Biomechanics of Joints and Implants

Sanjay Gupta · Ceby Mullakkara Saviour ·
Bidyut Pal · Souptick Chanda · Kaushik Mukherjee

Biomechanics of Joints and Implants

Concepts to Applications

Springer

Sanjay Gupta
Department of Mechanical Engineering
Indian Institute of Technology Kharagpur
Kharagpur, West Bengal, India

Bidyut Pal
Department of Mechanical Engineering
Indian Institute of Engineering Science
and Technology Shibpur
Howrah, West Bengal, India

Kaushik Mukherjee
Department of Mechanical Engineering
Indian Institute of Technology Delhi
New Delhi, Delhi, India

Ceby Mullakkara Saviour
Department of Mechanical Engineering
Indian Institute of Technology Kharagpur
Kharagpur, West Bengal, India

Souptick Chanda
Department of Biosciences
and Bioengineering
Indian Institute of Technology Guwahati
Guwahati, Assam, India

ISBN 978-981-96-0585-9 ISBN 978-981-96-0586-6 (eBook)
https://doi.org/10.1007/978-981-96-0586-6

This Springer imprint is published by the registered company Springer Nature Singapore Pte Ltd.
The registered company address is: 152 Beach Road, #21-01/04 Gateway East, Singapore 189721, Singapore

Preface

Biomechanics is a field of science that helps to gain insight into the mechanics of living systems. This interdisciplinary subject helps to understand the relationship between the structure and function of human joints, predict changes in the bone tissue due to alterations in the mechanical and biochemical environment, and propose methods of artificial interventions. The fundamental concepts of biomechanics encompass joint kinematics and kinetics, gait analysis and motion capture, tissue mechanics, implant design and analysis, bone remodelling, and bone fracture healing, as well as state-of-the-art techniques of modelling and simulation of biomechanical systems.

This book is derived from materials and lecture notes developed for the course "Mechanics of the Human Body", taught at the Indian Institute of Technology Kharagpur. It has been further enriched by the NPTEL online course *Biomechanics of Joints and Orthopaedic Implants*. Although primarily aimed at undergraduate and postgraduate engineering students, the course has also attracted medical practitioners with a keen interest in biomedical engineering. This book aims to be a valuable resource for students in biomedical and biomechanical engineering, highlighting recent advancements in the biomechanics of joints and orthopaedic implants.

The authors wish that the book accomplishes five goals. The first goal is to introduce the anatomy, physiology, and movements of the human body and its mechanical analogy to the students in order to evoke interest in the relationship between the structure and functions of the musculoskeletal system, in particular the main joints of the upper and lower limbs, as well as the spine. The second goal is to gain an insight into the structure, functions, movements, and forces acting on the joints during daily activities. The biomechanical concepts of the specific joints, the hip, knee, shoulder, elbow, and spine using the concept of static equilibrium have been presented.

The third goal is to quantitatively analyse the human gait cycle as a complex biomechanical system. Herein, focus on the details of gait analysis, measurement techniques, gait abnormalities, and motion capture system. Moreover, insights into the applications of mechanics of rigid bodies in the field of biomechanics have been provided. The estimation of musculoskeletal forces (joint forces and moments, muscle forces) using the fundamental concepts of joint dynamics (kinematics and kinetics) has been presented with relevant solved examples and exercises to impart basic understanding. The quantitative study of the skeletal system is envisaged to provide a comprehensive understanding of the applications of rigid body dynamics.

The fourth goal is to revisit the basic concepts of mechanics of deformable bodies relevant to bone tissue as well as implanted bone structure. The efficacies of joint replacement (hip, knee, shoulder) and spinal surgery have been explored in detail. The concepts of composite beams in the bone-implant structure have been presented to enhance the understanding of stress/strain shielding. Common failure mechanisms of implanted bone structures and biomaterials used in orthopaedic implants have been discussed. Moreover, state-of-the-art modelling techniques of biomechanical systems and experimental validation have been presented. The students/instructors are expected to be benefitted from the extensive discussions on state-of-the-art finite element modelling techniques using CT/MRI scan data.

The fifth goal is to gain insight into the theory and concepts of bone remodelling and fracture healing. The theory, mathematical formulations, and computational framework of bone adaptation have been explored. Different mathematical models of tissue differentiation algorithms, such as mechanoregulatory, bioregulatory, and mechanobioregulatory algorithms have been presented. A brief overview of artificial intelligence and its application in orthopaedic biomechanics research has been discussed.

The contributions of several people have culminated in the development of this manuscript. We sincerely thank Dr. Rahul Gautam Talukdar for his feedback on spine-related content, Dr. Nirmal Kumar Som for his insights into human anatomy, and Mr. Gowtham Reddy for his review on artificial intelligence content. We also thank Mr. Pedada Jagdish for reviewing the manuscript, Mr. Rounak Bhattacharya for his feedback on motion analysis, Mr. Saranjit Sarkar and Ms. Rupa Mukherjee for their contributions to figures and artworks. Lastly, we express our heartfelt gratitude to our families and friends for their unwavering support and encouragement throughout this project.

We welcome corrections and suggestions for improving this textbook, by email to sangupta@mech.iitkgp.ac.in.

Kharagpur, India
September 2024

Sanjay Gupta
Ceby Mullakkara Saviour
Bidyut Pal
Souptick Chanda
Kaushik Mukherjee

Contents

1 Musculoskeletal System: Structure and Function ... 1
1.1 Skeletal System ... 1
1.2 Bone ... 4
1.3 Muscle ... 7
1.4 Ligament ... 7
1.5 Tendon ... 8
1.6 Mechanical Analogy of Anatomical Elements ... 9
1.7 Joints ... 10
1.7.1 Classifications of Joints ... 10
1.8 Structure and Functions of Synovial Joints ... 13
1.8.1 Movements of Synovial Joints ... 14
1.8.2 Mechanical Analogy of Synovial Joints ... 16
1.8.3 Factors Influencing Joint Stability ... 19
1.8.4 Joint Disorders ... 19
1.9 Hip Joint ... 21
1.9.1 Functions of the Hip Joint ... 21
1.9.2 Femoral Neck Shaft Angle and Femoral Neck Anteversion Angle ... 22
1.9.3 Muscles of the Hip Joint ... 23
1.9.4 Factors Affecting Stability of the Hip Joint ... 25
1.9.5 Common Problems of the Hip Joint ... 27
1.10 Knee Joint ... 29
1.10.1 Anatomy of the Knee Joint ... 30
1.10.2 Functions of the Knee Joint ... 31
1.10.3 Patella ... 31
1.10.4 Menisci ... 32
1.10.5 Factors Affecting Stability of the Knee Joint ... 33
1.10.6 Muscles of the Knee Joint ... 33
1.10.7 Ligaments of the Knee Joint ... 34
1.10.8 Knee Deformities ... 36

1.10.9 Common Problems Associated with Knee Joint 36
1.11 Ankle Joint ... 37
1.11.1 Anatomy of the Ankle Joint ... 39
1.11.2 Functions of the Ankle Joint ... 40
1.11.3 Muscles of the Ankle Joint ... 40
1.11.4 Factors Affecting Stability of Ankle Joint ... 40
1.11.5 Common Problems Associated with Ankle Joint ... 43
1.12 Shoulder Joint ... 44
1.12.1 Anatomy of the Shoulder Joint ... 44
1.12.2 Movements Associated with Shoulder Joint ... 45
1.12.3 Shoulder Joint Muscles ... 46
1.12.4 Shoulder Joint Ligaments ... 48
1.12.5 Common Problems in the Shoulder Joint ... 48
1.13 Elbow Joint ... 51
1.13.1 Anatomy of the Elbow Joint ... 51
1.13.2 Elbow Joint Movements ... 52
1.13.3 Elbow Joint Muscles ... 52
1.13.4 Elbow Joint Ligaments ... 53
1.13.5 Common Elbow Joint Problems ... 53
1.14 Spine ... 55
1.14.1 Anatomy of the Spine ... 55
1.14.2 Ligaments of the Spine ... 59
1.14.3 Spine Movements ... 59
1.14.4 Disc Degenerative Disease ... 61
Multiple Choice Questions (MCQ) ... 62
References ... 65

2 Basic Biomechanics of Human Joints and Spine ... 67
2.1 Biomechanics of the Hip ... 67
2.1.1 Basic Biomechanics of the Hip Joint ... 68
2.1.2 Hip Joint Reaction Forces During Various Activities ... 76
2.2 Biomechanics of the Knee Joint ... 77
2.2.1 Instantaneous Centre of Rotation of the Knee Joint ... 78
2.2.2 Influence of the Patella on Knee Joint Mechanics ... 80
2.2.3 Screw–Home Mechanism of the Knee Joint ... 81
2.2.4 Basic Biomechanics of the Knee Joint: Sample Problems ... 82
2.2.5 Knee Joint Reaction Forces During Various Activities ... 88
2.3 Biomechanics of the Ankle Joint ... 88
2.3.1 Basic Biomechanics of the Ankle Joint: Sample Problems ... 89
2.4 Biomechanics of the Shoulder Joint ... 92

2.4.1 Closed Chain Mechanism 93
2.4.2 Degrees-of-Freedom of the Shoulder Mechanism 94
2.4.3 Glenohumeral Joint Reaction Force During Abduction 95
2.4.4 Basic Biomechanics of the Shoulder Joint: Sample Problems 95
2.5 Biomechanics of the Elbow Joint 98
2.5.1 Carrying Angle 98
2.5.2 Basic Biomechanics of the Elbow Joint: Sample Problem 99
2.6 Biomechanics of the Spine 108
2.6.1 Lumbar Spinal Load for Various Postures 108
2.6.2 Basic Biomechanics of the Lumbar Spine: Sample Problems 109
MATLAB Code for Solving 3D Force System Numerical Problem 112
Multiple Choice Questions (MCQ) 114
Practice Problems 116
References 120

3 Biomechanics of Gait 121
3.1 Gait Cycle 121
3.1.1 Phases of Gait Cycle 122
3.1.2 Gait Cycle Parameters 124
3.1.3 Basic Differences Between Running and Walking Cycles 125
3.1.4 Ground Reaction Forces 125
3.1.5 Centre of Gravity 126
3.1.6 Moment Arm of Ground Reaction Force 126
3.1.7 Determinants of Gait Cycle 129
3.2 Gait Analysis 133
3.2.1 Variation of Ground Reaction Force During Gait Cycle 134
3.2.2 Gait Abnormalities 137
3.3 Gait Measurement Techniques 138
3.3.1 Image Processing Systems 139
3.3.2 Floor Sensor Systems 140
3.3.3 Wearable Sensors 141
3.3.4 Motion Capture Systems: Marker-Based Versus Marker-Less 142
Multiple Choice Questions (MCQ) 147
References 149

4 Joint Kinematics and Kinetics 151
4.1 Joint Kinematics 151
4.1.1 Estimation of Joint Angles 151
4.1.2 Lower-Limb Joint Angles During Gait Cycle 152
4.1.3 Linear Variables 156
4.2 Joint Kinetics 156
4.2.1 Anthropometry Data 157
4.3 Biomechanical Models 162
4.4 Inverse Dynamics and Forward Dynamics 164
4.5 Link Segment Analysis: Joint Forces and Moments 165
4.5.1 Assumptions of the Link Segment Model 165
4.5.2 Bone-on-Bone Contact Force 166
4.6 Static Optimisation 173
4.7 Three-Dimensional Motion Analysis 178
4.7.1 Kinetic Analysis 196
4.8 Use of Musculoskeletal Models for Force Estimation 201
4.9 Postural Stability 204
4.10 Swaying During Quiet Stance 206
4.11 Inverted Pendulum Model 207
MATLAB Code for Solving the Numerical Problem Related to 3D Motion Analysis 210
Multiple Choice Questions (MCQ) 215
References 218

5 Bone Tissue Mechanics 219
5.1 Fundamentals of Stress and Strain 219
5.1.1 Stress and Strain 219
5.1.2 Stress–Strain Curve 221
5.1.3 Shear Stress due to Torsion 222
5.1.4 Pure Bending 223
5.1.5 Generalized Hooke's Law 228
5.1.6 Stress Tensor 229
5.2 Stress Transformation 232
5.3 Bone Structure 239
5.4 Cortical Bone 239
5.4.1 Mechanical Behaviour of Cortical Bone 240
5.5 Cancellous Bone 241
5.5.1 Mechanical Behaviour of Cancellous Bone 242
5.5.2 Density-Modulus Relationship of Cancellous Bone 245
5.5.3 Loading and Deformation of Bone Structure 249
5.5.4 Viscoelastic Property of Bone 250
5.6 Bone Remodelling 252
5.7 Trabecular Architecture 255
5.8 Anisotropic Properties of Bone 256

5.8.1 Micro-FE Models ... 258
5.8.2 Fabric Tensor Method ... 259
5.8.3 Strain-Based Algorithm ... 259
Multiple Choice Questions (MCQ) ... 260
Practice Problems ... 262
References ... 263

6 Biomechanics of Orthopaedic Implants Part I: Hip and Knee ... 267
6.1 Joint Replacement ... 267
6.1.1 Classification of Implants: Fixation Techniques ... 268
6.1.2 Implant Failure Mechanisms ... 270
6.2 Structural Analysis of Implant-Bone Structures ... 273
6.2.1 Composite Bar Subjected to Axial Loading ... 274
6.2.2 Composite Beam Subjected to Pure Bending ... 276
6.2.3 Bending of Unsymmetrical Beam ... 280
6.2.4 Bending of Unsymmetrical Composite Beam ... 284
6.3 Biomechanics of Hip Replacement ... 287
6.3.1 Total Hip Replacement ... 287
6.3.2 Common Classification of Stem Designs ... 292
6.3.3 Hip Resurfacing ... 293
6.4 Biomechanics of Knee Replacement ... 296
6.4.1 Effect of Implant Alignment ... 298
6.4.2 Effect of Implant Materials ... 298
6.4.3 Implant Stability: Role of Geometric Features ... 299
Practice Problems ... 300
References ... 301

7 Biomechanics of Orthopaedic Implants Part II: Shoulder and Spine ... 305
7.1 Biomechanics of Shoulder Arthroplasty ... 305
7.1.1 Conventional Total Shoulder Arthroplasty ... 306
7.1.2 Implant Stability: Role of Geometric Features ... 308
7.1.3 Reverse Total Shoulder Replacement ... 308
7.2 Biomechanics of Spinal Surgery ... 311
7.2.1 Spinal Fusion ... 312
7.2.2 Non-fusion Surgery ... 314
7.3 Biomaterials ... 315
7.3.1 Primary Requirements ... 316
7.3.2 Types of Biomaterials and Applications ... 317
Multiple Choice Questions (MCQ) for Chaps. 6 and 7 ... 319
References ... 323

8 Biomechanical Modelling, Analysis and Design of Orthopaedic Implants ... 325
8.1 The Need for Finite Element Modelling for Biomechanical Analysis ... 325
8.2 Development of FE Model ... 326
8.2.1 Image Processing and Geometrical Reconstruction of Bone ... 326
8.2.2 FE Modelling of Implanted Bone ... 329
8.2.3 Assignment of Bone Material Property ... 331
8.2.4 Musculoskeletal Loading and Boundary Conditions for FE Models ... 334
8.2.5 Population-Based Model ... 337
8.3 Verification and Validation of FE Models ... 338
8.3.1 Validation Using Strain Gauge Method ... 339
8.3.2 Validation Using Digital Image Correlation Technique ... 342
8.3.3 Validation Using Digital Volume Correlation Technique ... 344
8.3.4 Validation Using Linear Displacement Sensor ... 344
8.4 Steps in the Design of Orthopaedic Implants ... 346
8.4.1 Design Feasibility ... 346
8.4.2 Concept and Detailed Design ... 347
8.4.3 Design Verification ... 348
8.4.4 Manufacturing ... 348
8.4.5 Design Validation ... 349
8.4.6 Design Transfer ... 349
Multiple Choice Questions (MCQ) ... 349
References ... 352

9 Adaptive Bone Remodelling ... 355
9.1 Biology of Bone Remodelling ... 355
9.1.1 Different Phases of Bone Remodelling ... 356
9.1.2 Bone Turnover ... 357
9.1.3 Haversian Bone Remodelling ... 359
9.1.4 Trabecular Bone Remodelling ... 359
9.2 Theory of Adaptive Bone Remodelling ... 361
9.3 Bone Remodelling Models: Mathematical Formulations and Computational Schemes ... 363
9.3.1 SED-Based Bone Remodelling Model ... 363
9.3.2 Orthotropic Strain-Based Bone Remodelling Model ... 366
9.3.3 Mechanobiochemical Bone Remodelling Model ... 369
Multiple Choice Questions (MCQ) ... 376
References ... 378

10 Bone Fracture Healing 381
10.1 Types of Bone Fractures 381
10.2 Stages of Bone Fracture Healing 382
10.3 Cellular Mechanisms Involved in Fracture Healing 384
10.4 MSC Differentiation Pathway 387
10.5 Primary and Secondary Bone Healing 388
10.6 Need for Immobilisation of Fracture Site 389
10.7 Factors Affecting Bone Fracture Healing 389
10.8 Common Fracture Fixation Techniques 390
10.9 Secondary Fracture Healing Models 391
10.9.1 Mechanoregulatory Models 392
10.9.2 Bioregulatory Models 399
10.9.3 Mechanobioregulatory Models 400
10.10 Bone Ingrowth Around Porous Coated Implants 400
Multiple Choice Questions (MCQ) 401
References 404

Appendix: Artificial Intelligence in Orthopaedic Biomechanics 407

Solution to Practice Problems 425

About the Editors

Sanjay Gupta is a Professor in the Department of Mechanical Engineering at the Indian Institute of Technology Kharagpur and is leading the Biomechanics Research Group. He earned his Ph.D. from the Delft University of Technology, The Netherlands. Subsequently, he worked as a senior research fellow at the University of Southampton, UK, and as a research associate at the Imperial College London, UK. He completed his bachelor's degree in mechanical engineering from Bengal Engineering College, University of Calcutta, in 1989 and master's degree in mechanical engineering from Jadavpur University, Calcutta, in 1992. He has 26 years of teaching and research experience. His major areas of research include bone and joint mechanics, pre-clinical analysis of implant design, bone remodelling, and mechanobiology. He has 50 international journal publications in this domain of research. He was awarded the INSA Teachers Award 2022 by the Indian National Science Academy, New Delhi.

Ceby Mullakkara Saviour is pursuing his doctoral study in the field of Biomechanics at the Indian Institute of Technology Kharagpur. He completed his bachelor's degree in mechanical engineering at Cochin University of Science and Technology in 2016 and master's degree in mechanical engineering from the Indian Institute of Technology Guwahati in 2019. During his master's degree, he secured the DAAD-IIT sandwich fellowship and conducted his thesis work at Leibniz University Hannover, Germany, from September 2018 to March 2019. His major areas of research include tissue mechanics, orthopaedic implant design, bone remodelling, and mechanobiology.

Bidyut Pal is an Assistant Professor in the Department of Mechanical Engineering at the Indian Institute of Engineering Science and Technology (IIEST), Shibpur. He obtained his Ph.D. from Indian Institute of Technology Kharagpur in 2010. Subsequently, he worked as a postdoctoral research associate in the Biomechanics Research Group (Mechanical Engineering Department) at Imperial College London, UK, and as a postdoctoral research fellow in the Medical Engineering Research Group at Anglia Ruskin University, UK. Dr. Pal also served as a senior lecturer and lecturer in Mechanical Engineering at the University of Portsmouth, UK. He earned his master's degree in biomedical engineering from Jadavpur University, Kolkata, and his bachelor's degree in mechanical engineering from Kalyani Government Engineering College (Kalyani University), West Bengal, India. His broad research area is computational and experimental orthopaedic biomechanics.

Souptick Chanda is an Assistant Professor in the Department of Biosciences and Bioengineering at the Indian Institute of Technology Guwahati. He completed his bachelor's degree in mechanical engineering at the Indian Institute of Engineering Science and Technology, Shibpur (Erstwhile Bengal Engineering and Science University, Shibpur) in 2003 and master's degree in the same discipline from Indian Institute of Technology Guwahati in 2006. His doctoral study in biomechanics at the Indian Institute of Technology Kharagpur lasted until August 2015. Subsequently, he joined Nanyang Technological University, Singapore, as a postdoctoral research fellow. Dr. Chanda was awarded the Fulbright-Nehru Academic and Professional Excellence Fellowship in 2019 to carry out research on orthopaedic biomechanics at Harvard Medical School, Boston. His areas of research include biomechanics, implant design and optimisation, surgical simulations, and soft computing.

Kaushik Mukherjee is an Assistant Professor in the Department of Mechanical Engineering at the Indian Institute of Technology Delhi. He is leading the orthopaedic biomechanics and implant design research group as well as the Sports Biomechanics research lab. His research lies across orthopaedic biomechanics, implant design, musculoskeletal modelling, and mechanics of human movement. Dr. Mukherjee earned his doctoral and master's degrees from the Indian Institute of Technology Kharagpur, whereas his undergraduate degree was from West Bengal University of Technology (Kalyani Government Engineering College). Dr. Mukherjee pursued his postdoctoral research in the Department of Bioengineering, Imperial College London, UK, where he was also a recipient of the Marie Skłodowska-Curie Research Fellowship.

Chapter 1
Musculoskeletal System: Structure and Function

In this chapter, our focus will be directed towards the fundamental aspects of the human musculoskeletal system. A detailed understanding of the musculoskeletal system, particularly the intricate relationship between the structure and its function, is necessary prior to discussions on the biomechanics of the human body. Musculoskeletal elements, e.g. bone, muscle, ligament, and tendon, are discussed in the following articles.

1.1 Skeletal System

The skeletal system is a basic framework of bones and connective tissues that provides the human body with essential structure, support, and protection. Additionally, this intricate network of bones and joints allows us to perform a range of movements and activities essential for daily living. The skeletal system is classified into the appendicular skeleton and the axial skeleton, as presented in Fig. 1.1. The group of bones or the skeleton marked in yellow corresponds to the axial skeleton, encompassing the skull, thoracic cage (rib cage), and vertebral column (Fig. 1.1a). The skeletal segment marked in blue colour is knowns as the appendicular skeleton, which includes the upper limbs, lower limbs, shoulder girdle, and pelvic girdle (Fig. 1.1b).

In order to accurately describe the position and orientation of the anatomical structures in space, it is necessary to define anatomical reference planes and anatomical directions. In Fig. 1.2, three mutually perpendicular planes, i.e. the coronal plane, the sagittal plane, and the axial plane, are presented. These planes are referred to as anatomical planes of reference, and they are essential for precisely determining the position of structures in space. The sagittal plane is a vertical plane that passes through the body longitudinally, dividing the body into a left section and a right section. The coronal plane is another vertical plane, perpendicular to the sagittal plane, that divides the body into a front section and a back section. The transverse

S. Gupta et al., *Biomechanics of Joints and Implants*,
https://doi.org/10.1007/978-981-96-0586-6_1

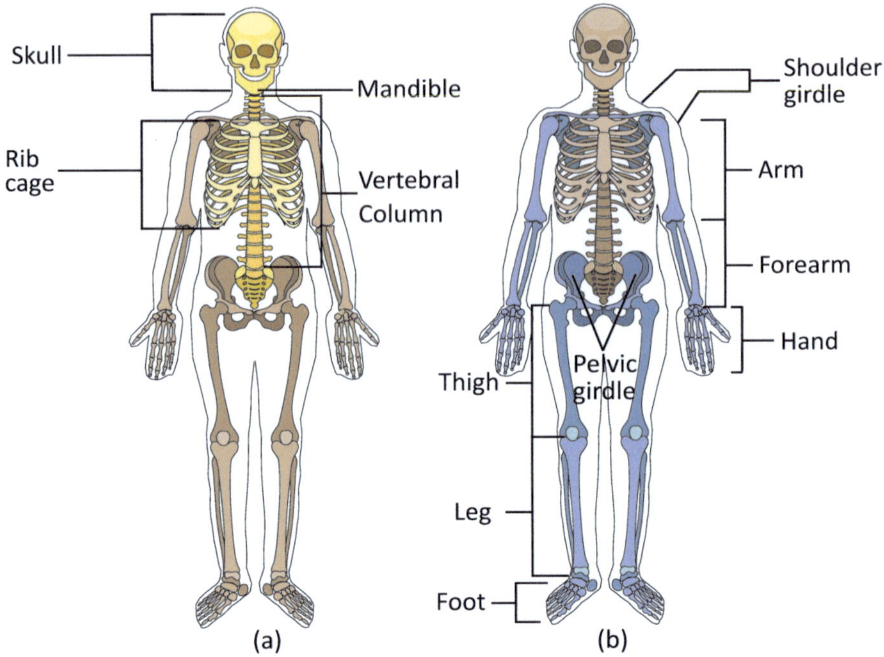

Fig. 1.1 The skeletal system; **a** axial skeleton, **b** appendicular skeleton

plane, also known as the axial plane, is an anatomical plane perpendicular to the above two planes that divides the body into top and bottom sections. Additionally, anatomical directions describe the position of a structure relative to another structure. Specific directions have been designated, including superior, inferior, anterior, posterior, proximal, distal, medial, and lateral, as illustrated in Fig. 1.3. For instance, the anterior and posterior directions are employed to describe structures located toward the front (anterior side) or toward the back (posterior side) of the body. An example is the toe being described as anterior to the heel. Part 'A' is said to be superior to part 'B', if part 'A' lies above part 'B'. Whereas, if part 'A' lies below the reference part 'B', we can say that part 'A' is inferior to part 'B'. For example, the pelvis is inferior to the abdomen. In addition to superior and inferior, anatomical terms such as proximal and distal are commonly used to denote relative positions. As the name suggests, while distal refers to a position far or at a certain distance, proximal indicates proximity. In precise terms, proximal indicates a position closer to the trunk or the point of attachment or origin of a structure, while distal refers to a location further away from it. For example, the foot is distal to the knee joint, whereas the knee is proximal to the foot. The knee joint is located closer to the trunk, while the foot is located farther away from it, making it distal to the knee joint. The femoral head is proximally located in the femur, whereas the femoral condyles are distally located.

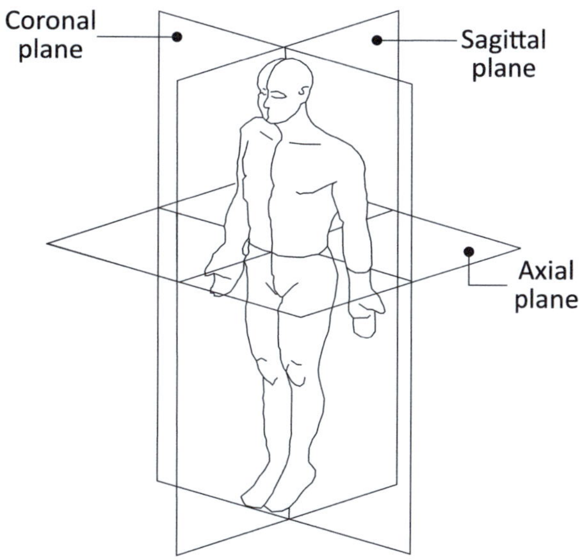

Fig. 1.2 The anatomical planes of references

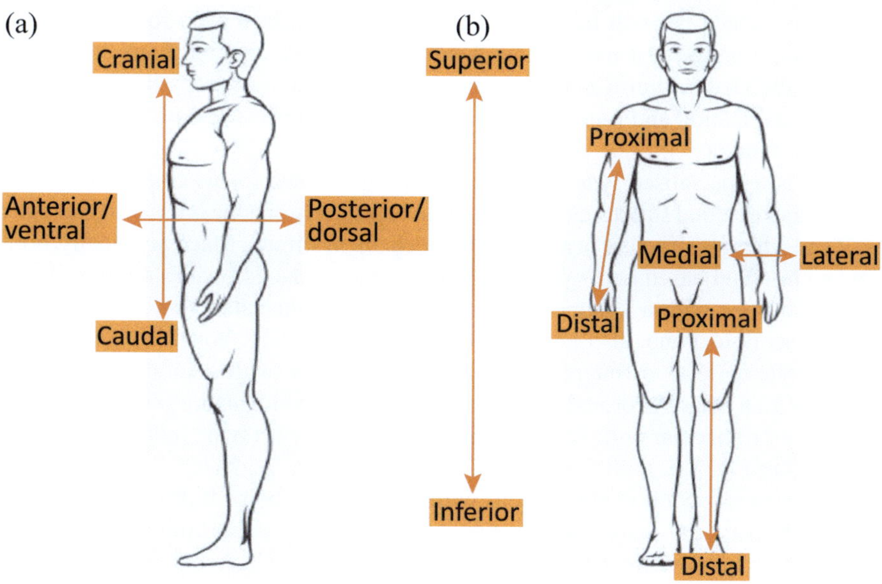

Fig. 1.3 Anatomical directions; **a** lateral (side) view, **b** anterior (front) view

The terms medial and lateral are used to describe positions relative to the midline of the body. A position closer to the midline is termed medial, while a position farther away from the midline is called lateral. For instance, when we hold our hand with the palm facing forward, the thumb is away from the midline of our body, making it the lateral-most finger. Meanwhile, the other fingers are positioned medially to the thumb. Additionally, ventral and dorsal are anatomical terms utilised to describe the position of structures. Ventral refers to structures located towards the front, while dorsal refers to structures positioned towards the back of the body. Similarly, cranial and caudal describe structures situated towards the head end of the body and away from the head, respectively.

1.2 Bone

The skeletal system comprises various types of bones, namely long bones, short bones, flat bones, irregular bones, and sesamoid or round bones, as illustrated in Fig. 1.4. The human skeleton in an adult consists of 206 bones. However, this count increases, if the commonly found sesamoid bones are considered. These sesamoid bones are located beneath the head of the first metatarsal in each foot, a common occurrence among most individuals. A sesamoid bone is a small bone commonly found embedded within a muscle or tendon near joint surfaces, existing as focal areas of ossification and functioning as a pulley to alleviate stress on that particular muscle or tendon.

Bone is a specialised type of connective tissue known for its complexity in the internal and external structures. One remarkable property of bone is its ability to repair itself and adapt its structure in response to mechanical stimuli. Macroscopically, bone can be classified based on its porosity. Cortical bone, also known as compact bone, is a dense bone with a volume fraction of a solid greater than 70%. On the other hand, cancellous bone, also known as spongy or trabecular bone, is characterised by a porous network of interconnected rods and plates, with a volume fraction of solid less than 70%. A coronal section of the femur provides an understanding of the structure of trabecular bone (cancellous bone) and compact bone (cortical bone), as presented in Fig. 1.5.

Bone serves several vital functions within the human body. It provides structural support, bearing the body's weight and ensuring stability. Additionally, bones act as protective barriers for critical organs, such as heart, lungs, and brain, shielding them from potential injuries. Bone also plays a pivotal role in blood cell formation through hematopoiesis. Hematopoiesis is the formation of blood cellular components within the bone marrow, where hematopoietic stem cells give rise to various types of blood cells, ensuring a continuous supply of blood cellular components. Moreover, bone acts as a mineral reservoir, storing essential minerals that play significant roles in various physiological processes. These minerals include calcium, phosphate, magnesium, sodium, and potassium, and their storage in bone contributes to maintaining proper mineral balance in the body. Another crucial role of bone is enabling

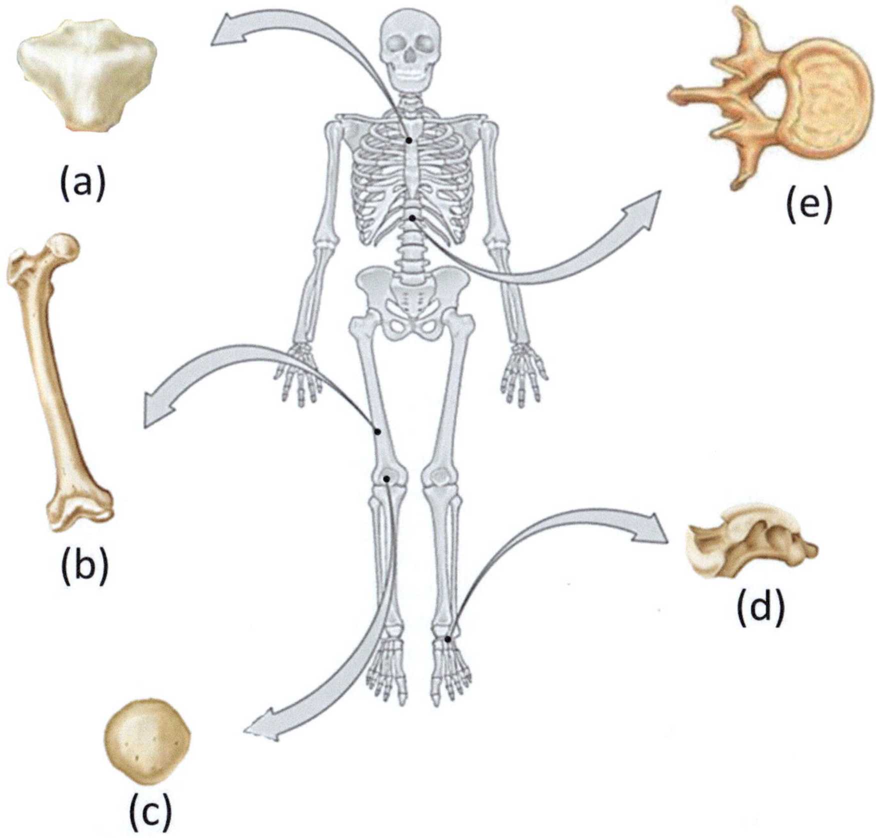

Fig. 1.4 Different types of bone; **a** flat bone, **b** long bone, **c** sesamoid bone, **d** short bone, **e** irregular bone

Fig. 1.5 A coronal section of the proximal femur

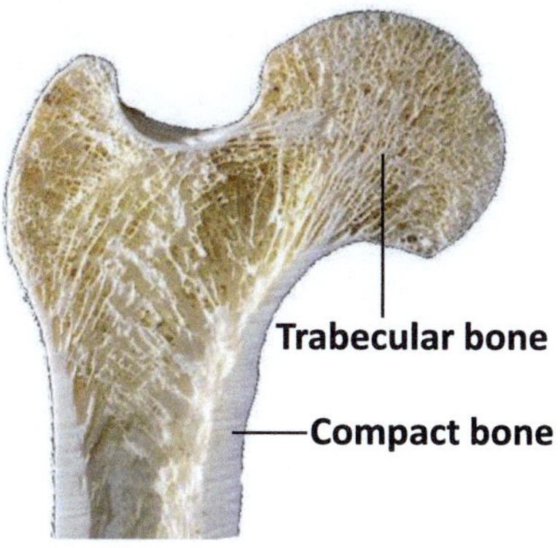

movements. Limbs can move and articulate due to the interaction between bones and muscles, granting mobility to the human skeleton.

Now, let us specifically focus on the anatomy of femur, the longest and strongest bone in the human body (Fig. 1.6). A long bone, such as the femur, can broadly be divided into three primary regions: the epiphysis, metaphysis, and diaphysis. The epiphyses, known as the connector, connect the femur with other bones to form two important joints, i.e. hip and knee joint. The metaphyseal regions consist of cancellous bone covered by a thin layer of cortical bone, providing attachment sites for muscles in the proximal and distal parts of femur. The diaphysis of the femur, on the other hand, comprises a thick layer of dense cortical bone, playing an active role in load transmission from the proximal end to the distal end of the femur. Beginning with its proximal end, we come across the spherical-shaped femur head (capital femoral epiphysis). Next, we encounter the neck of the femur, a critical bony structure within the femur. It is worth noting the sudden reduction of cross-sectional area in the femoral neck region as it extends from the femoral head. There are two bony landmarks in the proximal femur, the greater trochanter and the lesser trochanter (Fig. 1.6). The greater trochanter is on the lateral side, and the lesser trochanter is on the postero-medial side. As one moves away from proximal towards distal, the diaphysis (shaft) of the femur can be located. Further towards the distal side, the femur forms the metaphysis region. Metaphysis is a transitional zone between the diaphysis and the epiphysis. At the distal end of the femur, the medial and lateral condyles are located, which constitute the distal epiphysis of the femur.

Fig. 1.6 Anatomy of the femur

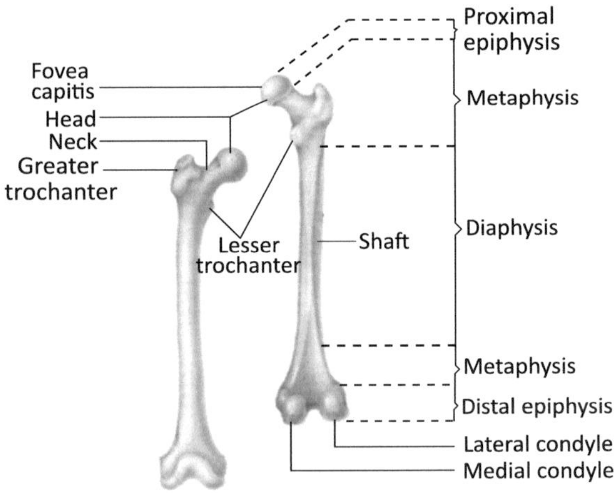

1.3 Muscle

The human body possesses nearly 700 muscles, categorised into three major types: skeletal muscles, cardiac and smooth muscles. The skeletal muscles are connected to the bones and are primarily responsible for the movement of the skeleton. Commonly they are designated as voluntary muscles because their activation and control are under conscious control. Cardiac muscles, on the other hand, are unique to the heart and are involuntary muscles. They play a crucial role in the rhythmic contractions of the heart, pumping blood throughout the body. The smooth muscles are found in the walls of blood vessels, the intestine, bronchi, uterus, and bladder. Commonly, the smooth muscles are involuntary and responsible for various functions within the internal organs.

Now, let us gain an insight into the structure of skeletal muscles. Skeletal muscles are composed of numerous subunits or bundles known as fascicles. Each fascicle is enveloped by connective tissue called the perimysium, and within each fascicle, there are multiple muscle fibres. In Fig. 1.7, we can observe that muscle fibres are enclosed by a sheath of endomysium, which consists of many fibrils or myofibrils. Typically, skeletal muscles attach to bones by means of tendons. The epimysium is a dense connective tissue that surrounds the entire muscle tissue and is continuous with the tendons, making it thicker and more collagenous. In other words, tendon and muscles are connected through the epimysium, providing structural integrity and support.

The functions of muscles are fundamental to human functioning. Muscles are responsible for generating motion, allowing us to perform various movements, such as walking, running, and grasping objects. Additionally, muscles play a critical role in maintaining posture, enabling us to stand and sit upright. Another essential function of muscles is heat production. Muscles generate heat during their contraction and relaxation processes, regulating the body temperature.

1.4 Ligament

Ligaments are tough bands of dense fibrous connective tissue that connect one bone to another bone. They play a vital role in providing stability to joint capsules. Ligaments and tendons have similar structure and composition, although subtle differences exist owing to different functions. The ligament contains collagen fibrils (70–80% by dry weight), elastin (1–15% by dry weight) and proteoglycan (1–3% by dry weight) and about 60% water. The hierarchical structure of ligament (similar to tendon) is shown in Fig. 1.8. The ligaments are flexible structures having viscoelastic properties; they permit natural movements of the bones on which they are attached. However, ligaments are strong and inextensible so as to offer resistance to bending and applied forces.

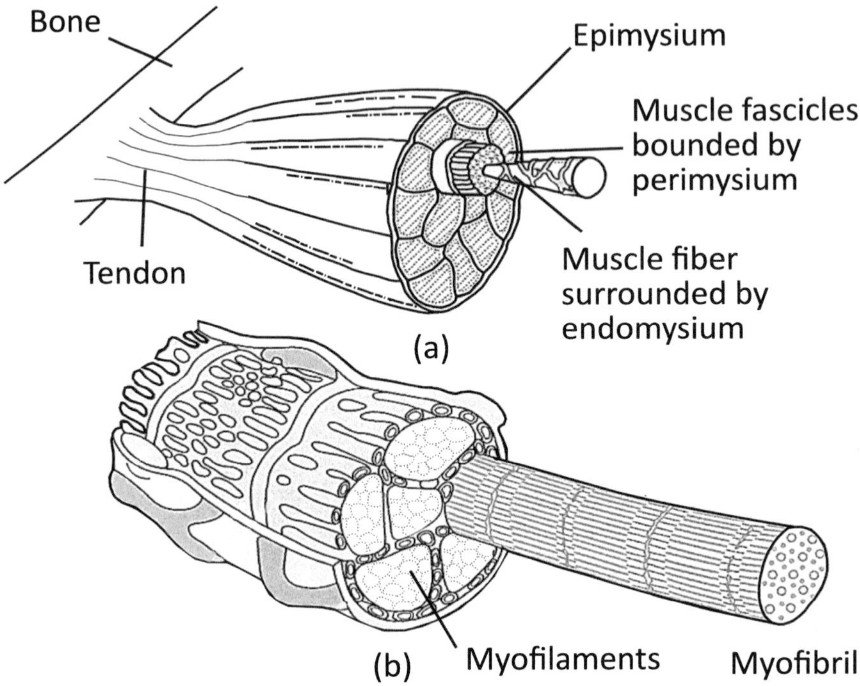

Fig. 1.7 The structures of **a** muscle, **b** muscle fibre

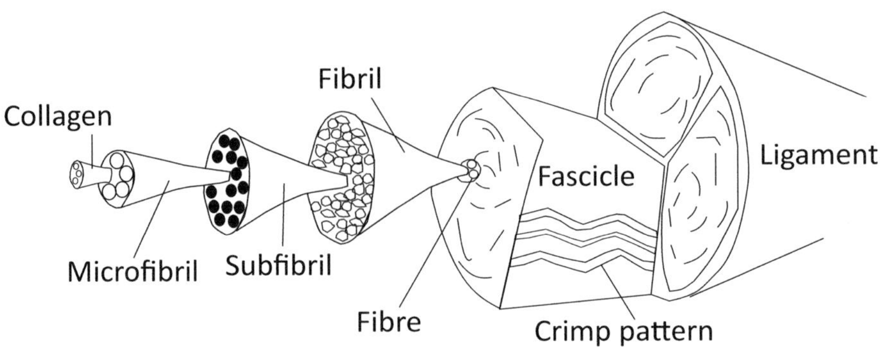

Fig. 1.8 Hierarchical structure of a ligament

1.5 Tendon

Tendons attach muscles to bones and transmit forces during movement, as shown in Fig. 1.9. Tendons can withstand tension and efficiently transmit the mechanical forces generated by muscle contractions to the skeletal system. They play a crucial role in

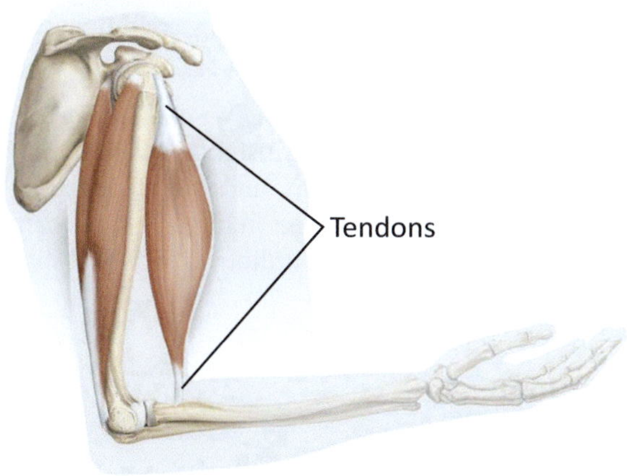

Fig. 1.9 Tendons attaches muscle to bone

enabling movement and locomotion in the body. Tendons and ligaments, having similar composition and structures, are made of collagen and exhibit viscoelastic material properties. Qualitatively, the mechanical behaviour of tendons and ligaments are similar, exhibiting the same non-linear stress–strain behaviour. However, tendon is stronger and stiffer than ligament since its collagen orientation is more aligned. The arrangement of collagen fibres differs to some extent in tendons and ligaments and is suited to the respective functions. The fibre composition in tendons has a well-ordered (aligned), parallel arrangement, which allows the tendon to sustain high uniaxial tensile force during physiological activities. In contrast, the fibres of ligaments are not entirely parallel but are closely interwoven with one another in order to sustain tensile loads along different directions.

1.6 Mechanical Analogy of Anatomical Elements

Comparing an essential anatomical element with an analogous mechanical element helps to comprehend the relationship between the structure and function of the musculoskeletal element.

- **Bone**: Comparable to rigid links in mechanical systems, bones provide structural support and act as a framework, serving as rigid components that provide shape and stability to the body.
- **Joints**: Analogous to standard mechanical joints, such as hinges or ball-and-socket joints. The joints enable movement and articulation between bones, permitting a wide range of motion.
- **Muscle and Tendon**: These are similar to actuators in mechanical systems. Muscles contract and generate force, while tendons efficiently transmit this force to bones, enabling controlled movements and locomotion.

- **Ligaments**: Similar to controllers or springs in mechanical systems, ligaments monitor and adjust the operating conditions of joints, providing stability and controlling the range of motion.

By drawing these comparisons, we can better appreciate the biomechanical aspects of the human body and gain insights into how its various elements work together to enable movement, support, and overall function. Understanding these mechanical analogies is useful in the fields of biomechanics and bioengineering, for the development of prosthetics, orthotics, and rehabilitation techniques.

1.7 Joints

Joints, also known as articulations, are crucial components of the musculoskeletal system, which serve as the meeting points of two or more bones. Within the skeletal system, joints are considered to be the weakest parts or links. Understanding the structure of joints is essential for comprehending their function, stability, and susceptibility to injuries and pathologies.

1.7.1 Classifications of Joints

Joints can be classified both functionally as well as structurally. Let us first look into the functional classification, which is based on the mobility of the joint. This classification results in three types of joints, namely synarthrosis, amphiarthrosis, and diarthrosis. Synarthrosis refers to immovable joints (e.g. sutures, gomphosis) and amphiarthrosis corresponds to slightly movable joints (e.g. radio-ulnar joint, pubic symphysis), whereas diarthrosis designates fully movable joints (e.g. hip, knee).

The structural classification of joints is based on the dominant type of connective tissue present. Three categories fall under this classification: fibrous joints, cartilaginous joints, and synovial joints. Fibrous joints are connected by collagenous fibres, and they are further subcategorised into sutures, syndesmosis, and gomphosis. Cartilaginous joints are united by cartilage and are further classified as synchondrosis and symphysis. Synovial joints are enclosed within a fibrous capsule, allowing free movement.

1.7.1.1 Fibrous Joints

Fibrous joints, as the name suggests, are bound together by collagenous fibres. They lack a joint cavity and depending on their type may allow limited or no movement. The subcategories of fibrous joints include the following:

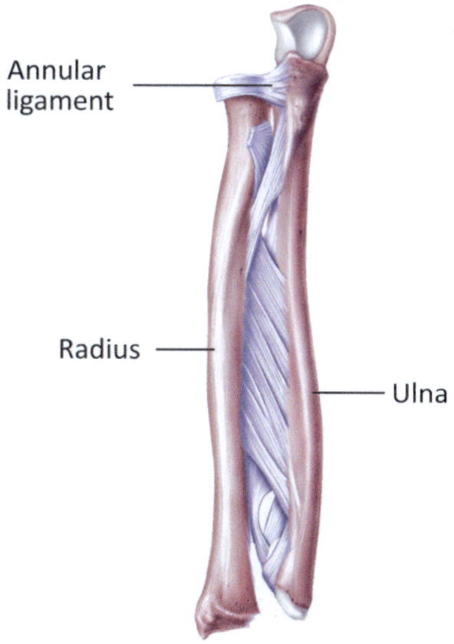

Fig. 1.10 A syndesmosis joint: connection between the radius and ulna via annular ligament

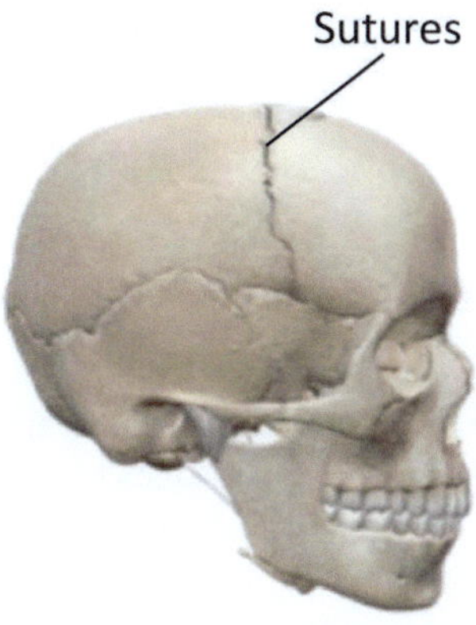

Fig. 1.11 Sutures found in the skull

- Syndesmosis: These are slightly movable fibrous joints. Example: The connection between the radius and ulna via annular ligament (Fig. 1.10).
- Gomphosis: An immovable joint that anchors teeth into their sockets. Example: The periodontal ligament secures the tooth in its bony socket.
- Sutures: These are immovable fibrous joints found in the skull as shown in Fig. 1.11.

1.7.1.2 Cartilaginous Joints

In cartilaginous joints, the adjacent bones are connected by cartilage and are classified into two types: synchondrosis and symphysis.

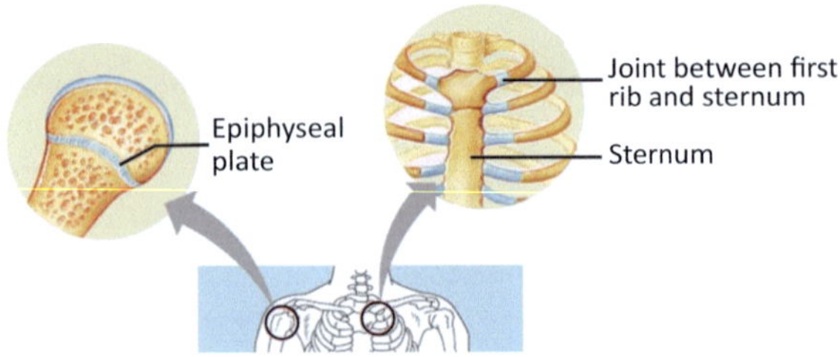

Fig. 1.12 Synchondrosis cartilaginous joints

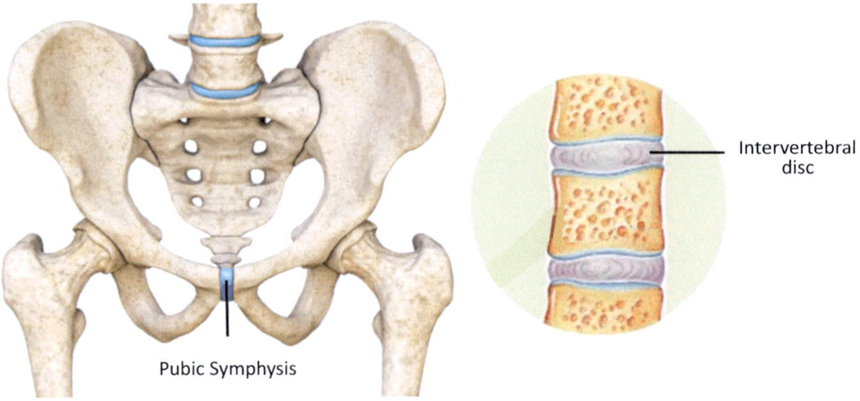

Fig. 1.13 Symphysis cartilaginous joints: **a** pubic symphysis, **b** intervertebral disc

- Synchondrosis: These are immobile cartilaginous joints, where hyaline cartilage completely bridges two adjacent bones. Examples include the epiphyseal plate in the humerus bone, and the joints between the first rib of the sternum as shown in Fig. 1.12.
- Symphysis: These are slightly movable cartilaginous joints connected by broad flattened discs of fibrocartilage. Examples include (a) the pubic symphysis, which connects the two hemi pelvises, and (b) the intervertebral disc as presented in Fig. 1.13.

1.7.1.3 Synovial Joints

Synovial joints are the most sophisticated type of joint in the skeletal system, allowing free movement within a fluid-filled synovial cavity (Fig. 1.14). These joints offer various degrees of mobility and are crucial for human locomotion and functional activities. The synovial fluid inside the joint cavity plays a vital role in reducing friction and maintaining joint health and function. In the following sections, we

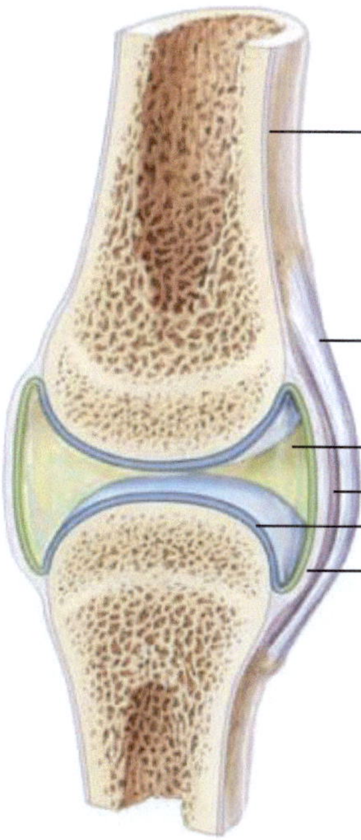

Fig. 1.14 General structure of a synovial joint

will explore the intricate structure of synovial joints while analyzing their unique functions within the musculoskeletal system.

1.8 Structure and Functions of Synovial Joints

Synovial joints are a type of joint that allows a wide range of movement. These joints consist of several key elements to facilitate smooth articulation and stability, which are briefly described as follows:

1. **Joint cavity and capsule**: Joint cavity is formed by a tough, and flexible capsule. The capsule surrounds and encloses the joint, holding the articulating bones together while providing a space for relative motion.
2. **Synovial membrane and synovial fluid**: The inner lining of the joint capsule is known as the synovial membrane. This membrane is responsible for secreting synovial fluid, a viscous fluid quite similar in consistency to a raw egg white. Synovial fluid can be described as the filtered solution of blood plasma and is primarily composed of hyaluronic acid. It plays a vital role in reducing friction between the articulating bones, ensuring smooth movement, and nourishing the joint's cartilage through diffusion.

3. **Articular cartilage**: The articulating surfaces of the bones within a synovial joint are covered with articular cartilage. This smooth and resilient cartilage cushions the ends of the bones and helps to distribute forces evenly during movement. Articular cartilage is a type of hyaline cartilage, which is different from the other two types of cartilage, elastic cartilage and fibrocartilage. It is generally considered a triphasic material composed of a porous matrix, water and ions. However, it does not contain any nerves or blood supply. The porous matrix consists of type II collagen and proteoglycans. Articular cartilage contains chondrocyte cells that help to maintain its structural integrity and biochemical balance. In most human joints, the thickness of cartilage layers typically varies in the range of 2–4 mm (Fox et al. 2009). Articular cartilage exhibits anisotropic material properties as well as heterogeneity, owing to variations in collagen, proteoglycans and water content.
4. **Ligaments**: Outside the joint capsule, ligaments reinforce and stabilise the joint. Ligaments are tough bands of fibrous connective tissue that connect bone to bone, providing additional support and preventing excessive movement that could lead to injury. These ligaments are either extracapsular ligaments located outside the capsule or intracapsular ligaments located inside the capsule.

The synovial joint is richly supplied with sensory nerves that can detect pain and monitor the extent to which the capsule is being stretched. Certain synovial joints exhibit unique structural adaptations known as articular discs. The articular discs are specialised fibrocartilaginous structures located between the articulating bones. They are found in cases where the articular bones have slightly different shapes, thus enhancing the joint's function and stability. Examples of synovial joints with articular discs are the temporomandibular joint (TMJ) and the knee joint.

1.8.1 Movements of Synovial Joints

Synovial joints provide a wide range of movements essential for the mobility and function of the human body. These movements are as follows.

- **Flexion**: A bending movement that decreases the angle between bones at the joint. Examples include bending the knee or shoulder (Fig. 1.15a).
- **Extension**: A straightening movement that increases the angle between bones. Examples include straightening the knee or shoulder after bending (Fig. 1.15b).
- **Abduction**: Movement away from the midline of the body. For instance, raising the arm or leg away from the centre of the body (Fig. 1.16a).
- **Adduction**: Movement towards the midline of the body. This action brings the arm or leg closer to the body's centre as presented in Fig. 1.16b.
- **Internal Rotation**: Rotating a limb towards the midline of the body (Fig. 1.17). For example, turning the arm or leg inwards.
- **External Rotation**: Rotating a limb away from the midline of the body (Fig. 1.17). It involves turning the arm or leg outwards.

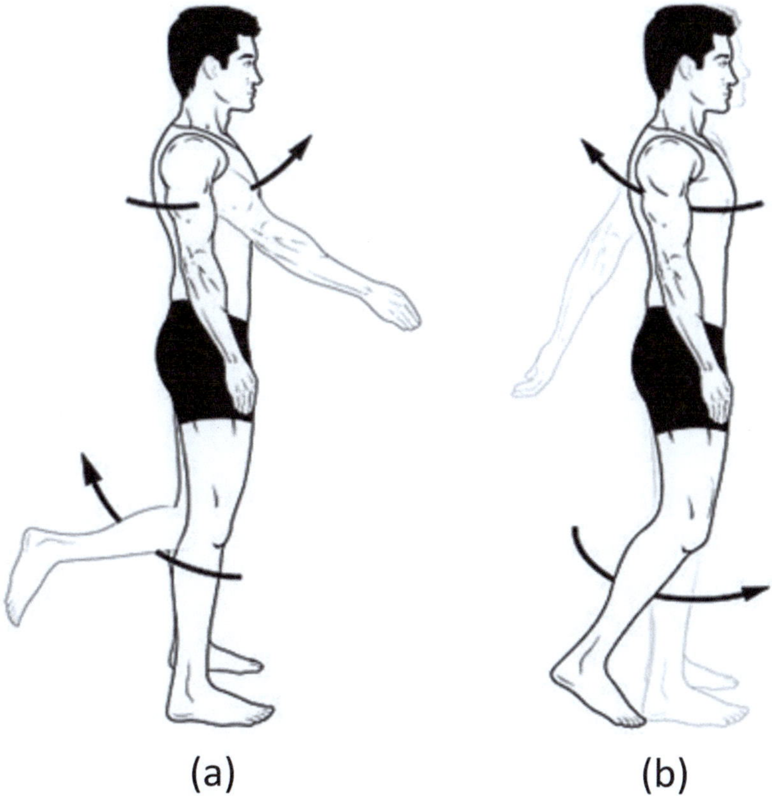

Fig. 1.15 Joint movements, **a** flexion, **b** extension

Additionally, there are few specific movements of the hands and feet as illustrated in Fig. 1.18:

- Palmar Flexion: The bending of the hand and wrist towards the palm side of the forearm.
- Dorsal Flexion: The upward bending of the hand and wrist, pointing the fingers towards the back of the hand.
- Pronation: Rotation of the forearm such that the palm faces downward.
- Supination: Rotation of the forearm such that the palm faces upward.
- Dorsiflexion: Flexion of the foot at the ankle joint, bringing the foot upwards towards the shin.
- Plantar Flexion: Extension of the foot at the ankle joint, pointing the foot downward.
- Eversion: Outward tilting of the sole of the foot away from the midline of the body.
- Inversion: Inward tilting of the sole of the foot towards the midline of the body.

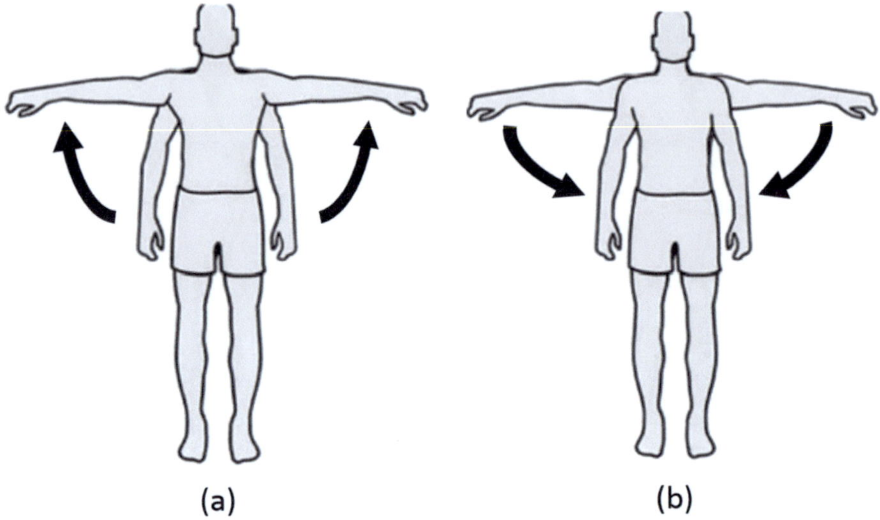

Fig. 1.16 Joint movements, **a** abduction, **b** adduction

Fig. 1.17 Internal and external rotation

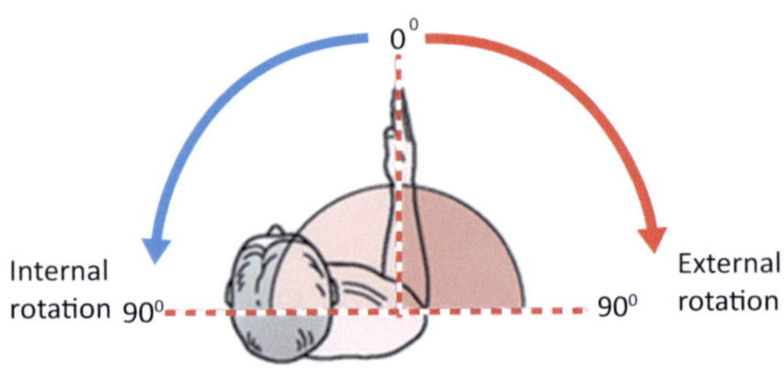

Understanding these various movements is crucial in assessing joint function, diagnosing joint disorders, and developing appropriate treatment plans to maintain joint health and mobility.

1.8.2 Mechanical Analogy of Synovial Joints

Synovial joints within the human skeletal system can be compared with the mechanical joint types, as shown in Fig. 1.19. Let us explore these analogies, as discussed in the following.

(a) **Pivot joint**: The pivot joint involves a rounded bone surface articulating with a ring formed by the second bone and ligament. It offers uniaxial movement, permitting rotation around the longitudinal axis. An example of a pivot joint

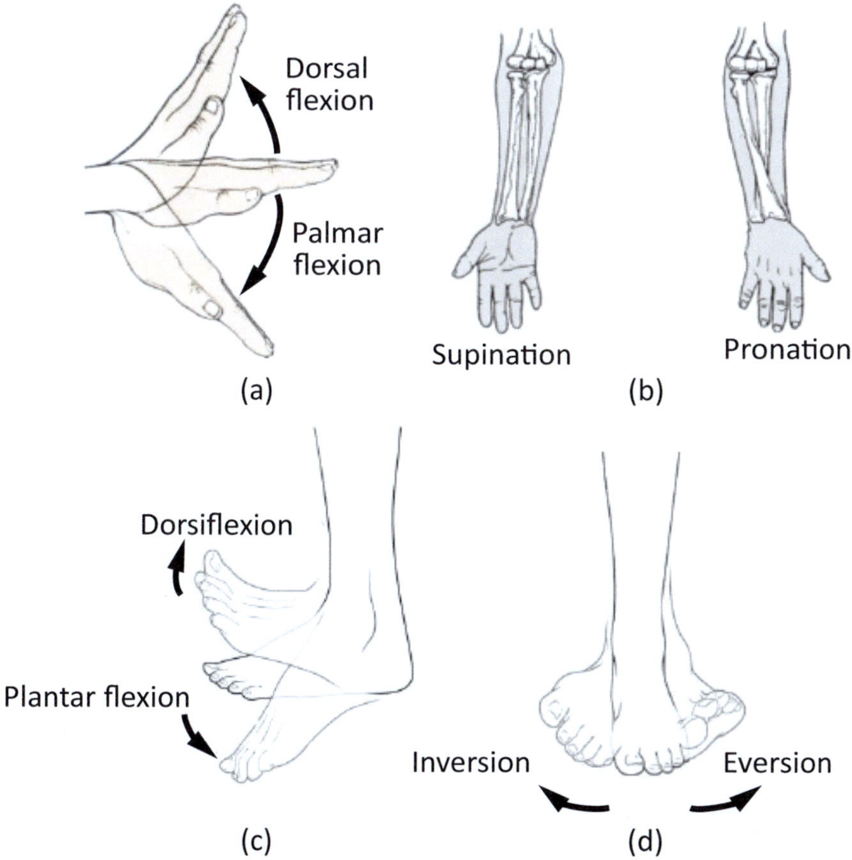

Fig. 1.18 Specific movements of the hands and feet; **a** palmer and dorsal flexion, **b** supination and pronation, **c** plantarflexion and dorsiflexion, **d** inversion and eversion

is the proximal end of the radioulnar joint, enabling supination (palm up) and pronation (palm down) movements of the forearm.

(b) **Hinge joint**: The hinge joint is similar to a door hinge, allowing rotation around one axis. It is formed by a convex bone surface fitting into a concave surface of the second bone. Examples of hinge joints in the body include the knee, elbow, ankle, and interphalangeal joints of the fingers. The primary movements produced in hinge joints are flexion, extension, and hyperextension.

(c) **Saddle joint**: A saddle joint occurs when one articulating bone surface is saddle-shaped, allowing the other bone to fit like a rider on a saddle. It offers biaxial movement, including circumduction (circular movement) and opposition (thumb to touch the fingertips). Examples of saddle joints are found between the trapezium bone and the metacarpal bone of the thumb.

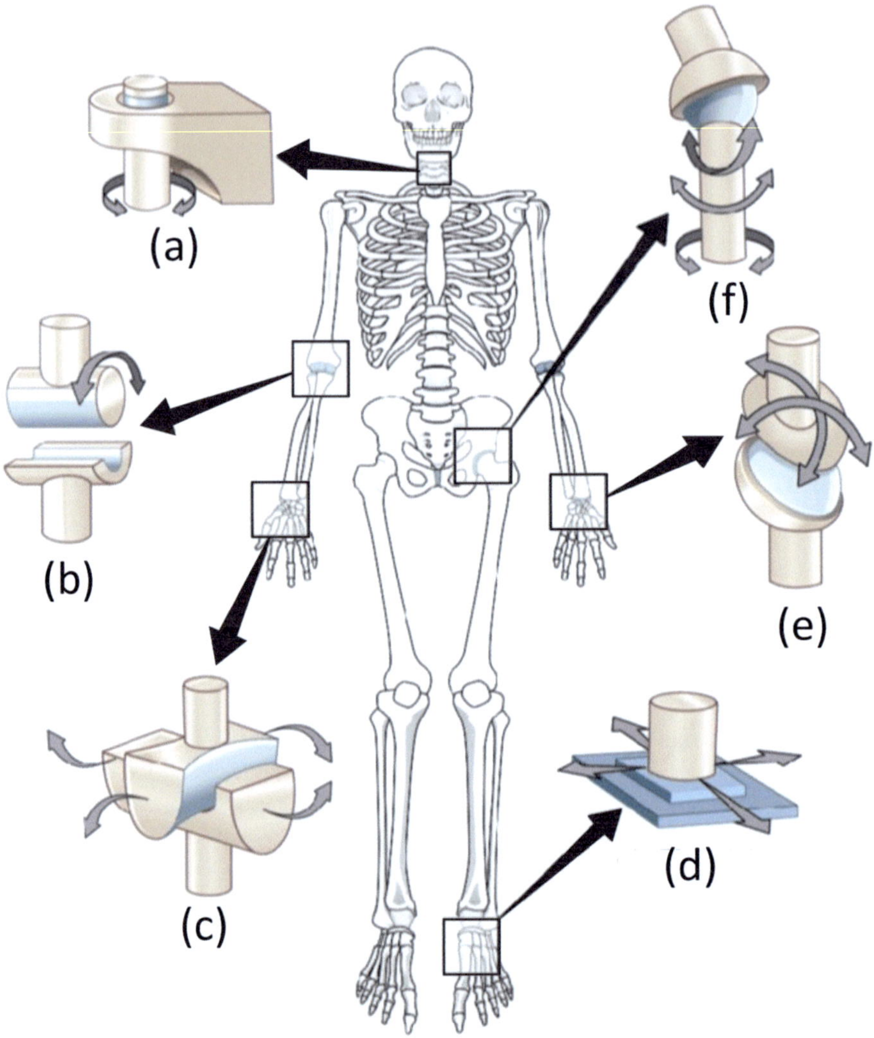

Fig. 1.19 Various types of synovial joints; **a** pivot joint, **b** hinge joint, **c** saddle joint, **d** planar joint, **e** condyloid joint, **f** ball and socket joint

(d) **Planar joint**: A planar joint is formed by two flat or slightly curved bone surfaces that allow only side-to-side movements while preventing rotation through ligament support. Examples of planar joints are the intercarpal and intertarsal joints in the hands and feet, the sternoclavicular joint in the shoulder, and the vertebrocostal joints.

(e) **Condyloid (Ellipsoid) joint**: The condyloid joint features a spherically shaped bone fitting into an oval depression, allowing biaxial movement. The movements possible in condyloid joints include flexion, extension, abduction, and adduction. Examples of condyloid joints are radiocarpal joints found in the wrists.

(f) **Ball and socket joint**: In this joint type, a ball-shaped bone fits into a cup-like depression, allowing multi-axial movements, such as flexion, extension, abduction, adduction, and rotation. Examples of ball and socket joints are the glenohumeral joint in the shoulder and the hip joint.

1.8.3 Factors Influencing Joint Stability

Joint stability is influenced by several factors that may be anatomical, biomechanical, or neuromuscular. In this section, the anatomical factors associated with joint stability will only be discussed. Ligaments play a crucial role in stabilising joints by connecting bones and preventing excessive movement. The shape of joint surfaces also determines joint stability; e.g. congruent articulations enhance stability. Additionally, the muscles surrounding a joint play a dynamic role in joint stability. Weak or imbalanced musculature can compromise stability, while well-conditioned muscles provide support and control joint movements.

1.8.4 Joint Disorders

Joint disorders can considerably affect an individual's daily life and overall well-being. Bone degenerative diseases, particularly osteoarthritis and rheumatoid arthritis, are major contributors to joint disorders. These conditions affect a large number of patients, leading to sustained pain and reduced mobility. Arthritis, in particular, leads to worn cartilage, resulting in irregular and rough articulating surfaces within a joint. As the cartilage wears away, the joint space between the bones decreases, causing bone-to-bone contact and pain.

Joint disorders may also be caused owing to traumatic injuries, leading to post-traumatic arthritis. Fractures are relatively common, occurring in approximately 12% of the global population. In order to address joint disorders, medical interventions like arthroscopy and arthroplasty are commonly employed. Arthroscopy is a minimally invasive procedure using an arthroscope to inspect and treat joint damage. It involves small incisions through which specialised instruments are inserted into the joint. The advantages of arthroscopy over traditional open surgery include reduced recovery time and less trauma to the surrounding tissue. Arthroplasty, or joint replacement, on the other hand, is a reconstructive surgical procedure for joint replacement using an implant or prosthesis. In this procedure, the joint is opened, and the affected bone fragments are removed or resected. Thereafter the implants are fixed in place, creating a new joint articulation. In specific fractures, such as intertrochanteric fractures in the femur and femoral neck fractures prevalent among elderly patients (Fig. 1.20), joint replacement is usually undertaken.

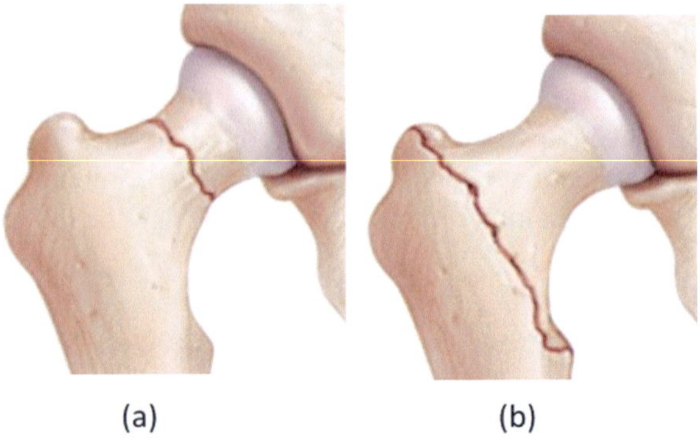

Fig. 1.20 Common femoral fractures; **a** femoral neck fracture, **b** intertrochantric fracture

Arthroplasty or joint replacement can be classified into total joint replacement, resurfacing arthroplasty, and hemi-arthroplasty. Total hip arthroplasty (THA) and hip resurfacing arthroplasty (HRA) are common surgical interventions for hip joint disorders (Fig. 1.21). Hemi-arthroplasty is a type of partial joint replacement, where half of the joint is replaced, and is performed only under special circumstances. For example, owing to specific traumatic injuries of the hip, a fractured femoral head or neck may be removed and replaced by a femoral implant, leaving the acetabular socket intact. Whereas, THA and HRA involve the replacement of bone mass of femur and pelvic bone with femoral and acetabular implants, respectively (Fig. 1.21). Unlike the THA, the medullary canal does not intervene in the HRA. Instead, only the articular surface of the femur is replacement by a resurfacing implant (Fig. 1.21a).

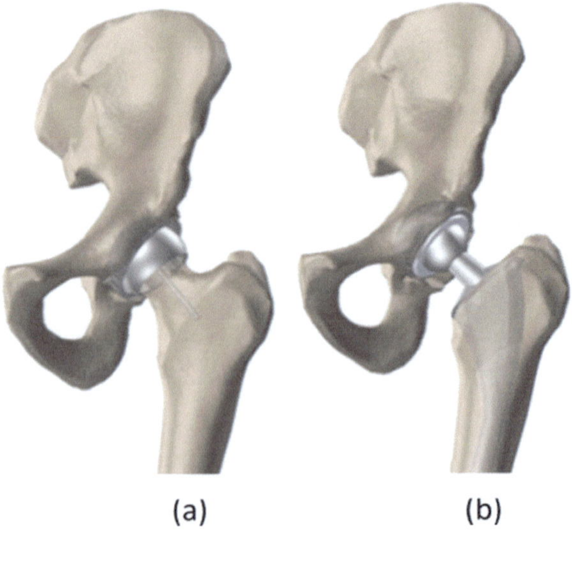

Fig. 1.21 Types of hip replacement surgeries; **a** hip resurfacing arthroplasty, **b** total hip arthroplasty

Prominent synovial joints are the hip, knee, and ankle joints in the lower limb, and the shoulder and elbow joints in the upper limb. A detailed discussion of these prominent human joints and the spine are presented in the following sections.

1.9 Hip Joint

The hip joint is a major load-bearing joint and involves the articulation between the pelvic bone and the femur. In this section, we will explore its structure, function, and common disorders. The hip joint is a synovial joint that forms a ball and socket joint. It consists of two main components: the femoral head (ball) and the acetabulum (socket), as shown in Fig. 1.22. The femoral head is the rounded part of the femur bone, resembling a ball. It is covered by hyaline cartilage, except for the fovea capitis, a small depression on the femoral head. The acetabulum is part of the pelvic bone and serves as the socket of the hip joint. It is a horseshoe-shaped articular surface covered by hyaline cartilage.

1.9.1 Functions of the Hip Joint

The hip joint plays the following vital roles in the human body:

- The hip joint is a major weight-bearing joint, supporting the body weight during static and dynamic postures.
- It facilitates the transfer of forces and moments between the upper and lower limbs.
- The hip joint allows a wide range of movement while maintaining stability during these motions.

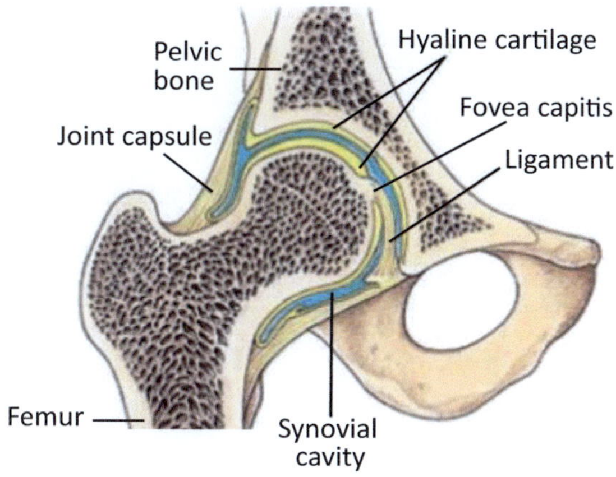

Fig. 1.22 A coronal section of the hip joint

1.9.2 Femoral Neck Shaft Angle and Femoral Neck Anteversion Angle

The femoral neck shaft angle and the femoral neck anteversion angle are essential measurements that help evaluate the alignment and orientation of the femur within the hip joint.

1.9.2.1 Femoral Neck Shaft Angle

The femoral neck shaft angle, also known as the CCD (caput-collum-diaphyseal) angle, is the angle formed by the intersection of the femoral head's axis (caput) and the femoral neck's axis (collum) with the shaft (diaphysis). The normal value is approximately 150° for new-borns and 126° for adults.

Coxa vara is a varus deformity that occurs when the neck shaft angle is reduced to less than 120° (Fig. 1.23a). Coxa vara can be classified into several subtypes as congenital, acquired, and developmental. Patients with coxa vara often exhibit Limb length discrepancy and Trendelenburg limp. If the disorder is present in both femurs, a "duck-waddle" gait can be observed in the subject in which the body sways from one side to the other.

Coxa valga is a valgus deformity that occurs when the neck shaft angle typically exceeds 135° (Fig. 1.23b). It can occur as a result of a lack of physiologic stress to the femur caused by skeletal abnormalities or neuromuscular hip dysplasia. Patients with coxa valga may present with symptoms such as reduced hip stability and an increased likelihood of joint pain.

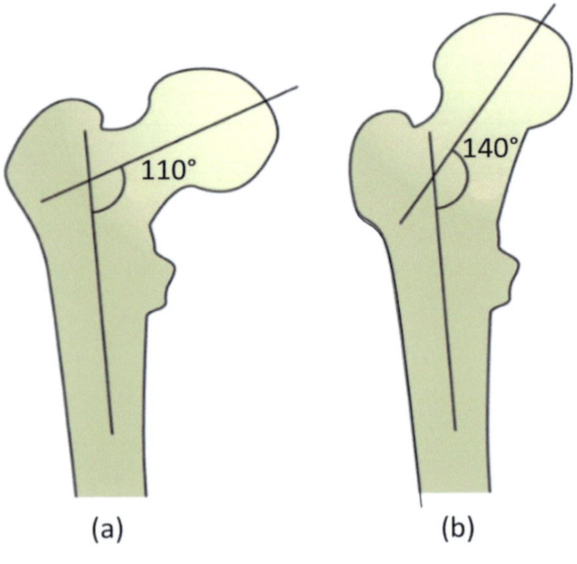

Fig. 1.23 Femoral deformities; **a** coxa vara, **b** coxa valga

1.9.2.2 Femoral Neck Anteversion Angle

The femoral neck anteversion angle is the angle between an imaginary transverse line through the knee joint and one through the centre of the femoral head and neck. The normal range varies between 15° and 20°. Retroversion deformity is characterised by a reduced femoral neck anteversion angle (Fig. 1.24a). Retroversion deformity can result in limited hip rotation, gait abnormalities (such as a "toeing in" gait), and, in some cases, hip pain. As the name suggests, "toeing in" gait is a gait abnormality in which the toes point inward or toward each other while walking. Anteversion deformity is due to increased femoral neck anteversion angle (Fig. 1.24b), and it may result in a pigeon-toed appearance of the feet. Retroversion and anteversion deformities damage the articular cartilage and acetabular labrum, which might lead to early osteoarthritis.

1.9.3 Muscles of the Hip Joint

The muscles surrounding the hip joint, presented in Fig. 1.25, have often been categorised into six muscle groups, according to their function. These muscle groups: (1) flexor, (2) extensor, (3) abductor, (4) adductor, (5) medial rotator and (6) lateral rotator. The role of each muscle group in specific movement is summarized in the following.

Flexor Muscles: These muscles are responsible for the flexion of the hip joint. Flexion involves moving the leg forward, bringing the knee towards the chest, with a maximum angle of 140°. The flexor group includes iliopsoas, rectus femoris, sartorius, adductor longus, adductor brevis, adductor magnus, tensor fasciae latae, pectineus, gluteus medius and gluteus minimus muscles.

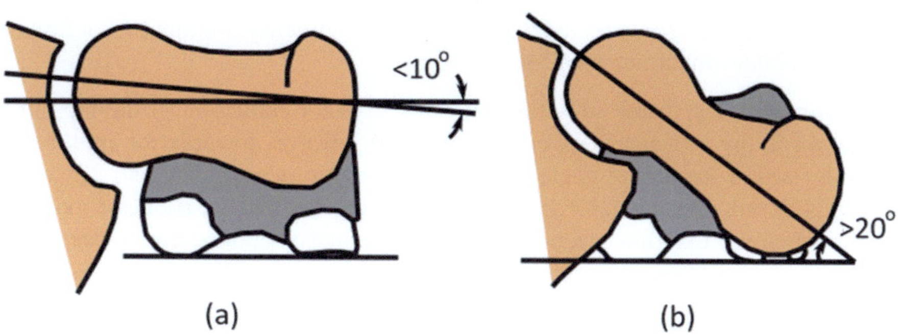

Fig. 1.24 Deformities associated with femoral neck anteversion angle; **a** retroversion deformity, **b** anteversion deformity

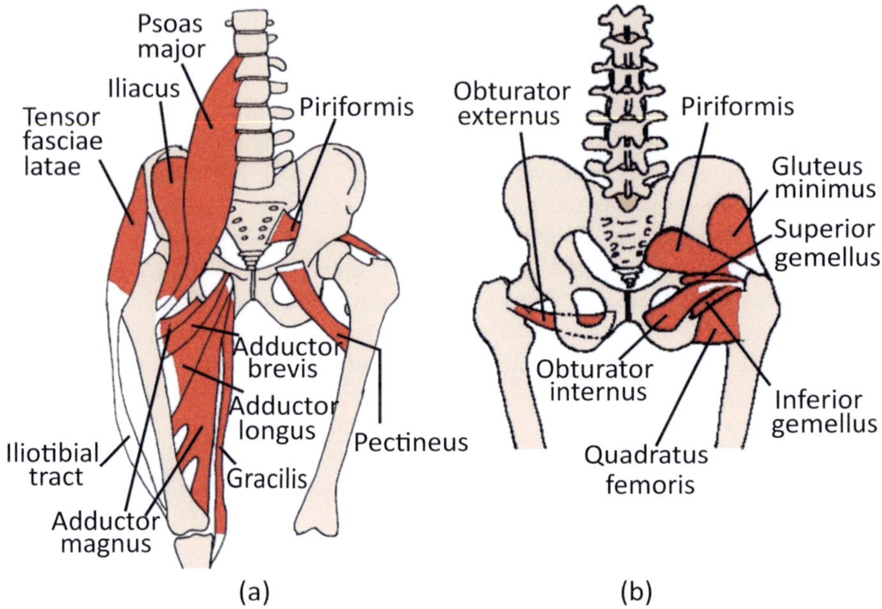

Fig. 1.25 Muscles associated with the hip joint; **a** anterior view, **b** posterior view

Extensor Muscles: These muscles are responsible for the extension of the hip joint. The extension involves moving the leg backward, straightening it from a flexed position, with a maximum angle of 20°. The extensor group includes gluteus maximus and hamstring (biceps femoris, semitendinosus, semimembranosus) muscles.

Abductor Muscles: These muscles are responsible for the abduction of the hip joint. Abduction is the movement of raising the leg away from the midline of the body, with a maximum angle of 50° when the hip is extended and 80° when the hip is flexed. The abductor group includes gluteus medius, gluteus minimus, tensor fasciae latae, and gluteus maximus muscles.

Adductor Muscles: These muscles are responsible for adduction of the hip joint. Adduction is the movement of bringing the leg inwards, towards the midline of the body, with a maximum angle of 30° when the hip is extended, and 20° when the hip is flexed. The adductor group includes adductor longus, adductor magnus, gracilis, obturator externus, pectineus, quadratus femoris, semimembranosus, semitendinosus, biceps femoris muscles.

Lateral Rotator Muscles: These muscles are responsible for lateral rotation of the hip joint. Lateral (External) rotation involves rotating the leg away from the midline of the body, typically with a maximum angle of around 30°–50°. The lateral rotator group includes quadratus femoris, piriformis, obturator externus, obturator internus, pectineus, adductor magnus, gluteus medius, gluteus maximus, biceps femoris, sartorius, iliopsoas, gemellus superior and gemellus inferior muscles.

Medial Rotator Muscles: These muscles are responsible for the medial rotation of the hip joint. Internal rotation involves rotating the leg towards the midline of the body, typically with a maximum angle of around 30°–50°. The medial rotator group includes tensor fasciae latae, gluteus medius, gluteus minimus, semimembranosus, and semitendinosus muscles.

1.9.4 Factors Affecting Stability of the Hip Joint

The stability of the hip joint is crucial for functioning and joint movements. The factors that contribute to the stability of the hip joint are as follows.

1.9.4.1 Shape of the Acetabulum

The acetabulum, the socket-like structure of the hip joint, has a deep and concave shape. This deep shape allows it to cover a significant portion of the spherical femoral head, providing a larger articular surface for the femur. This increased coverage enhances the stability of the joint, reducing the risk of dislocation and promoting smooth movement.

1.9.4.2 Ligaments of the Hip Joint

The hip joint is supported and stabilised by the extracapsular and intracapsular ligaments, as discussed in the following.

Extracapsular Ligaments

As the name indicates, these ligaments are seen outside the joint capsule, as shown in Fig. 1.26. The extracapsular ligaments comprise the iliofemoral, pubofemoral, and ischiofemoral ligaments.

- **Iliofemoral Ligament** (Ligament of Bigelow): It is also known as the "ligament of big yellow". This Y-shaped ligament is the strongest ligament in the body. Located on the anterior side of the hip joint, it resists the trunk falling backward in a standing posture.
- **Pubofemoral Ligament**: Pubofemoral Ligament is a triangular-shaped ligament adjacent to the iliofemoral ligament. It supports the joint inferomedially.
- **Ischiofemoral Ligament**: Ischiofemoral Ligament corresponds to a triangular band of strong fibres. It restricts internal rotation of the hip, regardless of the hip

Fig. 1.26 Extracapsular ligaments of the hip joint; **a** anterior view, **b** posterior view

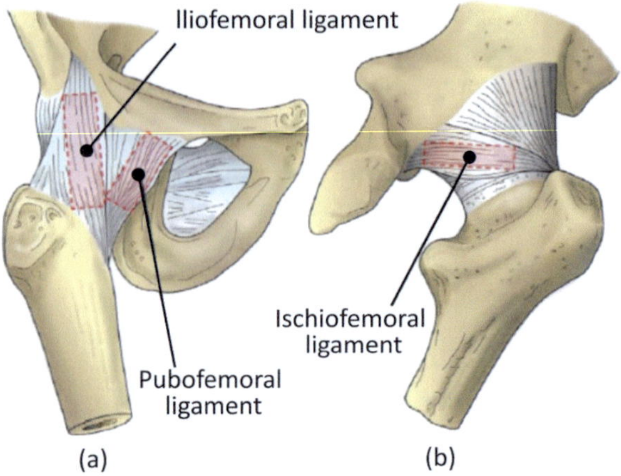

position (flexed, extended, or neutral). It is located on the posterior side of the hip joint.

Intracapsular Ligament

As the name indicates, these ligaments exist inside the joint capsule (Fig. 1.27).

The intracapsular ligaments consist of ligamentum teres and transverse ligament of the acetabulum.

- **Ligamentum Teres** (Ligament of the Head of the Femur): Ligamentum Teres is a relatively small ligament that runs from the acetabular fossa to the head of the

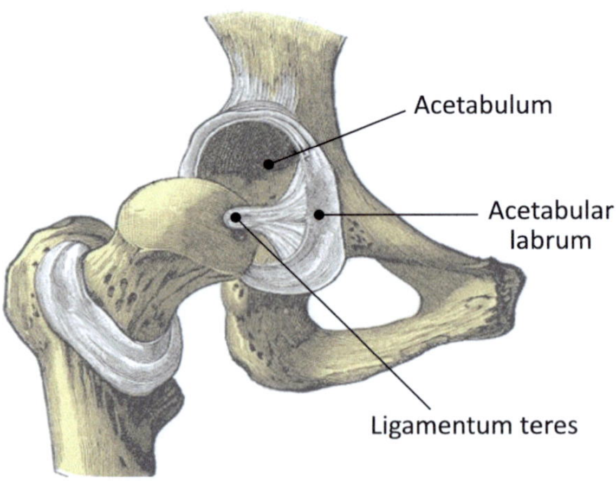

Fig. 1.27 Intracapsular ligaments of the hip joint

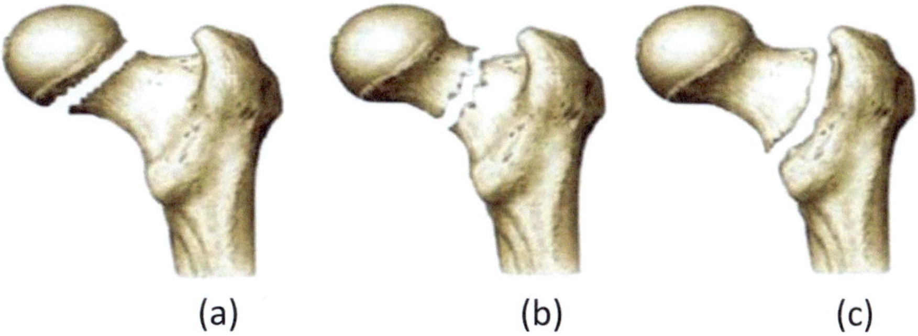

Fig. 1.28 Types of intracapsular fractures: **a** Subcapital, **b** transcervical, **c** basicervical

femur. It provides some stability to the hip joint and carries a small artery that supplies blood to the femoral head.

- **Transverse Ligament**: Transverse Ligament of the acetabulum is a portion of the acetabular labrum without cartilage cells within its fibres. It reinforces the acetabular labrum and helps maintain the stability of the joint.

The acetabular labrum is a ring of fibro-cartilaginous tissue that surrounds the acetabulum. It enhances the stability of the joint by deepening the socket and providing an articulating surface for the femoral head. The labrum also acts as a seal, maintaining a negative pressure within the joint, similar to a vacuum seal. This negative pressure further enhances joint stability. In a summary, the shape of the acetabulum, the strength and the arrangements of ligaments, and the presence of the acetabular labrum collectively contribute towards the stability of the hip joint.

1.9.5 Common Problems of the Hip Joint

Let us now discuss the common problems associated with the hip joint, including fractures and arthritis.

1.9.5.1 Fractures

Fractures of the hip joint can be classified into intracapsular and extracapsular fractures, depending on the location of the fracture.

Intracapsular Fractures: These fractures occur within the joint capsule of the hip. There are three main types of fractures: (a) subcapital, (b) transcervical, and (c) basicervical, as shown in Fig. 1.28. Subcapital fracture is characterised by a fracture occurring below the femoral head. Transcervical fracture is characterised by a fracture line that is transverse across the neck. Basicervical fracture is defined as a fracture of

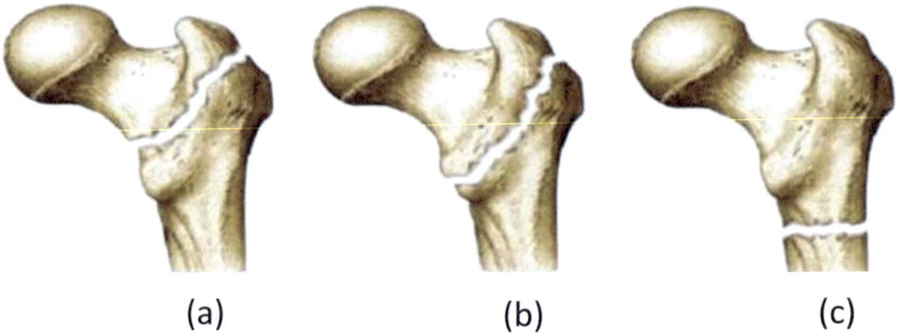

Fig. 1.29 Types of extracapsular fractures: **a** Intertrochanteric, **b** peritrochanteric, **c** subtrochanteric

the base of the femoral neck that occurs medially from the intertrochanteric line above the lesser trochanter. Intracapsular fractures are more common in older patients, typically in the age group between 60 and 80 years.

Extracapsular Fractures: These fractures occur outside the joint capsule of the hip. There are mainly three types of extracapsular fractures: (a) intertrochanteric, (b) peritrochanteric, (c) subtrochanteric, as shown in Fig. 1.29. An intertrochanteric fracture occurs between the trochanters of the femur. Peritrochanteric fracture extends along the trochanters. Subtrochanteric fractures are defined as fractures of the proximal femur that occur within 5 cm of the lesser trochanter. Extracapsular fractures are more common in younger adults and are often caused due to severe traumatic injuries.

1.9.5.2 Arthritis

Arthritis is a degenerative joint disease that affects the hip joint. The two most common forms of arthritis seen in the hip joint are osteoarthritis and rheumatoid arthritis.

Osteoarthritis: This is the most common form of arthritis in the hip joint. It is a degenerative condition in which the protective cartilage that covers the ends of bones gradually wears away (Fig. 1.30). As the cartilage wears away, bone-to-bone contact occurs, leading to the development of bone spurs. This often causes pain, stiffness, and limited range of motion in the hip joint. Osteoarthritis, typically associated with aging, is often seen in older adults or can otherwise be a manifestation of trauma-induced wear and tear of the joint.

Rheumatoid Arthritis: Rheumatoid arthritis is an autoimmune disorder in which the immune system attacks the synovial membrane that lines the joint capsule, causing inflammation and swelling. The inflammation can damage the cartilage, ligaments, and bones in the hip joint, leading to pain, stiffness, and reduced joint function.

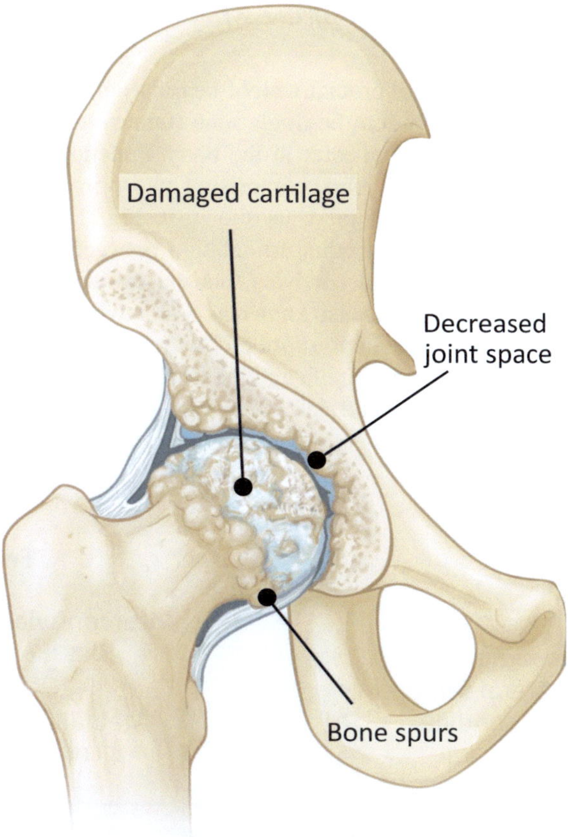

Fig. 1.30 Osteoarthritis in the hip joint

Rheumatoid arthritis can affect people of any age and is a chronic condition that requires ongoing management.

Both fractures and arthritis can significantly impact the function and mobility of the hip joint, leading to pain, restricted movement, and impaired quality of life. In some cases, surgical interventions may be necessary, particularly for severe fractures or advanced arthritis.

1.10 Knee Joint

In this section, we will gain an insight into the intricate structure and essential functions of the knee joint along with the common problems associated with this joint. Understanding the anatomy and mechanics of the knee is fundamental to appreciating its role in human locomotion and the challenges faced during physical activities.

1.10.1 Anatomy of the Knee Joint

The knee joint, a crucial weight-bearing joint in the lower extremity, serves as the connection between the thigh bone (femur) and the leg bone (tibia). Notably, it is the largest synovial joint in the body. Functionally, the knee joint is classified as a modified hinge joint, comprising two primary articulations: tibiofemoral joint and patellofemoral joint, as shown in Fig. 1.31. The tibiofemoral joint consists of the lateral and medial articulation between the condyles of the femur and the corresponding tibial condyles (plateaus). Owing to the presence of two condyles, tibiofemoral joint is also known as bi-condyloid joint. The patellofemoral joint is an intermediate articulation formed by the femur and the patella (kneecap). This joint exists between the patella and the trochlear groove of the femur, contributing to the overall movement and stability of the knee joint. The tibiofemoral joint is an example of a hinge joint, whereas the patellofemoral joint is an example of a gliding joint.

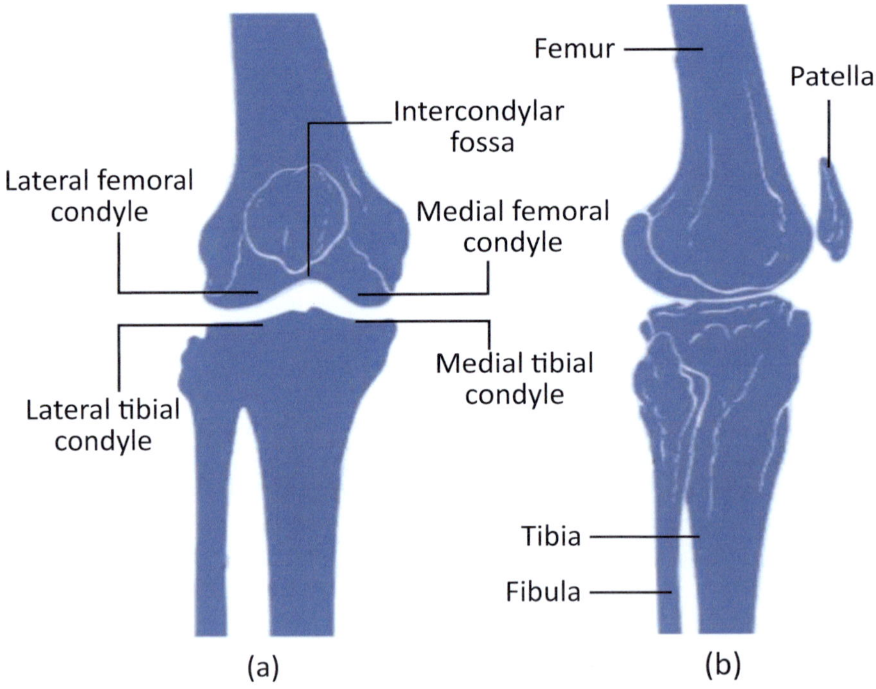

Fig. 1.31 Simplified sketch of knee joint; **a** anterior view, **b** lateral view

1.10.2 Functions of the Knee Joint

The primary functions of the knee joint are the following:

- Supporting heavy loads and body weight under both static and dynamic conditions.
- Facilitating the transfer of forces and moments from the upper body to the ankle and foot.
- Enabling human locomotion by facilitating a broad range of movements.
- Maintaining stability during various movements.

1.10.3 Patella

The patella, commonly known as the knee cap, is a sesamoid bone situated within the knee joint complex, as shown in Fig. 1.32. The patella, or knee cap, serves as a protective barrier for the knee joint, shielding it from potential impacts and injuries from the front. It offers mechanical advantages during knee extension by causing an anterior displacement of the quadriceps tendon, thereby increasing the moment arm of the quadriceps muscle force. This mechanical advantage enhances the effectiveness of knee extension during various movements. The patella aids in joint stability by facilitating a wider distribution of compressive stress on the femur. By increasing the contact area between the patellar tendon and the femur, the patella contributes to a more even distribution of the forces across the joint, reducing localised stress and promoting overall stability.

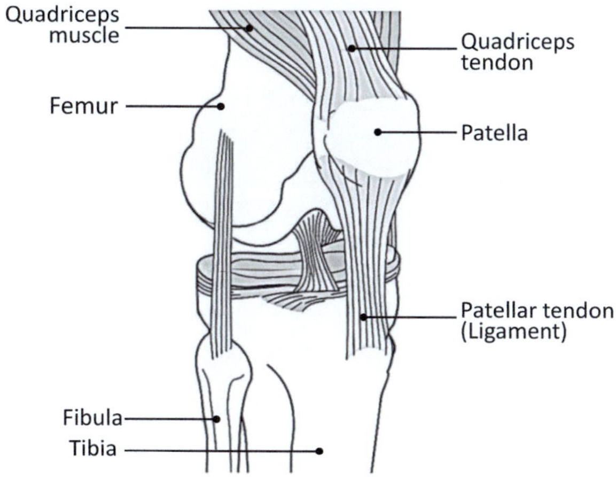

Fig. 1.32 Detailed sketch of the knee joint

1.10.4 Menisci

The menisci are fibrocartilage tissue located within the knee joint. The two C-shaped medial and lateral menisci consist of connective tissue with extensive collagen fibres containing cartilage-like cells. These menisci are attached at both ends to the intercondylar area of the tibia, forming a vital component of the knee-joint structure (Fig. 1.33). The menisci serve several crucial functions within the knee joint as stated below:

- Deepening the articular surface: The menisci deepen the articular surface of the tibia, thereby increasing the stability of the joint. This helps to ensure a better fit between the femur and tibia during movement.
- Improved joint congruence: By enhancing the congruence of the joint surfaces, the menisci promote smooth and coordinated movements, reducing the risk of dislocation or misalignment.
- Load distribution: The menisci play a vital role in distributing the forces more evenly across the tibial plateau. This even distribution of forces helps to prevent excessive stresses on specific areas, safeguarding the joint from wear and tear

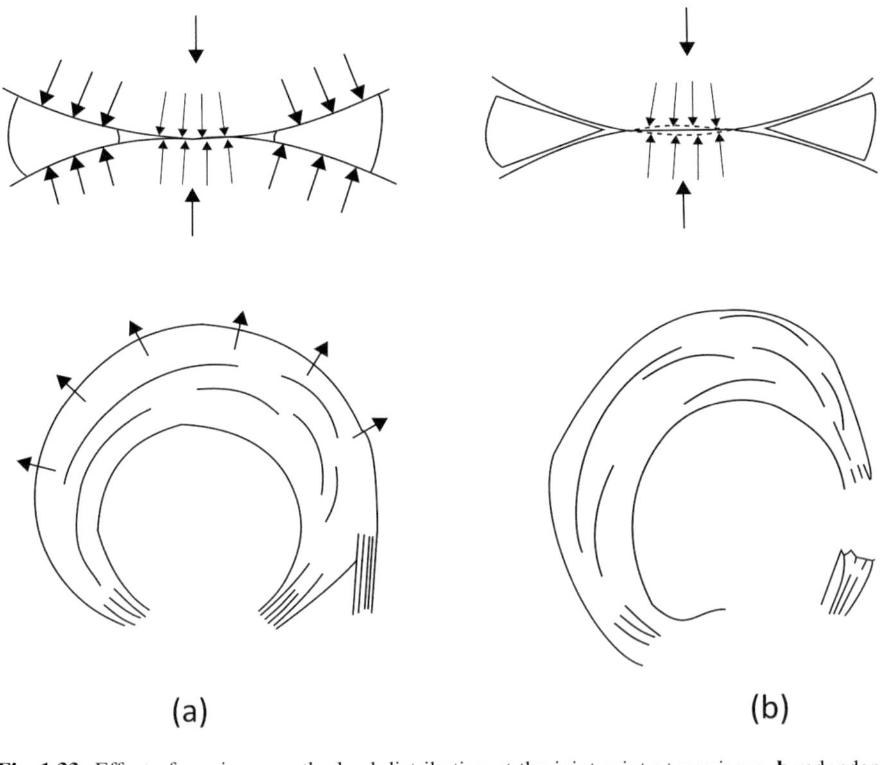

Fig. 1.33 Effect of meniscus on the load distribution at the joint **a** intact meniscus, **b** redundant meniscus

(Fig. 1.33). In a normal knee, the loads are distributed over a wide area of the tibial plateau. If the menisci are redundant, the load will be distributed on a limited to smaller contact area in the centre of the tibial plateau. Hence, the removal of menisci results in elevated stress on the articulating bones.

- Friction reduction: Acting as buffers, the menisci reduce friction between the tibia and femur, thereby minimising the wear on the joint surfaces during motion.
- Shock absorption: One of the essential functions of the menisci is to act as shock absorbers. They dampen the impact of weight-bearing activities, safeguarding the knee joint from excessive force and potential damage.

1.10.5 Factors Affecting Stability of the Knee Joint

The stability of the joint is influenced by various factors, which are summarized as follows. The tissue structure plays a crucial role in providing stability to the knee joint. The joint capsule, menisci, and ligaments collectively contribute to maintaining optimal knee stability. The ligaments play the primary role in stabilising the knee joints and providing static stability, whilst the muscles and tendons contribute to dynamic stability by ensuring smooth and controlled movements.

1.10.6 Muscles of the Knee Joint

The muscles around the knee joint can be divided into the following four groups, depending on the movements offered by them. Some of these muscles are illustrated in Fig. 1.34. The following movements and the associated muscle actions are essential for various activities like walking, running, and other common movements involving the knee joint. The coordinated action of these muscles ensures the stability and mobility of the knee joint during various daily activities.

Flexor Group: These muscles are responsible for the knee flexion movement. Flexion refers to the bending of the knee joint. The flexor group comprises the set of hamstrings (semimembranosus, semitendinosus, biceps femoris muscles), gracilis, sartorius, gastrocnemius, plantaris, and popliteus muscles.

Extensor Group: These muscles are responsible for the knee extension movement, which involves straightening the lower leg or moving the lower leg anteriorly. The extensor group comprises quadriceps muscles (rectus femoris, vastus medialis, vastus lateralis, and vastus intermedius) and the articularis genu.

Medial Rotator Group: These muscles are responsible for the medial (internal) rotation, which is the inward rotation of the knee joint. The medial rotator group comprises primarily the popliteus muscle.

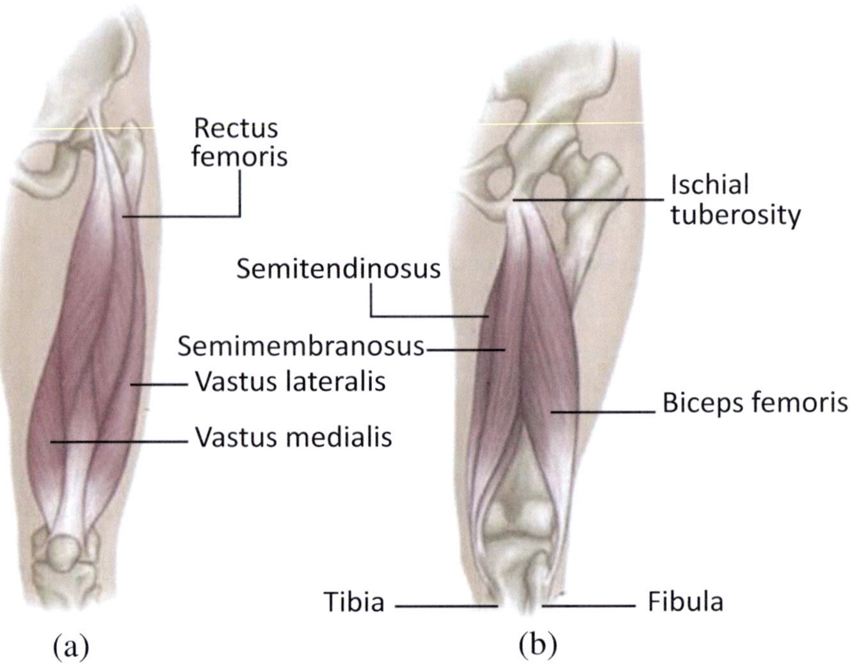

Fig. 1.34 Muscles around the knee joint in two views: **a** anterior, **b** posterior

Lateral Rotator Group: These muscles are responsible for lateral (external) rotation, which is the outward rotation of the knee joint. The lateral rotator group comprises primarily the biceps femoris and sartorius muscles.

1.10.7 Ligaments of the Knee Joint

There are five ligaments within the knee joint, which act as primary stabilisers. These ligaments play a crucial role in maintaining the stability and integrity of the knee joint during various movements, ensuring smooth and controlled motion. Damage or injuries to these ligaments can lead to instability and reduced functionality of the knee joint. A brief description of these ligaments and their function is presented below.

Medial Collateral Ligament (MCL): The MCL is a thick, wide, and flat ligament located on the medial (inner) side of the knee joint (Fig. 1.35). It attaches proximally to the medial condyle of the femur and distally to the medial condyle of the tibia. The MCL stabilises the knee during side-to-side movements and prevents excessive medial (inward) movements.

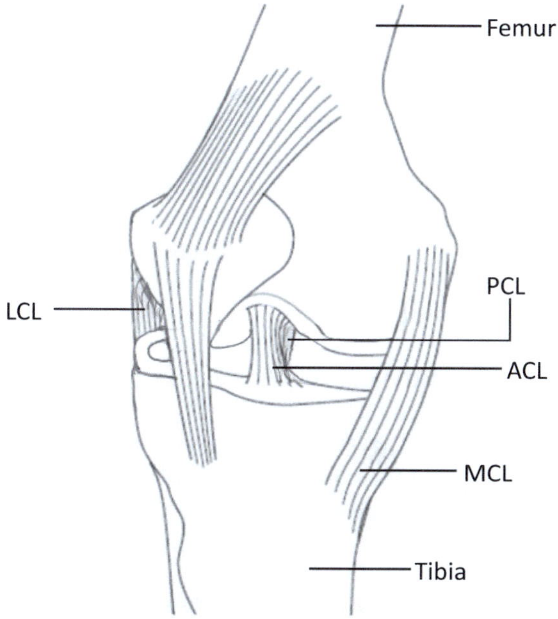

Fig. 1.35 Ligaments of the knee joint; MCL: Medial Collateral Ligament, LCL: Lateral Collateral Ligament, ACL: Anterior Cruciate Ligament, PCL: Posterior Cruciate Ligament

Lateral Collateral Ligament (LCL): The LCL is a thinner and rounder ligament situated on the lateral (outer) side of the knee joint (Fig. 1.35). It attaches proximally to the lateral femoral epicondyle and distally to a depression on the lateral surface of the fibula bone. The LCL provides stability during lateral (outward) movement and helps to prevent excessive lateral motion.

Patellar Ligament (Patellar Tendon): The patellar ligament is a continuation of the quadriceps tendon, which surrounds the patella (Fig. 1.35). It extends from the patella to the tibial tuberosity, a prominence on the tibia. The patellar ligament plays a vital role in knee extension and helps to maintain the position of the patella during movements.

Anterior Cruciate Ligament (ACL): The ACL is located within the knee joint and connects the anterior intercondylar region of the tibia to the femur in the intercondylar fossa (Fig. 1.35). It forms a cross-like structure with the posterior cruciate ligament (cruciate means 'cross' in Latin), providing rotational stability to the knee joint. The ACL prevents anterior dislocation of the tibia with respect to the femur.

Posterior Cruciate Ligament (PCL): The PCL also lies inside the knee joint, attaching to the posterior intercondylar region of the tibia and ascending anteriorly to attach to the anteromedial femoral condyle (Fig. 1.35). Together with the ACL, it forms the cruciate ligaments. The PCL prevents posterior dislocation of the tibia with respect to the femur.

1.10.8 Knee Deformities

In this section, we intend to discuss the knee deformities, known as varus and valgus knees (Fig. 1.36). A valgus deformity is manifested by inward bending of the knees (Fig. 1.36a). This misalignment can cause the knees to appear knock-kneed, affecting the distribution of forces within the knee joint. Conversely, a varus deformity is characterised by the outward bending of the knee, resulting in a bow-legged appearance (Fig. 1.36b). This misalignment results in additional pressure on the medial condyle (plateau) of the knee joint, potentially leading to various biomechanical problems. In order to quantitatively assess the severity of varus and valgus deformities, the angle of deformity is measured between the mechanical axis of the femur and the axis of the tibial shaft (Fig. 1.37). The mechanical axis of the femur is defined by a line from a point in the centre of the femoral head to a point in the centre of the knee joint. The axis of long bones, such as femur and tibia, is defined using a line passing through the intramedullary canal bisecting the long bone (Luo 2004). Understanding and measuring the varus and valgus deformities are vital in diagnosing and treating various knee conditions. An accurate assessment of these deformities enables healthcare professionals to devise appropriate treatment protocols, which may include bracing, physical therapy, or surgical interventions, depending on the severity of the misalignment.

1.10.9 Common Problems Associated with Knee Joint

The knee joint is a complex and crucial structure responsible for supporting our body weight, facilitating movement, and enabling various activities. Its intricate network of ligaments, muscles, and cartilage plays a vital role in offering a range of movements as well as maintaining stability during movements. However, it is susceptible to various injuries and degenerative conditions, which can significantly affect mobility and quality of life.

Ligament Injuries and Tears: Amongst the common problems associated with the knee joint, the cruciate ligament (ACL and PCL) tears are frequently encountered. The knee joint's vulnerability to injuries suffered during sports activities or accidents makes it susceptible to such tears (Fig. 1.38). Notably, the anterior cruciate ligament is the most commonly injured ligament in the knee.

Patellar tendinitis is characterised by inflammation of the patellar ligament, due to overuse and excessive stress on the patellofemoral joint, resulting in localised pain.

Meniscus tears are prevalent knee injuries. These C-shaped fibrocartilage pads can suffer longitudinal tears, often caused by forceful twisting or rotation of the knee. Athletes, especially those involved in contact sports, are at a higher risk of experiencing meniscus tears.

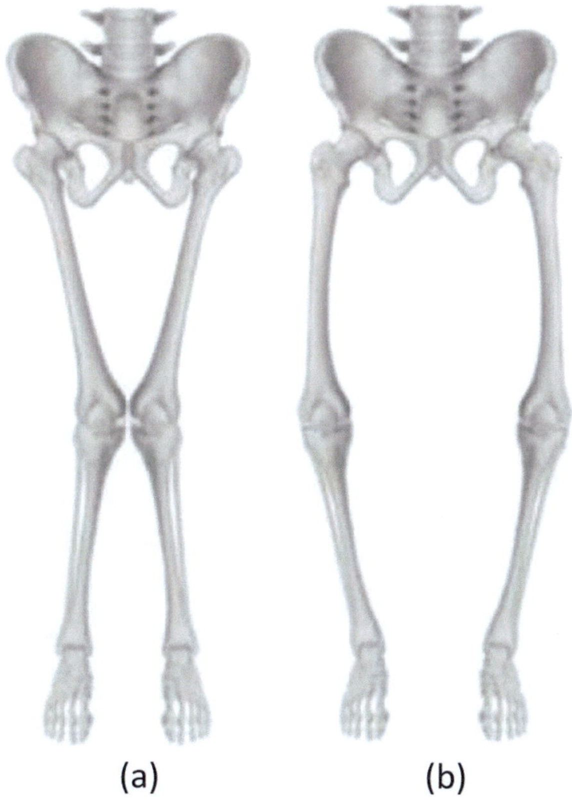

Fig. 1.36 Knee deformities; **a** valgus knee **b** varus knee

Knee arthritis is also a common problem affecting patients, manifested in the form of osteoarthritis and rheumatoid arthritis. Osteoarthritis is a degenerative bone disease that particularly affects the knee joint. Gradual wear of the cartilage leads to bone-to-bone contact and the formation of bone spurs, causing pain, stiffness, and restricted movement in the joint. In addition, rheumatoid arthritis, an autoimmune disorder, leads to inflammation and swelling of the synovial membrane, resulting in knee pain and stiffness. The attack of the immune system on normal tissues, including cartilage and ligaments, can lead to further complications in the knee joint.

1.11 Ankle Joint

The ankle joint is a complex structure in the human body, connecting the foot to the lower leg. It is an important structure since it forms the kinetic linkage allowing the lower leg to interact with the ground, a key requirement for many activities of daily living. Its remarkable design offers a wide range of movements, while

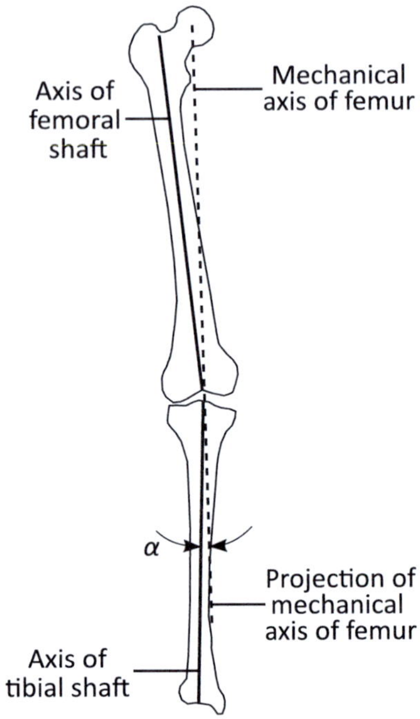

Fig. 1.37 The knee angle of deformity (α)

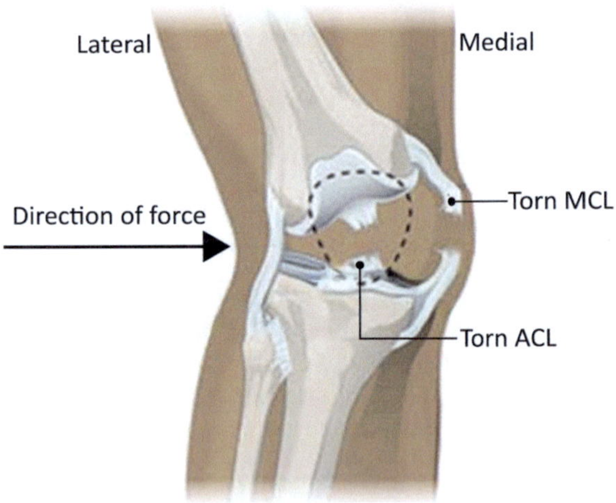

Fig. 1.38 Ligament tear in knee joint. ACL: anterior cruciate ligament, MCL: medial collateral ligament

providing stability and support during static and dynamic conditions. Understanding the anatomy and function of the ankle joint is required for appreciating its role in providing mobility and maintaining postural stability.

1.11.1 Anatomy of the Ankle Joint

The ankle joint is a synovial hinge joint connecting the lower leg (distal tibia and fibula) and the foot (talus). It is also known as the talocrural joint or tibiotalar joint. This joint primarily involves the articulation between the distal ends of the tibia and fibula with the talus, as shown in Fig. 1.39. The malleoli of the tibia and fibula constrain the talus, such that the joint functions as a hinge joint, and primarily contributes to the plantarflexion and dorsiflexion movements of the foot. Malleoli refers to the bony projections on either side of the ankle joint. There are lateral and medial malleolus associated with the tibia and fibula.

While the focus of this chapter excludes a detailed discussion on the foot, it is pertinent to understand the anatomy of the foot, owing to the pivotal role played by the talus in the ankle and foot. The foot comprises 28 bones and nearly 40 joints. It can be further categorised into three anatomical regions: forefoot, midfoot, and hindfoot. The forefoot encompasses the phalanges of five toes, five metatarsals, and the sesamoids of the first metatarsal head, along with interphalangeal and metatarsophalangeal joints. The midfoot includes the navicular, cuboid, and three cuneiform bones, incorporating naviculocuneiform, metatarsocuneiform, tarsometatarsal, and cubometatarsal joints. The hindfoot comprises the talus and calcaneus, featuring the subtalar, talonavicular, and calcaneocuboid joints. The subtalar joint allows for special foot movements, inversion, and eversion, which are already discussed earlier in Sect. 1.8.1.

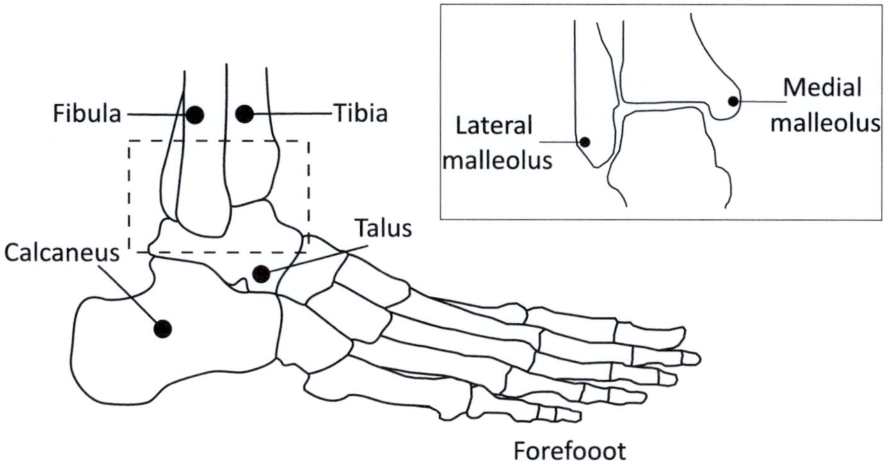

Fig. 1.39 Lateral view of the ankle joint. An enlarged posterior view of the ankle joint illustrating the malleoli is shown in the inset

1.11.2 Functions of the Ankle Joint

The ankle joint plays several vital roles, as stated in the following:

- It is a weight-bearing joint, supporting the body weight during different postures.
- It acts as a shock absorber, helping to dissipate the impact forces generated during weight-bearing activities. This function is crucial for reducing the risk of injury to the lower extremities.
- Positive power or work generated by the plantarflexor muscles about the ankle joint towards the end of the stance phase facilitates forward propulsion and swing initiation. This positive power is known as ankle push-off.
- It plays a key role in maintaining balance and proprioception, which is the body's ability to sense its position in space. This is essential for activities that require stability and coordination, such as standing on one leg or navigating uneven terrain.

1.11.3 Muscles of the Ankle Joint

The muscles of the ankle joint can be classified according to associated movements. The primary muscles of the ankle joint include the gastrocnemius, soleus, tibialis anterior, and the extensor digitorum longus.

Dorsiflexor Muscles: Dorsiflexion is the movement of the foot in an upward direction, bringing the foot closer to the shin bone. The tibialis anterior muscle is the major contributor to dorsiflexion of the ankle joint. This muscle is located on the front of the shin and runs down the leg to insert into the medial cuneiform and the base of the first metatarsal bone of the foot. When the tibialis anterior muscle contracts, it pulls the foot upward, causing dorsiflexion. Other dorsiflexor muscles having minor contributions are: extensor digitorum longus and extensor hallucis longus.

Plantarflexor Muscles: Plantarflexion is the movement of the foot in a downward direction, pointing the foot away from the shin bone. The main muscles contributing to plantarflexion are the gastrocnemius, soleus, and plantaris. These are called calf muscles, which insert at the calcaneous through the archilles tendon (Fig. 1.40). As the plantarflexor muscles contract, the ankle joint undergoes plantarflexion.

1.11.4 Factors Affecting Stability of Ankle Joint

The stability of the ankle joint is crucial for providing support during various movements. The factors that contribute to ankle stability are as follows.

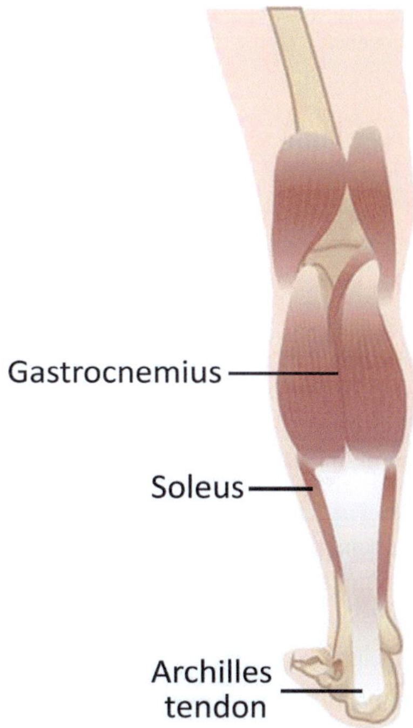

Fig. 1.40 The main calf muscles

1.11.4.1 Shape of the Articulating Surface

The superior surface of the talus, which articulates with the tibia and fibula, has a trochlear (pulley-like) shape. The trochlea of the talus fits into the mortise formed by the tibia and fibula, creating a stable hinge joint. This trochlear configuration helps to guide and limit the movement of the talus, thus contributing to the stability of the ankle joint during dorsiflexion and plantarflexion. The lower ends of the tibia and fibula create a mortise, or socket-like structure, that accommodates the trochlea of the talus. The mortise provides a stable bony structure that surrounds and supports the talus, enhancing joint stability.

1.11.4.2 Ligaments of Ankle Joint

There are several ligaments that surround the ankle joint. The lateral ligaments (anterior talofibular, posterior talofibular, and calcaneofibular) and the deltoid ligament on the inner side play key roles in stabilising the joint. The tibiofibular syndesmosis (fibrous joint) restricts movement between the tibia and fibula in everyday activities, preserving stability between the ends of the bones. This syndesmosis comprises three components: the anterior tibiofibular ligament, the posterior tibiofibular ligament, and the interosseous tibiofibular ligament. While the primary stability of the ankle joint

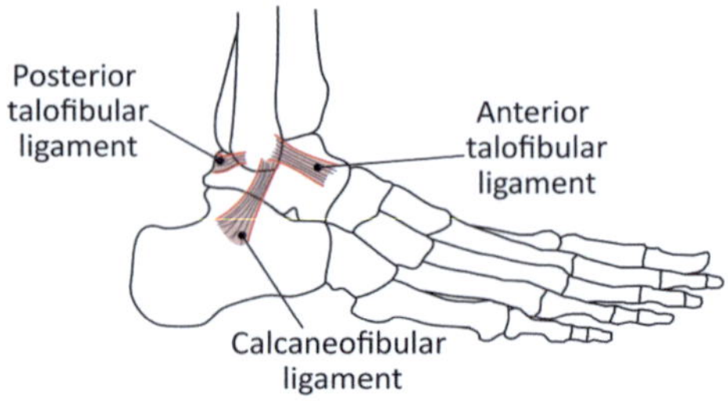

Fig. 1.41 Lateral ankle ligaments

is attributed to the ligaments around the ankle joint, the tibiofibular syndesmosis also contributes to the overall stability of the ankle complex. It works in conjunction with the ligaments on the medial and lateral sides to prevent abnormal movements and to maintain the structural integrity of the joint.

Lateral Ligaments: The lateral ligaments include the anterior talofibular ligament, the calcaneofibular ligament, and the posterior talofibular ligament (Fig. 1.41). These ligaments are essential for stabilising the lateral aspect of the ankle joint by resisting excessive inversion and providing stability during weight-bearing activities.

Anterior Talofibular Ligament (ATFL): This ligament runs from the anterior part of the lateral malleolus to the talus. The ATFL is the most commonly injured lateral ligament during ankle sprains. It provides stability against excessive inversion, preventing the foot from turning inward.

Posterior Talofibular Ligament (PTFL): This ligament runs from the posterior part of the lateral malleolus to the talus. The PTFL provides additional support against excessive plantarflexion and inversion of the ankle joint. The anterior and posterior ligaments withstand high tensile forces under plantarflexion and dorsiflexion, respectively.

Calcaneofibular Ligament (CFL): The CFL extends from the lateral malleolus to the lateral surface of the calcaneus (heel bone). The CFL provides lateral stability to the ankle joint and is particularly active during plantarflexion and inversion.

The stabilising role of the ankle joint is assumed by each of the lateral ligaments depending on the foot position. During dorsiflexion, the posterior talofibular ligament (PTFL) experiences maximum stress, while the calcaneofibular ligament (CFL) is taut, and the anterior talofibular ligament (ATFL) is loose. Conversely, in plantarflexion, the ATFL becomes taut and the CFL and posterior talofibular ligaments are loose (Knupp et al. 2015).

Deltoid Ligament: The medial aspect of the ankle joint is supported by the deltoid ligament (medial collateral ligament complex). The deltoid ligament is a strong, flat, triangular ligament located at the medial (inner) side of the ankle. Unlike the lateral ligaments on the outer side of the ankle, the deltoid ligament is a ligamentous complex that comprises of anterior tibiotalar ligament, tibionavicular ligament, tibiocalcaneal ligament, and posterior tibiotalar ligament. The deltoid ligament functions primarily to resist excessive eversion of the ankle, preventing the talus from moving away from the medial malleolus.

1.11.5 Common Problems Associated with Ankle Joint

Common issues associated with the ankle joint include ligament injuries, such as lateral ankle sprains, which can lead to chronic ankle instability and osteoarthritis. Ankle osteoarthritis can cause pain and functional limitations. Additionally, ankle impingement syndromes, posterior ankle pain, and ankle fractures are prevalent problems that can affect joint function and mobility.

1.11.5.1 Ankle Arthritis

Ankle arthritis can result from various factors, e.g. osteoarthritis, rheumatoid arthritis, and trauma. However, it is frequently associated with trauma, leading to a higher proportion of post-traumatic cases as compared to hip arthritis (Ewalefo et al. 2018). Post-traumatic osteoarthritis develops right after an injury, unlike other forms of arthritis that develop over time. It is characterised by injury-induced damage to the joint cartilage.

1.11.5.2 Fractures

Malleolar fractures specifically refer to fractures involving the malleoli, which are the bony protrusions on either side of the ankle joint. Fractures involving these structures are common and can occur due to various mechanisms, such as twisting injuries, direct trauma, or falls. The two main types of malleolar fractures are lateral and medial malleolus fractures. The lateral malleolus is more commonly fractured than the medial malleolus. It often results from inversion injuries, where the ankle is forced into twisting or rolling. This can happen during activities like walking on an uneven surface or playing sports that involve sudden changes in direction. It may also occur as a result of high-impact accidents. The majority of lateral malleolus fracture cases are stable, meaning they are non-displaced. However, impact injuries might lead to unstable fractures. In the case of a medial malleolus fracture, the medial malleolus is the fracture site. Medial malleolus fracture can result from direct trauma to the inner side of the ankle or from forces that cause excessive inward or outward rolling of the ankle.

1.12 Shoulder Joint

The shoulder joint is an example of a very complex musculoskeletal structure consisting of a chain of bones connecting the upper extremity to the trunk. The shoulder joint includes multiple bones, namely humerus, scapula, clavicle, as well as the thoracic cage.

1.12.1 Anatomy of the Shoulder Joint

The shoulder joint resembles a ball and socket joint, which facilitates rotation around different axes. It is a highly mobile joint with a large range of motion. The shoulder girdle comprises the scapula and clavicle, and functions as a movable but stable base for the motions of the humerus (Veeger et al. 1991). The shoulder consists of three synovial joints, which interconnect the bones.

- The sternoclavicular (SC) joint, connecting the sternum part of the thorax to the clavicle.
- The acromioclavicular (AC) joint, connecting the clavicle to the acromion of the scapula.
- The glenohumeral (GH) joint, usually represented by a ball and socket joint connecting the glenoid cavity of the scapula and the head of the humerus (Fig. 1.42).

The glenohumeral joint is formed by the articulation of the humeral head with the glenoid cavity, as shown in Fig. 1.43. The acromioclavicular joint is the joint formed by the acromion and the clavicle. The sternoclavicular joint is the joint formed by the sternum and the clavicle. One end of the clavicle forms the acromioclavicular joint, and the other end forms the sternoclavicular joint with the sternum. During movement of the arm, the medial border of the scapula glides over the dorsal side of the thoracic cage, pressed onto it by the combined action of muscles (Gupta 2002).

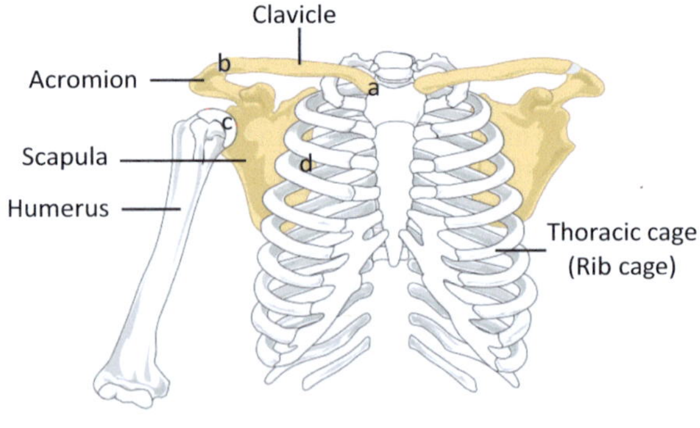

Fig. 1.42 The joints of the shoulder girdle: **a** SternoClavicular (SC) joint, **b** AcromioClavicular (AC) joint, **c** GlenoHumeral (GH) joint, **d** ScapuloThoracic Gliding Plane (STGP)

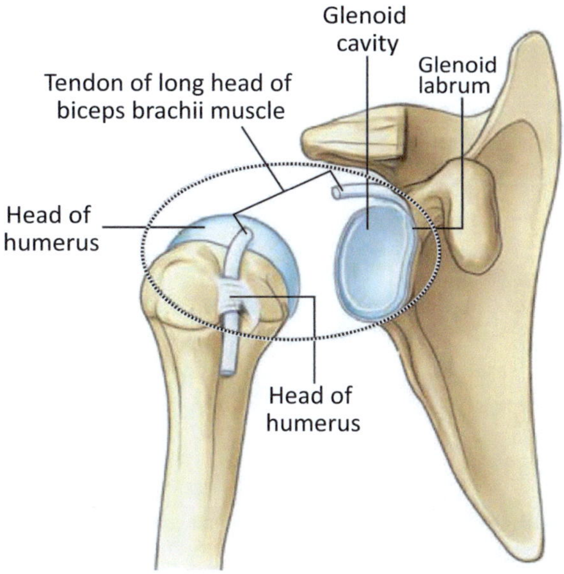

Fig. 1.43 The glenohumeral joint of the shoulder

This connection gives rise to a fourth joint, the Scapulo-Thoracic Gliding Plane (STGP).

Let us now look into the complex structure of the scapula (Fig. 1.44). The scapula is a large, flat, triangular bone consisting of solid bony ridges (glenoid, acromion, scapular spine, lateral border, medial border, and coracoid process) and two thin, hard laminated structures (infraspinous and supraspinous fossa) (Gupta 2002). The infraspinous fossa is surrounded by the glenoid, the scapular spine, the medial border, and the lateral border. The structure of the scapula, in ventral and dorsal views, is shown in Fig. 1.44a, b, respectively.

1.12.2 Movements Associated with Shoulder Joint

The shoulder joint offers the largest range of movements in the human body. The normal movements include abduction, adduction, flexion, extension, internal rotation, and external rotation. Raising the arm sideways is called abduction, in which the palm moves away from the midline of the body (Fig. 1.45a). Contrarily, lowering the arm back to the resting position is called adduction (Fig. 1.45b). Abduction and adduction movements take place in the coronal plane. In shoulder flexion, the arms are raised up straight above the head from the resting position by the sides of the body (Fig. 1.45c). An extension is the reverse movement of flexion when the arms are moved behind the body (Fig. 1.45d). The flexion and extension of the shoulder takes place in the sagittal plane. Flexion and extension can also be executed in the horizontal plane, and hence, they are termed horizontal flexion and horizontal extension.

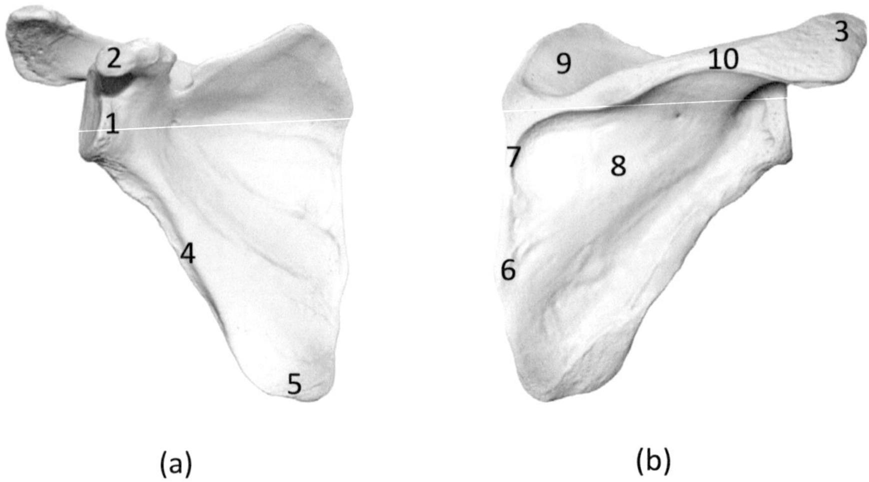

Fig. 1.44 Structure of the right scapula; **a** anterior view, **b** posterior view; 1. Glenoid, 2. Coracoid process, 3. Acromion, 4. Lateral border, 5. Angulus inferior, 6. Medial border, 7. Trigonum spinae, 8. Infraspinous fossa, 9. Supraspinous fossa, 10. Scapular spine

Here, the arms are held straight out to the side and the straight arms are then brought toward the midline of the body and across the chest. Apart from these, the movements of the shoulder are internal and external rotations. The internal rotation corresponds to the rotation of the arm inwards towards the midline of the body; whereas the movement of the arm away from the midline of the body is called external rotation (Fig. 1.45e). The range of motion of different movements is 180° for flexion, 45°–60° for extension, 180° for abduction/adduction and 90° for external/internal rotation.

1.12.3 Shoulder Joint Muscles

A large number of muscles are associated with the shoulder joint. These muscles either have an origin or insertion on the humerus, scapula, or clavicle (Fig. 1.46). The major muscles of the shoulder are the deltoid, trapezius, supraspinatus, infraspinatus, teres major, teres minor, rhomboideus, and latissimus dorsi. Other muscles include the long head of triceps, biceps brachii, pectoralis minor, serratus anterior, and subscapularis.

The rotator cuff group of muscles connects the upper arm to the shoulder blade. The primary function of the rotator cuff muscles is to enable rotation in the shoulder joint. This rotator cuff group consists of four muscles, teres minor, infraspinatus, supraspinatus, and subscapularis (Fig. 1.47). The subscapularis is attached to the anterior side of the scapula, whereas the infraspinatus and supraspinatus are attached to the posterior side of the scapula, one to the infraspinatus fossa, and the other

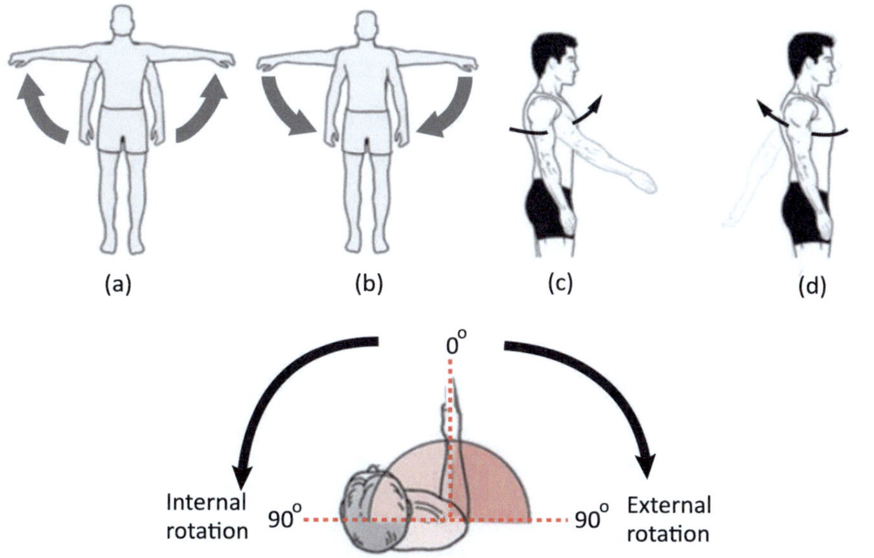

Fig. 1.45 Various movements of the shoulder joint; **a** abduction, **b** adduction, **c** flexion, **d** extension, **e** internal and external rotation

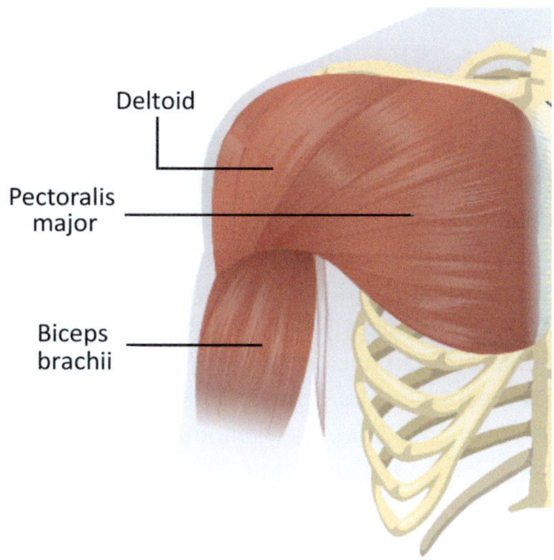

Fig. 1.46 Few major muscles around the shoulder joint

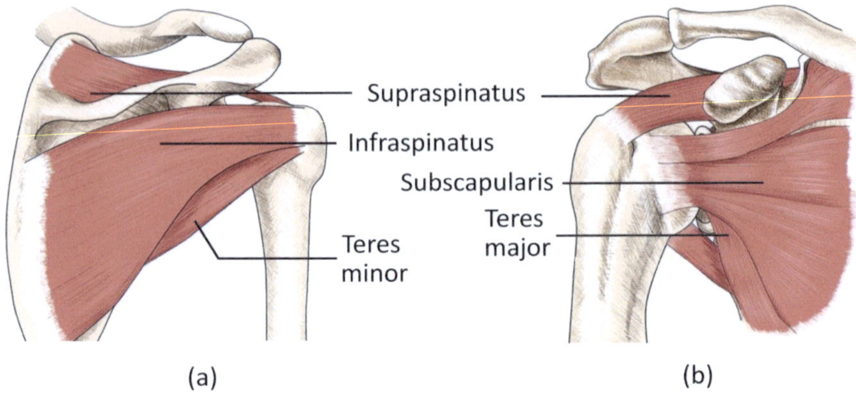

Fig. 1.47 Rotator cuff muscles; **a** posterior view, **b** anterior view

attached to the supraspinous fossa, respectively. The rotator cuff muscles insert at the scapula and attach to the humerus through a tendon. These muscles enable joint rotation and provide rotational stability to the shoulder. Tears in the tendons of these muscles are known as rotator cuff tears. The supraspinatus is the most commonly affected muscle within the group of rotator cuff muscles.

1.12.4 Shoulder Joint Ligaments

The major ligaments of the shoulder joint are coraco-acromial, coraco-humeral, glenohumeral, and coraco-clavicular ligaments. The coraco-clavicular ligament connects the coracoid process with the clavicle. The coraco-clavicular ligaments comprise of the conoid ligament and trapezoid ligament (Fig. 1.48). The coraco-humeral ligament connects the coracoid process with the humerus, Whereas the coraco-acromical ligament connects coracoid process with acromion. The gleno-humeral ligaments are a group of ligaments, which are a part of the glenohumeral joint capsule. The primary function of these ligaments is to provide stability to the joint. In particular, the glenohumeral ligament plays a major role in preventing the shoulder from dislocation.

1.12.5 Common Problems in the Shoulder Joint

Let us now look into the common problems associated with the shoulder joint. As indicated earlier, the shoulder joint offers the largest range of motion in the human body. However, this mobility may cause instability, which increases the potential risk of joint injury, often leading to degenerative processes, eventually causing pain

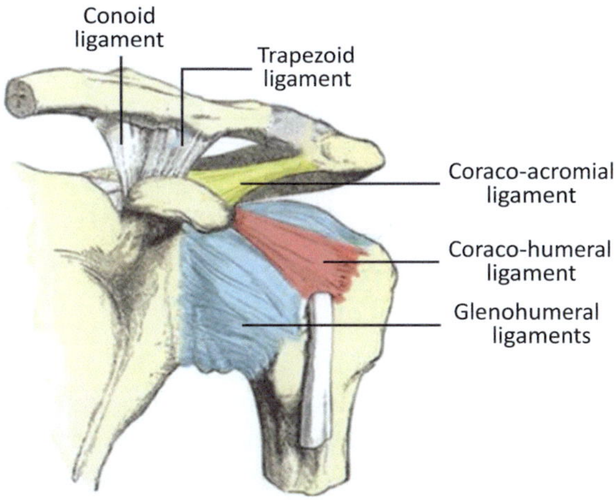

Fig. 1.48 The ligaments of the shoulder joint

and reduced mobility. It may also lead to impingement of the soft tissue or bony structures, resulting in pain. Major problems of the shoulder joint are as follows:

- Tendon inflammation or tendon tear,
- Shoulder instability,
- Arthritis,
- Bone fracture,
- Frozen shoulder.

The following is a brief discussion of the common problems associated with the shoulder.

- **Tendon inflammation or tendon tear**: The splitting and tearing of the tendons is a major problem that results from acute injury or degenerative changes in the tendons due to age, overuse, or a sudden injury. Under this category, the most common injuries are rotator cuff and biceps tendon injuries. Impingement occurs when the acromion pressurizes the underlying soft tissue during abduction. As the arm is lifted, the acromion impinges on the rotator cuff tendons and bursa. This leads to bursitis (inflammation of bursa) and tendonitis, causing pain and restrictive movement, as presented in Fig. 1.49. A bursa is a closed fluid-filled sac that functions as a gliding surface and cushion to reduce friction between tissues of the body. It is located adjacent to the glenohumeral joint towards the superior side.
- **Shoulder instability**: The shoulder instability occurs when the humeral head is forced out of the glenoid cavity or the shoulder cavity. Dislocation and subluxation are examples of shoulder instability (Ladd et al. 2021). These scenarios arise due to impact injury or overuse. As the ligaments, tendons, and muscles surrounding the joint cavity become loose or torn, dislocation may occur. Shoulder dislocations occur when the humeral head comes fully out of the glenoid cavity. A shoulder subluxation can be defined as partial or incomplete dislocation of the

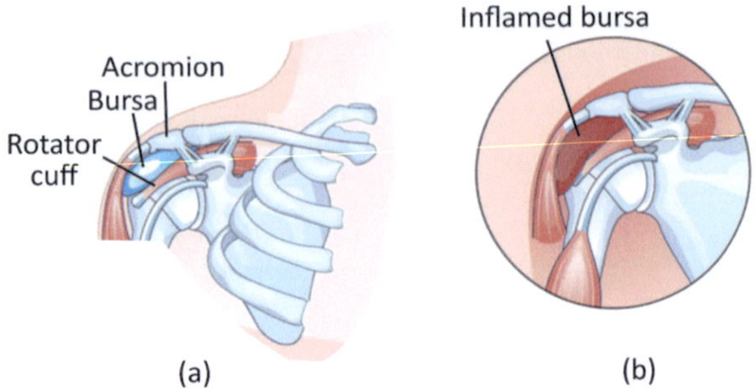

Fig. 1.49 Inflammation of the bursa: **a** normal bursa, **b** inflamed bursa

glenohumeral joint (Fig. 1.50). The repeated occurrence of the subluxations or dislocations can lead to arthritis.

- **Arthritis**: The shoulder arthritis is a degenerative joint disease common to all synovial joints. It is accompanied by shoulder pain and stiffness. Common arthritic disorders that affect the shoulder joint are osteoarthritis and rheumatoid arthritis. The glenohumeral and the acromial clavicular joints are typically affected, resulting in restricted movement of the shoulder joint.
- **Bone fracture**: Fracture can occur in the clavicle (collarbone), humerus (upper arm bone), and the scapula (shoulder blade) due to different reasons. In most cases, shoulder fracture is attributed to falls or traumatic injuries.

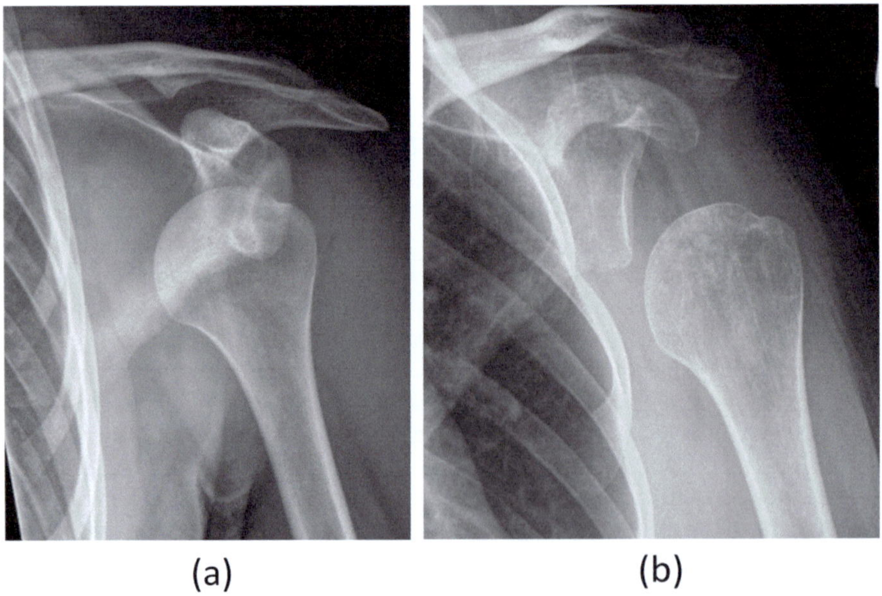

Fig. 1.50 Shoulder instability: **a** dislocation, **b** inferior subluxation

- **Frozen shoulder** (adhesive capsulitis): It is a condition that causes pain and stiffness in the shoulder and is quite common in patients. It is an inflammatory condition in the glenohumeral joint capsule that prevents normal movement of the shoulder. Eventually, the shoulder is frozen, the joint becomes stuck and its movement is limited.

1.13 Elbow Joint

The elbow joint consists of mainly three bones: humerus, radius, and ulna. In the following subsections, we will be discussing the anatomy, movements, muscles, ligaments, and common disorders associated with the elbow joint.

1.13.1 Anatomy of the Elbow Joint

The elbow joint resembles a hinge joint which facilitates a large range of motion. Let us now look into the articulation of the bony structures in the elbow joint. The articulation between the trochlea of the humerus and the trochlear notch of the ulna is called the humeroulnar joint (Fig. 1.51). The articulation between the head of the radius and the capitulum of the humerus is known as the humeroradial joint. Whereas, the articulation between the head of the radius and the radial notch of the ulna is called the radioulnar joint. The humeroulnar joint is an example of a hinge joint, whereas the humeroradial and radioulnar joints are examples of a pivot joint. The elbow joint primarily permits flexion, extension through humeroulnar and humeroradial joints. Whereas the movements, such as pronation and supination, are facilitated by the radio-ulnar joint.

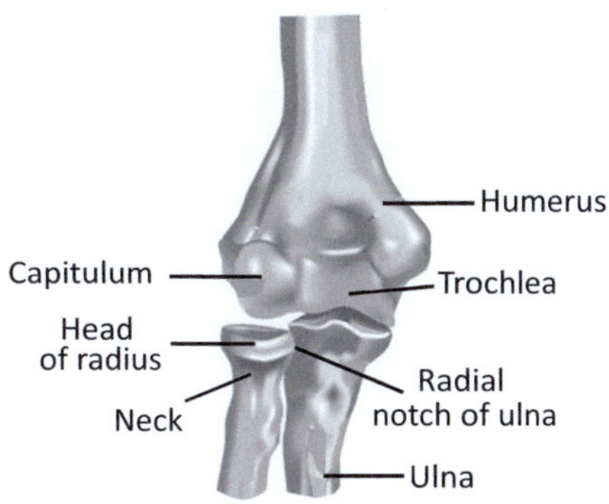

Fig. 1.51 Structure of the elbow joint

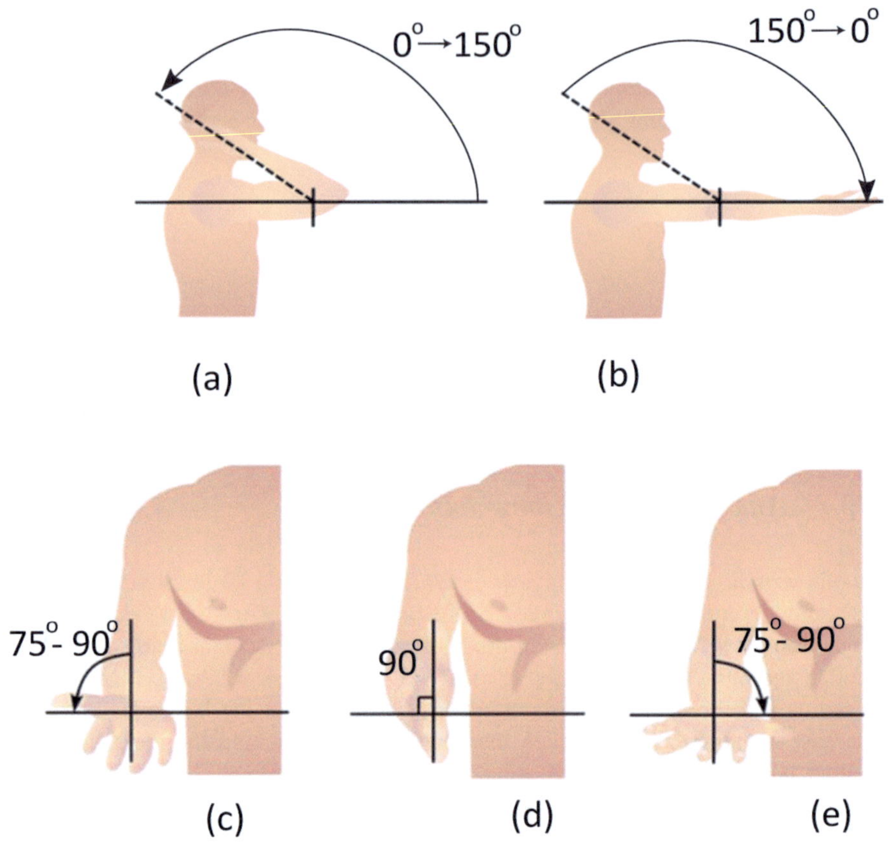

Fig. 1.52 Movements of elbow joints: **a** flexion, **b** extension, **c** supination, **d** neutral, **e** pronation

1.13.2 Elbow Joint Movements

Major movements of the elbow joints involve flexion and extension. The normal range of flexion or extension is up to ~150° (Fig. 1.52a, b). Apart from flexion and extension, some specialised movements, supination, and pronation, are also associated with the elbow joint. If we consider the arm in the neutral position as in Fig. 1.52a, opening the palm corresponds to supination with a normal range of 75°–90° (Fig. 1.52c), whereas closing the arm in the opposite direction corresponds to pronation with the same range (Fig. 1.52d).

1.13.3 Elbow Joint Muscles

The details of the muscles responsible for the elbow movements are as follows.

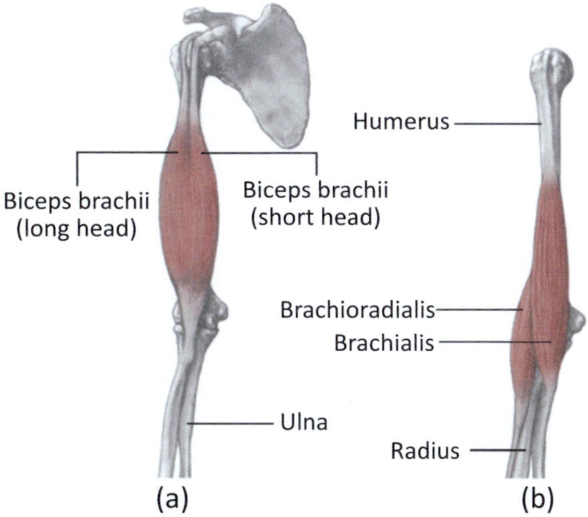

Fig. 1.53 Muscles associated with shoulder flexion: **a** biceps brachii, **b** brachialis and brachioradialis

Extension and Flexion Muscles: The primary extensor muscle responsible for the extension movement of the elbow is the triceps brachii. On the other hand, the primary flexor muscles that facilitate flexion of the elbow include the brachialis, biceps brachii, and brachioradialis (Fig. 1.53).

Pronation and Supination Muscles: The primary muscles responsible for pronation are the pronator quadratus and the pronator teres (Fig. 1.54). On the other hand, the muscles that enable supination are the supinator and the biceps brachii.

1.13.4 Elbow Joint Ligaments

The ligaments of the elbow joint are classified as the medial collateral ligament complex and the lateral collateral ligament complex (Fig. 1.55). The medial collateral ligament complex consists of the anterior bundle, posterior bundle, and transverse ligament (TL). Whereas, the lateral collateral ligament complex consists of the lateral ulnar collateral ligament (LUCL), radial collateral ligament (RCL), and annular ligament (AL).

1.13.5 Common Elbow Joint Problems

A variety of conditions can affect the elbow joint, leading to discomfort and functional limitations. Some common problems associated with the elbow joint include:

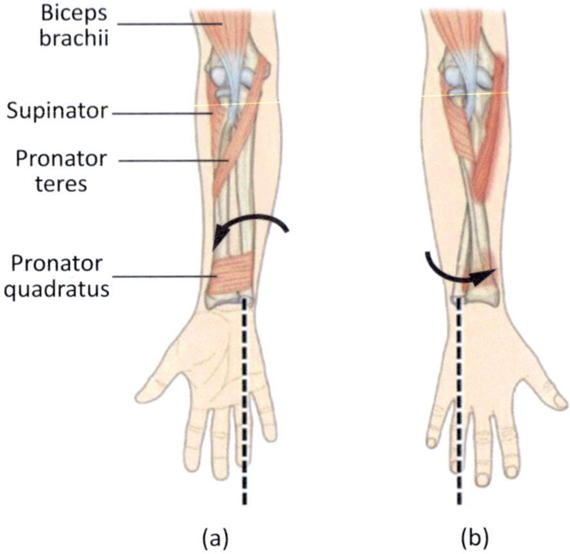

Fig. 1.54 Muscles associated with elbow movements; **a** supination, **b** pronation

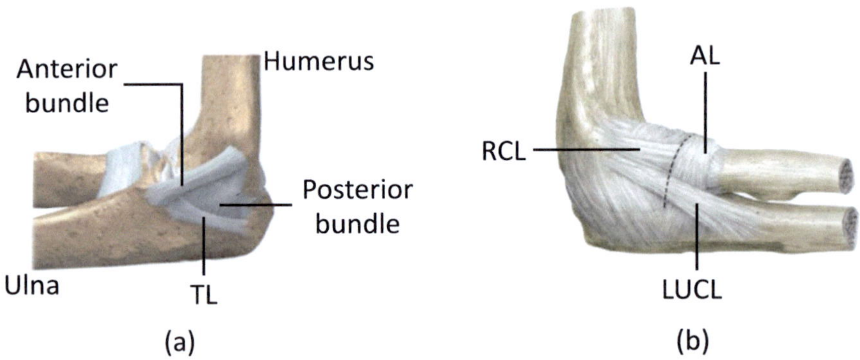

Fig. 1.55 Elbow joint ligaments; **a** Medial collateral ligament complex (MCL), **b** Lateral collateral ligament complex (LCL). TL: transverse ligament, RCL: radial collateral ligament, AL: annular ligament, LUCL: lateral ulnar collateral ligament

- **Arthritis**: Similar to other synovial joints, arthritis can afflict the elbow joint. Rheumatoid arthritis often affects the radio-ulnar joint, resulting in pain, stiffness, and joint deformities.
- **Tendonitis**: It is the inflammation (swelling) of the tendon, which results in pain and discomfort. Tennis elbow, also known as lateral epicondylitis, is a specific type of tendonitis that affects the tendons on the outside of the elbow. The pain may sometimes extend into the back of the forearm, accompanied by weakness in grip strength. It is commonly caused by repetitive gripping and wrist movements, such

as those used in playing tennis, hence the name 'tennis elbow'. Golfer's elbow, also known as medial epicondylitis, is another type of tendonitis that affects the tendons on the inside of the elbow. Activities, such as golfing, throwing, chopping wood using a hammer, can contribute to the development of golfer's elbow. The symptoms of golfer's elbow are pain and tenderness on the inside part of the elbow as well as stiffness and weakness in the affected arm.

1.14 Spine

The spine is the central support structure of the human body. Moreover, the spine is a crucial bony structure that provides support and protection for the spinal cord. It consists of a series of vertebrae that are interconnected by flexible discs, allowing for stability and mobility in the upper body.

1.14.1 Anatomy of the Spine

Commonly known as the backbone or the spine, the spinal or vertebral column consists of a series of bones (vertebrae), intervertebral discs, muscles, tendons, ligaments, and other tissues that span from the base of the skull to the tailbone. This column encloses the spinal cord and the fluid surrounding the spinal cord, passing through the spinal canal of each vertebra (Fig. 1.56). The major components of the spine are described briefly in the paragraphs below.

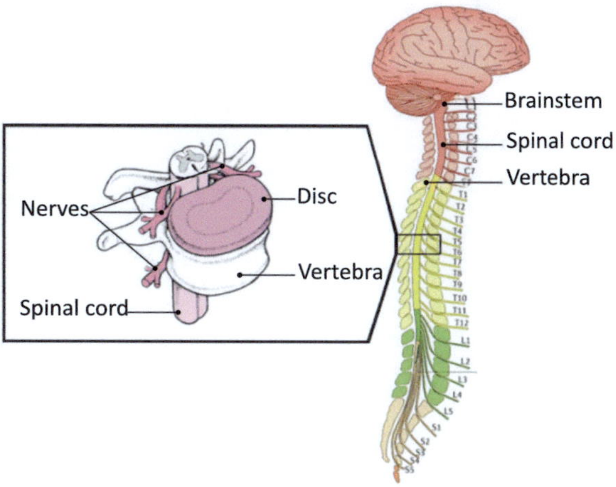

Fig. 1.56 Anatomy of the spine

1.14.1.1 Spinal Cord

The spinal cord is the main pathway of information transmission connecting the brain and the peripheral nervous system (PNS). The spinal cord is an integral part of the central nervous system (CNS) that supplies nerves and receives information via the PNS within the body (Fig. 1.56).

1.14.1.2 Vertebral Column

The human spine (vertebral column) is composed of 33 irregular-sized vertebrae. The main functions of the spine are to provide flexibility of motion, support for upper body weight, and protect the spinal cord. A healthy spine has three natural curves that resemble an S-shape in the sagittal plane. These curves provide resistance to axial loads on the spine and help to maintain its elasticity. The human spine comprises seven cervical vertebrae, twelve thoracic vertebrae, five lumbar vertebrae, one sacrum (five fused vertebrae), and 3–4 fused coccygeal segments (Fig. 1.57).

The cervical spine, which consists of seven cervical vertebrae, namely C1 to C7, is the most superior portion of the spine. The cervical spine provides mobility and stability to the head and is much more flexible than the other segments of the spine.

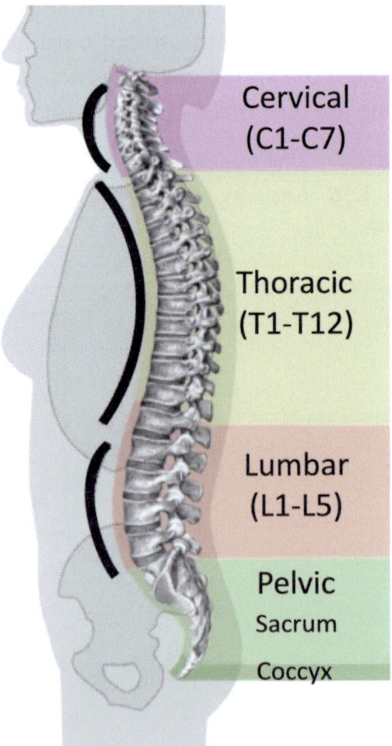

Fig. 1.57 Structure of the spine

The thoracic segment of the spine, comprising 12 thoracic vertebrae T1 to T12, holds the rib cage and protects internal organs, such as the heart and the lungs. The range of motion of the thoracic spine is very limited due to the restrictive nature of its joint articulations. The lumbar spine consists of five lumbar vertebrae, namely L1 to L5, and are specialised for weight bearing and strength. The primary function of the lumbar spine is to bear the weight of the upper body and transfer the load to the sacrum. The lumbar spine is more flexible than the thoracic spine and bears greater weight, thus making it the most susceptible region of the spine to injury.

The entire spine is connected to the hip bones (iliac) through the sacrum. The sacrum is a large, flat, triangular-shaped bone that is situated between the two pelvic bones and below the final lumbar vertebra (L5). Connection with the iliac bone forms the sacroiliac joint (on either side). The pelvic girdle comprises the sacrum along with the coccyx and the two sacroiliac joints. The coccyx, commonly known as the tailbone, is the last segment of the vertebral column.

1.14.1.3 Functional Spinal Unit

The functional spinal unit (FSU) or the motion segment is the smallest segment (unit) of the spine, which exhibits biomechanical characteristics similar to those of the entire spine. It comprises two consecutive vertebrae, the inter-vertebral disc, two facet joints, and seven surrounding ligaments, albeit without muscles.

1.14.1.4 Vertebrae

Let us now discuss the structure of the vertebrae in more detail. A typical vertebra is composed of the vertebral body and a posterior vertebral arch. The vertebral body is located on the anterior side of the vertebrae, whereas the vertebral arch is located on the posterior portion of the vertebra. The vertebral body has a large anterior middle part called centrum. The vertebral arch in the posterior portion consists of four articulating processes two superior and two inferior, two pedicles, two laminae, two transverse processes, and one spinous process (Fig. 1.58a). The pars interarticularis is a region between the superior and inferior articular process. The facet joints are symmetrical synovial joints that help to guide and restrain the movement of the vertebrae in all six degrees of freedom and act to transmit loads to the disc. The facets that form the facet joints are stabilising structures and carry almost one-third of the total compressive load borne by the spine segment.

The vertebral body is a cylindrical mass consisting of a core cancellous bone contained within a thin shell of cortical bone (Fig. 1.58b). It has an inferior surface and a superior surface that is slightly concave in shape, and they are known as vertebral endplates. Between two adjacent vertebrae, there is an intervertebral disc (anteriorly), and the articulating facets (posteriorly) that form the facet joints. The intervertebral disc, as presented in Fig. 1.58b, is a fibrocartilage cushion that lies between two vertebrae in the vertebral column. Each intervertebral disc (IVD) forms

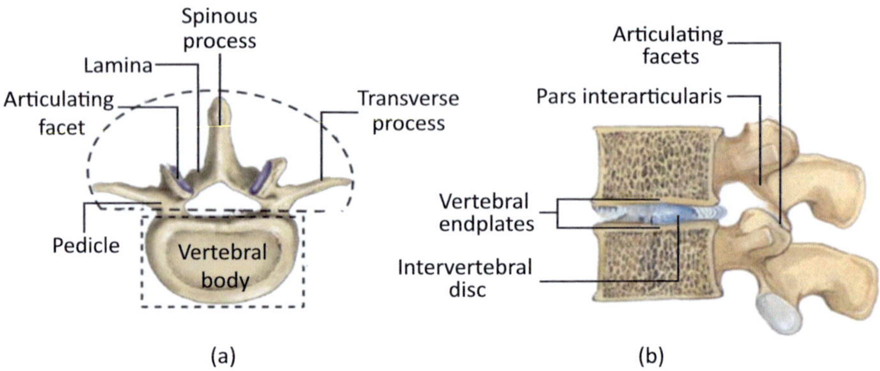

Fig. 1.58 Structure of the vertebra: **a** superior view, **b** sagittal section. The region within the dotted lines at the posterior end of vertebra represents the vertebral arch

a fibrocartilaginous joint which allows only a slight movement of the vertebrae. It also functions as a shock absorber within the spinal column. The intervertebral disc consists of two distinct parts; the annulus fibrosus and nucleus pulposus, as presented in Fig. 1.59. The nucleus pulposus is enclosed within the annulus fibrosus.

The annulus fibrosus consists of several concentric layers of fibrocartilage with annular fibres oriented in specific directions. The nucleus pulposus contains loose

Fig. 1.59 A simplified representation of the intervertebral disc; the fibre orientation is shown in the inset

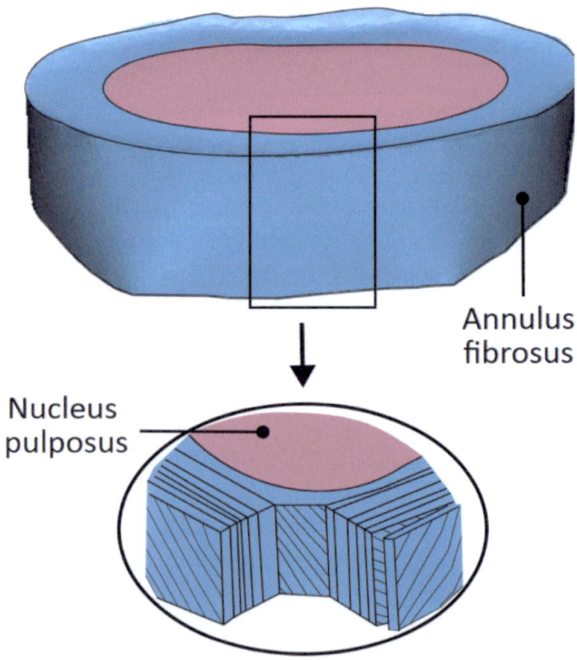

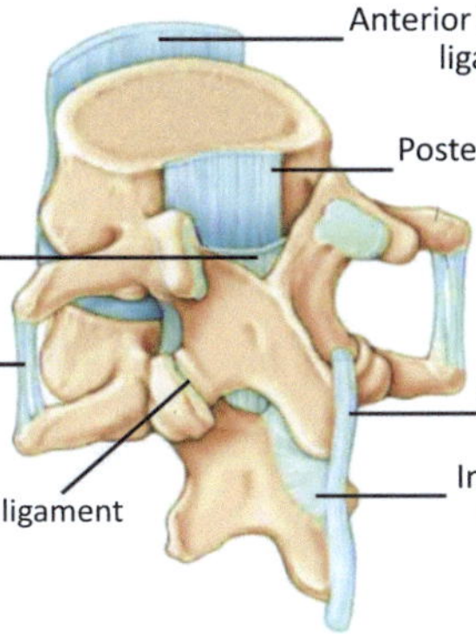

Fig. 1.60 Ligaments of the spine

fibres suspended in a gel-like substance. Hence, the composite formed by the annulus fibrosus matrix and the annular fibres enclosing the nucleus pulposus helps to distribute the pressure evenly throughout the disc.

1.14.2 Ligaments of the Spine

Let us look into the ligaments which play a major role in providing spinal stability. These ligaments are strong fibrous bands of connective tissue that hold the vertebrae together and provide stability to the spine within its physiologic range of motion. Excessive movements, such as hyper-extension or hyper-flexion are restricted by the ligaments. There are seven ligaments associated with spine, as shown in Fig. 1.60. These are anterior longitudinal ligament, posterior longitudinal ligament, supraspinous ligament, interspinous ligament, capsular ligament, intertransverse ligament, and ligamentum flavum.

1.14.3 Spine Movements

The movements associated with the spine are presented in Fig. 1.61. The movements are as follows:

- Flexion and extension in the sagittal plane.
- Lateral bending to the left or to the right in the coronal plane.
- Rotation of the spine towards the left or towards the right in the transverse plane.

We shall now look into the spinal muscles associated with the movements. The functions of the spinal muscles are to provide stability to the trunk at any posture and to produce movements during physiologic activity. The muscles located anterior to the vertebral column act as flexors, whereas the muscles located posterior to the

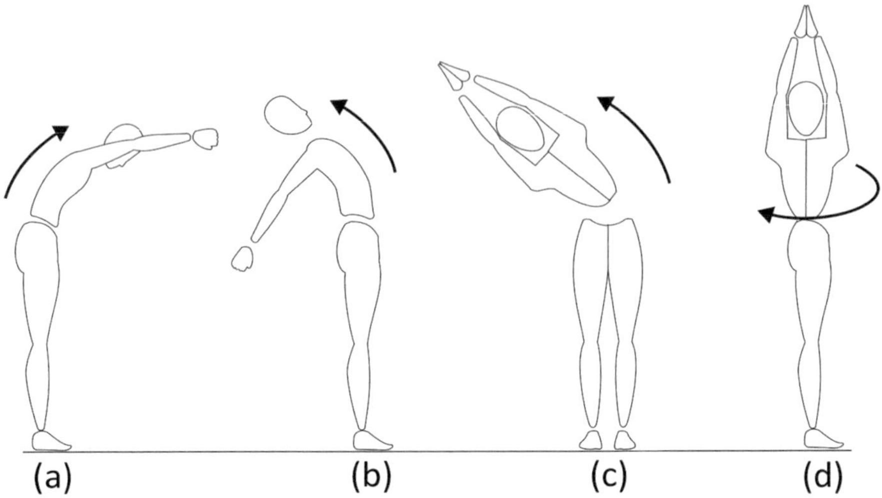

Fig. 1.61 Movements of spine: **a** flexion, **b** extension, **c** lateral bending, **d** rotation

vertebral column act as extensors. The major extensor muscles are erector spinae and multifidus. The erector spinae muscles generate the extensor force required for lifting, whereas the multifidus muscles stabilise individual lumbar motion segments. The major flexor muscles are the abdominal muscles and psoas. The major muscles responsible for the bending and rotation of the spine are the abdominal muscles, comprising the rectus abdominis, external oblique, internal oblique and the transverse abdominis muscles.

Let us now discuss the range of motion (ROM) of each part of the vertebral column. The cervical spine consists of the vertebrae C1 to C7, which is the most superior part of the vertebral column. It provides mobility and stability to the head while connecting it to the relatively immobile thoracic spine. The ROM for the cervical spine is as follows:

- Flexion/extension in the range of 80°–90°,
- Lateral bending in the range of 20°–45°,
- Rotation of 80°–90°.

The Atlas C1 is the first cervical vertebrae that articulates with the occiput (the lower portion of the head or the skull) forming the Atlanto-occipital joint. The second cervical vertebra (axis) C2 articulates with the atlas C1, forming the Atlanto-axial joint, as shown in Fig. 1.62.

The thoracic spine consists of vertebra T1 to T12 and plays a crucial role in providing structural support and protecting vital organs, particularly the heart and lungs, within the ribcage. In these thoracic vertebrae, the bones increase in size and mass as we move downward from T1 to T12. The increasing size and mass ensure more support for the weight of the body. Thoracic spine has very restricted ROM

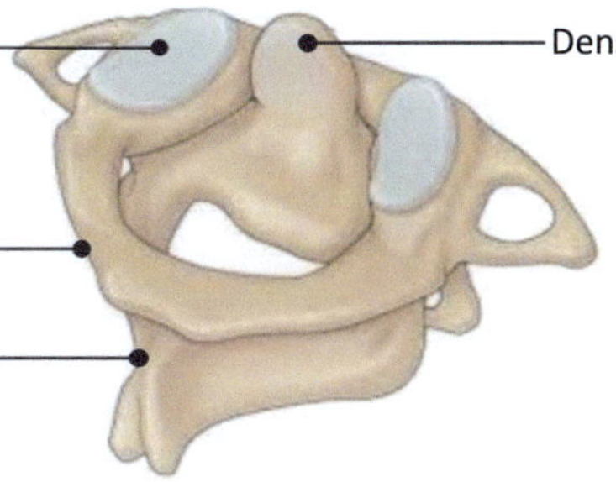

Fig. 1.62 Structure of the Atlanto-axial joint

during flexion and extension as compared to those of the cervical and lumbar spine, owing to the presence of the rib cage. The ROM of the thoracic spine is as follows:

- Flexion/extension in the range of 20°–40°,
- Lateral bending in the range of 30°–40°,
- Rotation of 30°–50°.

The lumbar spine is located towards the lower portion of the vertebral column and its main function is to bear the weight of the upper body, and transfer the forces and bending moments to the sacrum and further to the pelvic bone. The vertebrae in the lumbar spine are much larger in size as compared to those of the thoracic and cervical spine, in order to withstand heavy loads during various physiological activities. The lumbar spine is more mobile than the thoracic spine and also carries greater weight, making it the most likely region susceptible to injury in the spinal column. The ROM offered by the lumbar spine are flexion and extension in the range of 40°–90°, lateral bending of about 20°–45°, and rotation of 5°–15°.

1.14.4 Disc Degenerative Disease

Intervertebral disc degeneration is one of the main causes of spinal instability and low back pain. Degeneration means a specific injurious change in the composition, structure, and function of the spine. Factors, such as trauma, obesity, aging, and genetics have been found to be progenitors of disc degeneration. This is more common in the cervical and lumbar regions of the spine. Degeneration leads to disc dehydration, reduction in disc height, and disruption of outer annulus lamellae as well as the appearance of cracks and fissures (An and Masuda 2006) (Fig. 1.63). With more severe disc dehydration and reduction in disc height, greater load is carried by the facet joints as compared to a healthy spine, eventually leading to facet joint osteoarthritis.

Disc degeneration can cause disc herniation (abnormal bulging or rupture of the disc), which may further lead to nerve root or spinal cord compression and instability between vertebrae. Lumbar spinal stenosis (LSS) refers to a medical condition,

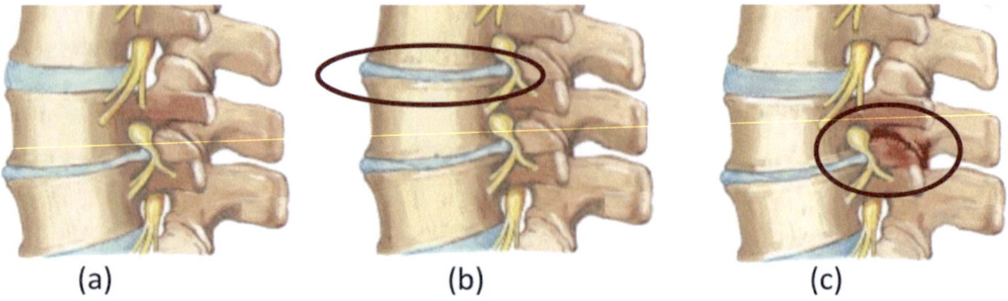

Fig. 1.63 Disc degenerative diseases: **a** intact spine, **b** reduced disk height, **c** osteoarthritis

wherein an anatomical narrowing of the spinal canal occurs. The nerves coming out from the spinal cord are irritated or pressed, thereby causing Sciatica, i.e. nerve root pain. Lumbar spondylolisthesis and spondylolysis are defects that are often identified during clinical evaluation of patients with low back pain (LBP). Spondylolisthesis is the forward displacement of a vertebra, especially the fifth lumbar vertebra. Spondylolysis is a stress fracture in the pars interarticularis and is most commonly observed in the lowest pair of the lumbar vertebrae. The disc degenerative diseases can be treated in two ways: conservative and surgical. The conservative type of treatment includes physiotherapy, medicine, or a combination of both. Within the surgical procedures, there are two methods: fusion and non-fusion surgeries. In fusion surgery, the surgical procedure involves fusion of two adjacent vertebrae using interbody cages and pedicle screws (Faizan et al. 2014). Whereas in the non-fusion type of surgery, the surgical procedure involves only the replacement of the normal intervertebral disc with an artificial intervertebral disc.

Multiple Choice Questions (MCQ)

1. Which of the following regions in a femur consist of hard cortical bone only?
   (A) Metaphysis
   (B) Diaphysis
   (C) Epiphysis
2. Which of the following is not a function of muscle?
   (A) Blood Cell Formation
   (B) Motion
   (C) Heat production
3. Which of the following attaches a skeletal muscle to the bone?
   (A) Tendon
   (B) Ligament
   (C) None of these

(D) All of these

4. Which is the weakest part of the musculoskeletal system?

(A) Muscles
(B) Joints
(C) Tendons
(D) Ligaments

5. Which of the following works as an actuator of the limb?

(A) Bone
(B) Joints
(C) Muscle + Tendon
(D) Ligaments

6. How many types of joints exist in our body based on the structural classification?

(A) 2
(B) 3
(C) 4
(D) 5

7. Which of the following joints can be found on the skull?

(A) Synarthroses
(B) Amphiarthroses
(C) Diarthroses

8. Which of the following joints contain fluid-filled cavities?

(A) Synarthroses
(B) Amphiarthroses
(C) Diarthroses

9. Which of the following is not a synovial joint?

(A) Sutures
(B) Shoulder joint
(C) Knee joint
(D) Elbow joint

10. Which of the following joints allows supination and pronation?

(A) Shoulder joint
(B) Knee joint
(C) Elbow joint
(D) Hip joint

11. Shoulder is an example of _____.

(A) Planar joint
(B) Saddle joint

(C) Hinge joint
(D) Ball and socket joint

12. What type of joint is the temporomandibular?

(A) Fibrous joint
(B) Cartilaginous joint
(C) Synovial joint

13. _____ is the largest synovial joint in the human body.

(A) Facet joint
(B) Knee joint
(C) Hip joint
(D) Shoulder joint

14. Condyloid joint belongs to the class of ______.

(A) Synarthroses
(B) Amphiarthroses
(C) Diarthroses

15. At gomphosis joint, bone ends are united by _____.

(A) Collagenic fiber
(B) Cartilage
(C) Dense irregular connective tissue

16. Which of the following factors influence joint stability?

(A) Shape of the articular surface
(B) Ligaments
(C) Muscles
(D) All of these

17. The rotator cuff is a group of ______ in the shoulder.

(A) Ligaments
(B) Muscles
(C) Bones
(D) Tendons

18. _____ is an auto-immune type of disorder.

(A) Osteoarthritis
(B) Rheumatoid arthritis
(C) Patellar tendinitis

19. Each vertebra has ___ sets of facet joints.

(A) 4
(B) 2
(C) 3

(D) 5

20. The main function of the lumbar spine is _____.

(A) protect vital organs such as the heart and lungs
(B) Bear the weight of the upper part of the body
(C) Hold the rib cage

Answer Key to MCQs

1. B; 2. A; 3. A; 4. B; 5. C; 6. B; 7. A; 8. C; 9. A; 10. C; 11. D; 12. C; 13. B; 14. C; 15. A; 16. D; 17. B; 18. B; 19. B; 20. B

References

An HS, Masuda K (2006) Relevance of in vitro and in vivo models for intervertebral disc degeneration. J Bone Joint Surg Am 88(Suppl 2):88–94

Ewalefo E, Fowokan O, Muchtar A, Raza M (2018) Management of posttraumatic ankle arthritis: literature review. Curr Rev Musculoskelet Med 11(4):546–557

Faizan A, Kiapour A, Kiapour AM, Goel VK (2014) Biomechanical analysis of various footprints of transforaminal lumbar interbody fusion devices. J Spinal Disord Tech 27(4):E118–E127

Fox SAJ, Bedi A, Rodeo SA (2009) The basic science of articular cartilage: structure, composition, and function. Sports Health 1(6):461–468

Gupta S (2002) Stress analysis of scapula, design considerations of glenoid prosthesis. Ph. D. thesis, Delft University of Technology, Netherlands

Knupp M, Lang TH, Zwicky L, Lötscher P, Hintermann B (2015) Chronic ankle instability (medial and lateral). Clin Sports Med 34(4):679–688

Ladd LM, Crews M, Maertz NA (2021) Glenohumeral joint instability: a review of anatomy, clinical presentation, and imaging. Clin Sports Med 40(4):585–599

Luo CF (2004) Reference axes for reconstruction of the knee. Knee 11(4):251–257

Veeger HE, Van der Helm FCT, Van der Woude LH, Pronk GM, Rozendal RH (1991) Inertia and muscle contraction parameters for musculoskeletal modelling of the shoulder mechanism. J Biomech 24(7):615–629

Chapter 2
Basic Biomechanics of Human Joints and Spine

This chapter is focussed on the mechanics associated with anatomical joints in the upper and lower limbs of the human body. The objective of this chapter is to gain insight into the structure, functions, movements, and forces acting on the joints during daily activities. We will unravel the biomechanical nuances of the specific joints, the hip, knee, shoulder, elbow, and spine using the concept of a free body diagram. These are the architectural keystones of our musculoskeletal system, enabling us to walk, run, jump, grasp objects, and execute countless other movements with precision and control.

2.1 Biomechanics of the Hip

The hip joint serves as the primary connection between the torso and lower limbs. The hip is analogous to a ball-and-socket articulation, with the femoral head as the spherical component and the acetabulum as the socket, as presented in Fig. 2.1.

The hip joint is a major load-bearing joint and plays a notable role in daily activities. Its primary function involves supporting body weight during both stationary postures like standing and sitting, as well as dynamic actions, such as walking, running, and jumping. Additionally, it aids in the transfer of forces and moments, generally referred to as load transfer, from the upper body to the lower extremities. The hip joint enables a wide range of movements, while maintaining stability throughout these motions.

S. Gupta et al., *Biomechanics of Joints and Implants*,
https://doi.org/10.1007/978-981-96-0586-6_2

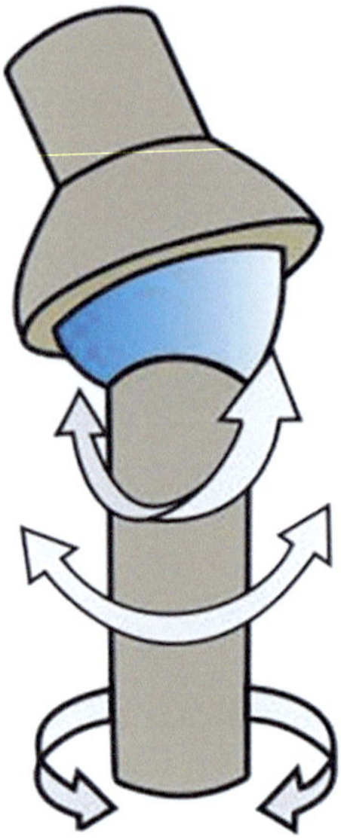

Fig. 2.1 Ball and socket articulation

2.1.1 *Basic Biomechanics of the Hip Joint*

In this section, basic biomechanics of the hip joint has been undertaken using the concept of free body diagram and static equilibrium. The forces acting on the hip joint are analysed for specific postures. We consider a one-legged stance under static conditions to evaluate the forces acting in the coronal plane. While this scenario is somewhat hypothetical, it could help us to comprehend the influence of joint configuration on its biomechanics. Likewise, as presented in Fig. 2.2, the forces acting on the hemi-pelvis are: the weight of the body except the lower limb on the same side $\left(\frac{6}{7}W\right)$, joint reaction force (***J***), and abductor muscle force (***A***). The force $\frac{6}{7}W$ acts at a lever arm, *b*, the distance between the body's centre of gravity and the centre of the femoral head (Fig. 2.2). The muscle force, ***A*** acts at a lever arm, *c* which extends from the lateral aspect of the greater trochanter to the centre of the femoral head. The joint reaction force (***J***) is the compressive force experienced between the femoral head and the acetabulum. This force passes through the centre of the femoral head (Q), which is the centre of the hip joint (Fig. 2.2).

Considering the known quantity to be the body weight (***W***), the lever arms (*b* and *c*) and the orientation of the muscle force (***A***), let us calculate the joint reaction force (***J***) and the magnitude of muscle force (***A***). The steps to determine these unknown forces, based on Fig. 2.2, are as follows.

(1) Using the equation of moment equilibrium about the joint centre Q, the abductor muscle force (***A***) is calculated in terms of the body weight (***W***) and the lever arms.

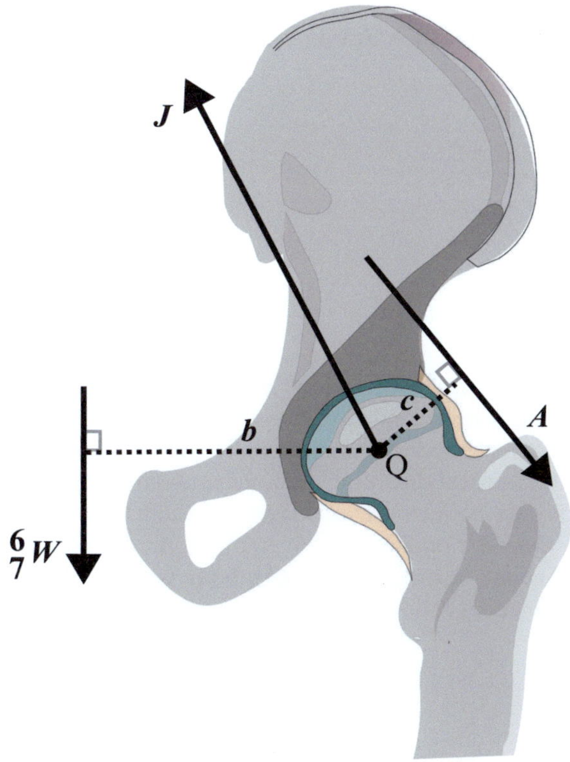

Fig. 2.2 Major forces acting on the pelvic bone. Femur is included in the figure for better visualization of the joint center Q

$$\frac{6}{7}Wb - Ac = 0$$

(2) It is evident from Fig. 2.2 that the forces are coplanar and concurrent, So the vector addition of all forces must be equal to zero for static equilibrium (Fig. 2.3).

$$\frac{6}{7}\boldsymbol{W} + \boldsymbol{A} + \boldsymbol{J} = 0$$

The final step involves the calculation of the joint reaction force, which is obtained from the force triangle (Fig. 2.3). Here, the length of the vector corresponds to its magnitude.

Let us discuss a detailed problem where the ground reaction force ($\boldsymbol{R}$) is considered. Figure 2.4 represents an instance of a single stance phase during walking. The gravitational force of a single leg is assumed to be one-seventh of the body weight, whereas the gravitational force acting on the upper part of the body is six-seventh of the body weight (6/7W), as shown in Fig. 2.4. In order to calculate the forces associated with the hip joint, let us examine the forces acting on the upper part of the body, as shown in Fig. 2.5. Considering the upper part of the body, there are

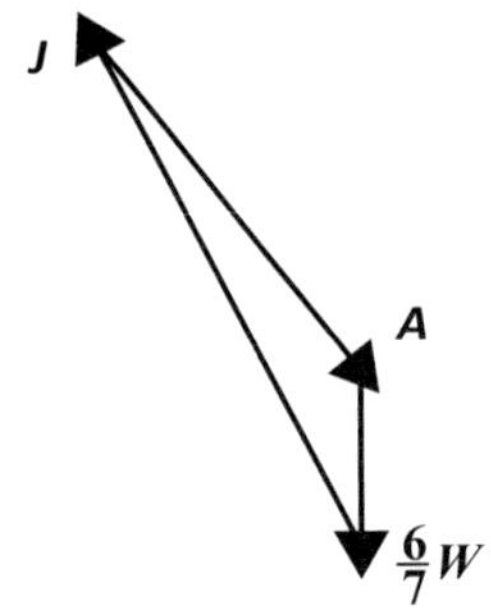

Fig. 2.3 Force triangle consisting of forces ***A***, ***J***, and $\frac{6}{7}\boldsymbol{W}$

two moments about the joint centre (Q). The first moment arises due to the abductor muscle. The second moment is caused by the gravitational force of the upper body. We assume abductor muscle force (***A***) to act at an inclination of 30° with respect to the vertical direction. Now, using the free body diagram of the upper part of the body, the abductor muscle force (***A***) can be estimated.

Considering the moment equilibrium of forces about Q, a relationship between the abductor muscle force (***A***), body weight (***W***) and the lever arms (b and c) is obtained, as follows:

$$\frac{6}{7}Wb - (A \cdot c) = 0$$

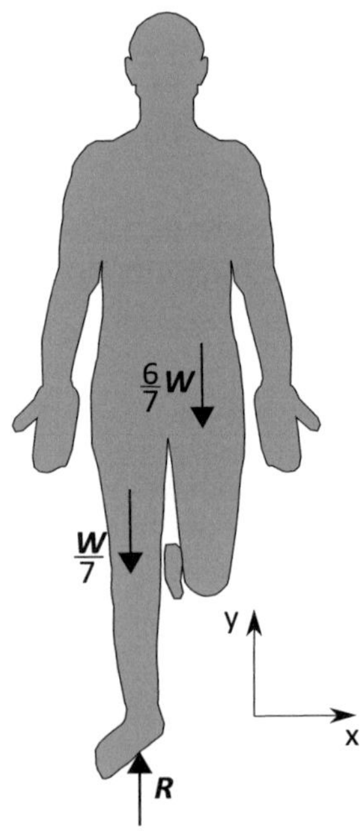

Fig. 2.4 Major forces acting at an instance of single stance phase

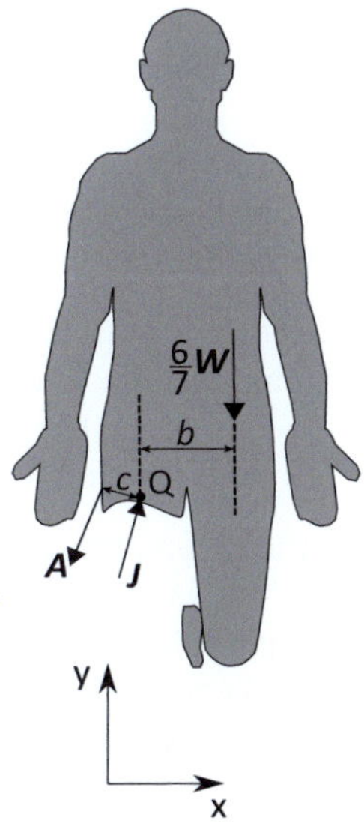

Fig. 2.5 Forces acting on the upper part of body

$$A = \frac{6Wb}{7c}$$

Assuming the ratio of lever arms (*c/b*) as 0.417, the abductor muscle is estimated to be nearly twice the body weight.

$$A = \frac{6Wb}{7c} = \frac{6W}{7 \times 0.417}$$

$$A \approx 2W$$

Now we can calculate the horizontal (A_x) and vertical components (A_y) of abductor muscle force, by resolving the force along the *x*- and *y*-directions (Fig. 2.6).

Horizontal component of abductor muscle force, $A_X = A\mathrm{Sin}\ 30°$

$$A_X = 0.5A = W$$

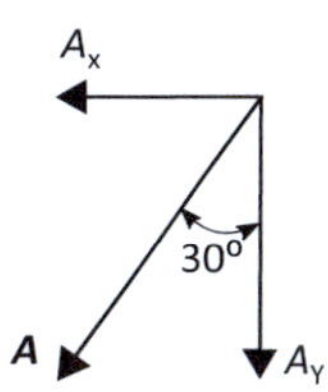

Fig. 2.6 Components of abductor muscle force

Vertical component of abductor muscle force, $A_Y = A\mathrm{Cos}\ 30°$

$$A_y \approx 0.87A = 1.74W$$

Considering the free body diagram of the lower limb, the forces acting on this limb are the following:

- Abductor muscle force, ***A***,
- Joint reaction force, ***J***,
- Weight of the lower limb, $\frac{W}{7}$,
- Ground reaction force, ***R***.

The ground reaction force is the reactive force which is equal to the weight of the body, obtained by considering the force equilibrium of the whole body (Fig. 2.4). Now, considering the free body diagram of the lower limb (Fig. 2.7), the joint reaction and abductor muscle forces are resolved along the horizontal and vertical directions. On considering the force equilibrium along the *x*-direction, we obtain the following:

$$A_x - J_X = 0$$
$$J_X = A_x$$

Since A_x was found to be equal to W (previous calculation), the horizontal component of the joint reaction force J_x is found to be equal to W.

In order to determine the vertical component of the joint reaction force, the following equation based on force equilibrium along the vertical direction is considered.

$$A_Y - J_Y - \frac{W}{7} + W = 0$$

Substituting the value of A_Y in the above equation results in

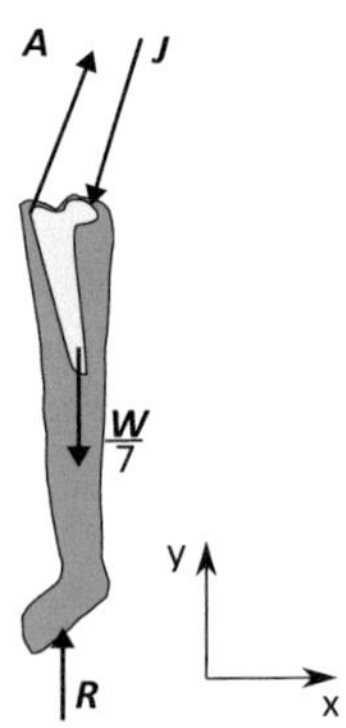

Fig. 2.7 Free body diagram of the lower limb

$$1.74W - J_Y - \frac{W}{7} + W = 0$$

Upon rearranging the above equation, the vertical component of the abductor muscle force is calculated as

$$J_Y = 1.74W - \frac{W}{7} + W = 0$$
$$J_Y = 1.74W + \frac{6W}{7} \approx 2.60W$$

Therefore, the vertical (J_y) and horizontal (J_x) components of the joint reaction force are equal to 2.60 times the body weight and the body weight, respectively. The resultant, J is calculated as

$$J = \sqrt{J_X^2 + J_Y^2} = W\sqrt{1^2 + 2.60^2} = 2.79W$$

Since we have found out the components of the abductor muscle forces, we proceed to estimate the inclination of the joint reaction force with respect to the x-direction (Fig. 2.8). The inclination angle (α) with respect to the vertical direction is determined as follows:

$$\tan\alpha = \frac{J_y}{J_x}$$
$$\tan\alpha = \frac{2.6W}{W} = 2.6$$
$$\alpha = 69^\circ$$

So far, we have considered simple 2D problems. Now, the hip joint reaction force and the muscle forces in a 3D force system are considered. For a 3D force system, the method of vector mechanics has been used to find out the unknown muscle forces, gluteus medius ($\boldsymbol{F}_1$), adductor longus ($\boldsymbol{F}_2$), adductor magnus ($\boldsymbol{F}_3$), and the joint reaction force ($\boldsymbol{J}$). The coordinates of the points in this problem are in centimetres (cm). The $\boldsymbol{J}$ is assumed to act on the centre of hip joint which is having coordinates as

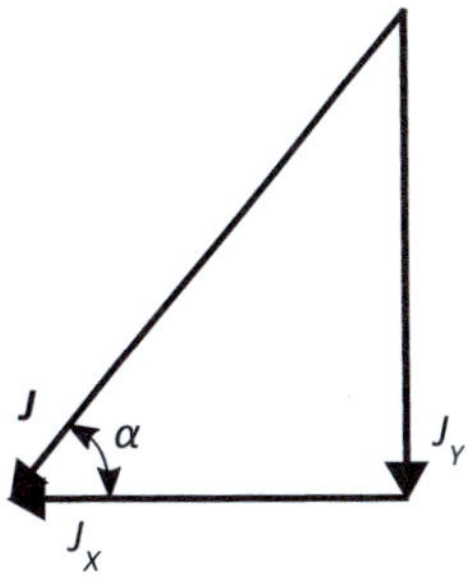

Fig. 2.8 Components of joint reaction force

Fig. 2.9 3D force system acting on the lower limb

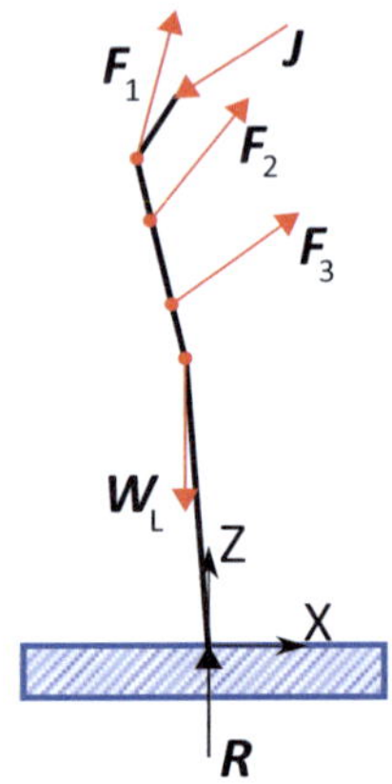

Table 2.1 Coordinates (in cm) of muscle attachments on the femur and the pelvic bone

Muscle force	Femur (x_1, y_1, z_1)	Pelvis (x_2, y_2, z_2)
$\boldsymbol{F_1}$	(−2.50, 2.50, 71.00)	(2.65, 2.52, 79.43)
$\boldsymbol{F_2}$	(−1.70, 1.70, 58.30)	(10.32, 5.78, 69.64)
$\boldsymbol{F_3}$	(−0.80, 0.80, 49.20)	(0.06, 10.0, 66.94)

(−0.8, 0.8, 73). The weight of the limb ($\boldsymbol{W_L}$) acts at the joint centre of knee, having a coordinate as (−0.4, 0, 40). The ground reaction force (***R***) acts at (0, 0, 0). The weight of the person is equal to 80 kg. The forces acting on the right femur, during a time instance of a single stance (only one limb is in contact with the ground), are shown in Fig. 2.9. The limb weight and the ground reaction force are denoted by $\boldsymbol{W_L}$ and ***R***, respectively. The origin of the coordinate system is considered at the point of contact of the limb on the ground. The coordinates of muscle attachment sites of various muscles on the femur and the pelvic bone are presented in Table 2.1.

The unit vector of muscle force F_1 is determined, using the position coordinates of muscle attachments on the femur and the pelvic bone (Table 2.1), as follows:

$$\widehat{\boldsymbol{F_1}} = \frac{(x_2 - x_1)\hat{i} + (y_2 - y_1)\hat{j} + (z_2 - z_1)\hat{k}}{\left|(x_2 - x_1)\hat{i} + (y_2 - y_1)\hat{j} + (z_2 - z_1)\hat{k}\right|}$$
$$= \frac{(2.65 + 2.50)\hat{i} + (2.52 - 2.50)\hat{j} + (79.53 - 71.00)\hat{k}}{\left|(2.65 + 2.50)\hat{i} + (2.52 - 2.50)\hat{j} + (79.53 - 71.00)\hat{k}\right|}$$
$$= 0.521\hat{i} + 0.002\hat{j} + 0.853\hat{k}$$

Similarly, the unit vectors corresponding to other muscle forces are obtained as

$$\widehat{\boldsymbol{F_2}} = 0.706\hat{i} + 0.240\hat{j} + 0.666\hat{k}$$
$$\widehat{\boldsymbol{F_3}} = 0.043\hat{i} + 0.460\hat{j} + 0.887\hat{k}$$

The muscle force vectors, having the magnitude and direction (unit vector) is obtained as follows:

$$\begin{aligned}
\boldsymbol{F_1} &= |\boldsymbol{F_1}|\widehat{\boldsymbol{F_1}} = |\boldsymbol{F_1}|\left(0.521\hat{\imath} + 0.002\hat{\jmath} + 0.853\hat{k}\right) \\
\boldsymbol{F_2} &= |\boldsymbol{F_2}|\widehat{\boldsymbol{F_2}} = |\boldsymbol{F_2}|\left(0.707\hat{\imath} + 0.240\hat{\jmath} + 0.667\hat{k}\right) \\
\boldsymbol{F_3} &= |\boldsymbol{F_3}|\widehat{\boldsymbol{F_3}} = |\boldsymbol{F_3}|\left(0.043\hat{\imath} + 0.460\hat{\jmath} + 0.887\hat{k}\right)
\end{aligned}$$

The segment weight of the limb ($\boldsymbol{W_L}$) is assumed to be one-seventh of the body weight ($\boldsymbol{W}$) and acting vertically downwards.

$$\boldsymbol{W_L} = -\frac{W}{7}\hat{k}$$

The hip joint reaction force ($\boldsymbol{J}$) can be expressed as

$$\boldsymbol{J} = J_X\hat{\imath} + J_Y\hat{\jmath} + J_Z\hat{k}$$

The ground reaction force ($\boldsymbol{R}$), which is equal to body weight, (only one limb is in contact with the ground, acting upwards), is written as

$$\boldsymbol{R} = W\,\hat{k}$$

The hip joint reaction force and muscle forces will be determined using static equilibrium equations. Considering force equilibrium along the x-, y-, and z-directions, the following equations are obtained:

$$\begin{aligned}
\Sigma F_x &= |\boldsymbol{F_1}|(0.521) + |\boldsymbol{F_2}|(0.706) + |\boldsymbol{F_3}|(0.043) + J_x \\
&= 0 \\
\Sigma F_y &= |\boldsymbol{F_1}|(0.002) + |\boldsymbol{F_2}|(0.240) + |\boldsymbol{F_3}|(0.460) + J_y \\
&= 0 \\
\Sigma F_z &= |\boldsymbol{F_1}|(0.853) + |\boldsymbol{F_2}|(0.667) + |\boldsymbol{F_3}|(0.887) \\
&\quad + J_z - |\boldsymbol{W_L}| + |\boldsymbol{R}| = 0
\end{aligned}$$

Considering the moment equilibrium about the hip centre, we can write

$$\begin{aligned}
\Sigma M &= \boldsymbol{r_1} \times \boldsymbol{F_1} + \boldsymbol{r_2} \times \boldsymbol{F_2} + \boldsymbol{r_3} \times \boldsymbol{F_3} \\
&\quad + \boldsymbol{r_4} \times \boldsymbol{W_L} + \boldsymbol{r_5} \times \boldsymbol{R} = 0
\end{aligned}$$

The moment arm vectors ($\boldsymbol{r_i}$) of forces are found using the position coordinate data, as discussed before.

$$\boldsymbol{r_1} = -1.70\hat{\imath} + 1.70\hat{\jmath} - 2.00\hat{k}$$

$$\boldsymbol{r_2} = -0.90\hat{i} + 0.90\hat{j} - 14.70\hat{k}$$
$$\boldsymbol{r_3} = 0\hat{i} + 0\hat{j} - 23.8\hat{k}$$
$$\boldsymbol{r_4} = 0.4\hat{i} - 0.8\hat{j} - 33.0\hat{k}$$
$$\boldsymbol{r_5} = 0.8\hat{i} - 0.8\hat{j} - 73.0\hat{k}$$

Upon carrying out the vector multiplication, the moment equilibrium equations are rewritten as follows:

$$\sum M_x = |\boldsymbol{F_1}|(1.45) + |\boldsymbol{F_2}|(4.12) + |\boldsymbol{F_3}|(10.95) + \frac{W}{7}(0.80) + W(-0.80) = 0$$
$$\sum M_y = |\boldsymbol{F_1}|(-0.41) + |\boldsymbol{F_2}|(-9.78) + |\boldsymbol{F_3}|(-1.02) + \frac{W}{7}(0.40) + W(-0.80) = 0$$
$$\sum M_z = |\boldsymbol{F_1}|(-0.89) + |\boldsymbol{F_2}|(-0.85) + |\boldsymbol{F_3}|(0) + \frac{W}{7}(0) + W(0) = 0$$

On solving the above set of six equilibrium equations (force and moment equilibrium), the unknown forces are determined as

$$|\boldsymbol{F_1}| = 3630.18\,\text{N};\ |\boldsymbol{F_2}| = -3793.96\,\text{N};\ |\boldsymbol{F_3}| = 995.76\,\text{N}$$
$$J_X = 743.88\,\text{N};\ J_Y = 444.06\,\text{N};\ J_Z = -2126.02\,\text{N}$$

The hip joint reaction force ($\boldsymbol{J}$) vector is found to be $\left(743.88\,\hat{i} + 444.06\,\hat{j} - 2126.03\,\hat{k}\right)$. A MATLAB code, which implements the steps involved in this problem, is presented at the end of this chapter. This code is useful to study the effect of changes in the muscle attachment sites as well as the body weight on the joint reaction force.

2.1.2 Hip Joint Reaction Forces During Various Activities

Direct in-vivo measurement of the joint reaction forces acting in a natural hip joint is almost impossible to acquire, without implantation. Technological advancements have introduced instrumented implants with telemetric features, permitting the real-time collection of data on hip joint reaction forces. The hip contact force has been estimated by several investigators using a force plate system and kinematic data with electromyography (EMG) for the normal hip joint (Crowninshield et al. 1978; van den Bogert et al. 1999) and using instrumented hip prosthesis for the implanted hip joint (Davy et al. 1988; Kotzar et al. 1991; Taylor et al. 1997; Bergmann et al. 1993, 2001) (Table 2.2). Most of these studies reported a large amount of inter-patient variability in the hip contact force. An in-depth analysis of the gait cycle and the joint reaction forces will be provided in Chap. 3.

Table 2.2 Hip joint reaction forces during various daily activities

Activity	Peak contact force (%BW)	Method
Walking	260–280[a]	Instrumented telemetric hip prosthesis
	270[b]	
	277[c]	
	211–285[d]	
	220–280[e]	EMG/force plate
	450–750[f]	
Stair climbing	260[a]	Instrumented telemetric hip prosthesis
	227–314[d]	
Jogging	550[g]	
Stumbling	870[g]	
Single leg stance	210–280[b]	
Standing up	181–220[d]	
Sitting down	149–176[d]	
Knee bend	117–177[d]	

Source [a] Davy et al. (1988)
[b] Kotzar et al. (1991)
[c] Taylor et al. (1997)
[d] Bergmann et al. (2001)
[e] van den Bogert et al. (1999)
[f] Crowninshield et al. (1978)
[g] Bergmann et al. (1993)

2.2 Biomechanics of the Knee Joint

As discussed previously in Chap. 1, the knee joint consists of two distinct articulations: the tibiofemoral joint and the patellofemoral joint. In terms of movement capabilities, the knee joint exhibits six degrees-of-freedom, delineating its capacity for motion in various directions. Among these, three are translational degrees-of-freedom, specifying movement along the anterior–posterior, medial–lateral, and superior-inferior axes (Fig. 2.10). Concurrently, the knee joint has three rotational degrees-of-freedom: flexion–extension, varus-valgus rotation, and axial rotation. While the knee joint is equipped for multifaceted movements, the most pronounced motion is the flexion–extension. This involves bending (flexion) and straightening (extension) of the knee joint.

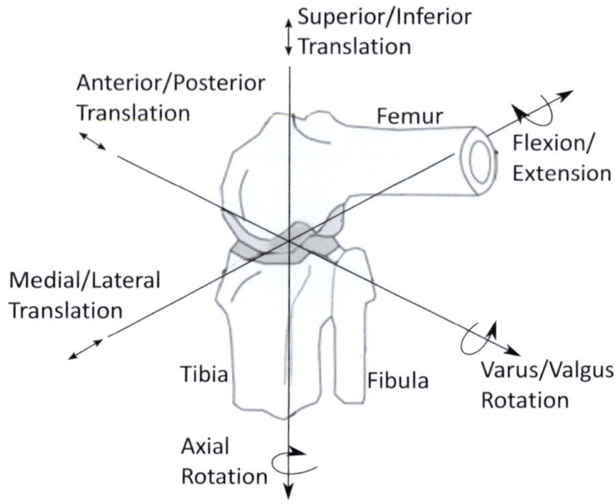

Fig. 2.10 Degrees of freedom of the knee joint

2.2.1 Instantaneous Centre of Rotation of the Knee Joint

A crucial aspect of knee joint kinematics is the trajectory of the knee's instantaneous centre of rotation (IC) throughout the entire range of knee flexion and extension, as presented in Fig. 2.11. In a normal knee, the trajectory follows a semi-circular path. As the femur rotates during flexion or extension, the instantaneous centre of rotation concurrently traverses along this semi-circular trajectory. This insight enhances our understanding of knee joint mechanics and its intricate dynamics during movement.

Let us thoroughly examine the trajectory traced by the instantaneous centre of rotation of the knee. In Fig. 2.12, point A moves to A′, whereas point B to B′ during the movement of the femur. The angle of rotation has been exaggerated for better clarity. The perpendicular bisectors of these lines AA′ and BB′ intersect at the

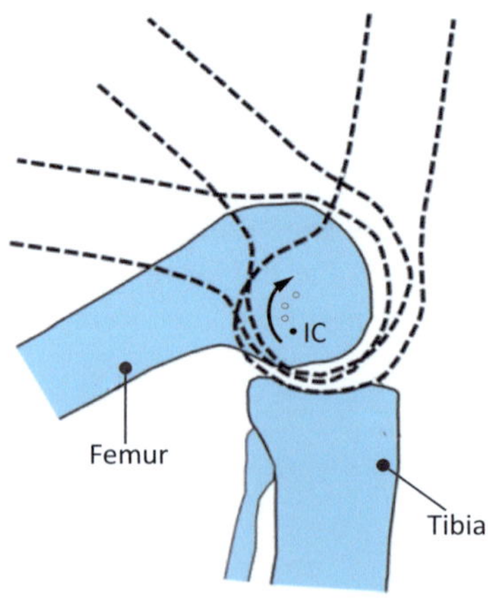

Fig. 2.11 Trajectory of the knee instantaneous centre of rotation

instantaneous centre of rotation (Fig. 2.12). Now the instantaneous centre of rotation is clinically determined by capturing successive radiograph images of the joint in different positions, usually in small intervals of about 10°, which needs to be obtained throughout the range of motion in a plane. Thereafter, the instantaneous centre of rotation can be located for each interval of rotation.

Let us focus our attention on knee flexion and the tibiofemoral joint. In Fig. 2.13, the changes in the instantaneous centre of rotation for various cases of relative motion between the tibia and femur have been presented. The first case corresponds to gliding (Fig. 2.13a). For ease of understanding, a line denoted by P is drawn from the instantaneous centre of rotation to the contact point and is perpendicular to the line, Q, which is tangential to the tibia surface. The arrow indicates the direction of the displacement of the contact points. Line Q remains tangential to the tibia surface, indicating that the femur glides on the tibia condyles during the interval of motion.

The next case, as shown in Fig. 2.13b, is pure sliding during knee flexion, where no change in the contact point of the tibia is observed. Figure 2.13c depicts the case of pure rolling, where changes in the contact points of femur and tibia are observed. Moreover, the instantaneous centre of rotation lies at the contact point during pure

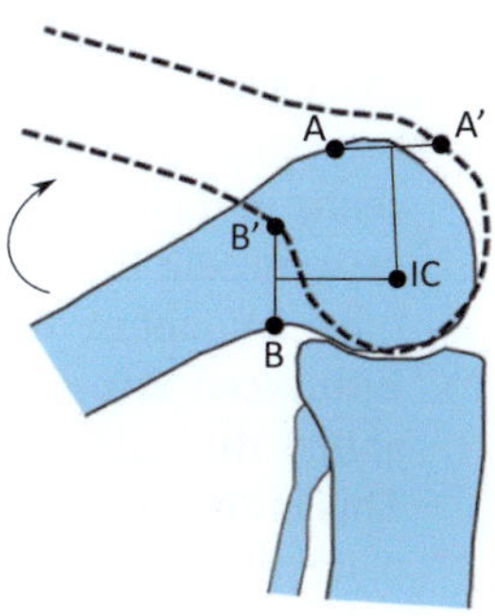

Fig. 2.12 The instantaneous center of rotation (IC) of the knee joint

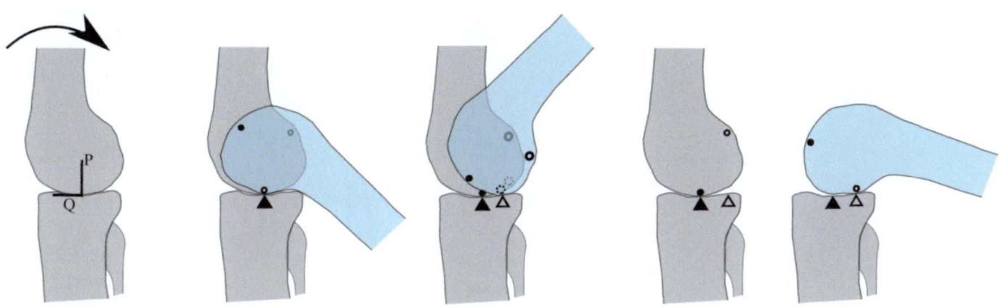

Fig. 2.13 Changes in the knee instantaneous center of rotation for various scenarios; **a** gliding, **b** pure sliding, **c** pure rolling, **d** combined sliding and rolling. Here, the triangles and circles indicate the contact points on tibia and femur, respectively

rolling. Now, let us discuss the case that corresponds to actual knee motion, that is, a combination of sliding and rolling as presented in Fig. 2.13d. The contact point of tibia and femur changes during this case. The interplay between sliding and rolling intricately influences the trajectory of the instantaneous centre of rotation throughout the flexion movement. As we know, the instantaneous centre of rotation changes during knee flexion.

2.2.2 Influence of the Patella on Knee Joint Mechanics

Two configurations of the knee joint, with and without the patella, can be distinguished by examining Fig. 2.14. Upon observation, a notable distinction arises in the lever arm between these two scenarios, particularly during the extension of the knee. This discrepancy is attributed to the influence of the patella. During knee extension, the lever arm of the quadriceps tendon experiences a substantial increase, owing to the presence of the patella. This mechanism highlights the mechanical advantages conferred by the patella in knee joint dynamics.

The functions of the patella are summarized as follows:

(1) **Increased quadriceps tendon moment arm**: The patella actively aids in knee extension by offsetting the quadriceps tendon with respect to the instantaneous centre of rotation. This offset effectively elongates the moment arm of the quadriceps muscle force during knee extension. As a result, in order to generate the same amount of moment, it requires a lesser muscle force.

(2) **Enhanced stress distribution**: Another notable function of the patella is its ability to facilitate a broader distribution of compressive stress on the femur. This is accomplished by expanding the contact area between the femur and the patellar tendon. By enlarging the contact region, the patella helps to minimise the chances of stress concentrations.

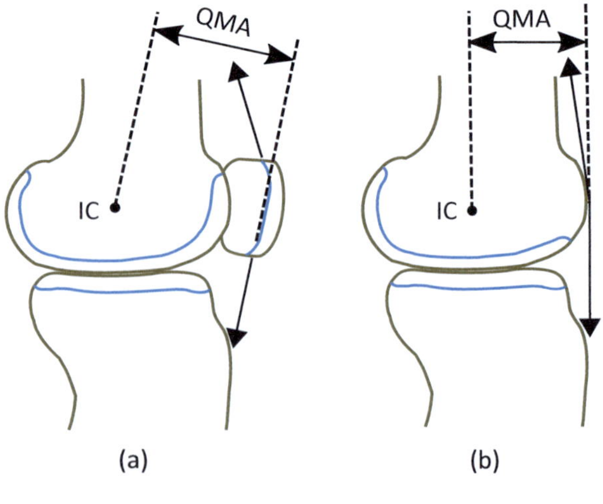

Fig. 2.14 Effect of patella on the moment arm of quadriceps (QMA); **a** knee with patella, **b** knee without patella

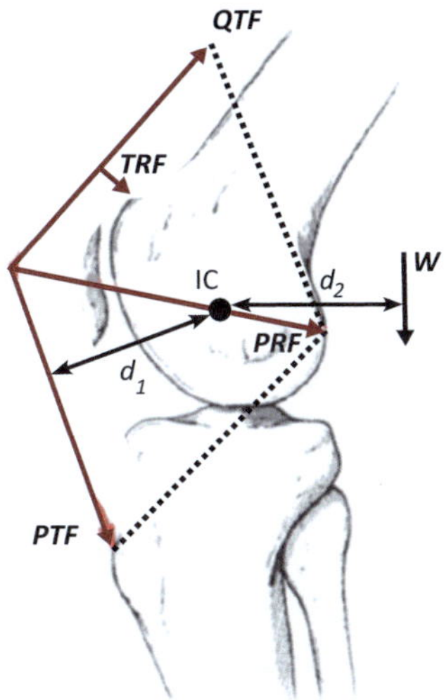

Fig. 2.15 Forces acting on a tibiofemoral joint. *W* is the body weight, *PRF*: patellar reaction force, *PTF*: patellar tendon force, *QTF*: quadriceps tendon force, IC: Instantaneous centre of rotation, d_1 and d_2 are moment arms of forces

Let us now focus on the forces acting on the patellofemoral joint, as presented in Fig. 2.15. The patellofemoral joint is subjected to compressive force, patellar reaction force (*PRF*), which is the resultant of patellar tendon force (*PTF*) and the quadriceps tendon force (*QTF*). The resultant force passes through the instantaneous centre of rotation (IC) of the patellofemoral joint. The *PTF* acts at a moment arm of d_1 from the IC. The tendon-femoral reaction force (*TRF*) acts between the quadriceps tendon and the trochlea. The weight of the body (*W*) is acting vertically downwards along the line of action of the centre of gravity, at a distance of *d*2 from the IC. As evident from Fig. 2.15, changes in postures will alter the distance d_2, which will have a pronounced effect on the force transmission at the patellofemoral joint.

2.2.3 Screw–Home Mechanism of the Knee Joint

The screw-home mechanism in the knee joint is considered to be vital for a stable standing and fully extended leg posture. It may be noted that there is a slight rotation of the knee during flexion and extension. During the final 20° of extension, femur medially rotates (similar to a screw), about 10°, relative to the tibia. This rotation results in tightening of the cruciate ligaments, which locks the knee and provides better joint stability. The knee is unlocked by lateral rotation of the femur relative to tibia. This complex movement during the terminal stage of extension is owing to the difference in the articular surface of the medial condyle and lateral condyle, as shown in Fig. 2.16.

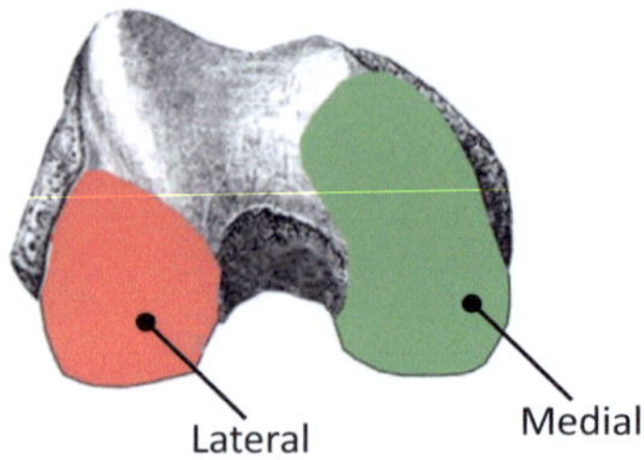

Fig. 2.16 Difference between articular surface geometry of medial, and lateral condyles of femur

2.2.4 *Basic Biomechanics of the Knee Joint: Sample Problems*

Let us now solve a numerical problem involving a simplified 2D force system, which includes the ground reaction force, ***R***, the patellar tendon force, ***P***, and the joint reaction force ***J***, as presented in Fig. 2.17. Here, the direction and point of application of the patellar tendon force (***P***) are the known quantities, whereas its magnitude is unknown. In the case of joint reaction force, both magnitude and direction are unknown; only the point of application of this joint reaction force is known in terms of instantaneous centre of rotation. For the ground reaction force (***R***), the direction, point of application and magnitude are known. Now, since the limb segment is in static equilibrium, the line of action of all three forces should intersect at one point.

The intersection point (I) is found out by extending the lines of action of ***P*** and ***R***, as indicated in Fig. 2.18. Based on the simplified system of coplanar forces, the line of action of ***J*** is determined by drawing a line between the point of application and the intersection point. Since the force system is coplanar and in static equilibrium, the magnitude of joint reaction force ***J*** can be found out graphically, using the force

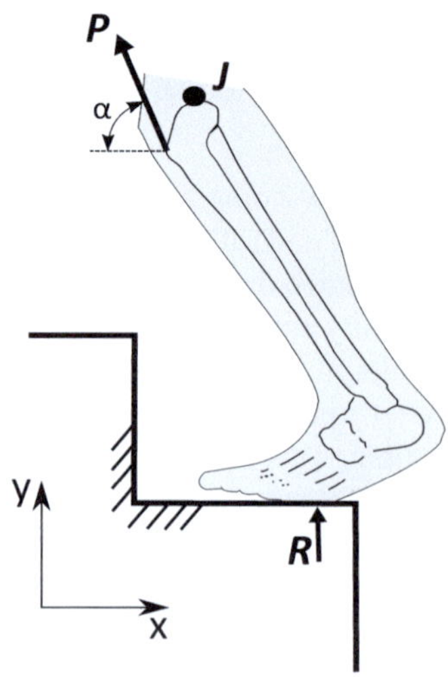

Fig. 2.17 A simplified 2D force system acting on a lower limb

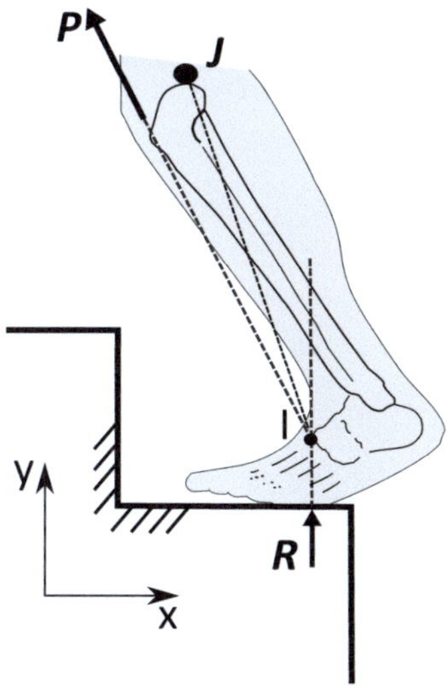

Fig. 2.18 Location of the intersection point (I) of the forces ***R*** and ***P*** in the lower limb

Fig. 2.19 Graphical method to determine force ***J*** using the force triangle

triangle, as indicated in Fig. 2.19. In the graphical method, length of each side indicates the magnitude of the corresponding forces.

Alternatively, the unknown forces can be determined using the equilibrium equations. In this problem, the ground reaction force (***R***) is equal to the body weight (***W***), since only one limb is in contact with the ground (Fig. 2.18). The magnitude of the patellar tendon force, ***P*** can be obtained by considering the moment equilibrium about the joint centre (O) (Fig. 2.20), as stated below.

$$\sum M_O = 0$$

$$Ra - Pb = 0$$

$$P = \frac{Wa}{b}$$

The ground reaction force acts at a lever arm a, which creates a counter-clockwise moment. Whereas, a force ***P*** acts at the lever arm b, creating a moment in the clockwise direction. Thereafter, using force equilibrium equations, the joint reaction force (***J***) can be determined (Fig. 2.21).

Considering the force equilibrium along the x-direction,

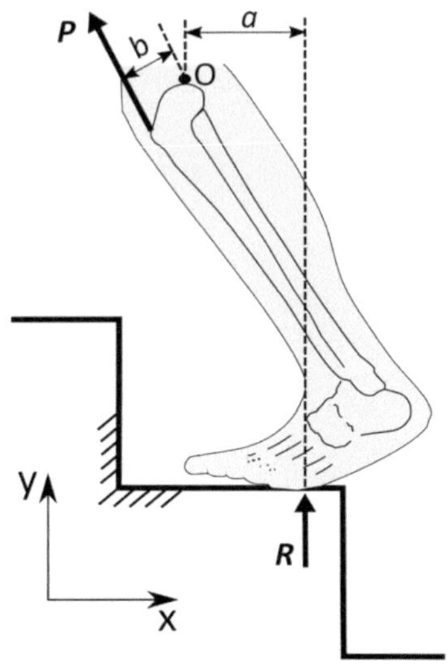

Fig. 2.20 The moment arms of the coplanar forces about the joint center (O)

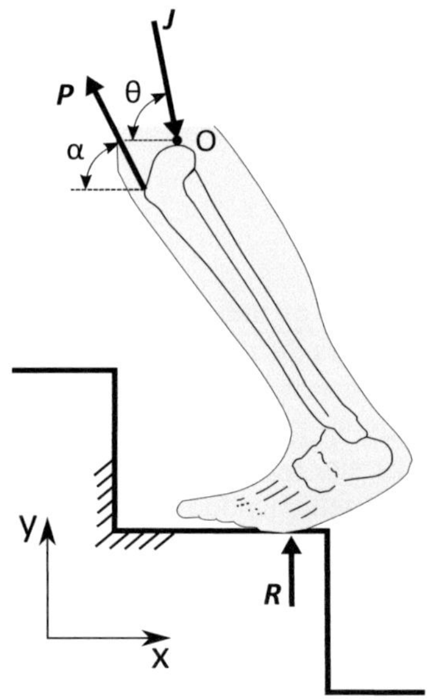

Fig. 2.21 Orientation of the joint reaction force (J) and the patellar tendon force (P), acting at angles θ and α, respectively with the horizontal

$$\Sigma F_x = J\cos(\theta) - P\cos(\alpha) = 0$$
$$J\cos(\theta) = P\cos(\alpha) \tag{2.1}$$

Considering the force equilibrium along the x-direction,

$$\Sigma F_y = -J\sin(\theta) + P\sin(\alpha) + R = 0$$

$$J\sin(\theta) = R + P\sin(\alpha) \tag{2.2}$$

Using Eqs. 2.1 and 2.2, we get

$$\tan(\theta) = \frac{R + P\sin(\alpha)}{P\cos(\alpha)}$$

This equation is further simplified to

$$\tan(\theta) = \frac{W + \frac{Wa}{b}\sin(\alpha)}{\frac{Wa}{b}\cos(\alpha)}$$

$$\theta = \tan^{-1}\left\{\frac{W + \frac{Wa}{b}\sin(\alpha)}{\frac{Wa}{b}\cos(\alpha)}\right\} \tag{2.3}$$

Upon substituting θ in the Eq. 2.3, joint reaction force (***J***) is obtained as

$$J = \frac{P\cos(\alpha)}{\cos\left[\tan^{-1}\left\{\frac{W + \frac{Wa}{b}\sin(\alpha)}{\frac{Wa}{b}\cos(\alpha)}\right\}\right]}$$

Let us look into another example regarding the calculation of patellar tendon force. Figure 2.22 shows the flexed knee in a static position. In this particular problem, we need to calculate the quadriceps muscle force necessary to maintain the knee flexion at a particular angle. The patellar tendon force, ***P*** acts at a perpendicular distance d_m from the joint centre, whereas the bodyweight acts with a lever arm of d_w. These lever arms are given as: $d_m = 50$ cm, $d_w = 60$ cm. The patellar tendon force, ***P*** and the bodyweight, ***W*** of 60 kg. Considering the forces acting on a flexed knee and moment equilibrium about the joint centre (C), the patellar tendon force (***P***) is determined in the following steps:

$$\sum M_C = 0$$

$$Pd_m - Wd_w = 0$$

On substituting the values of lever arms and the weight, we obtain the patellar tendon force (***P***):

$$P = \frac{60 \times 60}{50} = 72\,\text{kg}$$

In the next problem, we consider the leg segment of an individual in a sitting position, with the knee partially flexed such that the foot is not in contact with the ground, as shown in Fig. 2.23. The subject is wearing a boot of weight, $\boldsymbol{W_b}$. The weight of the lower leg is indicated by $\boldsymbol{W_l}$. Here, we need to find out the tibia-femoral joint contact force ($\boldsymbol{F_j}$) and the tensile force in the quadriceps muscle ($\boldsymbol{F_m}$). The joint

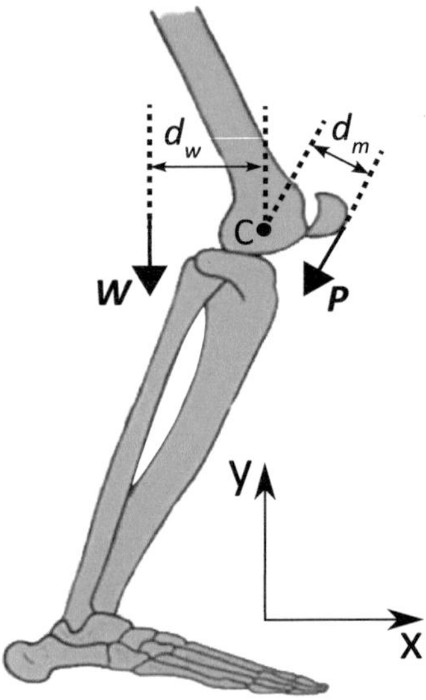

Fig. 2.22 Forces acting on a flexed knee. The leg segment is shown in the figure for better visualization

reaction force makes an angle α with the horizontal. The centre of gravity of the leg and boot are located at B and C, respectively. With respect to the joint centre (O), the points A, B, and C are at a distance of a, b, and c, respectively. The line of action of the muscle force is oriented at θ with the mechanical axis (Fig. 2.23). The mechanical axis of a bone segment is a straight line that connects the centre points of the proximal and distal joint regions in the frontal or sagittal plane. The mechanical axis is oriented at an angle, β with the horizontal.

Considering moment equilibrium about the joint centre O, the quadriceps muscle ($\boldsymbol{F_m}$) can be determined.

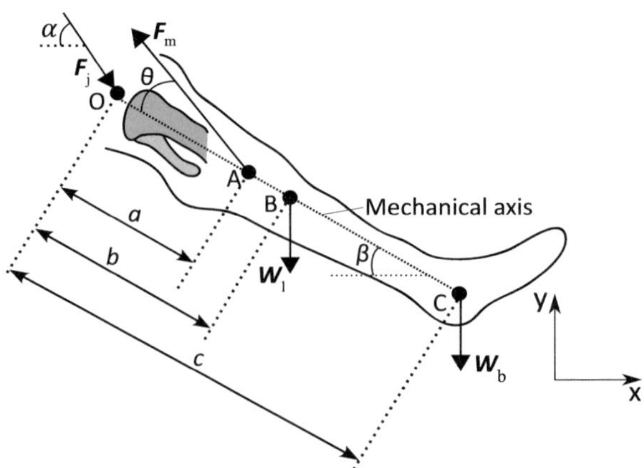

Fig. 2.23 Forces acting on a partially flexed knee in a sitting posture

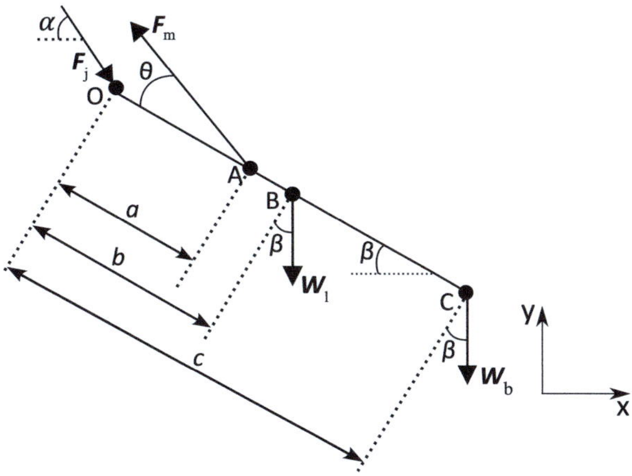

Fig. 2.24 Free body diagram of the flexed knee in sitting posture

$$\sum M_o = 0$$

$$F_m a \sin\theta - W_l b \cos\beta - W_b c \cos\beta = 0$$

$$F_m = \frac{(W_l b + W_b c)\cos\beta}{a \sin\theta} \tag{2.4}$$

In order to find out the unknown joint reaction force, we need to consider the force equilibrium equations. Resolving the forces along *x*-direction (Fig. 2.24) and considering equilibrium, we obtain:

$$\sum F_X = 0$$

$$F_j \cos\alpha - F_m \cos(\beta + \theta) = 0$$

$$F_j \cos\alpha = F_m \cos(\beta + \theta) \tag{2.5}$$

Similarly, resolving the forces along *y*-direction (Fig. 2.24) and considering equilibrium, we obtain

$$\sum F_y = 0$$

$$F_j \sin\alpha - F_m \sin(\beta + \theta) + W_l + W_b = 0$$

$$F_j \sin\alpha = F_m \sin(\beta + \theta) - (W_l + W_b) \tag{2.6}$$

Combining the Eqs. 2.5 and 2.6, we get

$$\tan\alpha = \frac{F_m \sin(\beta + \theta) - (W_l + W_b)}{F_m \cos(\beta + \theta)} \tag{2.7}$$

The values of *a*, *b*, and *c* are assumed to be 0.08, 0.14, and 0.28 times the body height (*h*). The weight of the limb and the boot are considered as 0.085 and 0.065

times the bodyweight (W), respectively. The angles θ and β are assumed to be 18° and 47°, respectively. On substituting these values in Eqs. 2.4, 2.5, and 2.7, we obtain the unknown forces and the angle as follows:

$$F_m = 0.830W;\ \alpha = 59.78^\circ;\ F_j = 0.697W$$

2.2.5 *Knee Joint Reaction Forces During Various Activities*

The variation of knee joint reaction forces during common daily activities provides valuable insights into knee joint biomechanics. The peak resultant forces during walking generally range between two to three times the body weight (BW). Table 2.3 presents the variation in peak knee joint reaction forces for various activities. Activities like normal walking, involving minimal flexion, incur lower joint forces. In comparison, activities that require greater flexion, such as stair ascent and descent, cause an abnormal increase in the joint reaction force, ranging from 3.3 to 5 times body weight (BW). In contrast, jogging results in a high joint force that is 4.5 times BW. This meticulous analysis emphasises the multifaceted nature of knee joint biomechanics, demonstrating how various activities and flexion patterns intricately influence the joint reaction forces.

2.3 Biomechanics of the Ankle Joint

The ankle joint is a synovial hinge joint, which is sometimes referred to as a "mortise-tenon" joint, commonly used in woodworking. In woodworking, the mortise-tenon joint involves inserting a projection (tenon) of one wood piece into a hole (mortise) of another, establishing a robust joint. Likewise, the ankle joint features the talus bone (tenon) fitting into the mortise shaped by the tibia and fibula, forming a stable and weight-bearing joint.

Table 2.3 Peak knee joint reaction forces during common daily activities

Activities	Maximum knee joint force (BW)	References
Level walking	2.6–2.8	Taylor and Walker (2001)
Downhill walking	4.4	Paul (1976)
Uphill walking	3.7	Paul (1976)
Stairs down	2.9–3.1	Taylor and Walker (2001)
Stair ascend	2.4–2.5	Taylor and Walker (2001)
Jogging	4.1–4.5	D'Lima et al. (2008)

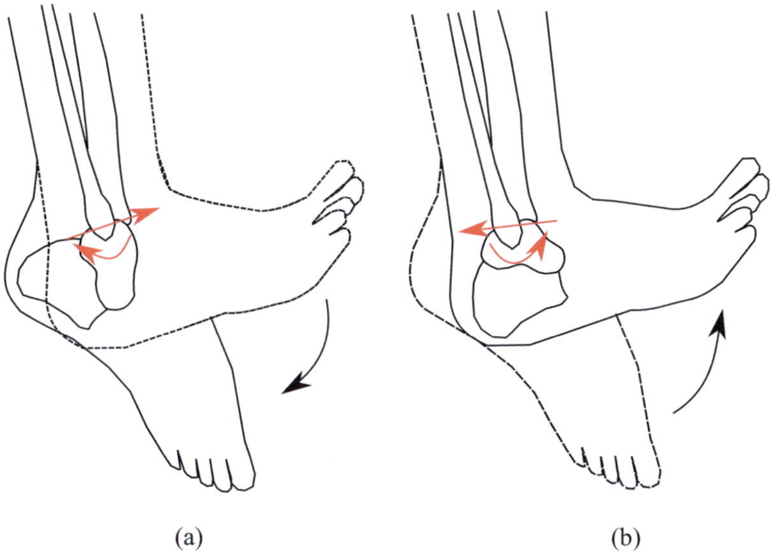

Fig. 2.25 Movements of the ankle joint; **a** plantarflexion, **b** dorsiflexion. The red arrows indicate gliding and rolling

The primary movements of the ankle joint are plantar and dorsiflexion, which occurs in the sagittal plane, about the talocrural axis (axis of rotation of the ankle joint). Plantarflexion involves pointing the toes downward away from the shin. As the foot moves downward, the talus bone rolls posteriorly and glides anteriorly on the tibiofibular surface (Fig. 2.25). Dorsiflexion involves lifting the foot upward, bringing the toes closer to the shin bone. During dorsiflexion, the talus bone within the ankle joint undergoes a specific movement pattern. As the foot moves upward, the talus bone rolls anteriorly and glides posteriorly on the tibiofibular surface (Fig. 2.25). During dorsiflexion, the talus locks against the sides of the ankle mortise, resulting in better joint stability. Dorsiflexion is crucial for activities requiring controlled foot descent, such as descending stairs or walking downhill. Plantarflexion, on the other hand, is vital for activities requiring propulsion and push-off, like jumping or accelerating during running.

2.3.1 Basic Biomechanics of the Ankle Joint: Sample Problems

Let us solve a numerical problem involving a simplified 2D force system representing an instance of toe-off (Fig. 2.26). The ground reaction force (***R***) and achilles tendon force (***P***) are given as 400 N and 1200 N, respectively. In addition, the ***P*** acts at an

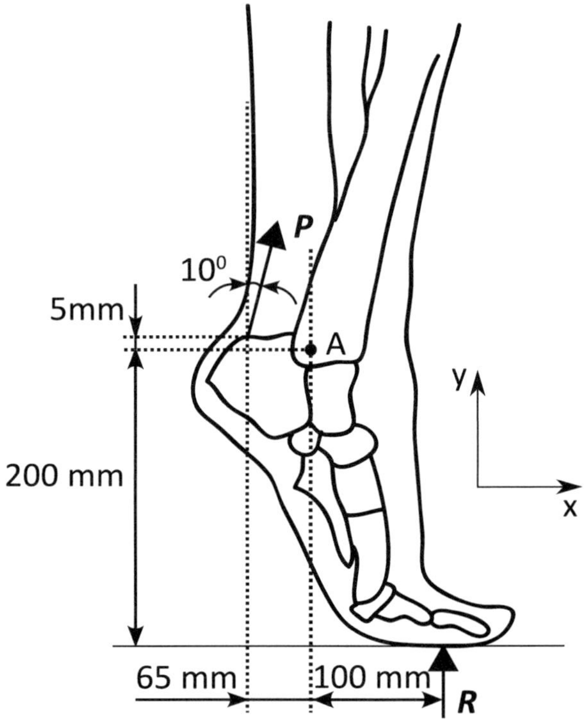

Fig. 2.26 Simplified illustration of the forces acting on the ankle joint during toe off

angle of 10° with the vertical. The objective is to determine the ankle joint force (***J***) acting at point *A*.

Let us assume that ***J*** acts as an angle α with the vertical as presented in Fig. 2.27. Since the foot is in static equilibrium, the sum of the forces along the *x*- and *y*-direction is equal to zero.

$$\Sigma F_x = P\sin(10^\circ) - J\sin(\alpha) = 0$$
$$\Sigma F_y = R + P\cos(10^\circ) - J\cos(\alpha) = 0$$

Equating the above two equations, we obtain α as

$$\tan(\alpha) = \frac{P\sin(10^\circ)}{R + P\cos(10^\circ)} = \frac{1200\sin(10^\circ)}{400 + 1200\cos(10^\circ)}$$
$$\alpha = 7.41^\circ$$

The joint reaction force (***J***) can be obtained after substituting α in the equilibrium equation, as follows:

$$J = \frac{1200\ \sin(10^\circ)}{\sin(7.41^\circ)} = 1615.72\,\text{N}$$

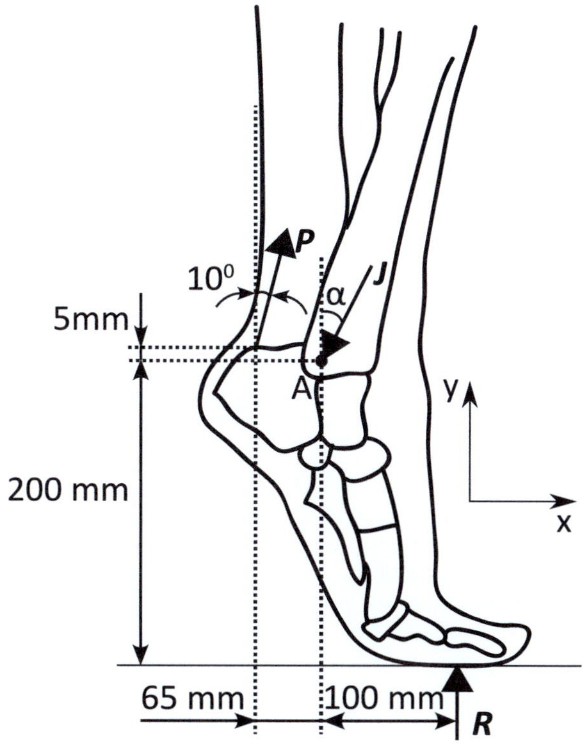

Fig. 2.27 Joint reaction force acting through the ankle joint center during toe off

In the next problem, let us consider the mass of the foot segment, and find out the joint reaction force (***J***) during the instance of mid-stance. In this problem, the foot is in the neutral position as shown in Fig. 2.28. The weight of the foot segment ($\boldsymbol{W_f}$), achilles tendon force (***P***), and total body weight is given as 10.5 N, 2100 N, and 720 N, respectively. Unlike the previous example, the ***P*** acts vertically upwards in this case.

Let us assume that the angles made by ***J*** with the vertical is equal to α. Similar to the previous example, let us consider the force equilibrium equations (Fig. 2.29).

$$\Sigma F_x = -J \sin(\alpha) = 0$$
$$\Sigma F_y = R - W_f + P - J \cos(\alpha) = 0$$

Here, the ground reaction force (***R***) is equal to the total body weight (BW), since we consider the instance of mid-stance. From the above equations, it is clear that ***J*** acts vertically downwards ($\alpha = 0°$) and is expressed as

$$J = \frac{R - W_f + P}{\cos(\alpha)} = \frac{R - W_f + P}{\cos(0°)}$$
$$= 2100 - 10.5 + 720 = 2809.5\,\text{N}$$

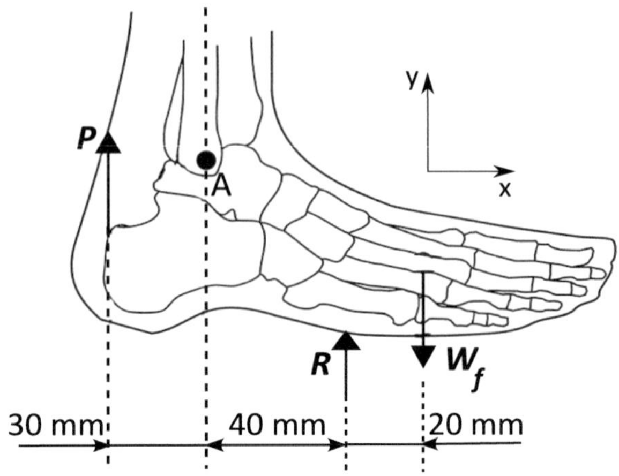

Fig. 2.28 Forces acting on the ankle joint during mid-stance

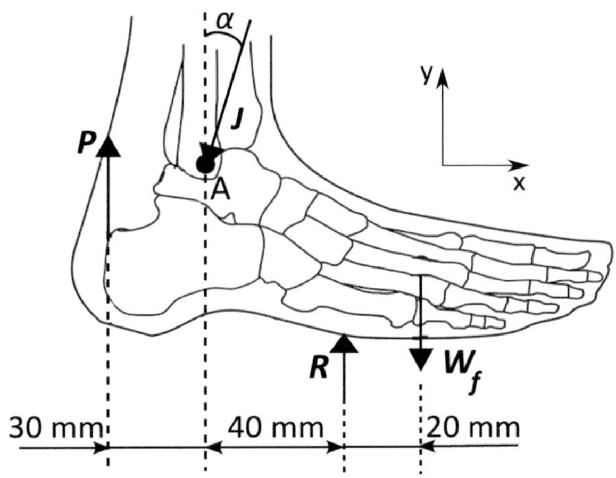

Fig. 2.29 Joint reaction force acting through the ankle joint centre during mid-stance

Here, $\boldsymbol{J}$ is equal to 2809.5 N $\approx$ 3.9 BW. It may be noted that the peak ankle joint reaction forces during walking generally range from three to six times the body weight (Stauffer et al. 1977; Steineman et al. 2021).

2.4 Biomechanics of the Shoulder Joint

The shoulder joint consists of a chain of bones connecting the upper extremity to the torso. In contrast to the pelvic joint, the shoulder joint offers a considerable range of motion (Sect. 1.12). The shoulder joint comprises three synovial joints: the Glenohumeral (GH-) joint, the Acromioclavicular (AC-) joint, and the sternoclavicular (SC-) joint. The articulating geometry of these synovial joints is complex; they can be regarded as ball joints for many purposes, neglecting the possibilities of small translations. During movement of the arm, the medial of the scapula slides over the dorsal side of the thoracic cage, pressed onto it by the combined action of muscles. This connection gives rise to a fourth joint, the scapula-thoracic gliding plane (STGP). The shoulder mechanism is an example of a very complex musculoskeletal system.

2.4.1 Closed Chain Mechanism

Before delving into the shoulder mechanism, it is important to understand the fundamental working principles of mechanisms. A kinematic chain is an assembly of rigid bodies connected by joints to provide constrained motion, whereas a mechanism is a kinematic chain with one fixed link. There are two types of mechanisms, the closed chain and the open chain, as shown in Fig. 2.30. In a closed chain mechanism, every link is connected through joints to two or more adjacent links (Fig. 2.30a). The terminal link experiences resistance enough to restrain its free motion. Hence, the terminal links are fixed. In case of an open chain mechanism, each link is only connected to a single joint (Fig. 2.30b). Moreover, the terminal link in the chain is not restrained and is freely movable. The salient features of the closed chain mechanism as encountered in the human body are: relatively small joint movements, low joint accelerations, large resistance forces, joint compression, decreased joint shear and translation, and enhanced dynamic stabilisation through muscle activation.

The shoulder joint represents a closed chain mechanism. The connection between the clavicle and the thorax occurs through the SC-joint, while the scapula is connected

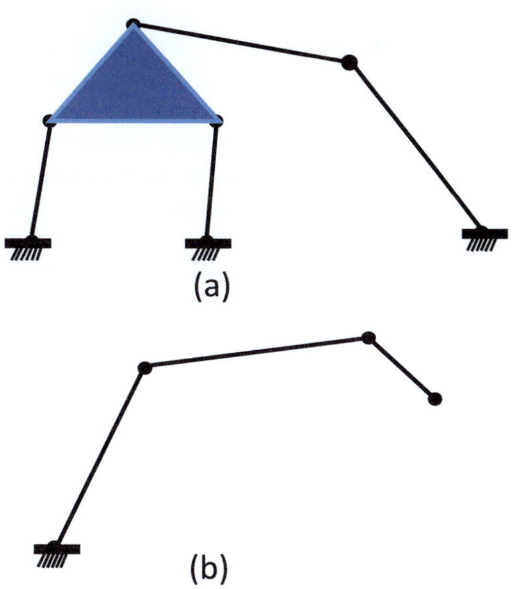

Fig. 2.30 Basic types of mechanisms; **a** closed chain mechanism, **b** open chain mechanism

to the clavicle through the AC-joint. Moreover, the scapula's medial border glides over the dorsal side of the thorax, restricting its movements. The bony process of the scapula, acromion, moves in a spherical manner around the SC-joint, while the medial border of the scapula slides along the ellipsoidal shape of the thorax, guided by muscles, m. serratus anterior and m. rhomboideus. The double connection of the scapula to the thorax, one by the clavicle, and the other by the scapula thoracic gliding plane, results in a closed chain mechanism. Due to the closed chain, the translational and rotational degrees-of-freedom of the scapula and clavicle are restricted. Simultaneous motions of the shoulder girdle and humerus are called scapulohumeral rhythm. This rhythm ensures proper and efficient movement of the shoulder joint. As we raise our arm, especially beyond a certain angle, the scapula starts to rotate to allow enough space for the arm bone to move smoothly. This coordinated movement helps prevent impingement or discomfort in the shoulder joint. In addition, the large motion range of the humerus is attributed to the simultaneous motions of the SC-, AC-, and GH-joints.

2.4.2 Degrees-of-Freedom of the Shoulder Mechanism

Let's now discuss the degrees-of-freedom (DOF) of the shoulder joint. The shoulder complex comprises several key joints, including the glenohumeral (GH), acromioclavicular (AC), and sternoclavicular joints (SC). The Delft shoulder model was originally developed to estimate muscle and joint contact forces throughout the entire shoulder complex (Van der Helm et al. 1994a, b), offering a comprehensive framework for understanding the complexities of these joints. In this model, GH-, AC-, and SC-joints were visualized as spherical joints, and offered three rotational degrees-of-freedom each. Additionally, the scapulothoracic joint's movement is conceptualized by modelling it in connection with the thorax. The thorax was modelled as an ellipsoid with two points, Thorax-Angulus inferior (Thorax-AI) and Thorax-Trigonum spinae (Thorax-TS), on the medial border of the scapula undergoing sliding movements (Fig. 2.31). This unique mechanism results in the constraint of two degrees-of-freedom (DOF) by the scapulothoracic gliding plane. The conoid ligament was assumed as a rigid structure, which reduced the total DOF by one. This comprehensive model provides a useful framework for understanding the complexities of shoulder and elbow mechanics. The DOF of the upper extremity containing thorax, shoulder, elbow, and wrist joint may be summarized as follows.

(1) **Thorax**:

- Three rotational DOF,
- Three translational DOF.

(2) **Sternoclavicular, Acromioclavicular, and Glenohumeral Joints**:

- Each of these joints offers three rotational DOF.

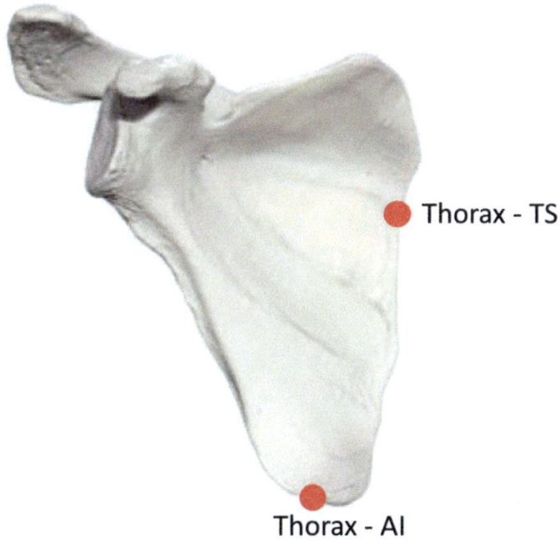

Fig. 2.31 The connection points between the scapula and the thorax

(3) **Scapulothoracic Joint**:

- The scapulothoracic gliding plane introduces constraints through the points Thorax-AI and Thorax-TS, which limits two DOF.

(4) **Conoid Ligament**:

- The rigid structure assumption of the conoid ligament further restricts one degree of freedom (axial rotation).

Considering the thorax and the shoulder joint complex, the total DOF is calculated as 12.

2.4.3 Glenohumeral Joint Reaction Force During Abduction

Let us explore how the GH-joint force varies with the movements of the arm. The GH-reaction force increases with an increase in abduction angle until 90° and thereafter, reduces (Table 2.4). A peak GH joint reaction force of ~393N, corresponding to unloaded abduction of 90° was reported by van der Helm (1994a, b).

2.4.4 Basic Biomechanics of the Shoulder Joint: Sample Problems

Let us discuss a few simple problems, for a better understanding of mechanics associated with the shoulder joint. The primary objective of these problems is to find the predominant GH-joint reaction force in the shoulder complex.

Table 2.4 Resultant glenohumeral joint reaction forces during unloaded humeral abduction

Abduction angle (°)	Glenohumeral joint reaction force (N)
0	19.26
30	165.84
60	325.84
90	392.94
120	346.01
150	192.83
180	82.79

Problem 1 In this problem, the subject is in the position of 90° abduction (Fig. 2.32). It is assumed that only the deltoid muscle force ($\boldsymbol{F_d}$) is involved and acts at a distance of 0.035 m from the centre of GH joint (O), so as indicated in Fig. 2.32. The weight of the arm ($\boldsymbol{W_a}$) is assumed to be 0.05 times the body weight (BW) and is acting at a distance of 0.35 m from O. We intend to calculate the GH-joint reaction force ($\boldsymbol{F_g}$).

Solution 1: The joint reaction force and the deltoid muscle create a couple since they are equal and acting at a distance in opposite directions. This couple opposes the moment produced by the weight of the arm. Considering the moment equilibrium about the centre of joint, the deltoid muscle force is calculated as

$$\sum M_O = 0$$
$$(0.035 \times W_a) - (F_d \times 0.0035) = 0$$
$$F_d = 10 \times W_a = 0.5 \text{ BW}$$

Considering the force equilibrium along the x-direction, the GH-joint reaction force is determined.

$$\sum F_x = 0$$

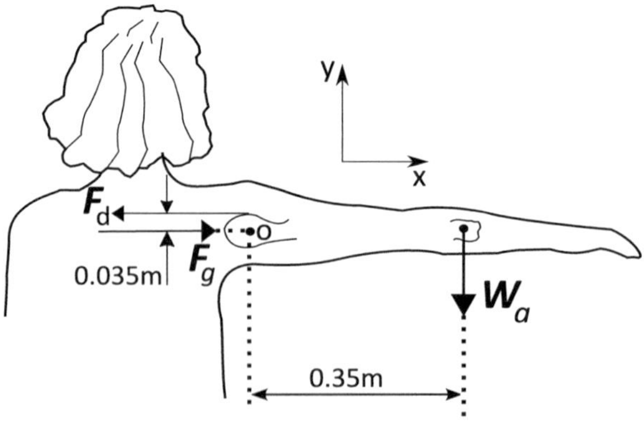

Fig. 2.32 Subject with 90° abducted arm

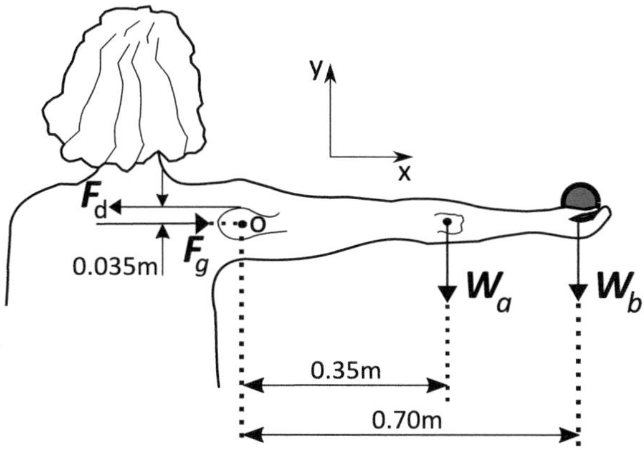

Fig. 2.33 Subject with 90° abducted arm and a ball in the hand

$$F_g - F_d = 0$$
$$F_g = F_d = 0.5\ \text{BW}$$

Hence, the GH-joint reaction force is calculated to be 0.5 BW.

Problem 2 In this problem, a ball having a weight ($\boldsymbol{W_b}$) equal to 0.025 times the body weight is carried at the end of the forearm (Fig. 2.33). The rest of the problem definition remains the same as stated in Problem 1. The location of the ball carried at the end of the forearm is 70 cm from the centre of rotation of the GH-joint.

Solution 2: Similar to the previous problem, considering the moment equilibrium, the deltoid muscle force is calculated as

$$\sum M_O = 0$$
$$(0.35 \times W_a) + (W_b \times 0.70) - (F_d \times 0.035) = 0$$
$$F_d = 1\ \text{BW}$$

Considering the force equilibrium along the *x*—direction, one can determine the GH-joint reaction force.

$$\sum F_x = 0$$
$$F_g - F_d = 0$$
$$F_g = F_d = 1\ \text{BW}$$

Hence, the GH-joint reaction force is equal to 1 BW.

2.5 Biomechanics of the Elbow Joint

The elbow joint may be considered a non-weight bearing joint, since it does not carry much of the body weight. However, this joint can support external loads and plays a crucial role in facilitating essential movements of the forearm, such as flexion, extension, pronation, and supination. Let us initiate our discussions on the basics of elbow joint mechanics and the degrees-of-freedom. The elbow joint provides two DOFs, owing to the articulations of humeroulnar and ulnar-radial joints, each providing one DOF through flexion–extension and pronation-supination movements, respectively.

2.5.1 *Carrying Angle*

Carrying angle is prescribed between the anatomical axis of the extended forearm and the anatomical axis of the arm, as shown in Fig. 2.34. It opens laterally at about 170° in males and 167° in the case of female subjects. However, it disappears when the elbow is flexed. The carrying angle permits the forearm to clear the hips in swinging movements during walking and is crucial when carrying objects.

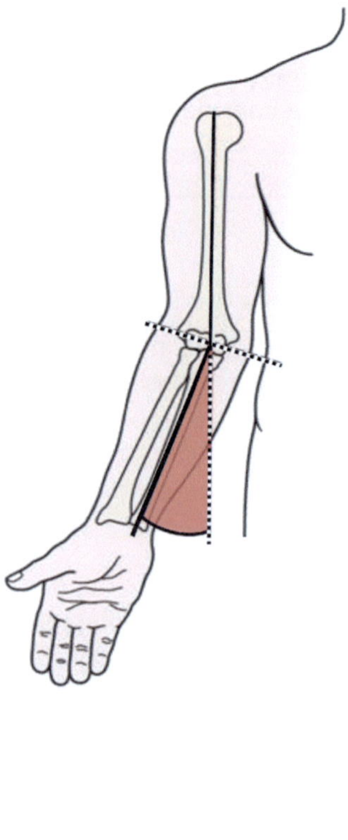

Fig. 2.34 Illustration showing the carrying angle in the elbow joint

2.5.2 *Basic Biomechanics of the Elbow Joint: Sample Problem*

Let us gain an insight into the basic biomechanics of the elbow using a numerical problem. Dynamic analysis is not considered because heavily loaded movements of the joint tend to be slow. As shown in Fig. 2.35, the subject is carrying a ball weighing 300 N with a 90° flexed elbow position. The hand is considered to be in a neutral position of rotation. Hence, there are no rotational movements, like pronation or supination. The biceps, brachialis, and brachiaradialis are the main flexor muscles acting to resist the external load. In order to simplify the analysis, the muscles of the wrist or the finger flexors are not considered in the analysis. The muscle forces are assumed to act along straight lines. The main steps involved in the analysis may be summarized sequentially as follows:

(1) Define the geometry and find the moment arms of the muscles about the joint axis.
(2) Using these moment arms, estimate the muscle force.
(3) Using estimated muscle forces, calculated the ligament force and the joint reaction forces.

Initially, in this analysis, we consider a unified bone that accounts for both the radius and ulna together. However, later, we will consider them as two separate bones. The insertion of m. biceps on the radius is 0.038 m from the joint axis, O, as shown in Fig. 2.36. The m. brachialis is also located at a similar distance, but the insertion is on the ulna (Fig. 2.37). The brachiaradialis muscle insertion is on the radius, at a distance of 0.24 m from O (Fig. 2.38). The limb weight ($\boldsymbol{W}$) of 32 N is assumed to act at a distance of 0.16 m from O. The wrist axis is denoted as B. The ball weight ($\boldsymbol{W_a}$) of 300 N is assumed to be acting at the distal end of the metacarpals which is 0.35 m away from O.

Using the geometric data, we can define the lines of action of the muscles and hence, calculate their moment arms about the joint axis passing through joint centre O. Let us consider the brachialis muscle and determine its moment arm (Fig. 2.37). The moment arm of the brachialis muscle about joint centre O is denoted by L_{Br}. Considering the ΔOEF, we can calculate the ∠OEF as follows:

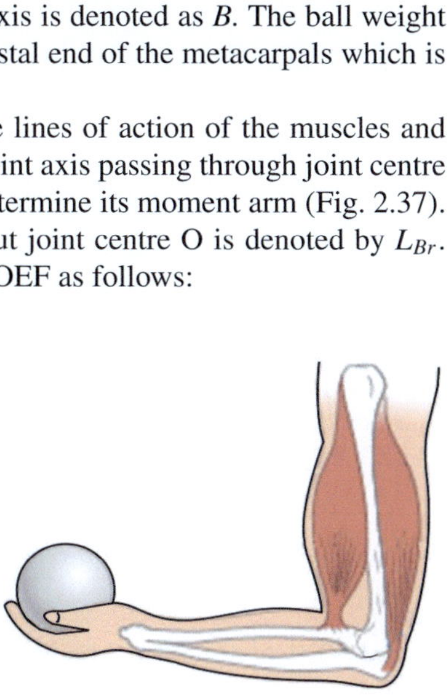

Fig. 2.35 Subject carrying a ball with elbow flexed at 90°

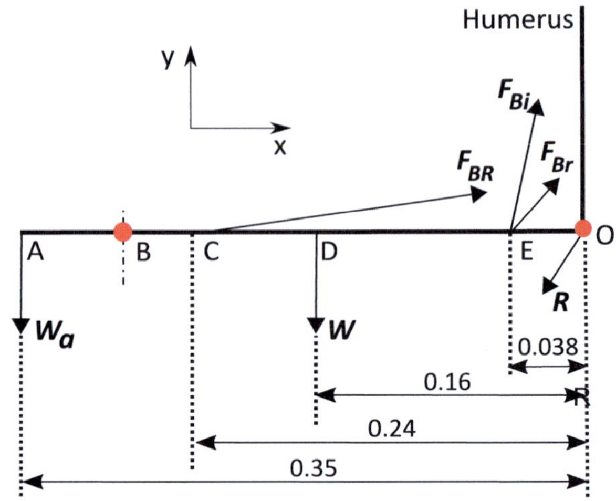

Fig. 2.36 Forces acting on the forearm carrying an external load during elbow flexion. The dimensions are in meters

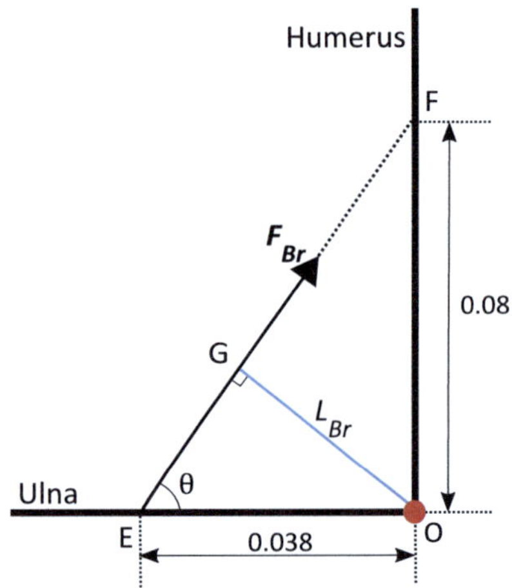

Fig. 2.37 Moment arm of m. brachialis

$$\tan \angle \mathrm{OEF} = \frac{0.080}{0.038} = 2.105$$
$$\theta = \angle \mathrm{OEF} = 64.6^\circ$$

Since $\angle \mathrm{OGE} = 90^\circ$ and the sum of all angles in a triangle is equal to 180°, the $\angle \mathrm{EOG}$ is 25.4°.

$$\begin{aligned} L_{Br} &= 0.038 \cos \angle \mathrm{EOG} \\ &= 0.038 \cos 25.4^\circ \approx 0.034\,\mathrm{m} \end{aligned}$$

The moment arm of m. brachialis, L_{Br} is found to be 34 mm.

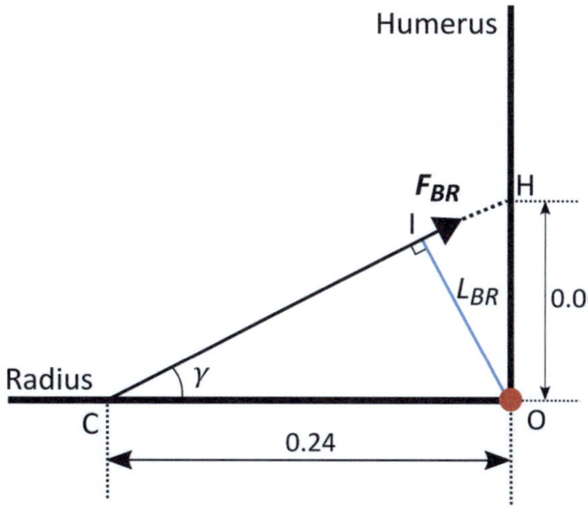

Fig. 2.38 Moment arm of m. brachiaradialis

Let us now calculate the moment arm of m. brachiaradialis, from Fig. 2.38. Following a similar procedure as earlier, we can calculate the moment arm of the m. brachiaradialis about O, as follows:

$$\begin{aligned} L_{BR} &= 0.24\,\mathrm{Cos}\angle \mathrm{COI} \\ &= 0.24\mathrm{Cos}\;74.84^\circ \\ &\approx 0.063\,\mathrm{m} \end{aligned}$$

From Fig. 2.38, we get

$$\gamma = \angle \mathrm{OCI} = 15.16^\circ$$

Following a similar procedure, the moment arm (L_{Bi}) of m. biceps is found out from the geometrical data in Fig. 2.39:

$$\begin{aligned} \tan\angle \mathrm{EOK} = \tan\angle \mathrm{ELJ} &= \frac{0.038 - 0.020}{0.320} \\ \angle \mathrm{EOK} &= 3.2^\circ \end{aligned}$$

Therefore,

$$\beta = \angle \mathrm{OEK} = 86.8^\circ$$

$$\begin{aligned} L_{Bi} &= 0.038\,\mathrm{Cos}\angle \mathrm{EOK} \\ &= 0.038\,\mathrm{Cos}\;3.2^\circ \approx 0.038\,\mathrm{m} \end{aligned}$$

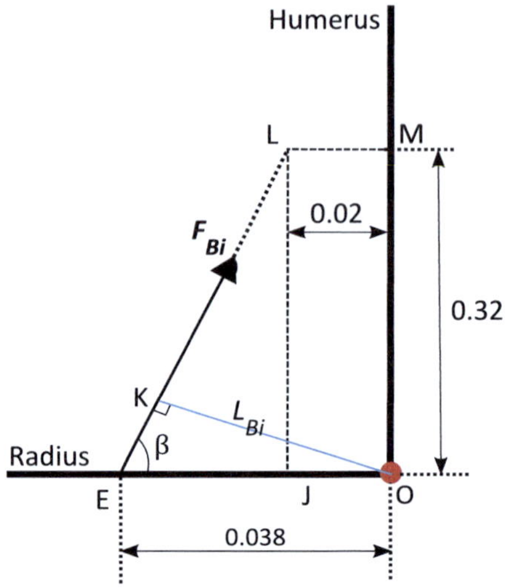

Fig. 2.39 Moment arm of m. biceps

The next step is the estimation of muscle forces. It is assumed that all three muscles are stressed to the same intensity. Hence, the forces produced in each muscle will be proportional to its physiological cross-sectional area. The typical values of the cross-sectional area of the muscles are given below:

$$A_{Bi} = 518 \times 10^{-6}\,\text{mm}^2$$
$$A_{Br} = 493 \times 10^{-6}\,\text{mm}^2$$
$$A_{BR} = 120 \times 10^{-6}\,\text{mm}^2$$

The muscle forces can be obtained by multiplying the cross-sectional area with the stresses generated in each muscle. In order to find out the muscle forces, we use the moment equilibrium about the joint centre O. Limb weight and external load produce moment about the joint centre O, and are expressed as follows:

$$\text{Moment due to external load} = 300 \times 0.35 = 105\,\text{Nm}$$
$$\text{Moment due to limb weight} = 32 \times 0.16 = 5.12\,\text{Nm}$$

The moment equilibrium about the joint axis O can be expressed as follows:

$$\text{Moment due to limb weight} + \text{Moment due to external load} \\ - \text{Moment due to muscles} = 0$$

$$105 + 5.12 - \boldsymbol{\sigma_m} \times [(A_{Bi} \times L_{Bi}) \\ +(A_{Br} \times L_{Br}) + (A_{BR} \times L_{BR})] = 0$$

$$105 + 5.12 = \boldsymbol{\sigma_m} \times [(A_{Bi} \times L_{Bi}) \\ +(A_{Br} \times L_{Br}) + (A_{BR} \times L_{BR})]$$

$$\sigma_m = \frac{105 + 5.12}{(518 \times 0.038 + 493 \times 0.034 + 120 \times 0.063) \times 10^{-6}} \approx 2.502386 \times 10^6\,\mathrm{Nm}^{-2}$$

Upon substituting the known values, we find the stress value to be $2.502386 \times 10^6\,\mathrm{Nm}^{-2}$, which is generated by the flexor muscles under the given problem description. Using this value of stress and the corresponding cross-sectional area, we calculate the individual muscle forces.

$$\begin{aligned} F_{Bi} &= \sigma \times A_{Bi} \\ &= 2.502386 \times 10^6 \times 518 \times 10^{-6}\,\mathrm{N} \\ &= 1296.23\,\mathrm{N} \\ F_{Br} &= 1233.67\,\mathrm{N} \\ F_{BR} &= 300.29\,\mathrm{N} \end{aligned}$$

It is assumed that each of the radius and the ulna carry a share of the external load equivalent to the strength of their respective muscle arrangement. Hence, the individual joint reaction forces can be estimated. Using the muscle forces, let us estimate the joint reaction forces of humeroradial and humeroulnar articulations. So far, we assumed that the radius and ulnar be a unified bone articulating with the humerus.

Estimation of Humeroradial Reaction Force: Let us, first, consider the humeroradial joint. We assume that the two bones (radius and ulna) carry an equal share of the total limb weight. Figure 2.40 illustrates the muscle forces, the share of limb weight ($W/2$), the share of the ball weight (W_R), and the humeroradial joint reaction force (R_R). However, the ball weight carried by the radius (W_R) and humeroradial joint reaction force (R_R) are unknown. For determining the ball weight shared by the radius, we can use the moment equilibrium about O, as follows:

$$\begin{aligned} & W_R \times 0.35 + \frac{W}{2} \times 0.16 \\ & \quad = F_{Bi} \times L_{Bi} + F_{BR} \times L_{BR} \\ & W_R \times 0.35 + \frac{32}{2} \times 0.16 \\ & \quad = 0.038 \times 1296.23 + 0.063 \times 300.29 \\ & W_R = 187.47\,\mathrm{N} \end{aligned}$$

W_R is the external load carried by the radius bone. The remaining load, that is $300 - 187.47 = 112.53\,\mathrm{N}$, is carried by the ulnar.

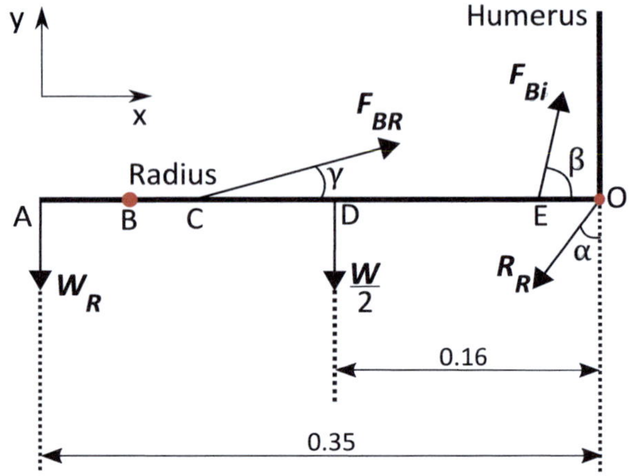

Fig. 2.40 Forces acting on the radius and the humeroradial joint carrying a portion of external load during elbow flexion

In order to find out the humeroradial joint reaction force ($\boldsymbol{R_R}$), the force equilibrium along the horizontal and vertical direction is considered. The α is the angle of inclination of the humeroradial joint reaction force with respect to the humeral shaft axis. We can write down the force equilibrium equations as follows:

$$\begin{aligned}\Sigma F_x &= F_{BR}\cos(\gamma) \\ &\quad + F_{Bi}\cos(\beta) - R_R\sin(\alpha) \\ &= 0\end{aligned} \tag{2.8}$$

$$\begin{aligned}\Sigma F_y &= F_{BR}\sin(\gamma) \\ &\quad + F_{Bi}\sin(\beta) - W_R - \frac{W}{2} - R_R\cos(\alpha) \\ &= 0\end{aligned} \tag{2.9}$$

Combining the Eqs. 2.8 and 2.9, we can calculate the tan(α) as given below:

$$\begin{aligned}\tan(a) &= \frac{F_{BR}\,\mathrm{Cos}(\gamma) + F_{Bi}\,\mathrm{Cos}(\beta)}{F_{BR}\,\mathrm{Sin}(\gamma) + F_{Bi}\,\mathrm{Sin}(\beta) - W_R - \frac{W}{2}} \\ &= \frac{300.29\,\mathrm{Cos}(15.16^\circ) + 1296.23\,\mathrm{Cos}(86.8^\circ)}{300.29\,\mathrm{Sin}(15.16^\circ) + 1296.23\,\mathrm{Sin}(86.8^\circ) - 187.47 - 16} \\ \tan(a) &= 0.3096 \\ \alpha &= 17.21^\circ\end{aligned}$$

The humeroradial joint reaction force ($\boldsymbol{R_R}$) is obtained upon substituting value of angle α in Eq. 2.8.

$$R_R = \frac{F_{BR}\cos(\gamma) + F_{Bi}\cos(\beta)}{\sin(\alpha)} = \frac{300.29\cos(15.16^\circ) + 1296.23\cos(86.8^\circ)}{\sin(17.21^\circ)}$$
$$= 1223.97\,\text{N}$$

Now that we have calculated the humeroradial joint reaction force, its effect on humeroradial contact can be considered by resolving the joint reaction force into normal and transverse components. In this calculation, the normal component of the humeroradial joint reaction force (R_N) is assumed (arbitrarily) to make an angle of 60° with the humeral shaft axis (Fig. 2.41). We have already calculated the inclination of R_R, which is 17.21° with the humeral shaft axis (Fig. 2.42). On resolving the humeroradial joint reaction force, we obtain the normal (R_N) and transverse (R_T) components:

$$\begin{aligned} R_N &= R_R\cos(60^\circ - 17.21^\circ) \\ &= 1223.97\cos(42.79^\circ) = 898.21\,\text{N} \end{aligned}$$

$$\begin{aligned} R_T &= R_R\sin(60^\circ - 17.21^\circ) \\ &= 1223.97\sin(42.79^\circ) = 831.62\,\text{N} \end{aligned}$$

Estimation of ligamentous force: Let us look into the contribution of the interosseous ligamentous force (Fig. 2.42). So far, this ligament was not considered in the analysis. In order to understand the contribution of this ligament, let us estimate the ligamentous force ($\boldsymbol{F_L}$), assuming a friction coefficient of 0.02 at the contact point (Fig. 2.41a). The frictional force at the point of contact is obtained by multiplying the frictional coefficient by the normal force. Now, the frictional force is directed along the tangential direction and is obtained as $0.02 \times 898.21 = 17.96$ N.

The sliding force can, therefore, be expressed as

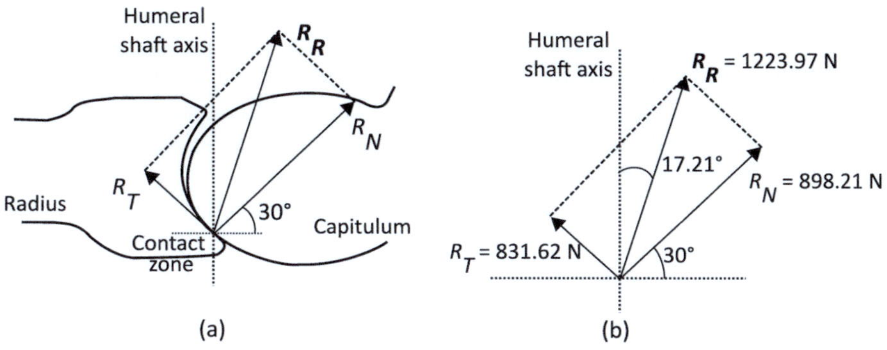

Fig. 2.41 Humeroradial joint reaction force: **a** components of the joint force, **b** estimates of the force components

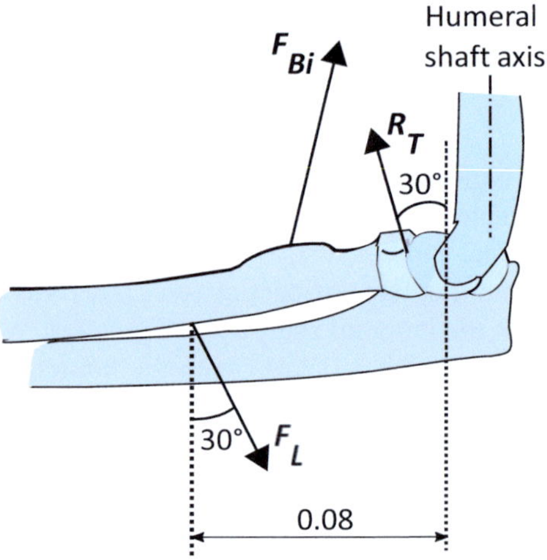

Fig. 2.42 Estimation of ligamentous force

$$F_s = R_T - 17.96 = 831.62 - 17.96 = 813.66\,\text{N}$$

This unbalanced sliding force (F_s) along the transverse direction will tend to move the radial head sideways across the capitulum (Fig. 2.41a). However, this is not the case in reality, because this force is balanced by the ligamentous system, which restores the radial contact to equilibrium. The force imposed by the ligament $\boldsymbol{F_L}$ is assumed to act at an angle of 30° with the vertical, as shown in Fig. 2.42. This force is equal in magnitude but opposite in direction to the sliding force (F_s) at the point of contact. Since these two forces are not collinear but assumed to be coplanar, they will impose an extension moment on the radius. Now, it is assumed that this ligamentous force acts at a distance of 0.08 m from the joint axis and along the radial shaft (Fig. 2.42).

Recalculation of humeroradial reaction force: Let us recalculate the humeral radial joint reaction force, considering the effect of the ligamentous force (Fig. 2.43). The resultant force needs to be recalculated since the ligamentous system balances a portion of the transverse component of the reaction force. The normal component of the force R_N remains unchanged, as 898.21 N acts at a 60° to the humeral shaft axis (Fig. 2.43). Although the transverse component has its direction intact, the magnitude is equal to the frictional force equal to 17.96 N (Fig. 2.43).

The revised humeroradial joint reaction force ($\boldsymbol{R'_R}$) is equal to

$$R'_R = \sqrt{R_N^2 + R_T'^2} = \sqrt{898.21^2 + 17.96^2} = 898.38\text{N}$$

The angle (x) with which $\boldsymbol{R'_R}$ is inclined can be found out as follows:

$$\tan(60 - x) = \frac{17.96}{898.21} = 0.02$$

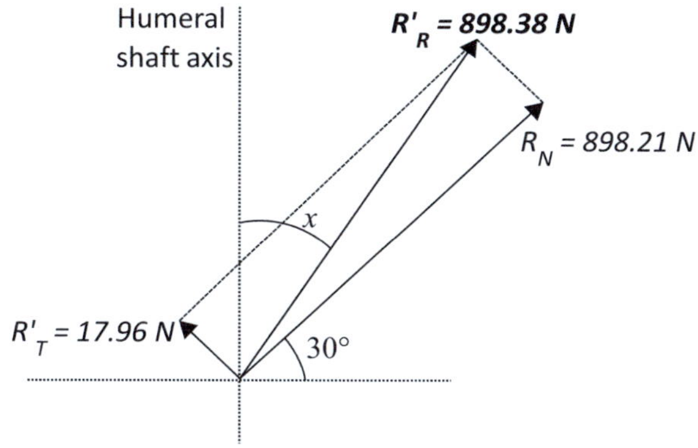

Fig. 2.43 Revised components of the humeroradial joint reaction force

$$x = 60 - \tan^{-1}(0.02) = 58.85^\circ$$

The moment imposed on the radius by the ligament is calculated as: $0.08\cos(30^\circ) \times 813.66 = 56.37\,\text{Nm}$ (Fig. 2.42). This moment opposes the flexion muscle arrangement and, therefore, decreases the load that can be carried. For the radial bone, the load carried was estimated to be 187.47 N with a moment arm of 0.35 m, which resulted in a moment of 65.61 Nm. Subtracting the moment imposed by the ligament on the radius, 56.37 Nm, resulting in a moment of 9.24 Nm. This moment is divided by the distance of 0.35 m to obtain the revised load carried by the radial bone, 26.40 N. For the ulna bone, the load was earlier estimated to be 112.53 N acting at a moment arm of 0.35 m, resulting in a moment of 39.39 Nm. Upon adding the moment of 56.37 Nm imposed by the ligament on ulna, we get the revised moment as 95.76 Nm, which corresponds to a revised load (W_U) of 273.60 N.

Estimation of Humeroulnar Reaction Force: Let us estimate the humeroulnar joint reaction force ($\boldsymbol{R_U}$). The ligament acts between the radius and ulnar as shown in Fig. 2.44. Considering the force equilibrium along the vertical direction, $\Sigma F_x = 0$,

$$F_{Br}\cos(\theta) - F_L\sin(30^\circ) - R_U\sin(\alpha) = 0$$

Considering the force equilibrium along the vertical direction, $\Sigma F_y = 0$,

$$F_{Br}\sin(\theta) + F_L\cos(30^\circ) - W_U - \frac{W}{2} - R_U\cos(\alpha) = 0$$

Equating the above two equations, results in the following:

$$\tan(\alpha) = \frac{F_{Br}\cos(\theta) - F_L\sin(30^\circ)}{F_{Br}\sin(\theta) + F_L\cos(30^\circ) - W_U - \frac{W}{2}}$$

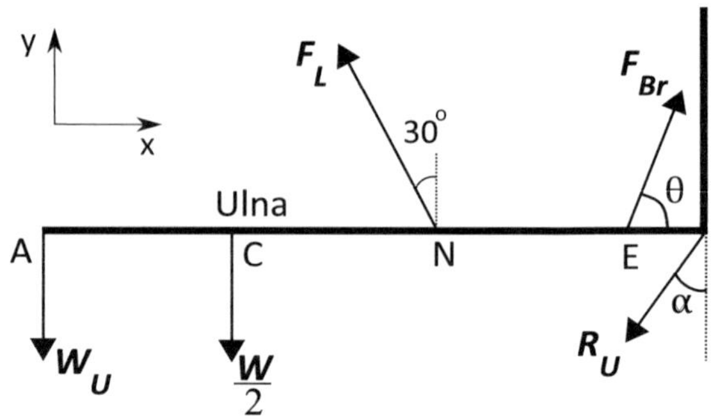

Fig. 2.44 Estimation of humeroulnar joint reaction force

$$= \frac{1233.67\cos(64.6°) - 813.66\sin(30°)}{1233.67\sin(64.6°) + 813.66\cos(30°) - 273.51 - 16} = 0.08$$

$$\alpha = \tan^{-1}(0.08) = 4.57°$$

Substituting the value of α in any of the above equilibrium equations, we obtain the revised humeroulnar joint reaction force ($\boldsymbol{R_U}$).

$$R_U = \frac{F_{Br}\cos(\theta) - F_L\sin(30°)}{\sin(\alpha)} = \frac{1233.67\cos(64.6°) - 813.66\sin(30°)}{\sin(4.57°)} = 1535.38\,\mathrm{N}$$

Hence, the revised humeroulnar joint reaction force is calculated as 1535.38 N.

2.6 Biomechanics of the Spine

The spine can be viewed as a modified elastic rod due to its flexible spinal column, the shock-absorbing properties of the discs, and the stabilising role of the ligaments. The lumbar spine segment bears the majority of the weight from the upper part of the body. Hence, the discussion of this section will be focused on the lumbar spine, which is located at the lower part of the vertebral column. The primary function of the lumbar spine is to bear the weight of the upper part of the body and to transfer the forces and bending moments to the sacrum. The lumbar spine is more flexible than the thoracic spine and carries more weight making it the most likely region susceptible to injury within the entire spinal structure.

2.6.1 Lumbar Spinal Load for Various Postures

Let us discuss the loads acting on the lumbar spine. Figure 2.40 presents the normalised spinal loads in various body postures (Wilke et al. 1999). These are

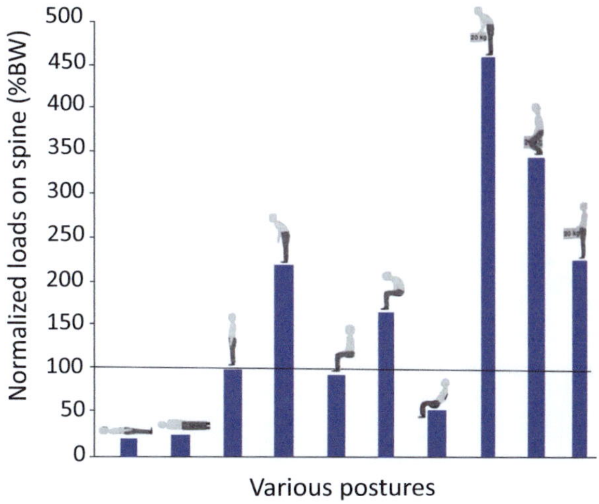

Fig. 2.45 Spinal load for various postures

calculated with respect to the load during upright standing. The highest loads on the lumbar spine are generally produced while lifting heavy objects. The factors influencing the spinal loads are as follows:

- the size, weight and position of the object lifted with respect to the centre of motion of the spine;
- the posture, position of the centre of gravity of the upper part of the body with respect to the centre of motion of the spine.

As evident from Fig. 2.45, the loads on the spine are dependent on the body position. In the supine position (lying down), the load on the spine is minimal because the body weight is eliminated. The forward bending movement produces a bending moment on the lumbar spine which is equal to the product of the upper body weight and the lever arm. Additionally, the forward bending movement generates increased tensile and compressive stresses in the intervertebral disk, owing to the bending action.

2.6.2 *Basic Biomechanics of the Lumbar Spine: Sample Problems*

Now, let us explore how the bending moment on the lumbar spine is calculated. The numerical problem is relatively simple, but it illustrates how load position and posture can influence the forces and bending moment on the lumbar spine.

Consider a subject lifting an object having weight ($\boldsymbol{W_O}$) equals 100 N using three different postures. The weight of the upper part of the body ($\boldsymbol{W_U}$) is taken as 300

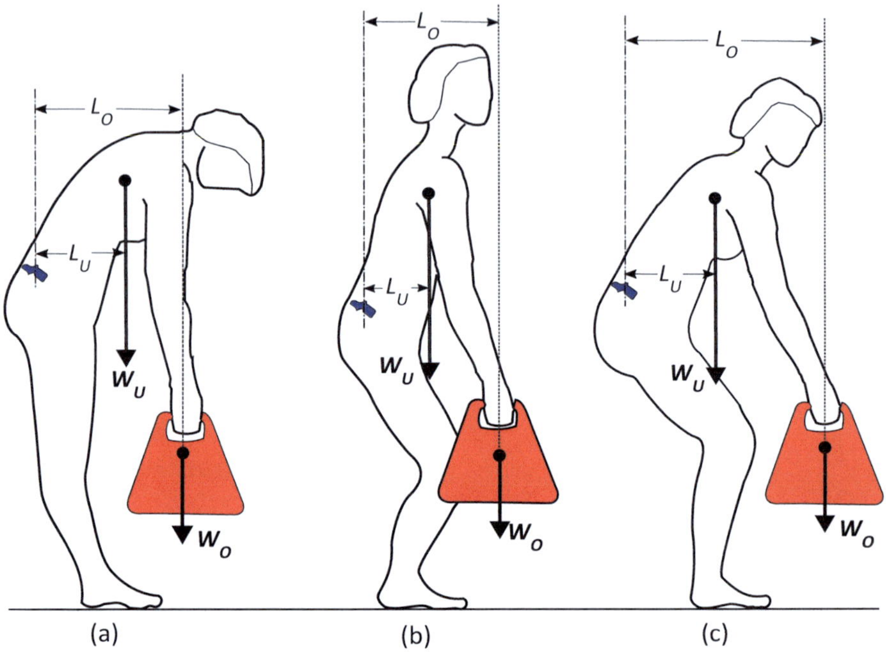

Fig. 2.46 Various postures of lifting a weight; **a** case 1: forward bending, **b** case 2: knees bend, **c** case 3: bending forward with knees bend

N. The first case is forward bending, which is presented in Fig. 2.46a. In the second case, the weight is carried closer to the body with knees bent (Fig. 2.46b). In the third case (Fig. 2.46c), the subject is bending forward with knees bend. The lever arm of the weight of the upper part of the body (L_U) and the length of the lever arm of the weight of the object (L_O) are different for three different cases. Now, let us determine the bending moment for each of these scenarios Fig. 2.46.

The first case involves the calculation of the bending moment during forward bending (Fig. 2.46a). Assuming L_O as 0.40 m, the bending moment of $\boldsymbol{W_O}$ about the centre of lumbar disk is equal to $\boldsymbol{W_O}$ multiplied by its lever arm (L_O), which is equal to 40 Nm. If L_U is assumed to be 0.25 m, then the bending moment produced by the upper part of the body is 75 Nm (300 N multiplied by 0.25 m). Therefore, the total forward bending moment (M_1) is the sum of these two bending moments, which is equal to 115 Nm, owing to $\boldsymbol{W_U}$ and $\boldsymbol{W_O}$. In the second case, with the object positioned closer to the body and the knees bent (Fig. 2.46b), we assume Lo and L_U to be 0.35 m and 0.20 m, respectively. The total forward bending moment is calculated as the sum of the bending moment produced by the object and the upper part of the body.

$$
\begin{aligned}
M_2 &= 0.35 \times 100 + 0.20 \times 300 \\
&= 35 + 60 = 95\,\text{Nm}
\end{aligned}
$$

This value is less than the value calculated in the first case. In the third case, we are calculating the bending moment during the forward bending with the knees bent. In this case, we assume the lever arm of the object (L_O) to be 0.50 m, since the object is held away from the trunk (Fig. 2.46c). Since the subject is forward bending, similar to the first case, L_U is equal to 0.25 m. The bending moment is calculated as follows:

$$\begin{aligned} M_3 &= 0.50 \times 100 + 0.25 \times 300 \\ &= 50 + 75 = 125\,\text{Nm} \end{aligned}$$

The third case involves the highest bending moment among all the cases. A higher bending moment leads to increased stress within the functional spinal unit (FSU).

In the next numerical problem, which is an extension of the previous, we will find the static loads on the lumbar segment. Similar to the third case of the previous problem, the subject with an upper part body weight of 300 N is lifting an object of 100 N (Fig. 2.46c). We assume that the spine is flexed approximately at 35° (Fig. 2.47). In this case, we identify three main forces acting on the lumbar segment:

(1) the force produced by the upper part of the body weight ($\boldsymbol{W}_U$) is 300 N, assuming approximately 50% of the total body weight;
(2) the force produced by the weight of the object ($\boldsymbol{W}_O$) is 100 N;
(3) the erector spinae muscle force, ***Fe*** with known direction and unknown magnitude.

Let us examine the free body diagram of the lumbar segment, as illustrated in Fig. 2.47. The axial force ($\boldsymbol{F}$) and tangential or shear ($\boldsymbol{T}$) force are acting at the centre of the lumbar segment (C), as shown in Fig. 2.47b. The erector spinae muscle force $\boldsymbol{F}_e$ is acting with a lever arm L_e. The lumbar segment is oriented at 35° to the horizontal axis (Fig. 2.47b). The lever arms of the object, upper body weight and muscle, about the centre of the lumbar segment (C), are 0.40, 0.25, and 0.05 m, respectively (Fig. 2.47a). Now, for the body to be in static equilibrium, the sum of the moments must be equal to 0. Considering the moment equilibrium about the centre of the lumbar segment (C),

$$\begin{aligned} \Sigma M_c &= 0 \\ W_O \times 0.4 + W_U \times 0.25 - F_e \times 0.05 &= 0 \\ 100 \times 0.4 + 300 \times 0.25 - F_e \times 0.05 &= 0 \end{aligned}$$

Based on the above equations, the force applied by the erector spinae muscle is calculated to be

$$F_e = 2300\,\text{N}$$

Let us now calculate the $\boldsymbol{F}$ and $\boldsymbol{T}$, as indicated in Fig. 2.47b. We can determine the values of these unknown forces by analysing the forces in the x- and y-directions and applying the principles of static equilibrium, as outlined below.

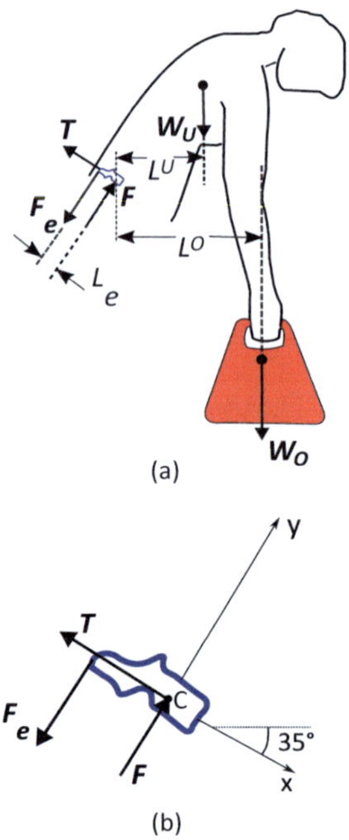

Fig. 2.47 Diagrammatic representation of the forces acting on a lumbar segment. **a** Action of major forces, **b** enlarged view of the lumbar segment showing the axial and shear loads

$$\Sigma F_x = 0$$
$$W_U \sin(35^\circ) + W_O \sin(35^\circ) - T = 0$$
$$300 \sin(35^\circ) + 100 \sin(35^\circ) - T = 0$$
$$T = 229.4\,\text{N}$$

$$\Sigma F_y = 0$$
$$W_U \cos(35^\circ) + W_O \cos(35^\circ) + F_e - F = 0$$
$$300 \cos(35^\circ) + 100 \cos(35^\circ) + 2300 - F = 0$$
$$F = 2627.6\,\text{N}$$

The axial and the shear forces acting on the lumbar segment was found to be 2627.6 N and 229.4 N, respectively.

MATLAB Code for Solving 3D Force System Numerical Problem

```
% Define symbolic variables for unknowns
    syms F1 F2 F3 Jx Jy Jz real
```

```
% Coordinates for forces
% Joint Reaction Force (J)
J = [Jx; Jy; Jz];
% Coordinates of muscle attachment points
F1_femur = [-2.50; 2.50; 71.00];
F1_pelvis = [2.65; 2.52; 79.43];
F2_femur = [-1.70; 1.70; 58.30];
F2_pelvis = [10.32; 5.78; 69.64];
F3_femur = [-0.80; 0.80; 49.20];
F3_pelvis = [0.06; 10.0; 66.94];
% Coordinates for the hip joint center
hip_joint_center = [-0.8; 0.8; 73];
% Define the ground reaction force (R) and limb weight (WL)
total_weight = 784.8; % Total weight of the person
limb_weight = total_weight / 7; % Weight of the limb in N
ground_reaction_force = total_weight; % Total ground reaction force
% Coordinates for the limb weight and ground reaction force
WL_coord = [-0.4; 0; 40];
R_coord = [0; 0; 0];
% Calculate vectors for muscle forces
F1_vector = F1_pelvis - F1_femur;
F2_vector = F2_pelvis - F2_femur;
F3_vector = F3_pelvis - F3_femur;
% Normalise force vectors for equilibrium equations
F1_normalised = F1_vector / norm(F1_vector);
F2_normalised = F2_vector / norm(F2_vector);
F3_normalised = F3_vector / norm(F3_vector);
% Force vectors for limb weight and ground reaction force
WL_vector = [0; 0; -limb_weight];
WL_vector_norm=WL_vector/norm(WL_vector);
R_vector = [0; 0; ground_reaction_force];
R_vector_norm=R_vector/norm(R_vector);
% Calculate moments about the hip joint center
moment_F1 = cross(F1_femur - hip_joint_center, F1_normalised);
moment_F2 = cross(F2_femur - hip_joint_center, F2_normalised);
moment_F3 = cross(F3_femur - hip_joint_center, F3_normalised);
moment_WL = cross(WL_coord - hip_joint_center, WL_vector_norm);
moment_R = cross(R_coord - hip_joint_center, R_vector_norm);
% Set up equilibrium equations for forces
eq1 = F1 * F1_normalised(1) + F2 * F2_normalised(1) + F3 * F3_normalised(1)
+ WL_vector(1) + R_vector(1) + J(1) == 0;
eq2 = F1 * F1_normalised(2) + F2 * F2_normalised(2) + F3 * F3_normalised(2)
+ WL_vector(2) + R_vector(2) + J(2) == 0;
eq3 = F1 * F1_normalised(3) + F2 * F2_normalised(3) + F3 * F3_normalised(3)
+ WL_vector(3) + R_vector(3) + J(3)== 0;
```

```
% Set up equilibrium equations for moments
moment_eq1 = F1 * moment_F1(1) + F2 * moment_F2(1) + F3 * moment_F3(1) + limb_weight * moment_WL(1) + ground_reaction_force* moment_R(1) == 0;
moment_eq2 = F1 * moment_F2(2) + F2 * moment_F2(2) + F3 * moment_F3(2) + limb_weight * moment_WL(2) + ground_reaction_force* moment_R(2) == 0;
moment_eq3 = F1 * moment_F1(3) + F2 * moment_F2(3) + F3 * moment_F3(3) + limb_weight * moment_WL(3) + ground_reaction_force* moment_R(3) == 0;
% Solve the system of equations
[sol_F1, sol_F2, sol_F3, sol_Jx, sol_Jy, sol_Jz] = solve([eq1, eq2, eq3, moment_eq1, moment_eq2, moment_eq3], [F1, F2, F3, Jx, Jy, Jz]);
% Display results
F1_value = double(sol_F1);
F2_value = double(sol_F2);
F3_value = double(sol_F3);
J_value = [double(sol_Jx), double(sol_Jy), double(sol_Jz)];
disp(['Muscle Force F1: ', num2str(F1_value)]);
disp(['Muscle Force F2: ', num2str(F2_value)]);
disp(['Muscle Force F3: ', num2str(F3_value)]);
disp(['Joint Reaction Force J: ', num2str(J_value)]);
```

Multiple Choice Questions (MCQ)

1. Which movement results in a reduction in the angle between the forearm and upper arm (humerus) in the elbow joint?
   (A) Pronation
   (B) Extension
   (C) Flexion
   (D) Supination
2. Which of the following muscle is involved in elbow extension?
   (A) Biceps brachii
   (B) Brachialis
   (C) Brachioradialis
   (D) Triceps brachii
3. Carrying angle is the angle between the long axes of ________ and ________.
   (A) Femur, fibula
   (B) Femur, tibia
   (C) Extended forearm, upper arm

(D) Scapula, humerus

4. What is the degree of freedom of the acromio-clavicular joint?

   (A) 1
   (B) 2
   (C) 3
   (D) 4

5. What are the degrees of freedom of the shoulder joint with respect to the thorax?

   (A) 10
   (B) 6
   (C) 12
   (D) 3

6. The scapulohumeral rhythm is the simultaneous motion of the shoulder girdle and _________.

   (A) Scapula
   (B) Humerus
   (C) Clavicle
   (D) Thorax

7. The connection ________________ is an example of a closed chain mechanism.

   (A) Thorax-clavicle-humerus
   (B) Humerus-scapula-clavicle
   (C) Clavicle-scapula-humerus
   (D) Thorax-clavicle-scapula

8. Kinematic chain is an assembly of rigid bodies connected by joints to provide ______motion.

   (A) Constrained
   (B) Unconstrained

9. How many degrees of freedom does the radio-ulnar joint have?

   (A) 1
   (B) 2
   (C) 3
   (D) 4

10. Which muscle group provides stability to the shoulder?

    (A) Gluteal
    (B) Hamstring
    (C) Quadriceps
    (D) Rotator cuff

11. The patella allows a wider distribution of the compressive force on ____.

(A) Tibia
(B) Fibula
(C) Femur
(D) Humerus

12. The peak knee joint reaction forces during normal walking is in the range of _ ________ the body weight.

(A) 100–180%
(B) 200–300%
(C) 320–400%
(D) 450–550%

13. The mass of the vertebrae ________ from C1 to the sacrum.

(A) Increases
(B) Decreases
(C) Remains constant

14. Which of the following/(s) is/(are) affected by the femur-neck-shaft angle?

(A) Joint reaction force
(B) Lever arm of the muscle forces
(C) Joint reaction force and lever arm of the muscle force

15. Which of the following is NOT a movement of the spine?

(A) Extension
(B) Supination
(C) Bending
(D) Torsion

Answer Key to MCQs

1. C; 2. D; 3. C; 4. C; 5. B; 6. B; 7. D; 8. A; 9. A; 10. D; 11. C; 12. B; 13. B; 13. A; 14. C; 15. B.

Practice Problems

2.7.1 A subject weighing 80 kg is standing on one foot. Find the hip joint reaction force ($\boldsymbol{F_J}$). The abductor muscle group ($\boldsymbol{F_A}$) is assumed to make an angle of 70° with the horizontal, as given in Fig. 2.48. The weight of the limb and reaction force are denoted as $\boldsymbol{W_L}$ and $\boldsymbol{R}$, respectively. The $\boldsymbol{W_L}$ is equal to 13 kg. The dimensions in Fig. 2.48 are in meters.

2.7.2 A subject is lifting a dumbbell of 10 kg, as shown in Fig. 2.49. The head and trunk weigh 5 kg and 35 kg, respectively. Find the erector spinae muscle force ($\boldsymbol{F_E}$), given the body is flexed 60° to the vertical. The moment arm of the erector

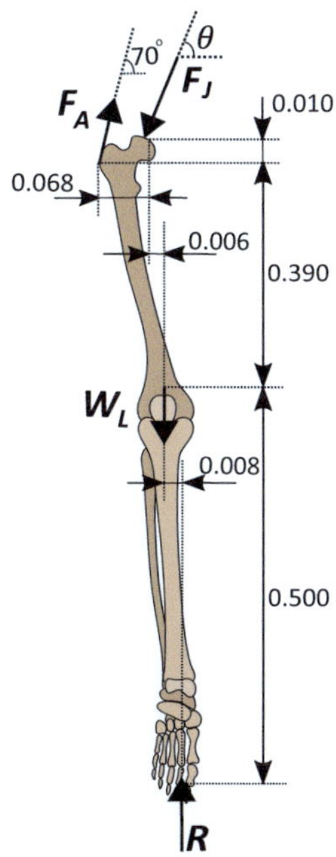

Fig. 2.48 Forces acting on a lower limb while standing

spinae muscle about the segment centre may be assumed as 0.04 m. The dimensions in Fig. 2.49 are expressed in meters.

2.7.3 A subject with a flexed elbow is holding a ball weighing 4 kg as shown in Fig. 2.50. The weight of the forearm is 2.5 kg. Find out the force exerted by the biceps muscle. You may ignore the contribution of the other muscles in this problem. The dimensions in Fig. 2.50 are expressed in meters.

2.7.4 The subject is about to rise from a squatting position as shown in Fig. 2.51. Determine the tensile force exerted by the patellar tendon. The weight of the whole body and lower leg are 70 kg and 5 kg, respectively. The dimensions in Fig. 2.51 are expressed in meters.

2.7.5 During an instance of humeral abduction, the muscle forces of the deltoid and supraspinatus were found to be 400 N and 200 N, respectively. The deltoid and supraspinatus are acting at an angle of 5° (with vertical) and 35° (with horizontal) as shown in Fig. 2.52. The weight of the arm (W_A) is equal to 50 N. Find out the Glenohumeral joint reaction force.

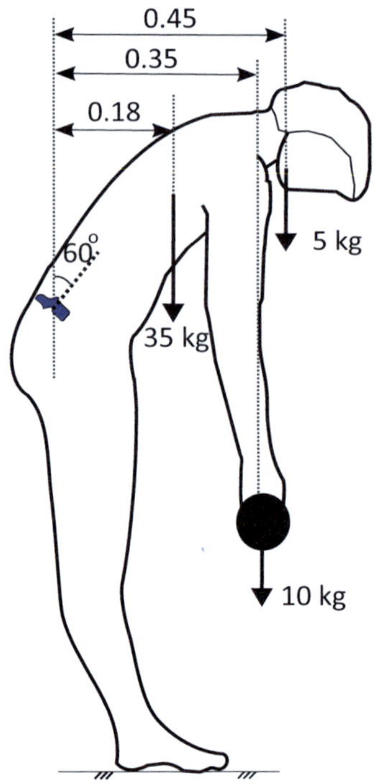

Fig. 2.49 A subject lifting a dumbbell of 10 kg

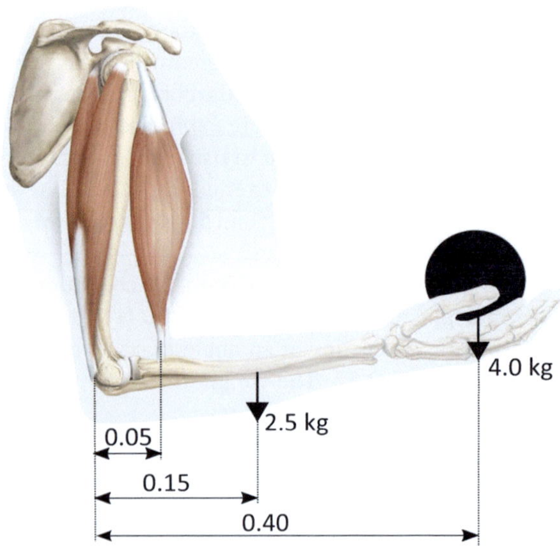

Fig. 2.50 A subject with flexed elbow holding a ball

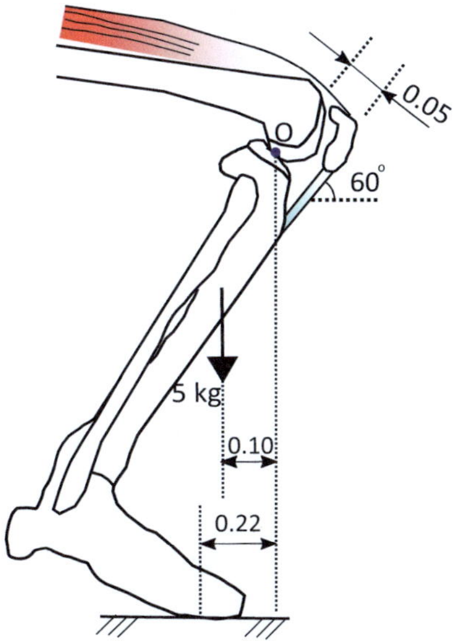

Fig. 2.51 Forces acting on the shank when a subject is about to rise from squatting position

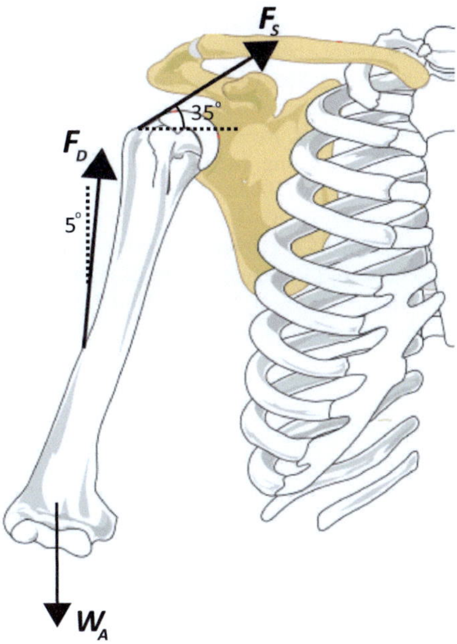

Fig. 2.52 Forces acting on the arm during an instance of humeral abduction

References

Bergmann G, Graichen F, Rohlmann A (1993) Hip joint loading during walking and running, measured in two patients. J Biomech 26:969–990

Bergmann G, Deuretzbacher G, Heller M, Graichen F, Rohlmann A, Strauss J, Duda G (2001) Hip contact forces and gait patterns from routine activities. J Biomech 34:859–871

Crowninshield RD, Johnston RC, Andrews JG, Brand RA (1978) A biomechanical investigation of the human hip. J Biomech 11(1–2):75–85

Davy D, Kotzar G, Brown R, Heiple K, Goldberg VM, Berilla J, Burstein A (1988) Telemetric force measurements across the hip after total arthroplasty. J Bone Joint Surg Am 70:45–50

D'Lima DD, Steklov N, Patil S, Colwell CW Jr (2008) The Mark Coventry Award: in vivo knee forces during recreation and exercise after knee arthroplasty. Clin Orthop Relat Res 466(11):2605–2611

Kotzar G, Davy D, Goldberg VM, Heiple K, Berilla J, Brown R, Burstein A (1991) Telemeterized in vivo hip joint force data: a report on two patients after total hip surgery. J Orthop Res 9:621–633

Paul JP (1976) Approaches to design—force actions transmitted by joints in human body. Proc R Soc Lond B Biol Sci 192:163–172

Stauffer RN, Chao EY, Brewster RC (1977) Force and motion analysis of the normal, diseased, and prosthetic ankle joint. Clin Orthop Relat Res 127:189–196

Steineman BD, Quevedo González FJ, Sturnick DR, Deland JT, Demetracopoulos CA, Wright TM (2021) Biomechanical evaluation of total ankle arthroplasty. Part I: Joint loads during simulated level walking. J Orthop Res. 39(1):94–102

Taylor SJ, Walker PS (2001) Forces and moments telemetered from two distal femoral replacements during various activities. J Biomech 34(7):839–848

Taylor S, Perry J, Meswania J, Donaldson N, Walker P, Cannon S (1997) Telemetry of forces from proximal femoral replacements and relevance to fixation. J Biomech 30:225–234

van den Bogert AJ, Read L, Nigg BM (1999) An analysis of hip joint loading during walking, running, and skiing. Med Sci Sports Exerc 31:131–142

van der Helm FCT (1994a) A finite element musculoskeletal model of the shoulder mechanism. J Biomech 27(5):551–569

van der Helm FCT (1994b) Analysis of the kinematic and dynamic behavior of the shoulder mechanism. J Biomech 27(5):527–550

Wilke HJ, Neef P, Caimi M, Hoogland T, Claes LE (1999) New in vivo measurements of pressures in the intervertebral disc in daily life. Spine 24(8):755–762

Chapter 3
Biomechanics of Gait

Walking, although seemingly simple, is a complex and remarkable human movement that involves a series of coordinated actions. This activity is characterised by the continuous changes in the ground reaction forces, while maintaining balance and movement in a specific direction. While it might appear effortless for most people, walking poses significant challenges for individuals with disabilities or gait disorders, such as in patients with cerebral palsy. Learning to walk is a developmental milestone that takes months to master. Even for healthy individuals, walking can become challenging with age. Replicating the complexity of walking in robots, particularly maintaining balance, remains a formidable engineering challenge. While two-legged machines can be designed to mimic walking, ensuring their stability during movement is a persistent hurdle. Understanding and predicting human walking is essential for various reasons. It is critical for designing prostheses and orthoses that provide support to individuals with gait disorders, enhancing their mobility and quality of life. Moreover, it is essential for developing functional electrical stimulation methods, which help activate muscles in individuals with paralysis or spinal cord injuries. Controlled muscle activation at the right time and sequence can potentially help paralyzed patients regain the ability to walk.

3.1 Gait Cycle

Understanding the gait cycle is crucial for various fields, including sports science, physical therapy, orthopaedics, and ergonomics, since it provides valuable insights into human locomotion and can help diagnose and treat gait abnormalities or injuries. Essentially, normal gait is characterised by a rhythmic and alternating pattern of movements involving the trunk and limbs, resulting in the forward progression of the body's centre of gravity while having at least one of the feet on the ground at all times. In order to understand the mechanics of walking, we need to look into

S. Gupta et al., *Biomechanics of Joints and Implants*,
https://doi.org/10.1007/978-981-96-0586-6_3

the details of the gait cycle. The gait cycle represents a single sequence of functions performed by a lower limb during walking. It commences when the reference foot contacts the ground and concludes with the subsequent floor contact of the same foot.

3.1.1 Phases of Gait Cycle

The gait cycle can be divided into two main phases, the stance phase and the swing phase, as shown in Fig. 3.1. The cycle typically starts with the forward propulsion of the body from the standing position and thus propels one leg forward. During the stance phase, the foot remains in contact with the ground. In contrast, the "swing phase" involves the period when the foot is off the ground, moving forward in preparation for the next step. Each of these phases consists of specific events and sub-phases that contribute to the overall efficiency and stability of walking (Figs. 3.1 and 3.2).

Stance phase: The gait cycle commences with the heel strike, the instant when one foot (right) first touches the ground (Fig. 3.2). Simultaneously, the opposite foot (left) transitions into the swing phase, setting the stage for coordinated movement. Following the heel strike, the foot elegantly rolls forward, ensuring that the entire sole comes into contact with the ground. This phase, referred to as "foot flat" or "loading response," establishes a stable foundation for weight-bearing (Fig. 3.2). At midstance, the body's centre of gravity directly aligns over the supporting leg.

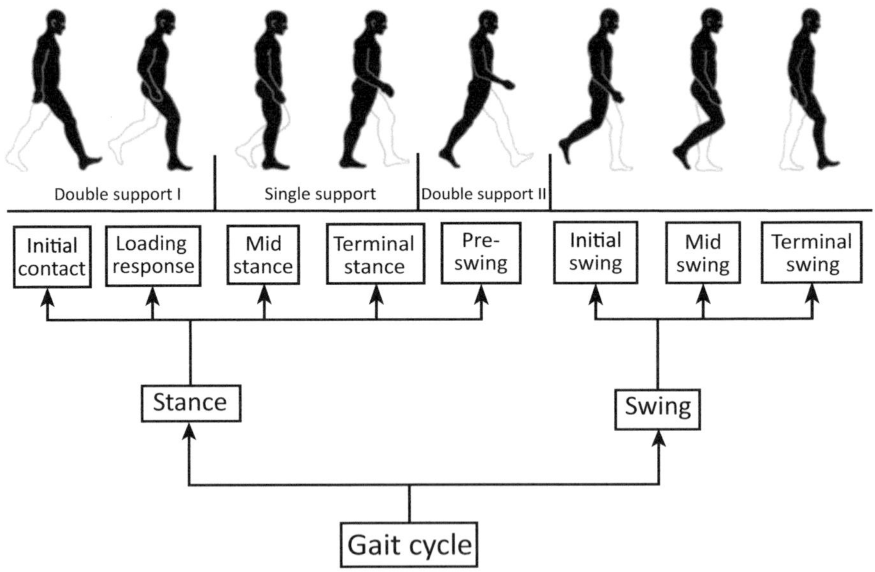

Fig. 3.1 Different phases of the gait cycle

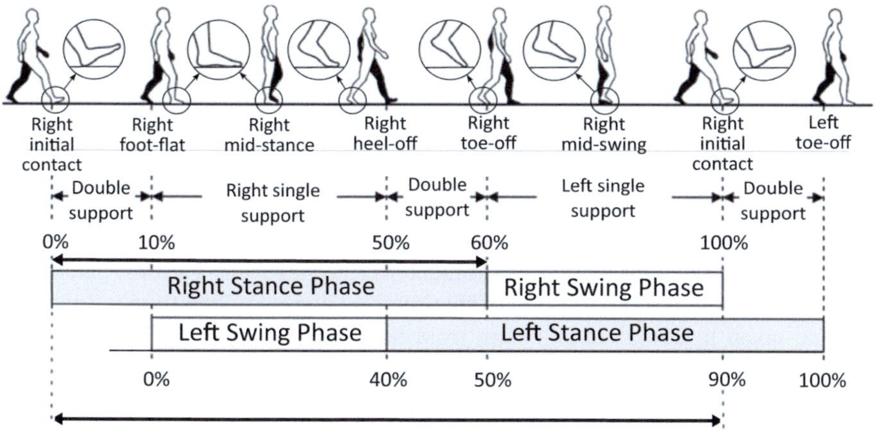

Fig. 3.2 Graphical representation of the gait cycle indicating the phases and instances

This midpoint in the stance phase signifies a moment of equilibrium and stability, crucial for the subsequent phases of walking. The next phase, heel-off, involves the lifting of the heel. This action contributes to the forward progression of the body. The toe-off phase marks the pivotal transition from the stance phase to the swing phase (Fig. 3.2). During this phase, the toe of the supporting foot firmly pushes off the ground, propelling the body forward and initiating the swing phase. The duration between heel-off and toe-off is known as push-off.

Swing phase: With the foot (right) leaving the ground, it enters the initial swing phase (Fig. 3.2). During this stage, the limb moves freely forward, preparing for the impending heel strike of the same foot. The mid-swing phase is distinguished by the limb's graceful and unhindered forward swing. The knee remains flexed during this part of the cycle, allowing continued progression of the limb. As the limb extends its forward swing, it prepares itself for the upcoming heel strike (Fig. 3.2). The knee gradually extends during this phase, culminating in the swing phase.

In a typical gait cycle, the stance phase accounts for approximately 60% of the gait cycle (Fig. 3.2). The swing phase constitutes about 40% of the gait cycle. Single support, when only one leg is in contact with the ground, lasts for roughly 40% of the gait cycle (Fig. 3.2). Double support, when both feet are in contact with the ground, takes up approximately 20% of the gait cycle (Fig. 3.2). The double support phase is succeeded by the right single support phase, during which only the right leg is in contact with the ground. This phase spans from 10 to 50% of the gait cycle (Fig. 3.2). This phase is followed by another double support phase, where both legs touch the ground, covering up to 60% of the cycle. The right leg then enters into the swing phase, moving freely through the air while the left leg supports the body's weight (Fig. 3.2). Finally, the right leg returns to the ground in the heel strike position, completing one full gait cycle. It may be noted, however, that the timespan of different phases within a gait cycle is not rigidly defined, since there can be variations in gait for individuals owing to factors, such as walking speed, terrain, and personal biomechanics. Motion capture systems, force plates, and other advanced instrumentations have been used to precisely measure and analyse the gait cycle, as discussed later.

3.1.2 Gait Cycle Parameters

The time and step parameters have an influence on the gait cycle. The time parameters are step time, double support time, stride time, and cadence. Step time is defined as the time taken to complete one step, from heel strike to toe-off. Double support time is the duration when both feet are in contact with the ground, simultaneously. Whereas, the stride time is defined as the time required to complete one full gait cycle, from one heel strike to the next. Cadence quantifies the number of steps taken per unit of time, usually expressed as steps per minute. Typically, the cadence during normal walking ranges from 100 to 115 steps per minute, which reflects the speed of walking.

The step parameters are step length, stride length, step width, and toe-out angle (Fig. 3.3). The step length is defined as the distance between successive heel contacts of opposite feet, indicating how far each foot travels during one step. In normal gait, the right step length is equal to the left step length. The stride length defined is the distance between successive heel contacts of the same foot, spanning the entire gait cycle. In normal gait, it is typically twice the step length. Step width measures the side-to-side distance between the lines formed by both feet, providing information about the width of the step and the individual's base of support. The toe-out angle quantifies the angle at which the foot is placed concerning the forward direction during walking, revealing the orientation of the feet throughout the gait cycle (Fig. 3.3).

Let us now discuss the effect of speed on the gait parameters. When analysing the influence of increased walking speed on various gait parameters, several observations become evident. As walking speed increases, individuals tend to take longer steps. This is reflected in the increased stride length. Simultaneously, the step width decreases to provide more balance, while walking at an increased speed. At faster walking speeds, individuals naturally aim to optimise their gait for efficiency. A narrower step width requires less lateral movement and reduces energy expenditure, making it more suitable for brisk walking and running. With an increase in walking speed, both the step period (the time to complete a step) and the double support period

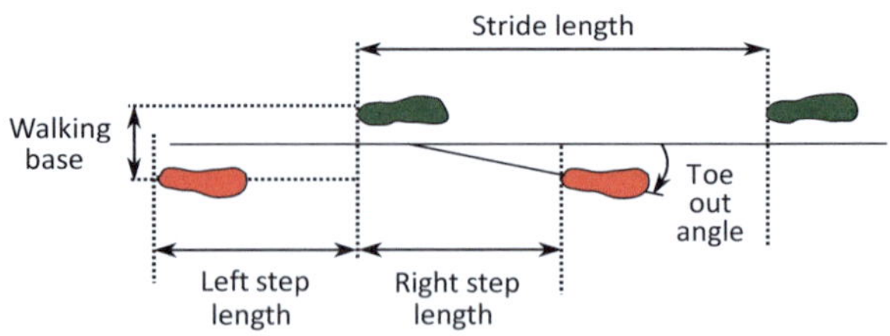

Fig. 3.3 The step parameters of a gait cycle

(the time when both feet are in contact with the ground) decrease. This reduction in time for double support is intuitive, as faster walking involves less time with both feet on the ground.

3.1.3 Basic Differences Between Running and Walking Cycles

The basic differences in phases between running and walking cycles are discussed in the following.

Double support phase: In walking, there is a distinct double support phase where both feet are in contact with the ground, simultaneously. This phase provides stability and is a key characteristic of walking gait. In running, the double support phase essentially does not exist. There are instances during the running gait cycle where both feet are off the ground. This absence of double support contributes towards the continuous nature of running.

Stance to swing phase ratio: In walking, the stance phase typically accounts for the majority of the gait cycle, roughly 60%. In this phase, one foot is on the ground while the other swings forward. In running, the ratio of stance to swing phases reverses. The stance phase becomes shorter, and the swing phase becomes longer.

Double swing phase: In running, there is a phenomenon known as the double swing phase. During certain instances of the gait cycle, both legs are off the ground, creating two separate swing phases as opposed to walking's single swing phase. This characteristic sets running apart from walking and allows for a faster and more dynamic mode of locomotion. In walking, at least one foot is in contact with the ground at all times.

Apart from these basic differences in phases stated above, there are a lot of differences in running and walking, such as the ground reaction force pattern, muscle excitation joint forces, and moments. In the analysis of human gait, the centre of gravity (CoG) and the ground reaction force (GRF) play a critical role in the mechanics of walking, details of which are discussed in the following.

3.1.4 Ground Reaction Forces

Ground Reaction Force (GRF) is the force exerted by the ground on an individual contact with it during activities like, walking or running. This force is essential for generating motion. The GRF varies during the gait cycle, and it primarily exists during the stance phase of walking (Fig. 3.4). There are three components for the GRF: vertical, anterior–posterior, and medial–lateral components. While the vertical component supports the body's weight, the anterior–posterior component aids in forward progression and braking during walking. The activation of muscles around each joint generates the medial–lateral ground reaction force.

Although there are variations in GRF in the anterior–posterior and medial–lateral directions during the gait cycle, these variations are relatively small as compared to the predominant vertical component of the GRF (Fig. 3.4). During normal walking, the first peak of the GRF typically occurs at around 15% of the gait cycle, representing the initial impact when the foot strikes the ground (Fig. 3.4). Subsequently the second peak, occurring at approximately 48%, corresponds to the push-off phase as the foot leaves the ground (Fig. 3.4). However, it may be noted that these values are subject-specific. The variation in the direction of the GRF vector will be discussed in Sect. 3.2.1.

3.1.5 Centre of Gravity

The centre of gravity (CoG) is a point in the human body through which the total body weight of a person or object is considered to act. It is typically located around the midsection of the body, approximately at the level of the navel or belly button (Fig. 3.5). In gait analysis, it is crucial to optimise the path of the CoG in order to minimise energy consumption. The most efficient gait pattern involves the CoG traveling in a straight line during the gait cycle. The line of gravity is an imaginary vertical line that extends from the CoG to the supporting surface, representing the force of gravity acting on the body. Maintaining the CoG within the base of support is preferred for stability and efficiency. During various postures and movements, such as, forward bending or leaning backward, the location of the CoG may shift outside the base of support, affecting balance and requiring additional muscular effort to maintain stability.

3.1.6 Moment Arm of Ground Reaction Force

GRF can create moments (torques) about the joints in the body, such as the hip, knee, or ankle. These moments are influenced by the distance (moment arm) between the joint centre and the line of action of the GRF. Muscles around the joints must generate opposing moments to counteract the moments created by GRF, thereby maintaining balance and control joint motion. For example, when the GRF is posterior to the ankle joint (at heel strike), it creates a clockwise moment about the ankle joint centre C, as shown by GRF_1 in Fig. 3.6. As the foot flattens and the GRF passes through the ankle joint centre C, the moment decreases (GRF_2 in Fig. 3.6). In contrast, during toe-off, the GRF is anterior to the ankle joint centre C, creating a counter-clockwise moment (GRF_3 in Fig. 3.6). The magnitude of these moments depends on the distance between the GRF and the ankle joint centre C, as presented in Fig. 3.6. Greater distances result in larger moments, requiring more muscular effort to counterbalance

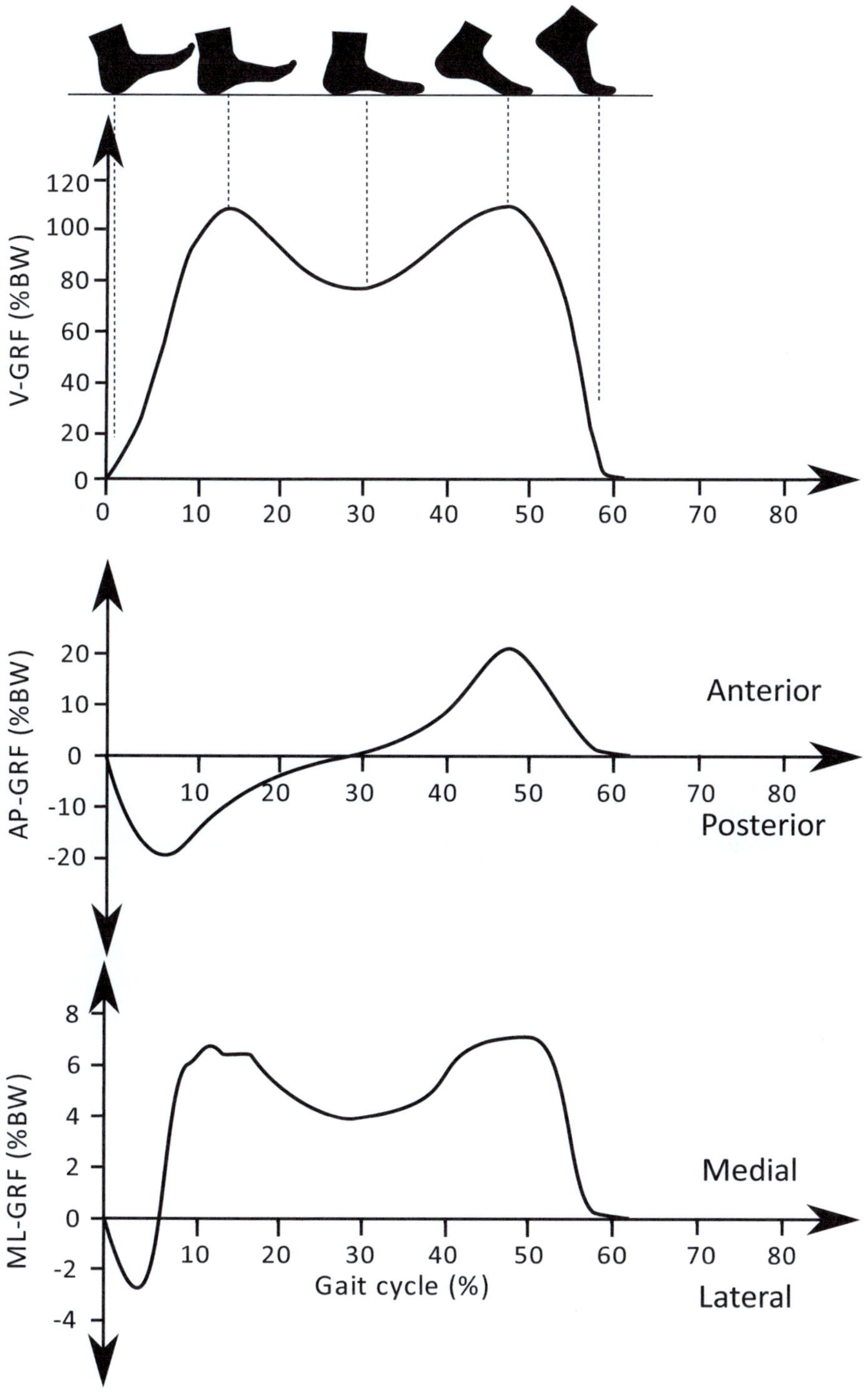

Fig. 3.4 Variations in the components of GRF during a gait cycle; vertical (V-GRF), anterior–posterior (AP-GRF), and medial–lateral (ML-GRF)

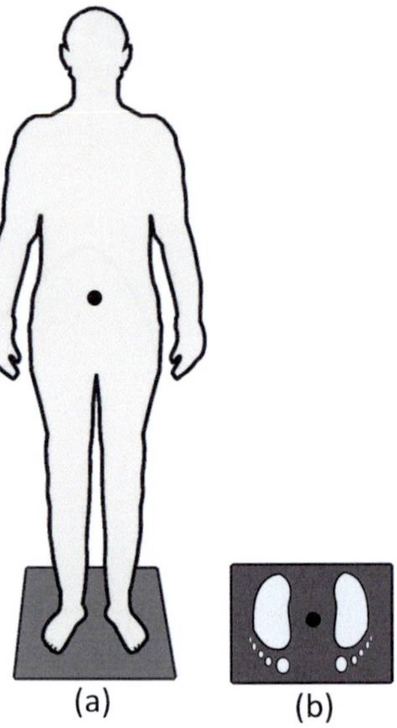

Fig. 3.5 Location of the centre of gravity; **a** actual, **b** projection on the base of support

and control joint motion. These variations in ankle moments during a gait cycle highlight the dynamic nature of walking and the continuous adjustments made by the musculoskeletal system to maintain balance and forward progression. Muscles around the ankle joint must generate appropriate forces to counteract these moments and ensure smooth and controlled motion throughout the gait cycle.

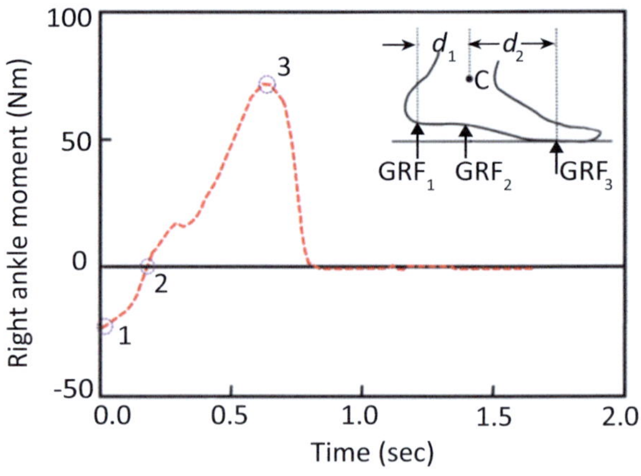

Fig. 3.6 Variation of ankle moment during a gait cycle

3.1.7 Determinants of Gait Cycle

Human locomotion is a complex phenomenon, involving numerous simultaneous movements in the 3D spatial coordinate system, making it challenging to analyse without a unifying principle. The unifying principle is based on the concept that locomotion involves the movement of the centre of gravity (CoG) along the most energy-efficient path. Moreover, this understanding allows for a qualitative analysis of the key factors that determine gait. The six main determinants are pelvic rotation, pelvic tilt, knee flexion during stance, knee and foot mechanisms, and lateral pelvic displacement. These determinants, work in a synergistic way to ensure that, during locomotion, the centre of gravity traverses through space along a smooth pathway requiring the least energy expenditure. Saunders et al. (1953) suggested that gait disorders can be analysed in terms of changes in these primary determinants so that therapeutic interventions can be applied to correct the disorders. Before we discuss the gait determinants, let us comprehend the meaning of "displacement of CoG".

The path described by the CoG, during normal walking of a healthy person, is similar to a smooth sinusoidal wave (Fig. 3.7). It is evident from Fig. 3.7a that the COG is displaced twice along the vertical direction (inferior–superior direction) and the total vertical displacement is approximately equal to 1.8 inches (Saunders et al. 1953). However, this value can vary for different individuals based on their body height. The peaks in the oscillation of this vertical displacement occur at mid-stance (30% of gait) and mid-swing (80% of gait). In contrast, during the double support (at 55% of gait cycle), the CoG achieves its lowest vertical position. Even at the peak of vertical displacement, the CoG lies slightly below the level corresponding to that of standing. In simpler words, one may appear to be slightly shorter during walking than during standing. Similar to vertical displacement, the CoG traverses along the medial–lateral direction (Fig. 3.7b) and along the anterior–posterior direction (Fig. 3.7c). The total medial–lateral displacement is approximately equal to the total vertical displacement (Saunders et al. 1953). However, the frequency of the variation in medial–lateral displacement is equal to that of gait whereas it is twice in the case of vertical displacement and anterior–posterior direction (Fig. 3.7).

In order to understand the effect of various determinants on the path traversed by CoG, we need to first discuss a hypothetical gait known as compass gait. Compass gait is characterised by minimal knee bending (knee remains relatively straight), where the pelvis moves through a series of arches, centred at the ground contact point and with a radius equal to the leg's length (Fig. 3.8). In Fig. 3.8, sharp inflexion points exist at the intersection of the arcs, which would lead to an abrupt change in the direction of forward acceleration. Consequently, according to Newton's first law of motion, greater force will be required for sudden changes in acceleration, resulting in increased energy expenditure. Moreover, the vertical displacement of the CoG associated with the compass gait is almost double as compared to the normal gait.

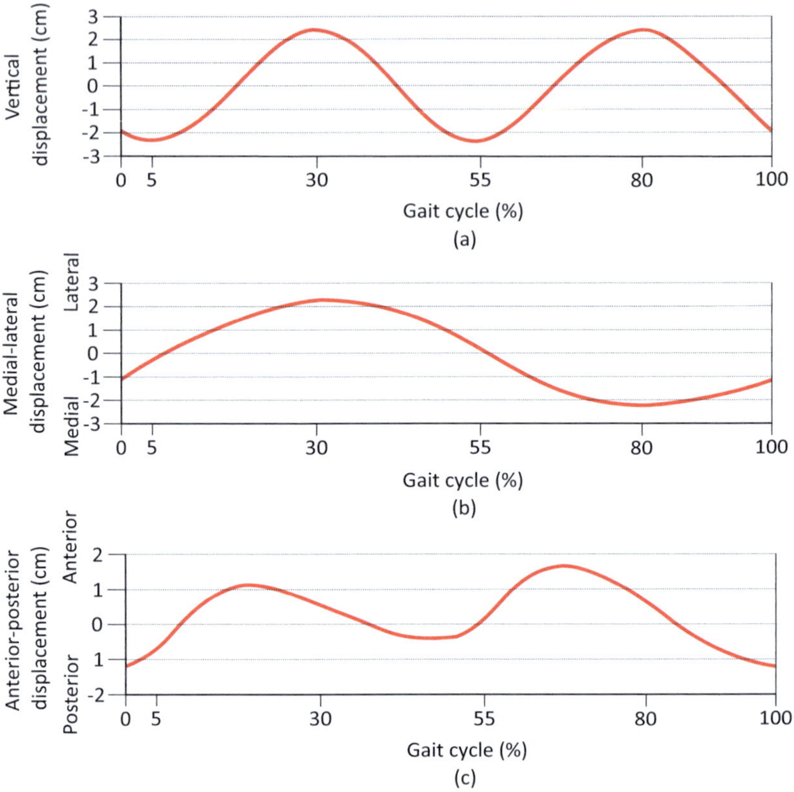

Fig. 3.7 Variation in the displacement of CoG along different directions during a gait cycle; **a** vertical, **b** medial–lateral, **c** anterior–posterior. Adapted from Saunders et al. (1953) and Jansen et al. (2014). The values in this figure are for the sake of representation only and can vary for different subjects

First Determinant: Pelvic Rotation

Pelvic rotation is associated with hip movement, specifically internal and external rotation during the stance phase. On average, the magnitude of this rotation is approximately 4° on either side of the central axis. The pelvic rotation results in the flattening of the path of CoG by elevating the lower extremities of the arc (Fig. 3.9). Consequently, the angle of inflection at the intersection of arcs is reduced resulting in lesser energy expenditure.

Second Determinant: Pelvic Tilt

In the gait cycle of a healthy person, the pelvis is tilted downwards on the contralateral limb, which is known as positive Trendelenburg. Since the pelvis is tilted downwards, the knee joint of the contralateral limb must flex to provide enough clearance for the swing phase. The pelvic tilt results in a relative adduction of the limb in the stance

Fig. 3.8 Illustration of compass gait. The pelvis is denoted by a single bar with a small cuboid representing CoG of the body. The pathway of CoG is a series of interconnecting arcs. The legs are represented by rigid bars articulating only at the hip

Fig. 3.9 Effect of pelvic rotation on path of CoG. The dotted line within the box denotes the path of compass gait

phase and relative abduction of the limb in the swing phase. Unlike the effect of pelvic rotation, the peak of the arc is lowered due to the pelvic tilt, resulting in flattening of the path of CoG.

Third Determinant: Knee Flexion in the Stance Phase

During a heel strike, the knee joint executes full extension. Thereafter, the knee joint undergoes flexion and continues until foot flat. Later, the knee undergoes extension again, which is immediately followed by flexion during heel-off. This extension—flexion–extension–flexion period occurs during the stance phase and is also called a "double knee lock". The peaks in the path of CoG is lowered by the knee flexion, resulting in reduced energy expenditure during gait.

Fourth and Fifth Determinants: Foot and Knee Mechanisms

In order to comprehend the role of foot and knee mechanisms in smoothing the pathway of CoG, let us draw two intersecting arcs as shown in Fig. 3.10. The effect of the foot rotations on the displacement patterns of the knee and CoG is presented in Fig. 3.10. The angular displacements of the ankle, foot, and knee are related and they have a major role in reducing the energy expenditure during a gait cycle. The first arc is defined with respect to the initial stages of the stance phase (Fig. 3.10). This arc is traced as the ankle rotates with respect to the point of heel strike. The second arc is made by the foot rotating about a point at the fore-foot, during the later stages of the stance phase. During heel contact, the foot is dorsiflexed and the knee joint is fully extended, such that the lower limb is as long as possible. Thereby, the centre of gravity (CoG) reaches its lowest vertical position. The rapid plantarflexion of the foot, along with the initiation of knee flexion, helps to maintain the CoG at roughly the same level as it moves forward, slightly flattening and subsequently reversing the curve (path traced by the CoG). Similarly, the knee flexion during heel-off facilitates the flattening and slight reversal of the curve. By smoothing out the abrupt inflexions at the point of intersection of the arcs, a sinusoidal path is established that reduces energy expenditure and results in a smoother gait.

Sixth Determinant: Lateral Displacement of Pelvis

The CoG of the body displaces laterally during the gait cycle, as discussed earlier (Fig. 3.7b). This displacement is produced by the lateral shift of the pelvis or by

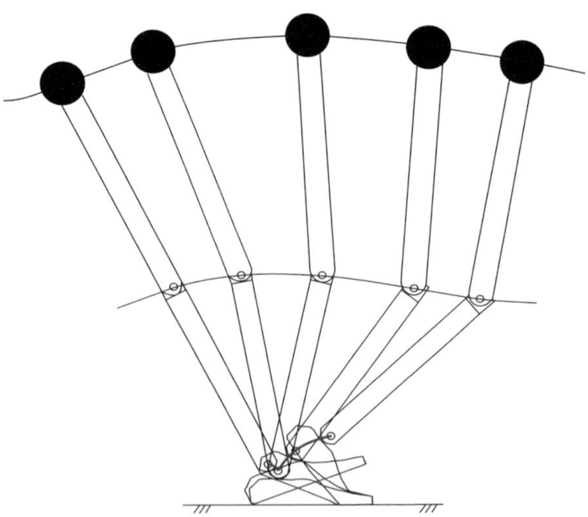

Fig. 3.10 Effect of foot rotation on the displacement patterns of knee and CoG

relative hip adduction (Saunders et al. 1953). The amount of lateral displacement would have been more if the limbs in the human body were parallel to each other. The tibiofemoral angle, along with the relative hip adduction, avoids the case of parallel limbs and thereby reduces the lateral displacement (Saunders et al. 1953). This, in turn, smoothens the gait as well as reduces the energy expenditure.

3.2 Gait Analysis

Let us now explore the details of gait analysis. Gait analysis is a critical component of biomechanics and is used to assess and study human walking patterns. It provides valuable insights into the forces and movements that influence walking. Several key factors are important in gait analysis, as outlined below:

- Gravity Force: This force represents the weight of the body and acts vertically downward. It is a constant force that influences the body's motion during walking.
- Muscle Forces: Muscles play a crucial role in generating the forces needed for walking. Muscle controls the joint movements and provides propulsion during each instance of the gait.
- Inertia Forces: Inertia is the resistance of an object to changes in its state of motion. In gait analysis, inertia forces come into play as the body accelerates and decelerates during walking.
- Mass of the Body: The mass of each body segment, including the arms, legs, and trunk, affects how the body moves and responds to external forces. It is a key factor in calculating moments and accelerations during walking.
- Ground Reaction Forces: These forces are exerted by the ground on the body and are essential for maintaining balance and forward motion during walking. Ground reaction forces act in the opposite direction of the gravity force and vary throughout the gait cycle.

When a person is standing, the ground reaction force is equal in magnitude to the person's weight but acts in the opposite direction to the force of gravity. This balance of forces is crucial for maintaining an upright stance. During walking, the ground reaction forces change continuously as each foot alternates between the stance and swing phases of the gait cycle. Gait analysis involves measuring and quantifying these forces and movements to understand how they contribute to normal and abnormal walking patterns. It is used in different fields, including orthopaedics, sports science, physical therapy, and biomechanical research, to diagnose and treat gait abnormalities, design assistive devices, and improve rehabilitation strategies.

3.2.1 Variation of Ground Reaction Force During Gait Cycle

Understanding the variation of ground reaction force (GRF) throughout a gait cycle is essential for comprehending human walking. As discussed in Sect. 3.1.4, the vertical component is the largest component which has a peak value more than body weight during the gait cycle. It is worth mentioning that gait analysis often focuses on the two force components in a two-dimensional plane, neglecting the medial–lateral component.

The line of action of the GRF vector varies during the stance phase of a gait cycle, as presented in Fig. 3.11. In contrast, the GRF is not present during the swing phase, since the leg is not in contact with the ground. Understanding the timing and magnitude of these forces is important for evaluating and diagnosing gait abnormalities and optimising walking patterns. Moreover, the alignment of the GRF vector is crucial in maintaining balanced and efficient movement during walking. In healthy individuals, the GRF vector passes close to the body's centre of gravity, contributing to smooth and coordinated gait. However, in individuals with gait disorders or abnormalities, the GRF vector deviates from the ideal alignment, passing outside the body, as shown in Fig. 3.12. This deviation increases the moment arm or lever arm, which is the perpendicular distance from the joint axis to the line of action of the force vector. This misalignment leads to a considerable increase in moment or torque around the joints owing to the deviation of the GRF vector from its normal path through the body.

A larger moment arm leads to a greater moment or torque created by the GRF. In order to maintain balance and prevent falls, the body's muscles must counteract the increased moments, caused by the misalignment of the GRF. However, the muscles have relatively small moment arms, meaning they exert their force closer to a joint. In order to counteract the larger moments created by the misaligned force vector, muscles need to exert considerably greater force. Hence, gait disorders that cause deviations in the alignment of the GRF vector can lead to increased joint moments. In order to compensate for these moments and maintain stability, muscles must generate more force, which can be challenging and may result in fatigue and difficulty in walking. The deviations in GRF and joint moments in individuals with pathological

Fig. 3.11 Variation of ground reaction force vector during the stance phase of gait cycle

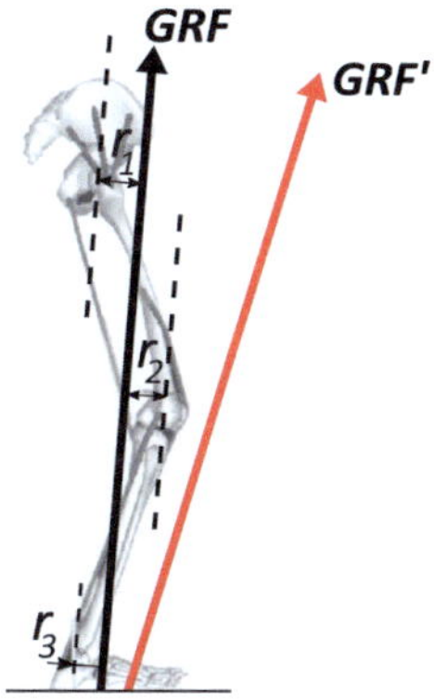

Fig. 3.12 Effect of the alignment of GRF on the lever arm; Moment arms are denoted by r_1, r_2, and r_3, respectively. The GRF' vector corresponds to the ground reaction force vector of a person with gait abnormality

gait, in particular for cerebral palsy patients, are notable and can have a substantial impact on their gait patterns and overall mobility.

Ground Reaction Force in Cerebral Palsy Patients: In patients with cerebral palsy, the vertical component of the GRF during the gait cycle differs markedly from that of normal individuals. This discrepancy is evident in Fig. 3.13, where different gait patterns of cerebral palsy patients (crouch, recurvatum, jump, and mild) are compared with the ground reaction force of a normal gait (blue line). Cerebral palsy patients often exhibit higher peak values and deviations in the vertical component of the GRF as compared to normal gait. Different gait patterns in cerebral palsy patients result in distinct deviations in the GRF. For example, the "jump" pattern (represented by a triangle) shows a considerable difference in the GRF in comparison to the normal gait (Fig. 3.13).

Joint Moments in Cerebral Palsy Patients: Joint moments, specifically plantar dorsiflexion moments, are also affected in patients with cerebral palsy. These joint moments play a crucial role in controlling movement during walking. In patients with cerebral palsy, misalignment of the GRF leads to longer moment arms, which in turn increases the joint moments around the hip, knee, or ankle joints. In order to maintain balance and stability while walking, muscles must generate additional force to counteract the increased joint moments. This compensation is necessary to prevent falls and maintain mobility. Different gait patterns of cerebral palsy patients lead to varying degrees of deviations from the normal joint moment curve (Fig. 3.13).

The common gait abnormalities of the knee in cerebral palsy patients can be classified into four types: jump, crouch, recurvatum, and mild (Sutherland and Davids 1993; Lin et al. 2000). Jump gait was characterised by increased knee flexion in the early stance phase, through initial double support, with correction of the knee motion to normal or near normal extension in mid-stance and late-stance. In crouch gait, there was increased knee flexion through the stance phase. Recurvatum gait described increased knee extension in the mid-stance and late-stance phases. The mild group was defined as those who walked better than the other three groups and were closer to the normal motion (Lin et al. 2000). The cerebral palsy patients

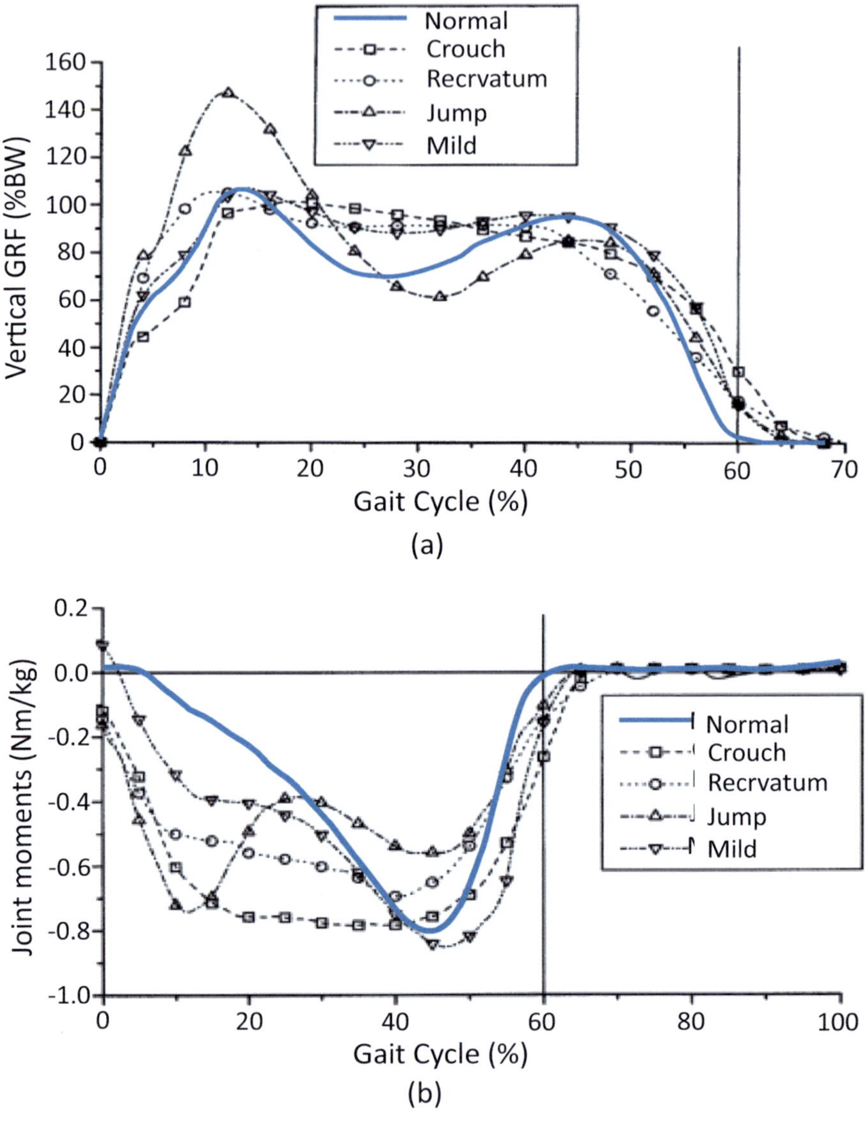

Fig. 3.13 Effect of gait abnormalities in cerebral palsy patients on, **a** vertical component of GRF, **b** ankle joint moments (Adapted from Lin et al. 2000)

experienced deviations in ground reaction forces and joint moments during walking due to the altered mechanics of their gait. These deviations require increased muscle forces to maintain stability, which can result in unique gait patterns and challenges in mobility.

3.2.2 Gait Abnormalities

Let us look into common forms of gait abnormalities and their associated characteristics:

(1) **Antalgic gait**: When an individual experiences pain in one limb, there is a tendency to adjust the gait to reduce the amount of weight-bearing on the affected side, thus minimising pain. In order to reduce the muscle activation surrounding the affected side and thereby minimise the joint compression, the patient tend to lean over the affected side. This altered gait pattern is called as antalgic gait. The term "antalgic" is derived from the Greek words "anti" meaning against and "algia" meaning pain. This gait abnormality shortens the stance phase of the affected side as compared to that of the unaffected. Common causes of this gait abnormality are osteoarthritis, rheumatoid arthritis, tendinitis.
(2) **Trendelenburg gait**: This gait abnormality is characterised by the trunk shift over. Let us assume that a patient is suffering from a weakened abductor muscle on his left side (Fig. 3.14). During the stance phase on the left lower limb, weakened abductor muscles allow the pelvis to tilt down on the right side. In order to compensate for this muscle weakness, the trunk rotates to the weakened side thereby reducing the demand on the abductor muscle to maintain a level pelvis (Fig. 3.14). The common cause of this gait abnormality is weak hip abductor muscles (gluteus medius and gluteus minimus).
(3) **Ataxic gait**: This gait abnormality causes irregular steps that affect one's ability to walk in a straight line when walking heel-to-toe. This gait is common in patients with cerebellar ataxia (damage to the cerebellum).
(4) **Propulsive gait**: Propulsive gait is a type of walking pattern characterised by a forward-leaning posture. The steps in a propulsive gait are often short and shuffling, with the feet being dragged along the ground. This gait abnormality is often associated with neurological conditions, such as Parkinson's disease, that affect movement and posture control.
(5) **Scissors gait**: As the name suggests, this gait abnormality is characterised by the scissor-like gait pattern. The knees and thighs tend to hit or cross each other while walking. This type of gait usually affects patients with spastic diplegia cerebral palsy.
(6) **Spastic gait**: Spasticity is abnormal muscle tightness due to prolonged muscle contraction. Spastic gait is a consequence of walking with a stiff leg. When attempting to lift this stiff leg while walking, it either drags or moves in a semi-circular motion (known as circumduction). This gait pattern is frequently observed in patients with cerebral palsy, multiple sclerosis, or hemiplegia.

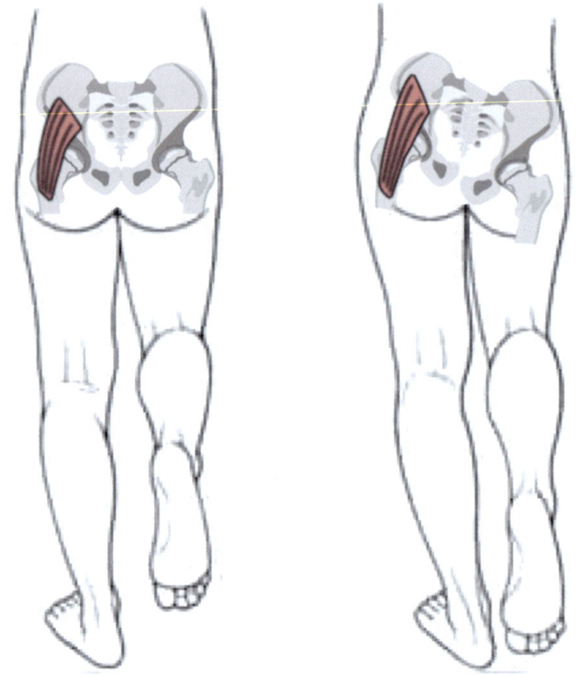

Fig. 3.14 Illustration to explain Trendelenburg gait **a** pelvic tilt in normal gait, **b** excessive pelvic tilt in trendelenburg gait

3.3 Gait Measurement Techniques

Gait analysis provides valuable insights into various aspects of gait, including its use in detecting and monitoring disorders and diseases. This is particularly relevant for identifying pathological gait owing to neurological conditions, such as Parkinson's disease or musculoskeletal problems like arthritis and tendonitis. Moreover, gait analysis is instrumental in addressing age-related changes in gait patterns, which affect a large portion of the population. Additionally, it plays a pivotal role in the development of prosthetics and orthotics, offering support and rehabilitation for individuals with gait disorders.

Clinical gait analysis serves as a comprehensive approach to understanding gait anomalies, enabling early disease diagnosis facilitating treatment of patients through rehabilitation, and evaluating pathologic gait and athletic performance. Traditionally, gait analysis relied on subjective methods involving therapist observation and judgment. While these methods provide valuable qualitative information about a patient's gait, they lack the precision and objectivity offered by modern techniques. Another traditional method involves plantar pressure monitoring, which offers insights into plantar pressure distribution but cannot capture gait parameters during the swing phase. Overall, these subjective methods raise concerns about accuracy and precision, posing challenges for precise diagnosis, follow-up, and treatment of gait disorders.

Advancements in technology have introduced objective techniques for gait analysis. These techniques offer enhanced measurement accuracy and precision, using non-wearable and wearable sensors to collect reliable data on various gait parameters. The availability of extensive data enables the diagnosis and tracking of gait abnormalities, effectively. Objective techniques are categorised into non-wearable sensors and wearable sensors, with the former including systems like imaging devices and force sensors. The non-wearable system requires the use of controlled research facilities, where the sensors are located in a laboratory setup to capture data on the patient's gait. Whereas, wearable sensor systems capture data on the gait of a patient's everyday activities outside a gait laboratory. Therefore, it can be carried out anywhere and the data can be analysed later. A third type of hybrid system uses a combination of both non-wearable sensor and wearable sensors. The following section provides an in-depth overview of non-wearable sensor systems used in gait analysis. These systems extract various gait parameters through digital image processing and floor sensors.

3.3.1 Image Processing Systems

A typical image processing system comprises one or more optical sensors employed to collect data pertinent to gait analysis. Various image processing methods, such as silhouette-based, optical flow analysis, and depth-based methods are employed for gait analysis. In this section, depth-based or range imaging system is discussed in detail. It encompasses techniques employed to compute and generate a distance map from a given viewpoint (Jain et al. 1995). These techniques enable the extraction of vital elements of the image with improved efficiency and real-time processing speed.

In gait analysis, range imaging systems contribute to the extraction of depth-related features and enhance the accuracy of gait parameter measurements. It includes methods like time of flight, stereoscopic 3D, and structured light to measure the distance of objects from a viewpoint, enabling the creation of 2D representations with depth information. Other methods are laser range scanners and infrared thermography.

Time-of-flight system: Utilising infrared light, this system determines the depth information by measuring the time it takes for an emitted signal to return after being reflected by an object. It calculates distance based on the speed of light and the time difference between signal emission and reception.

Stereoscopic vision system: This system uses a stereo camera setup with two or more image sensors to simulate human binocular vision, enabling depth perception. It relies on finding corresponding points in different images and calculates depth based on similar triangles between optical sensors, light emitters, and object in the scene.

Structured light system: This method involves projecting structured light patterns onto objects to capture their 3D topography using a 2D imaging camera. Well-known examples include the Microsoft Kinect sensor, which reconstructs object shapes based on 3D surface patterns.

3.3.2 *Floor Sensor Systems*

The floor sensor systems utilise sensors positioned along the floor on designated 'force platforms.' These platforms measure gait information through pressure sensors and ground reaction force sensors (GRF).

Force plates and force platform: These sensors measure GRF exerted by the feet during contact with the ground. They can also provide essential data on moments required for gait analysis. Force plates are often arranged as platform (Fig. 3.15).

Pressure measurement system: These systems involve sensors that quantitatively measure foot pressure distribution during the stance phase of walking (Fig. 3.16). They identify discrepancies in plantar pressure profiles between the left and right feet, aiding in the assessment of gait characteristics. However, they do not measure the horizontal or shear components of applied forces. Pressure measurement systems and force platforms can quantify the centre of pressure.

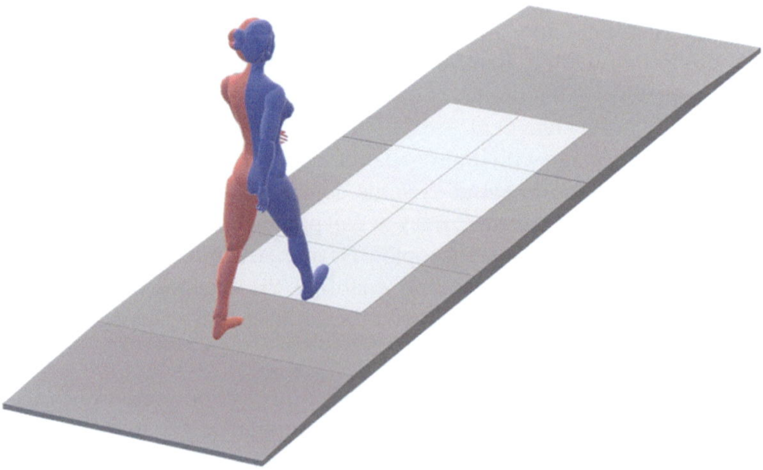

Fig. 3.15 Force platforms embedded in the walkway. Adapted from Baptista et al. (2023)

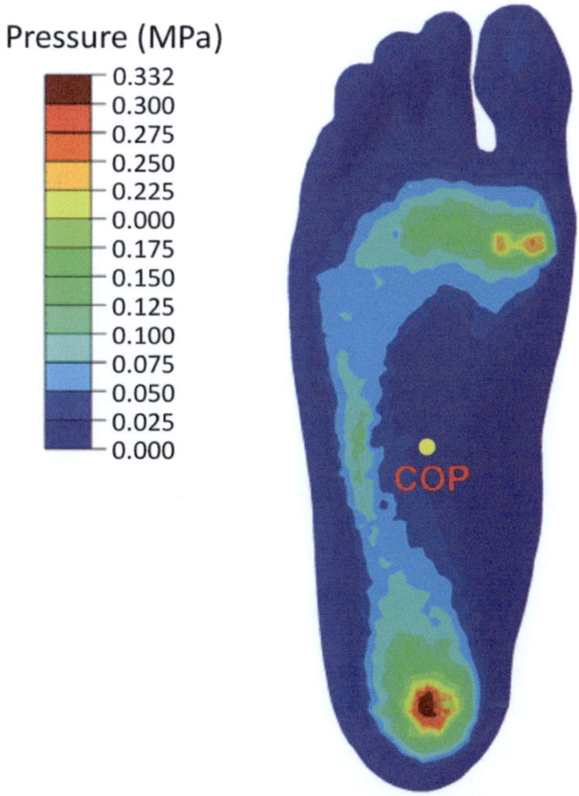

Fig. 3.16 Plantar pressure distributions and the location of centre of pressure (Wang et al. 2019)

3.3.3 Wearable Sensors

Gait analysis utilising wearable sensors involves the placement of these devices on different anatomical locations of the patient's body, such as the feet, knees, or hips, to quantify diverse characteristics of human gait. A brief overview of the most commonly used sensors are outlined below.

- Inertial Sensors: These sensors measure specific forces, angular rate of movement, and body orientation using a combination of accelerometers, gyroscopes, and magnetometers.
- Pressure sensing insoles: Pressure sensors are embedded within footwear to measure the distribution of foot pressure during the stance phase of walking, providing valuable insights into weight-bearing patterns and gait abnormalities.
- Electromyography (EMG): EMG sensors record the electrical activity generated by skeletal muscles. They can be used non-invasively with surface electrodes or invasively with wires or needle electrodes to analyse muscle activity.

- Ultrasound and Ultra-Wideband: These sensors are employed to measure the stride length and the distance of separation between feet, providing additional gait parameters.
- Goniometers: These sensors are used to measure joint angles. Several types of goniometers are available based on the working principle. One of them is strain gauge-based goniometers, where the change in resistance is based on flexion, thus providing information about joint movement.

Wearable sensor systems offer the advantage of capturing gait data in real-world environments, enabling a more comprehensive understanding of an individual's gait as compared to non-wearable sensors. These sensors can be particularly useful for monitoring patients' progress during rehabilitation, evaluating sports performance, and identifying gait abnormalities outside of a controlled laboratory setup.

3.3.4 Motion Capture Systems: Marker-Based Versus Marker-Less

The measurement of human motion stands as one of the most intriguing yet formidable challenges. This section delves into optical motion tracking solutions. Such solutions can be broadly classified into two categories: marker-based systems and marker-less systems.

Marker-based motion capture system: Marker-based systems use reflective markers placed strategically at specific anatomical landmarks of the human body (Fig. 3.17). These markers serve as tracking points for capturing motion data. The position of these markers is recorded using a calibrated multi-camera system. This approach enables the tracking of intricate body movements, including joint positions and body segment orientations. Due to factors like cost, complexity, and the requisite technical expertise required for operation, marker-based systems are primarily employed in specialised laboratories, particularly for clinical rehabilitation and other specific applications.

Marker-less motion capture system: Over the past decade, significant interest has been directed toward marker-less solutions. These systems aim to alleviate the cost and operational complexities associated with marker-based counterparts while simplifying the motion capture process. Marker-less systems typically employ depth-sensing camera systems, often referred to as RGB-D cameras, capable of capturing both colour and depth information. A prime example of such a camera system is the Microsoft Kinect. In Fig. 3.18, four cameras are presented, which are used in the motion capture system. In addition, the subject is trying to climb the rehabilitation stairs (Fig. 3.18). Marker-less systems find extensive use in clinical and biomechanical research, however, they do come with notable limitations. These limitations encompass a shorter measurement range, sensitivity to bright sunlight, and potential interference between multiple sensors. Most crucially, marker-less systems tend

Fig. 3.17 Marker-based motion capture system (*Source* Cerfoglio et al. 2021)

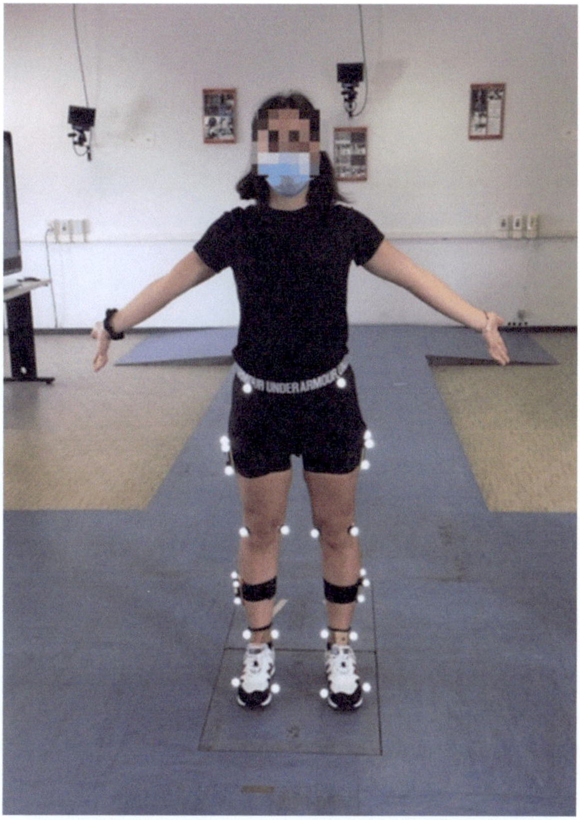

to exhibit a lower accuracy and precision in motion tracking as compared to the marker-based systems, which remain the gold standard in 3D gait measurement.

Let us now explore the details of a typical marker-based motion capture system. A comprehensive setup is illustrated in Fig. 3.19, showing two infrared cameras, mounted on the ceiling. In the central area of the walkway, four force plates are embedded into the floor (Fig. 3.19). Additionally, the system employs fourteen reflective markers, indicated by the white circles in Fig. 3.19. These markers are strategically placed on the subject's body to capture the movements accurately. Essentially, a typical marker-based motion capture system uses multiple cameras, reflective markers, and an integrated force platform or instrumented walkway. The marker data is meticulously recorded and stored in a file, comprising the x, y, and z coordinates of each marker at every sample point, offering precise gait analysis at any given instance during the data acquisition. The Cartesian Coordinate System (CCS) for each adjacent body segment (for a particular joint) involves defining axes based on palpable or X-ray identifiable bony landmarks. The common origin, shared by both the CCSs, serves as the reference point for determining the linear translation

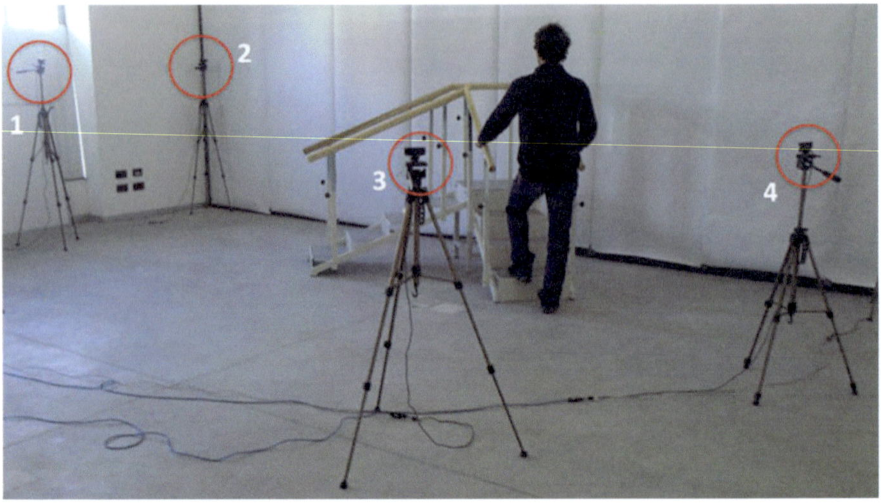

Fig. 3.18 Marker-less motion capture (mocap) system (*Source* Colombo et al. 2013)

in the joint (Fig. 3.20). Subsequently, the Joint Coordinate System (JCS) is established based on the two CCSs. Two of the JCS axes are fixed to the body and one remains 'floating'. Finally, the joint motion, encompassing three rotational and three translational components, is determined based on the JCS (Wu et al. 2002). In the case of a hip joint, the bony landmarks used are as follows:

- anterior superior iliac spine (ASIS),
- posterior superior iliac spine (PSIS),
- femoral epicondyle (FE).

Based on the bony landmarks, the coordinate system of the pelvic bone is defined as follows:

(a) The *X*-axis is formed by the line, parallel to a line lying in the plane defined by the two ASISs and the midpoint of the two PSISs (Fig. 3.20). The positive *X*-direction points anteriorly.
(b) The *Z*-axis is formed by the line connecting the two ASISs. Positive *Z*-direction points laterally.
(c) The *Y*-axis is formed by the line perpendicular to the *X*- and *Z*-axis. Positive *Y*-axis points cranially. The origin of the pelvic coordinate system coincides with the hip centre of rotation, which is generally considered to be the same as the centre of hip joint (Fig. 3.20).

In the case of femur, the coordinate system is defined as follows:

(d) The *y*-axis is formed by the line joining the midpoint between medial and lateral FEs and the origin. Positive y-direction points cranially.

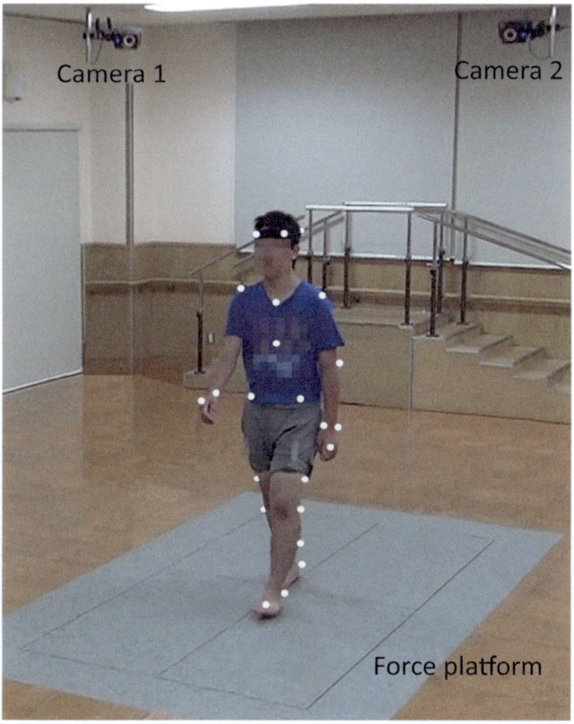

Fig. 3.19 Motion capture system with reflective markers (white), infrared cameras and force plates (Adapted from Naruse et al. 2017)

(e) The z-axis is formed by the line perpendicular to the y-axis, lying in the plane defined by the origin and the two FEs. Positive z-direction points laterally.
(f) The x-axis is formed by the line perpendicular to the y- and z-axis. The positive y-axis points anteriorly (Fig. 3.20).

The hip joint coordinate system is formed with two fixed axes (e_1 and e_3) and one floating axes (e_2). Here e_1 coincides with the Z-axis of the pelvic coordinate system, e_3 coincides with the y-axis of the femur coordinate system. The floating axis, e_2 coincides with the axis perpendicular to e_1 and e_2. The flexion/extension and mediolateral translations are defined using e_1, whereas the internal/external rotation and proximo-distal translation are defined based on e_3. The e_2 is used to define the abduction/adduction and antero-posterior translation. After the acquisition of the marker data using the motion capture system, kinematic and kinetic analyses are carried out, which has been discussed later in Chap. 4.

Marker-based motion capture, though regarded as the gold standard owing to its rich body of literature, non-invasive nature, and high accuracy, comes with its own set of challenges. One notable challenge is its susceptibility to soft tissue artefact. Soft tissue artefact corresponds to the error associated with the relative motion between the region of skin upon which the markers are attached and the underlying bone. This may arise due to muscular contractions, skin movement, and inertial effects (Cappozzo et al. 1996). The magnitude of error associated with soft tissue artefacts

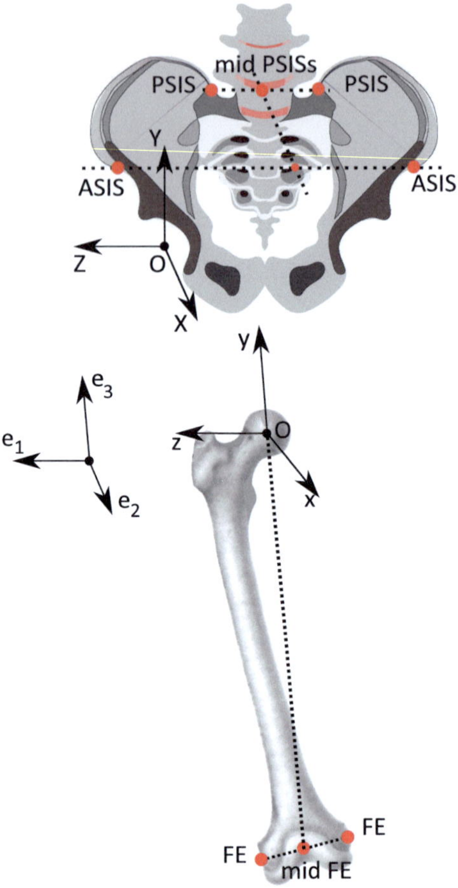

Fig. 3.20 Hip joint coordinate system; pelvic coordinate system (*XYZ*), femoral coordinate system (*xyz*), and the JCS for the right hip joint (e_1 e_2 e_3)

depends on the physical characteristics of the subject, marker locations, and the movement involved (Holden et al. 1997; Schwartz et al. 2004; Fuller et al. 1997).

Among other motion capture systems, the biplanar fluoroscopic imaging system stands out as an advanced technology that allows for the simultaneous capture of X-ray images in two planes. Moreover, this technique eliminates the skin movement artefacts associated with the marker-based movement capture systems. It permits the capture of high-quality digital images, providing detailed insights into skeletal kinematics and joint motion during dynamic physical tasks (Taylor et al. 2011). This system has been employed in a range of medical fields, including orthopaedics, spinal surgery, and joint kinematics analysis, because of its high precision in estimating joint kinematics during natural movements (Phan et al. 2018). Although this technology provides valuable insights in joint kinematics; the associated radiation risks have been a major limitation.

Multiple Choice Questions (MCQ)

1. The lateral distance between the lines of the two feet is called as ________.

   (A) Walking base
   (B) Step length
   (C) Stride length
   (D) Compass gait

2. The distance between successive instants of heel contact (strike) of the same foot is called _____________.

   (A) Step length
   (B) Stride length
   (C) Walking base

3. Swing phase constitute____% of the gait cycle.

   (A) 60
   (B) 40
   (C) 20
   (D) 15

4. What happens to the time duration of the swing phase when the walking speed increases?

   (A) Increases
   (B) Decreases
   (C) Remains constant

5. The energy consumption in the gait cycle is _________ when the centre of gravity (CoG) of the body follows a straight line.

   (A) Maximum
   (B) Minimum
   (C) Unaltered

6. The human gait has _______ determinants to minimise energy expenditure.

   (A) 3
   (B) 4
   (C) 6
   (D) 7

7. The ratio of the time frame of the stance phase to the swing phase reverses during running as compared to normal walking.

   (A) True
   (B) False

8. When the ground reaction force of gait disorder patients passes outside the body, the moment arm around the joints __________.

   (A) Increases
   (B) Decreases
   (C) Remains the same

9. Clinical gait analysis involves ____ measurement/(s).

   (A) Objective
   (B) Subjective
   (C) Both objective and subjective

10. _________ technique is used for recording the electrical activity produced by skeletal muscles.

   (A) Inertial measurement unit
   (B) Electromyography
   (C) Goniometry

11. Which of the following is a non-wearable sensor system?

   (A) Goniometer
   (B) Force plate
   (C) Wireless EMG
   (D) Accelerometer

12. Which is a major limitation of the marker-less system for optical motion tracking?

   (A) Short-range and inoperability in bright sunlight
   (B) Requiring a controlled environment
   (C) Artefacts from skin movement

13. The single support phase constitutes approximately____% of the gait cycle.

   (A) 60
   (B) 40
   (C) 20
   (D) 30

14. While running, _________ phase/(s) exist.

   (A) Double swing
   (B) Double support
   (C) Both double swing and double support

15. By reducing the walking base, the lateral displacement of the centre of gravity is _______.

   (A) Reduced
   (B) Increased

(C) Unaltered

Answer Key to the MCQs

1. A; 2. B; 3. B; 4. A; 5. B; 6. C; 7. A; 8. A; 9. C; 10. B; 11. B; 12. A; 13. B; 14. A; 15. A.

References

Baptista RR, Bravo MD, Heidner GS et al (2023) Age-related gait adaptations of ground reaction forces. Res. Biomed. Eng. 39:915–924

Cappozzo A, Catani F, Leardini A, Benedetti MG, Croce UD (1996) Position and orientation in space of bones during movement: experimental artefacts. Clin Biomech 11(2):90–100

Cerfoglio S, Galli M, Tarabini M, Bertozzi F, Sforza C, Zago M (2021) Machine learning-based estimation of ground reaction forces and knee joint kinetics from inertial sensors while performing a vertical drop jump. Sensors (Basel, Switz) 21(22):7709

Colombo G, Facoetti G, Rizzi C (2013) Virtual testing laboratory for lower limb prosthesis. Comput Aided Des Appl 10(4):671–683

Fuller J et al (1997) A comparison of lower-extremity skeletal kinematics measured using skin- and pin-mounted markers. Hum Mov Sci 16(2–3):219–242

Holden JP et al (1997) Surface movement errors in shank kinematics and knee kinetics during gait. Gait Posture 5(3):217–227

Jain RC, Kasturi R, Schunck BG (1995) Machine vision. McGraw-Hill

Jansen K, De Groote F, Duysens J, Jonkers I (2014) How gravity and muscle action control mediolateral centre of mass excursion during slow walking: a simulation study. Gait Posture 39(1):91–97

Lin CJ, Guo LY, Su FC, Chou YL, Cherng RJ (2000) Common abnormal kinetic patterns of the knee in gait in spastic diplegia of cerebral palsy. Gait Posture 11(3):224–232

Phan CB, Nguyen DP, Lee KM, Koo S (2018) Relative movement on the articular surfaces of the tibiotalar and subtalar joints during walking. Bone Joint Res 7(8):501–507

Naruse H, Fujisawa TX, Yatsuga C, Kubota M et al (2017) Increased anterior pelvic angle characterizes the gait of children with attention deficit/hyperactivity disorder (ADHD). PLoS One 12(1):e0170096

Saunders JB, Inman VT, Eberhart HD (1953) The major determinants in normal and pathological gait. J Bone Joint Surg Am 35-A(3):543–558

Schwartz M, Trost JP, Wervey R (2004) Measurement and management of errors in quantitative gait data. Gait Posture 20:196–203

Sutherland DH, Davids JR (1993) Common gait abnormalities of the knee in cerebral palsy. Clin Orthop Relat Res 288:139–147

Taylor KA, Terry ME, Utturkar GM et al (2011) Measurement of in vivo anterior cruciate ligament strain during dynamic jump landing. J Biomech 44(3):365–371

Wang Y, Wong DW, Tan Q, Li Z, Zhang M (2019) Total ankle arthroplasty and ankle arthrodesis affect the biomechanics of the inner foot differently. Sci Rep 9(1):13334

Wu G, Siegler S, Allard P, Kirtley C et al (2002) ISB recommendation on definitions of joint coordinate system of various joints for the reporting of human joint motion—part I: ankle, hip, and spine. J Biomech 35(4):543–548

Chapter 4
Joint Kinematics and Kinetics

Joint dynamics, comprising joint kinematics and kinetics, is the area of physics related to joint forces and the associated motions. Joint kinematics is the study of the spatial and temporal aspects of motion, without considering the forces causing that motion. Joint kinetics, however, is the study of the joint forces and moments that are associated with a specific motion. Together, these subjects provide an insight into the biomechanics of human movement.

4.1 Joint Kinematics

Joint kinematics is a fundamental subject in the field of human locomotion, serving as the cornerstone for understanding the motion of the human body. It encompasses the study of joint motion, angles, and the interplay between various skeletal segments, during a wide range of activities. In order to fully describe a body segment in space, we need several spatio-temporal variables as listed below:

- Position of the Centre of Mass (COM) of the segment (x, y, z),
- Joint angle $\left(\theta_{xy}, \theta_{yz}, \theta_{xz}\right)$,
- Linear velocity of the segment $(\dot{x}, \dot{y}, \dot{z})$,
- Angular velocity of segment $\left(\omega_{xy}, \omega_{yz}, \omega_{xz}\right)$,
- Linear acceleration of the segment denoted by $(\ddot{x}, \ddot{y}, \ddot{z})$,
- Angular acceleration of segment $\left(\alpha_{xy}, \alpha_{yz}, \alpha_{xz}\right)$.

4.1.1 Estimation of Joint Angles

In this section, the discussions are limited to the sagittal plane, for the sake of simplicity. In order to calculate the joint angle of a limb segment at a particular

S. Gupta et al., *Biomechanics of Joints and Implants*,
https://doi.org/10.1007/978-981-96-0586-6_4

time instance, we require the position data of two markers. Moreover, the line drawn between them needs to be parallel to the long-axis of the limb segment under consideration. Figure 4.1 presents the lower limb with, (a) segment angles and (b) joint angles, along with five anatomical markers representing the joints. The calculations of lower limb joint angles are stated in the following. Segment angles are measured in a counter clockwise direction (Fig. 4.1a), with respect to the horizontal (Winter 2009). The hip joint angle, α, is determined using the long axis of the femur and the pelvic axis, with flexion considered to be positive and extension to be negative (Fig. 4.1b). The angle α is calculated using the adjacent segment angles, θ_{54} and θ_{43}, as stated below:

$$\alpha = \theta_{54} - \theta_{43}$$

Here the segment angles, θ_{ij}, are calculated using the position data of the anatomical markers:

$$\theta_{ij} = \tan^{-1}\left(\frac{y_j - y_i}{x_j - x_i}\right)$$

The knee joint angle, β, is defined by the angular displacement of the long axis of the femur with respect to the long axis of the tibia (Fig. 4.1b). The knee angle (β) is calculated using the segment angles, θ_{43} and θ_{32}, as follows:

$$\beta = \theta_{43} - \theta_{32}$$

The ankle angle (γ) is calculated using the segment angles, θ_{32} and θ_{21}, as follows (Fig. 4.1b):

$$\gamma = \theta_{32} - \theta_{21} + 90^\circ$$

Joint movements are characterised by the magnitude and the sign of the angle value. For example, fully extended knee is described as 0° flexion. If $\theta_{43} > \theta_{32}$, the knee is said to be flexed (β is positive) and if $\theta_{43} < \theta_{32}$, the knee is said to be extended (β is negative). A similar convention is used for hip joint angle. The ankle joint angle, however, follows a slightly different rule. If $\theta_{32} > \theta_{21} - 90^\circ$, ankle is said to be plantarflexed, whereas if $\theta_{32} < \theta_{21} - 90^\circ$, ankle is said to be dorsiflexed.

4.1.2 Lower-Limb Joint Angles During Gait Cycle

Movements of the lower limb joints during a gait cycle play a pivotal role in our ability to walk effectively and efficiently. The ankle joint undergoes dorsiflexion and plantarflexion throughout the gait cycle (Fig. 4.2). During initial contact, the ankle is in the neutral position and later, the ankle plantarflexes (3°–5°) until the

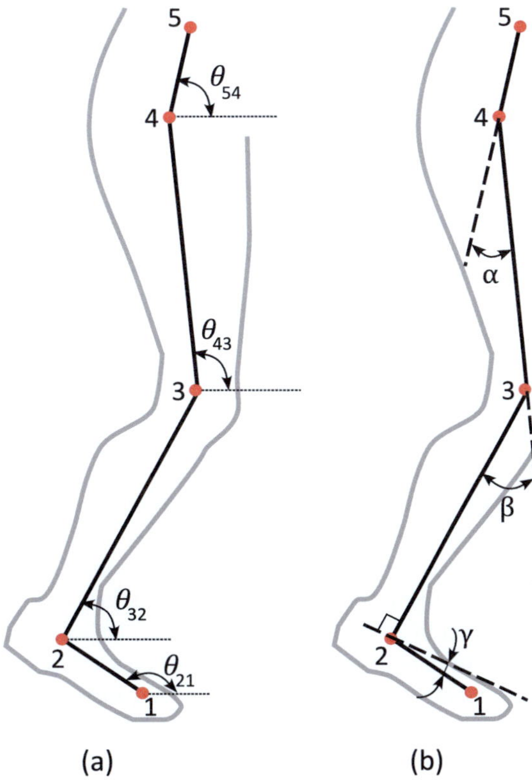

Fig. 4.1 Simplified lower limb model showing, **a** the segment angles (θ_{ij}); **b** the joint angles (α, β, γ). Red dots indicate markers

foot flat stage. This phase is sometimes referred to as the first rocker, since the foot pivots about the heel. In this phase, the dorsiflexor muscles anterior to the ankle joint are engaged in eccentric contractions, effectively controlling the plantarflexion. Eccentric contraction refers to the lengthening of muscles while generating force, whereas concentric contraction is related to the shortening of the muscles. These muscles function as shock absorber, facilitating a seamless transfer of weight to the lower limb. The foot remains stationary during the duration between foot flat and heel rise, whereas the tibia moves relative to the foot. The duration from foot flat to heel off is referred to as the second rocker, the pivoting motion about the ankle joint. During this phase, the tibia moves relative to the foot, and the plantar flexor muscles posterior to the ankle joint act concentrically to propel the body forward and initiate the heel-off. This concentric contraction is essential for generating the force required for forward propulsion.

Later in the gait cycle, the heel tends to lift off resulting in plantar flexion of about 20° at the end of the stance phase. The plantar flexor muscles contract concentrically to drive the foot into a plantar flexed position, propelling the body forward. This is referred to as the 'third rocker', with the pivot point under the metatarsal heads. In the swing phase, ankle dorsiflexors concentrically contract resulting in rapid dorsiflexion,

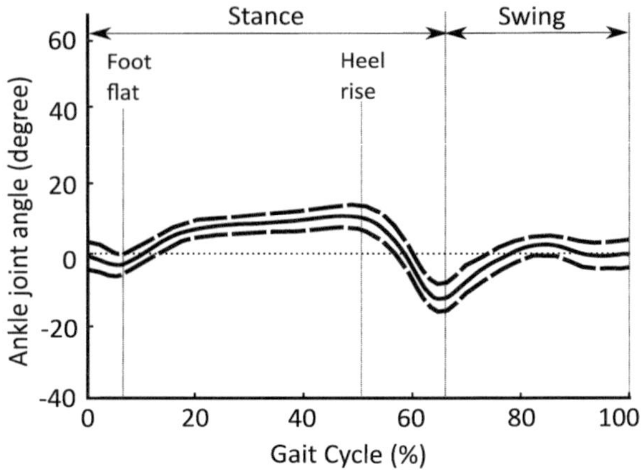

Fig. 4.2 Variation in ankle joint angle during a gait cycle (adapted from Hu et al. 2017)

facilitating the ground clearance of the foot. By mid-swing, a neutral position (0°) is achieved and maintained for the remainder of the swing phase.

Shifting our focus to the knee joint, which exhibits cyclic flexion and extension movements during the walking cycle, ranging from 0° to 70° (Fig. 4.3). After the initial contact in the stance phase, the eccentric contraction of knee extensor muscles occurs, resulting in flexion up to ~20°. The eccentric contraction of the knee extensors helps to absorb the impact forces generated during the heel strike phase. After the first peak of knee flexion, the knee joint almost fully extends. Later, during the pre-swing phase, the knee continues to flex in preparation for the swing phase. This flexion along with the ankle dorsiflexion allows for the toes to clear the ground. Throughout the mid-swing phase, the knee continues to flex, reaching a maximum of 65°. During the late swing phase, the knee rapidly extends in preparation for the second heel strike.

Let us discuss the variation of hip joint angle with the gait cycle (Fig. 4.4). The hip undergoes flexion upon heel strike to absorb shock and dissipate forces, followed by extension to propel the body forward. In the swing phase, the hip undergoes flexion

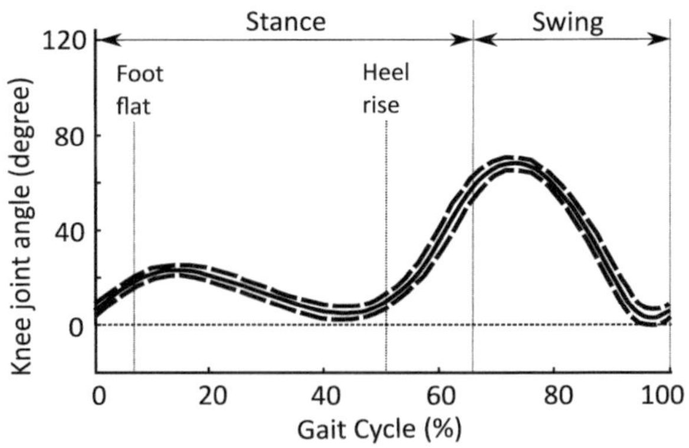

Fig. 4.3 Variation in knee joint angle during a gait cycle (adapted from Hu et al. 2017)

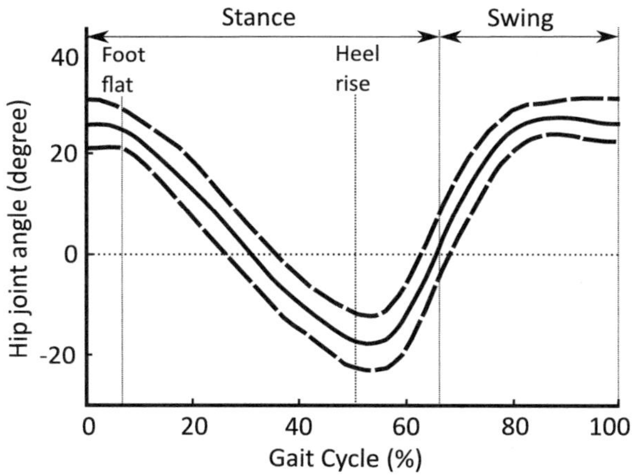

Fig. 4.4 Variation in hip joint angle during a gait cycle (adapted from Hu et al. 2017)

to facilitate limb advancement and extension to prepare for the subsequent stance phase. The peak hip extension (~20°) occurs immediately after the heel strike of the opposite foot. Subsequently, the weight shifts to the leading limb, and the trailing limb initiates hip flexion, marking the pre-swing phase. This flexion reaches its peak (~30°) during the terminal swing phase. Understanding these joint movements and their variations throughout the gait cycle is crucial in analysing normal walking patterns and addressing issues related to gait abnormalities and rehabilitation.

The finite difference method is useful to calculate angular velocity and acceleration from the angular displacement data. There are several approaches, such as forward and central difference approach. The angular velocity obtained using the forward difference approach is expressed as

$$\omega_{x_i} = \frac{\Delta x}{\Delta t} = \frac{x_{i+1} - x_i}{t_{i+1} - t_i}$$

The forward difference approach estimates the velocity at a time instance halfway between the two-time instances (i and $i + 1$). The angular velocity about the x-axis at ith time instance, calculated using central difference method, is written as follows:

$$\omega_{x_i} = \frac{\theta_{i+1} - \theta_{i-1}}{2\Delta t}$$

Similarly, the angular acceleration is derived from the angular velocity as follows:

$$\alpha_{x_i} = \frac{\omega_{i+1} - \omega_{i-1}}{2\Delta t}$$

where α_{x_i} denotes the angular acceleration about the x-axis, ω_{x_i} is the angular velocity at time i, and $2\Delta t$ represents the time difference between the successive data points of angular velocities. The central difference method is used to calculate linear and angular accelerations, since it is considered to be more accurate than the forward difference method, especially when dealing with time-dependent data and the need for precise calculations.

4.1.3 *Linear Variables*

In order to determine the displacement of the centre of gravity (COG) of a limb segment, we rely on position data of the markers, corresponding to two different time instances obtained from the gait analysis. Based on the displacement data, velocities and accelerations of the limb segment are calculated. A finite difference method is employed to compute linear velocities from displacement data. In this scheme, the position coordinates of the COG (x, y, z) with respect to the global coordinate system are used. The central difference approximation in order to calculate velocity involves a time interval of $2\Delta t$ rather than Δt.

$$V_{x_i} = \frac{x_{i+1} - x_{i-1}}{2\Delta t}$$

Accordingly, the velocities in y- and z-directions are calculated. Following a similar method, we can also determine the linear accelerations (a). In order to calculate the linear acceleration along the x-direction at the ith time instance (a_{x_i}), we can use the following expression:

$$a_{x_i} = \frac{V_{x_{i+1}} - V_{x_{i-1}}}{2\Delta t}$$

where a_{x_i} represents the linear acceleration along the x-axis, V_{x_i} is the linear velocity at time i, and $2\varDelta t$ is the time difference between successive data points of linear velocities.

4.2 Joint Kinetics

In this section, we discuss the forces and moments associated with the human movement. Using the kinematic and anthropometric data along with ground reaction force, it is possible to estimate the joint moments and forces.

4.2.1 Anthropometry Data

Anthropometry data, which encompasses measurements of body segment lengths and mass properties, plays a crucial role in human movement analysis. Anthropometry involves the study of physical measurements of the body to understand variations between individuals or groups. For comprehensive human movement analysis, it is essential to have accurate data on kinetic measures, including the masses of body segments, their moments of inertia, locations of joint centres, origins and insertions of muscles, anatomical angles for tendon pull, and the lengths and physiological cross-sectional areas of muscles. The quality and completeness of anthropometric data are paramount, as they significantly impact the accuracy of any biomechanical analysis.

The lengths of body segments can vary among individuals owing to factors, such as ethnic origin, sex, and overall body build. In order to standardize these measurements, average segment lengths are often expressed as a percentage of individual's total body height (Fig. 4.5). This method allows for a more universal and easily applicable representation of body segment lengths, making it a valuable component of anthropometric data in the study of human movement. The analysis of mass distribution, mass centres, and moments of inertia is essential for kinematic and kinetic studies in biomechanics. These properties help in understanding how mass is distributed throughout the body and how it influences movement. These data are often obtained through various methods, including cadaveric studies and imaging techniques. Table 4.1 provides information on the weight, centre of mass, and radius of gyration for different body segments, including the hand, forearm, upper arm, foot, leg, total leg, head and neck, shoulder mass, thorax, abdomen, and pelvis. This data is compiled from various references, reflecting the importance of a diverse range of sources in obtaining comprehensive information on mass and inertial properties of the body segments. These properties are crucial for accurate modelling and analysis in the field of biomechanics and contribute to a better understanding of human movement and stability.

Anthropometric data specific to individual muscles is essential for calculating muscle forces during normal movement. Muscles that belong to the same group, share the load in proportion to their relative cross-sectional areas. However, the mechanical advantage of each muscle can vary based on several factors, including the moment arm length at its point of origin and insertion on the bone. Muscles are connected to bones through tendons, and the precise locations of the muscle's origin and insertion determine the length of its moment arm. This length influences the muscle's mechanical advantage, affecting its ability to generate force and contribute to movements around a joint. The presence of bone structures beneath the muscle can also alter the angle at which the muscle pulls the bone. Moreover, tendons play a critical role in muscle function by attaching muscles to bones and affecting the angle of pull. All of these factors need to be considered when calculating muscle forces and analysing their contributions to human movement.

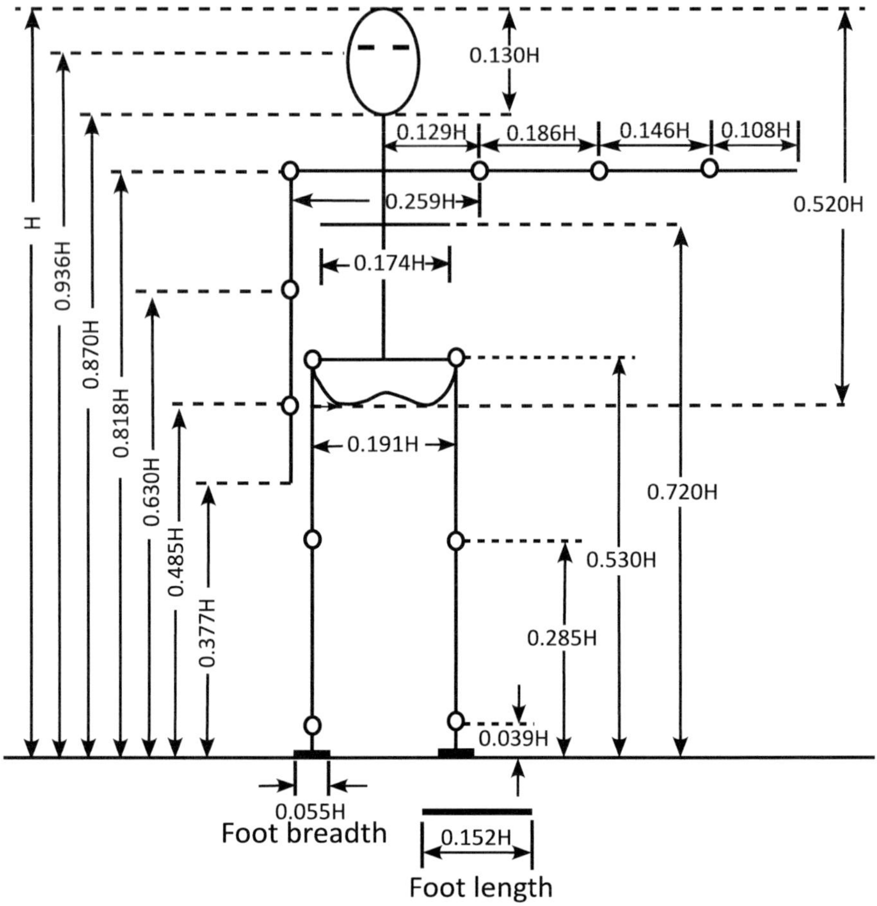

Fig. 4.5 Anthropometry data: body segment lengths in terms of body height (H)

Muscle Anthropometry: The process of determining the forces generated by individual muscles during normal movement typically requires obtaining muscle dimensions and considering the proportional sharing of load among muscles within the same group based on their relative cross-sectional areas. Additionally, the mechanical advantage of each muscle may vary, influenced by the moment arm length at its origin and insertion.

The physiological cross-sectional area (PCA) of a muscle is a critical measure that reflects the number of sarcomeres in parallel. Sarcomeres are the fundamental contractile units within muscle fibres. PCA is an indicator of the maximum possible muscular force. The angle between the long axis of a muscle and the muscle fibres is referred to as the pennation angle (Fig. 4.6). In parallel-fibred muscles, PCA is calculated as the mass of the muscle fibre (m in g) divided by the product of its density (d in g/cm^3) and length (l in cm), expressed as

Table 4.1 Anthropometric data on segment mass, location of centre of mass, radius of gyration and density (Winter 2009)

Segment	Definition	Segment weight/total body weight	Centre of mass/segment length		Radius of gyration/segment length			Density (g cm^3)
			Proximal	Distal	C of G	Proximal	Distal	
Hand	Wrist axis/knuckle II middle finger	0.006	0.506	0.494	0.297	0.587	0.577	1.16
Forearm	Elbow axis/ulnar styloid	0.016	0.430	0.570	0.303	0.526	0.647	1.13
Upper arm	Glenohumeral axis/elbow axis	0.028	0.436	0.564	0.322	0.542	0.645	1.07
Forearm and hand	Elbow axis/ulnar styloid	0.022	0.682	0.318	0.468	0.827	0.565	1.14
Total arm	Glenohumeral joint/ulnar styloid	0.050	0.530	0.470	0.368	0.645	0.596	1.11
Foot	Lateral malleolus/head metatarsal II	0.0145	0.50	0.50	0.475	0.690	0.690	1.10
Leg	Femoral condyles/medial malleolus	0.0465	0.433	0.567	0.302	0.528	0.643	1.09
Thigh	Greater trochanter/femoral condyles	0.100	0.433	0.567	0.323	0.540	0.653	1.05
Foot and leg	Femoral condyles/medial malleolus	0.061	0.606	0.394	0.416	0.735	0.572	1.09
Total leg	Greater trochanter/medial malleolus	0.161	0.447	0.553	0.326	0.560	0.650	1.06
Head and neck	C7-T1 and 1st rib/ear canal	0.081	1.00	–	0.495	1.116	–	1.11
Shoulder mass	Sternoclavicular joint/glenohumeral axis		0.712	0.288				1.04
Thorax	C7-T1/T12-L1 and diaphragm[a]	0.216	0.82	0.18				0.92
Abdomen	T12-L1/L4-L5[a]	0.139	0.44	0.56				
Pelvis	L4-L5/greater trochanter[a]	0.142	0.105	0.895				

(continued)

Table 4.1 (continued)

Segment	Definition	Segment weight/ total body weight	Centre of mass/ segment length		Radius of gyration/ segment length			Density ($g\ cm^3$)
			Proximal	Distal	C of G	Proximal	Distal	
Thorax and abdomen	C7-T1/L4-L5[a]	0.355	0.63	0.37				
Abdomen and pelvis	T12-L1/greater trochanter[a]	0.281	0.27	0.73				1.01
Trunk	Greater trochanter/ glenohumeral joint[a]	0.497	0.50	0.50				1.03
Trunk, head, and neck	Greater trochanter/ glenohumeral joint[a]	0.578	0.66	0.34	0.503	0.830	0.607	
Head, arms, and trunk (HAT)	Greater trochanter/ glenohumeral joint[a]	0.678	0.626	0.374	0.496	0.798	0.621	
HAT	Greater trochanter/ middle rib	0.678	1.142	–	0.903	1.456		

The radii of gyration are for rotation about the medial–lateral direction

[a] These segments are presented relative to the length between the greater trochanter and the glenohumeral joint

$$\text{PCA} = \frac{m}{d \times l}$$

Parallel fibred-muscles are those muscles with pennation angle equal to zero. In contrast, pennate muscles have muscle fibres oriented at an angle to the muscle's long axis and the PCA is calculated as

$$\text{PCA} = \frac{m \cos \alpha}{d \times l}$$

During muscle contraction, the pennation angle increases and vice-versa (Awad et al. 2021; Bolsterlee et al. 2017).

Muscle stress (σ) is defined as the force per unit cross-sectional area of a muscle (Brand et al. 1986). Stress values of muscles typically range between 20 and 100 N/ cm^2. Higher stress values are often observed in pennate muscles. Maximal isometric force is the force capacity of a muscle in the isometric condition (no change in muscle length) (Chen and Franklin 2023). It is expressed as the product of muscle stress and

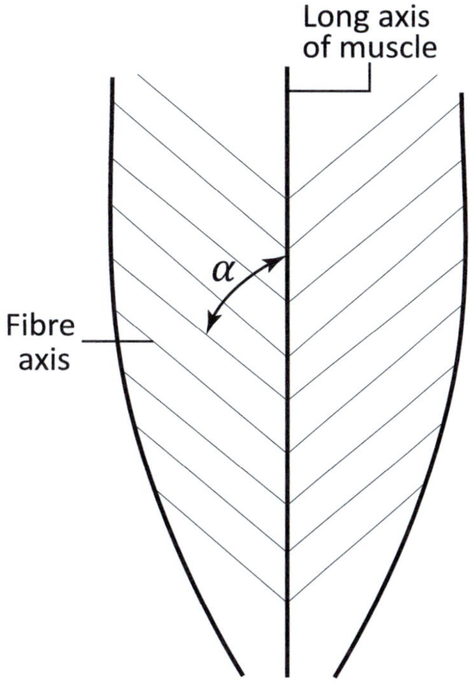

Fig. 4.6 Pennation angle (α) is the angle between the long axis of a muscle and the muscle fibre

corresponding PCA:

$$F_{\mathrm{iso,max}} = \sigma_{\mathrm{iso,max}}\mathrm{PCA}$$

The mechanical advantage of a muscle is determined by the origin and insertion points of the muscle, which define the angle of pull of the tendon on the bone. Each muscle has its unique moment arm length, which is the length of a line drawn perpendicular to the muscle that passes through the joint centre. This moment arm influences the mechanical leverage of a muscle around the joint centre, thereby affecting its contribution to joint movement. In the lower extremity, major biarticulate muscles such as, gastrocnemius, hamstring, and rectus femoris, have distinct moment arm lengths at both the proximal and distal ends. For example, the gastrocnemius has a typical moment arm of 5 cm at the ankle and 3.5 cm at the knee. Understanding these moment arm lengths and muscle properties is essential for comprehending joint kinetics.

4.3 Biomechanical Models

The fundamental components of a musculoskeletal model need to be discussed. Rigid bodies possess mass, centre of mass, and moment of inertia, collectively known as inertial parameters. Additionally, we require information about the geometry of these rigid bodies. Ideal joints are defined by their position and orientation with respect to a reference frame, accompanied by the constraints on joint angles, known as the kinematic constraints. Muscles are treated as actuators and are defined by their lines of action and specific properties. Kinematic variables are crucial for describing movement and are distinct from the forces responsible for causing that movement. These variables are linear and angular displacements, velocities, and accelerations.

In order to comprehensively model and analyse musculoskeletal systems, we need precise displacement data for key kinematic variables, including the centre of gravity of body segments, joint centres of rotation, and key anatomical landmarks that define the extremities of limb segments. These displacement data are indispensable for effective musculoskeletal modelling and analysis and are typically represented with regard to a spatial reference system, which is either relative or absolute. Kinetics refers to forces and moments that cause movements. These forces or moments can include both internal and external forces. The internal forces or moments arise from the muscles and ligaments. Whereas the external force is the ground reaction force, which arises due to the contact of foot on the ground, or from external loads carried by a subject.

According to Newton's second law, the force along the x-, y-, and z-directions produces an acceleration along respective directions on the bone segment having mass, m. While calculating moments, parameters such as the moment of inertia and angular acceleration are required. The moment is mathematically expressed as the product of the moment of inertia and angular acceleration. In two-dimension, the dynamic equilibrium equations are written as follows:

$$\sum F_x = m(a_{\text{CG}})_x;\ \sum F_y = m(a_{\text{CG}})_y;$$
$$\sum M_{\text{CG}} = I_{\text{CG}}\alpha$$

Here F_x and F_y are the external forces acting on the body along x- and y-directions, respectively. The M_{CG} denotes the moment about the centre of gravity. The m, a_{CG}, and α represents the mass, linear acceleration, and angular acceleration of the body, respectively.

Sample Problem 4.1

Now, let us look into a sample problem to demonstrate the effect of inertial forces on joint moments during knee flexion. Figure 4.7 presents a simplified lower extremity. We need to find out the magnitude and direction of the moment $\boldsymbol{M_K}$ acting at the knee joint. The knee muscles generate a moment that produce a tangential acceleration, $a_t = 20\ \text{m/s}^2$ at the ankle. The mass of the lower leg (m) is given as 4 kg and the moment of inertia I about the centre of gravity, I_{CG} is given as 0.08 kg m^2. We will tackle this numerical problem by first calculating the moment of inertia about the knee joint centre. Although the moment of inertia about the centre of gravity of the segment is given, we need to find out the moment of inertia about the knee joint centre using the parallel axis theorem as given below:

Moment of inertia about the knee joint centre (O),

$$\begin{aligned} I_{\text{O}} &= I_{\text{CG}} + md^2 = 0.08 + 4 \times (0.20)^2 \\ &= 0.24\,\text{kg}\,\text{m}^2 \end{aligned}$$

Angular acceleration, $\alpha = \frac{a_t}{r} = \frac{20}{0.5} = 40\,\text{rad s}^{-2}$.

The moment balance equation about the knee centre is expressed as follows:

$$\sum M_{\text{O}} = I_{\text{O}}\alpha$$

The summation of the external moments about the knee joint centre is equal to the product of the moment of inertia about the point O and the angular acceleration. The negative sign in the angular acceleration indicates a clockwise direction.

$$\begin{aligned} &4 \times (9.81) \times (0.2) - M_K = 0.24 \times (-40) \\ &4 \times (9.81) \times (0.2) - M_K = -9.6 \\ &7.85 - M_K = -9.6 \\ &M_K = 7.85 + 9.6 = 17.45\,\text{Nm} \end{aligned}$$

The positive sign indicates a counter-clockwise moment. When the lower leg is undergoing knee flexion, the total moment is more than double the moment produced

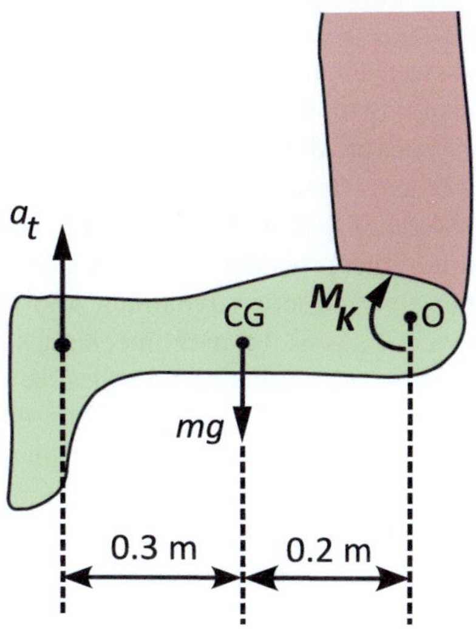

Fig. 4.7 A simplified diagram of the lower extremity showing joint moment and acceleration during knee flexion

by the segment mass. This means that the knee flexor muscles need to generate greater moments to accelerate the lower leg and hold the lower leg in position. This biomechanical insight highlights the activity of the knee flexor muscles, during knee flexion.

4.4 Inverse Dynamics and Forward Dynamics

Inverse dynamics and forward dynamics are two key concepts in biomechanics, particularly in the analysis of musculoskeletal systems. The following are brief descriptions of both methods.

Inverse dynamics: Inverse dynamics is a method of biomechanical analysis used to determine the forces and moments acting on the human body, particularly at the joints, based on measured kinematic data (position, velocity, and acceleration) and anthropometric information. In simpler terms, it helps to answer the question: "What forces and moments (torques) are required to produce an observed motion?"

Inverse dynamics typically starts with data collected from the motion capture systems, which provide information regarding the movement of body segments and joint angles. Using the measured kinematic data and knowledge of segment masses, lengths, and other relevant parameters, inverse dynamics calculations solve for the forces and moments (torques) acting on the joints. This includes estimating the joint reaction forces, muscle forces, and other internal loads. Inverse dynamics is widely used in biomechanics research, sports science, and clinical applications to understand the forces and moments (torques) involved during various activities, such as walking, running, and other activities.

Forward dynamics: Forward dynamics, on the other hand, is a biomechanical modelling approach that focuses on predicting the motion of the human body based on known inputs, such as muscle activations or joint torques. In other words, it answers the question: "Given certain forces and moments (torques), what will be the resulting motion?"

Forward dynamics typically starts with inputs like muscle activation patterns, joint moments (torques), and external forces. Using mathematical models of the musculoskeletal system, this method calculates the movement of the body segments over time. This procedure involves solving equations of motion considering factors like muscle contraction dynamics, joint mechanics, and segmental interactions. Forward dynamics is useful in biomechanics for designing simulations, understanding motor control, and predicting how the changes in muscle activations or external loads affect the movement. It is often used in robotics, rehabilitation, and sports science to optimise movement strategies or design assistive devices. In a summary, inverse dynamics helps to determine the forces and moments required to produce observed movements, while forward dynamics enables us to predict how the body will move when certain forces or moments (torques) are applied.

The role of a 3D motion capture system is crucial for understanding and analysing human movement. A 3D motion capture system consists of multiple cameras, and reflective markers. These markers are used to track the trajectories of body segments during movements. By recording the positions of these markers over time, the system provides kinematic data, including angular displacement, which describes the movement of the joint in three-dimensional space. Finite difference method applied to angular displacement data is useful for the calculation of angular velocity and angular acceleration. These derivatives provide critical information about the dynamic aspects of movement, such as the speed and direction of joint rotations. In addition to the kinematic data, a 3D motion capture system is often complemented by the force plates. These force plates, integrated into the floor, are used to measure the ground reaction forces acting on the feet during motion. Kinematic data from the motion capture system, along with anthropometric information, is fed into a musculoskeletal model, which represents the body's anatomy. These models help to simulate how muscles, joints, and bones interact during movements. The use of musculoskeletal model in joint force estimation will be discussed later in Sect. 4.7.

4.5 Link Segment Analysis: Joint Forces and Moments

In this section, we discuss the use of link segment analysis to calculate net joint forces and moments during a movement. The link segment model, an essential mathematical representation, plays a pivotal role in this process. Imagine an anatomical segment of the body, like the leg, which we can represent using rigid elements. This representation forms the basis of the link segment model. The following are the assumptions of the link segment model.

4.5.1 Assumptions of the Link Segment Model

1. **Fixed mass distribution**: Each segment within the model has a fixed point mass situated at its centre of gravity.
2. **Fixed centre of gravity**: The position of the centre of gravity for each segment remains fixed with respect to the local coordinate frame during movement.
3. **Joint types**: Joints connecting the segments are typically considered hinge joints or ball-and-socket joints.
4. **Constant moment of inertia**: The mass moment of inertia for each segment about its Centre of Mass is assumed to be constant during the movement.
5. **Constant segment length**: The length of each segment remains constant during movement, signifying that the distance between hinge or ball-and-socket joints remains fixed.

4.5.2 Bone-on-Bone Contact Force

The joint bone-on-bone force is the actual force acting across the articulating surfaces, since it accounts for the effect of muscle activity and the action of the ligaments. The articulating surfaces are pulled together by the actively contracting muscles, resulting in a compressive and shear force at the joint. In the simplest case possible, the bone-on-bone force equals the sum of the compressive force (due to the muscle and ligament) and the joint reaction force. The difference between the joint bone-on-bone force and the joint reaction force is explained with the help of two cases (Fig. 4.8). In both cases, the lower segment weighs 90 N (Fig. 4.8a). Here, the joint reaction force is equal to the weight of the lower segment, which is 90 N. In the first case, the lower segment weight of 90 N is hanging passively (Fig. 4.8a). This weight is balanced by the ligaments and muscle forces. In this case, the two muscles are not activated but assisted by ligaments pulling upwards by an equal and opposite force of 45 N each. In the second case, since two muscles are activated, the total upward force is 150 N. The bone-on-bone force is 60 N due to the compression, to satisfy the equilibrium condition. This means that a force of 60 N exists across the articulating surface when the muscles are activated (Fig. 4.8b).

In order to calculate intersegmental forces and moments, we utilise a systematic approach. First, we isolate each segment by removing it from the link segment model, similar to which we analyse a frame problem in engineering mechanics. This process entails drawing a free-body diagram for each isolated segment, illustrating the reaction forces and moments at the joints. Intersegmental forces and moments are computed using the equations of motion and an inverse dynamics approach. This meticulous analysis allows us to determine the net joint forces and moments at each

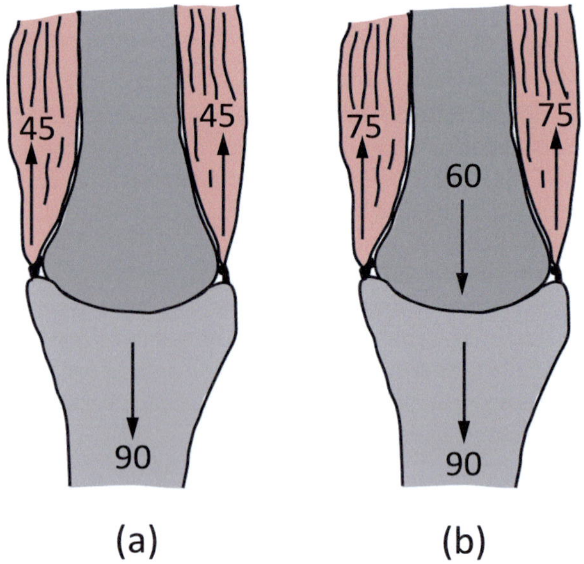

Fig. 4.8 Bone-on-bone contact force; **a** case 1, force = 0, **b** case 2, force = 60 N (compressive)

joint. When we perform this process for each individual segment and consider the collective free body diagrams and reaction forces for all segments, we can effectively reassemble the segments to obtain the complete link segment model. This model encapsulates the interplay of forces, moments, and kinematics within the system. In the subsequent sections, we will discuss the practical application of these principles, exploring how link segment analysis can provide valuable insights into human biomechanics.

Sample Problem 4.2

An individual is executing mid-stance instance of gait cycle on top of a force platform, as shown in Fig. 4.9. The dimensions in Fig. 4.9 are in centimetres (cm). The objective is to determine the joint reaction force and the net muscle moment acting around the ankle joint. Assume that the horizontal component of the ground reaction force during the mid-stance is zero. The mass of the subject and the foot alone are equal to 65 kg and 0.9 kg, respectively. Here point A is the ankle joint centre and point C is the centre of mass of the foot segment.

In this problem, the vertical ground reaction force (R_{y1}) must balance the body weight in order to satisfy the force equilibrium in the y-direction (Fig. 4.10). Therefore, R_{y1} is equal to the body weight and acts upward, counteracting the gravitational force. Using the subject's mass ($m = 65$ kg) and the acceleration due to gravity ($g \approx 9.81$ m s^{-2}), we get R_{y1} to be approximately 637.65 N. Considering force equilibrium along the x-direction,

$$\sum F_x = 0$$

$$R_{x2} + R_{x1} = 0$$

Here, R_{x1} and R_{x2} are the horizontal components of ground reaction force and ankle joint reaction force, respectively. Assuming the R_{x1} to be zero for simplicity, we can write

$$R_{x2} = -R_{x1} = 0$$

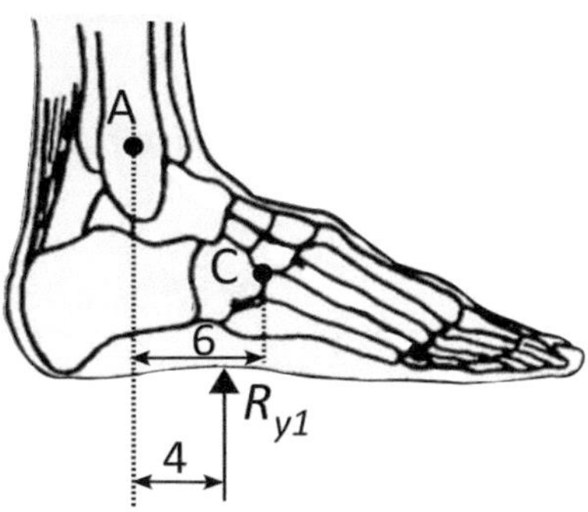

Fig. 4.9 A foot at mid-stance phase showing the position of centre of mass (C), ankle joint centre (A) and the verticle ground reaction force (R_{y1})

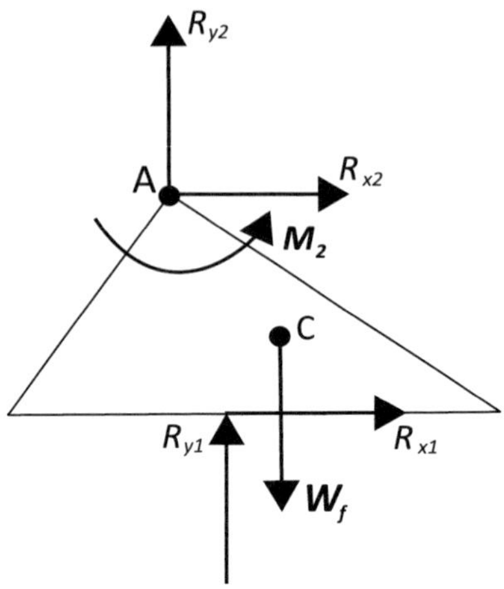

Fig. 4.10 Forces acting on the foot segment at mid-stance phase

For calculating the reaction force along the vertical direction (R_{y2}), we consider the force equilibrium along the y-direction.

$$\sum F_Y = 0$$
$$R_{y2} + R_{y1} - W_f = 0$$

Here, W_f and R_{y2} are the weight of the foot segment and vertical component of ground reaction force, respectively. The $W_f = m_f g$, where m_f denotes the mass of the foot segment.

$$R_{y2} = -R_{y1} + m_f g = -637.50 + 9.81 \times 0.9$$
$$R_{y2} = -628.82 \text{ N}$$

Considering moment equilibrium about foot centre of mass C,

$$\sum M_C = 0$$
$$M_2 - R_{y1} \times 0.02 - R_{y2} \times 0.06 = 0$$

Here, M_2 denotes the moment about ankle joint centre A.

$$M_2 = 637.65 \times 0.02 + (-628.82 \times 0.06)$$
$$M_2 = -24.98 \text{ Nm}$$

The joint reaction force at the ankle joint is equal to 628.82 N, in the downward direction. Whereas, the moment about the ankle joint is equal to 24.98 Nm. The negative sign in the moment signifies that the muscle moment acting on the foot at the ankle joint is in the clockwise direction.

Sample Problem 4.3

In this numerical problem, using the kinematic data during the swing phase of the foot, we need to calculate the muscle moment and the reaction forces at the ankle joint (Fig. 4.11). Let the mass of the subject (m) and foot segment length (l) of the subject be equal to 65 kg and 0.20 m, respectively. Based on the anthropometric relations, the foot segment mass (m_f) is obtained, as a fraction of the body weight, as 0.94 kg.

The angular acceleration about the Centre of Mass (C) of the foot, α is equal to 21.69 rad/s^2. The linear accelerations, along x and y directions denoted as, a_x and a_y are equal to 9.07 and −6.62 m/s^2, respectively. The inertial characteristics of the foot are calculated based on anthropometry relations. The radius of gyration (k) is related to the length of the limb segment (l) as follows:

$$k = 0.475 \times l = 0.475 \times 0.20 = 0.095 \text{ m}$$

The moment of inertia (I_C) is calculated as a product of the mass and the square of the radius of gyration as given below.

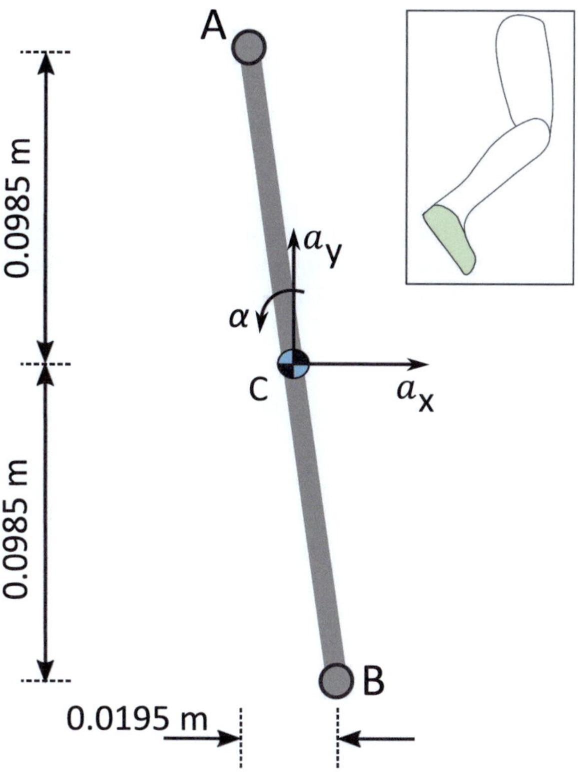

Fig. 4.11 Simplified representation of a foot segment during swing phase

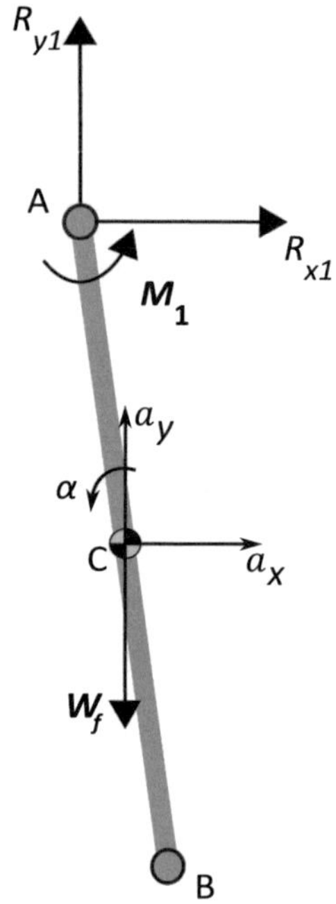

Fig. 4.12 Forces acting on a foot segment during swing phase

$$I_C = m_f k^2 = 0.94 \times 0.095^2 = 0.0085\ \text{kg m}^2$$

Since this is a dynamic problem, we will be considering dynamic equilibrium conditions. The ground reaction force is equal to zero, since the foot is not in contact with the ground during the swing phase. The forces acting on the foot segment are presented in Fig. 4.12. There are three unknowns, such as reaction forces (R_{x1} and R_{y1}) and moment (M_1). Let us consider dynamic equilibrium along the x-direction.

$$\sum F_x = m_f a_x$$
$$R_{x1} = m_f a_x = 0.94 \times 9.07$$
$$R_{x1} = 8.53\ \text{N}$$

Considering dynamic equilibrium along the y-direction:

$$\sum F_y = m a_y$$
$$R_{y1} - W_f = m \times (-6.62)$$

Here W_f denotes the weight of the foot segment.

$$R_{y1} - m_f g = m \times (-6.62)$$

$$R_{y1} = 0.94 \times 9.81 - 0.94 \times 6.62 = 3\text{ N}$$

Using the moment equilibrium about the Centre of Mass (C) of the foot, we can find out the ankle joint moment (M_1):

$$\sum M_C = I_O$$

$$M_1 - R_{x1} \times 0.0985 - R_{y1} \times 0.0195 = 0.0085 \times 21.69$$

$$M_1 = 0.0085 \times 21.69 + 8.53 \times 0.0985 + 3 \times 0.0195$$

$$M_1 = 1.08\,\text{Nm}$$

Let us now discuss some observations on the values of joint reaction forces and moments. The horizontal reaction force at the ankle joint is responsible for the horizontal acceleration calculated for the foot segment (Fig. 4.12). During the swing phase, following the toe-off instant, the foot is moving upwards against gravity. This means the foot is decelerating, and as a result, the vertical reaction force at the ankle joint is actually less than the static gravitational force acting on the foot. Another important observation is related to the ankle joint moment. This moment was found to be positive, implying that it acts in a counter-clockwise direction. Moreover, it indicates a net dorsiflexor activity which might be due to the action of the tibialis anterior muscle, resulting in dorsiflexion of the ankle joint.

Sample Problem 4.4

In this numerical problem, we consider the same time instant as that of sample problem 4.3 (Fig. 4.13). Here, we extend the problem to calculate the muscle moments ($\boldsymbol{M_2}$) acting at the knee joint centre (D). The limb segment is given as 43.5 cm long. The angular acceleration (α) is equal to 36.9 rad/s^2. The linear accelerations, a_x and a_y, are equal to -0.03 m/s^2 and -4.24 m/s^2, respectively. Based on the anthropometric data, the mass of the limb segment is calculated as 3.72 kg, considering a body mass of 65 kg. The radius of gyration, based on the length of the limb segment, equals to 0.131 m.

As mentioned earlier, if we know the distal joint moments and forces, the net moment and forces at the proximal end of the segment can be calculated. Since we have already calculated the ankle joint moments and forces earlier, we can now estimate the net joint forces and moments at the knee joint. So, this example demonstrates the estimation of the net forces and moments at the knee joint based on the prior knowledge of the ankle joint moments and forces. According to Newton's third law, the directions of the reaction forces and moments at the distal end are opposite

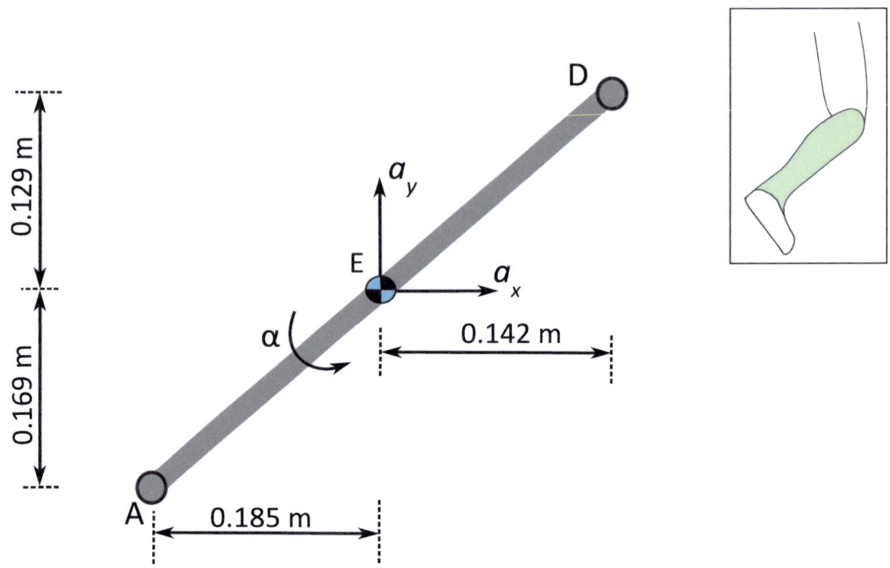

Fig. 4.13 Simplified representation of leg segment during swing phase

as compared to the previous problem. The three unknowns are joint reaction forces (R_{x2}, R_{y2}) and moment ($\boldsymbol{M_2}$).

Considering the dynamic equilibrium along the x-direction (Fig. 4.14):

$$\sum F_x = ma_x$$
$$R_{x2} - R_{x1} = ma_x$$
$$R_{x2} = 8.53 + 3.72x(-0.03)$$
$$R_{x2} = 8.42 \text{ N}$$

Considering dynamic equilibrium along the y-direction:

$$\sum F_y = ma_y$$
$$R_{y2} - R_{y1} - mg = ma_y$$
$$R_{y2} = R_{y1} + mg + ma_y$$
$$R_{y2} = 3.00 + 3.72 \times 9.81 + 3.72 \times (-4.24)$$
$$R_{y2} = 23.83 \text{ N}$$

Considering moment equilibrium about the centre of mass of the limb segment (E):

$$\sum M_c = I_c\alpha,$$

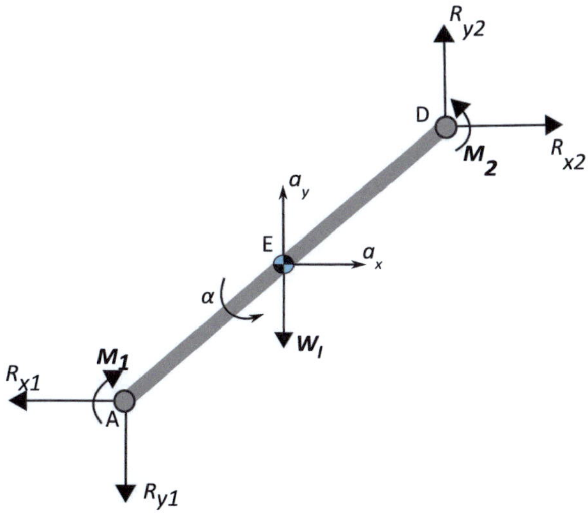

Fig. 4.14 Free body diagram of leg during swing phase

$$
\begin{aligned}
M_2 - M_1 - 0.169R_{x1} + 0.185R_{y1} \\
- 0.129R_{x2} + 0.142R_{y2} = I_c a \\
M_2 = M_1 + 0.169R_{x1} - 0.185R_{y1} \\
+ 0.129R_{x2} - 0.142R_{y2} + I_c\alpha \\
M_2 = 1.08 + (0.169 \times 8.53) \\
- (0.185 \times 3.00) + (0.129 \times 8.42) \\
- (0.142 \times 23.83) + (0.0638 \times 36.9) \\
M_2 = 1.08 + 1.44 - 0.56 \\
+ 1.09 - 3.38 + 2.35 \\
M_2 = 2.02\,\text{Nm}
\end{aligned}
$$

The joint moment ($\boldsymbol{M_2}$) is calculated to be positive, indicating a counter-clockwise moment around the knee joint. This means that $\boldsymbol{M_2}$ is trying to extend the knee joint. In this particular instance, the quadriceps muscles are actively involved in rapidly extending the swinging leg, which aligns with the counter-clockwise nature of $\boldsymbol{M_2}$. It is important to note that the angular acceleration of the leg is the combined effect of two reaction forces (R_{x2} and R_{y2}) and joint moment ($\boldsymbol{M_2}$).

4.6 Static Optimisation

Link-segment models assume that each joint functions as a ball and socket joint or hinge joint, in which the muscles generate moment (torque) about the joint centre. Accordingly, the estimated reaction force is equal to the bone-on-bone forces.

However, it is important to note that our muscles differ from motor torque. Instead, they function as linear motors which are associated with compressive and shear forces across the joint surfaces. Consequently, it becomes necessary to incorporate these muscle-induced forces into the free body diagram and recalculate the joint forces. For ease of understanding, the forces contributed by ligaments and anatomical constraints, which exist during the joint movements, have not been considered.

There are around fifteen major muscles that contribute to the joint moments across various joints in the lower limb. The line of action of each of these muscles is different and changes continuously with time. Some muscles like biceps femoris may contribute towards moments around the hip joint as well as the knee joint. In order to obtain accurate estimates of individual muscle forces, it is crucial to incorporate a detailed anatomical or kinematic model that includes information about the lines of action for each muscle relative to the joint centre. The extensor moment (M_j) at any joint and time instant (t) is mathematically expressed as the net algebraic sum of cross products between all force vectors and their respective moment arm vectors.

$$M_j(t) = \sum_{J=1}^{N_e} F_{ei}(t) \times d_{ei}(t) - \sum_{i-1}^{N_f} F_{fi}(t) \times d_{fi}(t)$$

In the above equation, N_e represents the number of extensor muscles, while N_f represents the number of flexor muscles. Each extensor muscle force at time t is denoted as $F_{ei}(t)$, and its corresponding moment arm as $d_{ei}(t)$. Similarly, for flexor muscles, the force in the ith muscle at time t is represented as $F_{fi}(t)$, and its moment arm as $d_{fi}(t)$.

Accurately measuring individual muscle forces is a challenging task, and there is no practical way to directly measure muscular forces for an entire musculoskeletal system or even specific parts like the upper or lower limbs. Alternatively, mathematical modelling offers the possibility to predict these individual muscle forces. Despite the dynamic nature of the musculoskeletal system, static optimisation has been used to obtain satisfactory results depending on the specific evaluation goals and purposes of the study. In this approach, the dynamic forces are essentially treated as static forces at every time instant.

Now, let us look into the indeterminacy in muscle force estimation. The mathematical modelling of a musculoskeletal system results in a set of equations that describe force or moment equilibrium. These equations take the form of $Ax = b$, where x represents the unknown muscular forces. In other words, this set of equilibrium equations, along with the constraints on muscle forces, helps to define the musculoskeletal system. In this equation, A is the coefficient matrix for the variable x, and b is a constant vector. When there are fewer equations than the unknowns in a system of linear or polynomial equations, it is considered to be undetermined. In order to solve an undetermined system, optimisation theory comes into play. The goal is to find an optimal solution, denoted as x, that satisfies the equations with a minimum cost. One can choose a suitable cost function $C(x)$ to guide this optimisation process. Since muscles can only generate positive forces, constraints are essential in the mathematical model.

We will now focus our discussion on the mathematical modelling involved in muscle force estimation, in particular the static optimisation technique based on the Lagrange multiplier method. An overview of the muscle force estimation using inverse dynamics and static optimisation is presented in Fig. 4.15. The experimental joint kinematics data is used to calculate generalized joint moments and forces, using the inverse dynamics method. Joint moments or torques are the net combined effect of individual muscle forces spanning the joint. The contribution of each muscle can now be calculated using an optimisation scheme, as discussed in the following. An initial guess of muscle forces is made, based on which the joint equilibrium is verified. However, an initial guess is not required for the Lagrangian multiplier optimisation method. Thereafter, changes in joint forces are made, if necessary. The procedure is repeated until an optimal muscle force distribution is achieved.

The optimisation problem is typically formulated as follows: minimise $C(x)$ subject to the equality constraints $h(x) = 0$, and the inequality constraints $g(x) \leq 0$. Here, $C(x)$ represents the cost function associated with the unknowns, $h(x)$ represents a set of equality constraints, and $g(x)$ represents a set of inequality constraints. This framework is useful to model and analyse the complex interactions of muscles and forces within a musculoskeletal system. The Lagrange multiplier method, used for solving the static optimisation problem, is a powerful numerical approach for identifying the local maxima and minima of a function, while considering constraints. The Lagrange multipliers transform the problem of constrained minimisation of $C(x)$ into an unconstrained minimisation problem. This transformation can be expressed in the form of the Lagrangian function, L.

$$L(x \cdot \mu) = C(x) + \mu(Ax - b)$$

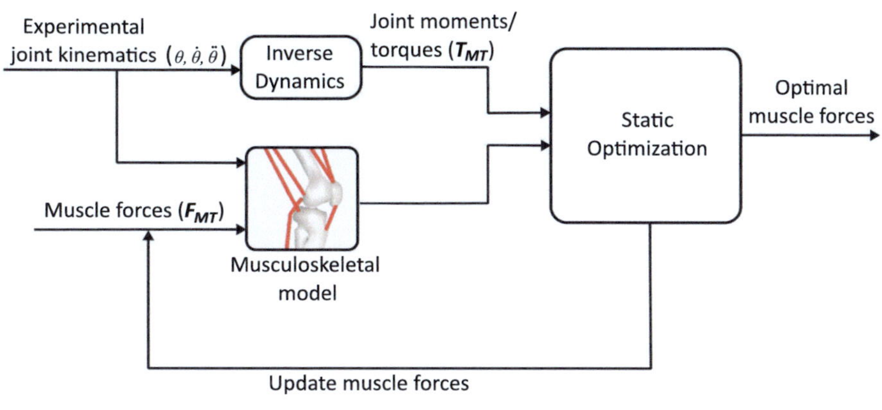

Fig. 4.15 Method for estimation of optimal muscle forces. Adapted from Erdemir et al. (2007)

Here, $Ax - b$ represents the residual of the equilibrium equation, and μ is a set of Lagrange multipliers. This method allows us to effectively optimise complex systems subject to various constraints.

Sample Problem 4.5

A sample problem may be considered to comprehend the optimisation scheme. Consider a link segment of length '*2l*' as indicated in Fig. 4.16 with three muscle forces, $\boldsymbol{f}_1$, $\boldsymbol{f}_2$, and $\boldsymbol{f}_3$. We need to find out the expressions for the muscle forces in terms of known quantities with the help of the Lagrange multiplier method. The known quantities are the area of cross section of each muscle, moment arm of each muscle about joint centre, *E*, the joint reaction forces, weight of the segment denoted as A_i, d_i, $\boldsymbol{R}_1$, and $\boldsymbol{Mg}$, respectively. The two endpoints of the links are *D* and *E*. The angle between the axis of the link and the vertical is denoted by θ. The distance of the centre of gravity (CG) from joint centre *E* is *l*, along the link axis. The moment equilibrium about *E* is expressed in the following equation:

$$M_g l \sin(\theta) - 2R_1 l \sin(\theta) = -f_1 d_1 + f_2 d_2 + f_3 d_3$$

The above equation is in the form $Ax = b$; where $Ax = -f_1 d_1 + f_2 d_2 + f_3 d_3$ and $b = Mgl\sin(\theta) - 2R_1 l \sin(\theta)$. The unknown muscle forces are denoted by x. The cost function for the segment model can be expressed as a function of muscle stresses as follows:

$$C = \left(\sigma_1^2 + \sigma_2^2 + \sigma_3^3\right)/2$$

The Lagrangian function (L) is expressed as

$$L(x, \mu) = C(x) + \mu(Ax - b)$$

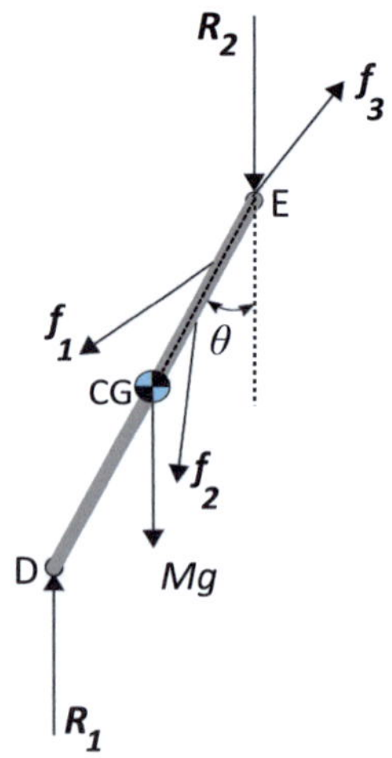

Fig. 4.16 Link segment model of length *2l*

$$L = \frac{\sigma_1^2 + \sigma_2^2 + \sigma_3^3}{2} + \mu(-A_1\sigma_1 d_1 + A_2\sigma_2 d_2 + A_3\sigma_3 d_3 - Mgl\sin(\theta) + 2R_1 l\sin(\theta)).$$

The muscle forces are replaced by the products of muscle stresses and the physiological cross-sectional area. The constrained minimal solution is obtained by equating the differentials of the L to zero.

Taking the partial derivative of L with respect to muscle stress, the following equations are obtained:

$$\frac{\partial L}{\partial \sigma_1} = \sigma_1 - \mu A_1 d_1 = 0$$
$$\frac{\partial L}{\partial \sigma_2} = \sigma_2 + \mu A_2 d_2 = 0$$
$$\frac{\partial L}{\partial \sigma_3} = \sigma_3 + \mu A_3 d_3 = 0$$

Using the above equations, the muscle stresses σ_2 and σ_3 is expressed in terms of σ_1 as follows:

$$\sigma_2 = -\frac{d_2 A_2}{d_1 A_1}\sigma_1$$
$$\sigma_3 = -\frac{d_3 A_3}{d_1 A_1}\sigma_1$$

Differentiating L with respect to μ results in the following:

$$\frac{\partial L}{\partial \mu} = -\sigma_1 A_1 d_1 + \sigma_2 A_2 d_2 + \sigma_3 A_3 d_3 - Mgl\sin(\theta) + 2R_1 \sin(\theta) = 0$$

Upon substituting σ_2, and σ_3 in terms of σ_1 in the above equation, we get

$$-\sigma_1 A_1 d_1 - \frac{d_2 A_2}{d_1 A_1}\sigma_1 A_2 d_2 - \frac{d_3 A_3}{d_1 A_1}\sigma_1 A_3 d_3 - Mgl\sin(\theta) + 2R_1 l \sin(\theta) = 0$$

Rearranging the above leads to

$$\sigma_1 = \frac{-MglA_1 d_1 \sin(\theta) + 2R_1 l A_1 d_1 \sin(\theta)}{A_1^2 d_1^2 + A_2^2 d_2^2 + A_3^2 d_3^2}$$

The muscle force f_1 is obtained by multiplying the stress with the area of cross-section A_1.

$$f_1 = \frac{-MglA_1^2 d_1 \sin(\theta) + 2R_1 l A_1^2 d_1 \sin(\theta)}{A_1^2 d_1^2 + A_2^2 d_2^2 + A_3^2 d_3^2}$$

Similarly, the unknown muscle stresses and forces f_2 and f_3 are found out to be

$$f_2 = \frac{MglA_2^2 d_2 \sin(\theta) - 2R_1 l A_2^2 d_2 \sin(\theta)}{A_1^2 d_1^2 + A_2^2 d_2^2 + A_3^2 d_3^2}$$

$$f_3 = \frac{MglA_3^2 d_3 \sin(\theta) - 2R_1 l A_3^2 d_3 \sin(\theta)}{A_1^2 d_1^2 + A_2^2 d_2^2 + A_3^2 d_3^2}$$

In order to ensure positive muscle forces, we need to introduce constraints $f_i \geq 0$, which means the individual muscle forces should always be greater than or equal to zero. In order to incorporate these constraints into the optimisation problem, we include an additional set of Lagrange multipliers (μ_i). The additional set of Lagrange multipliers along with the slack variable (g_i) are used to handle inequalities within the Lagrangian function, allowing us to incorporate constraints effectively into the optimisation problem.

$$L(x \cdot \mu) = C(x) + \mu(Ax - b) + \mu_i\left(g_i^2 - x_i\right)$$

Cost functions: The cost functions are often referred to as performance criteria. These performance criteria are designed to represent the objective of minimising the activation of the muscular system during specific activities. Over the years, various optimisation approaches have been developed to address the challenge of solving redundant muscle force systems. These approaches primarily differ in how the cost function is formulated and in the choice of solution methods. Several cost functions have been suggested to predict the muscle forces, including the principles of minimal muscle force (Seireg and Arvikar 1973, 1975; Pedotti et al. 1978), minimal muscle stress (Crowninshield et al. 1978; Crowninshield and Brand 1981; Dul et al. 1984a, b), and minimal energy expenditure or consumption (Hardt 1978; Davy and Audu 1987).

4.7 Three-Dimensional Motion Analysis

The output data from any marker-based motion capture system is a file consisting of marker coordinates corresponding to a time instant. These coordinates are defined according to a reference system, called global coordinate system (GCS), that is fixed in the laboratory. The purpose of this section is to discuss the essential steps required to transform the marker coordinate data into the anatomical axes of the body segments. This is required for kinematic analysis since the marker data defined with respect to the anatomical coordinate system (ACS) of the segment serves as the input data of this analysis.

The ACS pertinent to a limb segment with its centre of mass (COM) and its axes x–y–z is shown in Fig. 4.17. The GCS is denoted by axes X–Y–Z, which remain fixed, irrespective of camera setups. The marker axis system, denoted by x_m–y_m–z_m, for each segment, may change across different laboratories or even within the same laboratory across different experiments. In order to ensure accurate 3D analysis, a minimum of three independent markers per body segment is necessary, with no shared markers between adjacent segments. Additionally, markers on each segment should not align collinearly but instead, form a plane in 3D space. The tracking markers m_{T1}, m_{T2}, and m_{T3} define the tracking marker plane, as shown in Fig. 4.17.

The relation between the marker coordinate system (MCS), x_m–y_m–z_m, and the ACS, x–y–z is obtained from the anatomical calibration process. In this process, the subject assumes the standard anatomical position in which the subject stands upright, facing forward with each arm hanging on either side of the body, and palms facing anterior. In this position, the ACS can be defined by placing extra calibration markers temporarily on the well-known anatomical points (skeletal landmarks) of the segment. For example, in the case of the right leg segment, three tracking markers could be placed at the midpoint on the anterior surface of the tibia (m_{T1}), on the lateral malleolus (m_{T2}), and on the head of the fibula (m_{T3}). During calibration, medial malleolus and the medial epicondyle of the tibia can be considered the anatomical landmarks. The markers, m_{C1} and m_{C2}, corresponded to the medial malleolus and the medial epicondyle of the tibia, respectively. The coordinates of the three tracking and the two calibration markers are recorded and averaged over a calibration time of ~1 s, while the subject executes the anatomical position. The y-axis of the ACS of the leg segment can be defined as the line joining the midpoint between the medial (m_{C1}) and lateral malleolii (m_{T2}) and the midpoint between the head of the fibula (m_{T3}) and the medial epicondyle (m_{C2}) of the tibia. These midpoints are the ankle

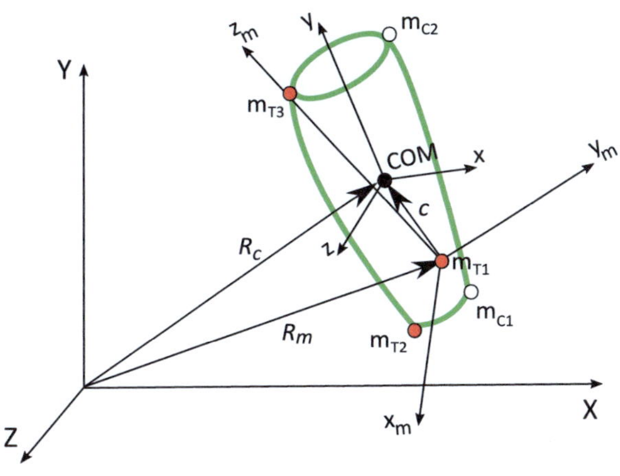

Fig. 4.17 Various coordinate systems involved in the transformation of marker data from global coordinate system (X–Y–Z) to anatomical coordinate system (x–y–z)

and knee joints, respectively. The leg x-axis is the line drawn perpendicular to the plane defined by the anatomical y-axis and the line joined by m_{C1} and m_{T2} (Fig. 4.17). According to the right-hand thumb rule, the direction of the leg z-axis is drawn as a line perpendicular to the anatomical x–y plane.

The ACS can now be defined, relative to the tracking markers (m_{T1}, m_{T2}, and m_{T3}). The **c** is defined as the vector originating from m, the origin of the tracking marker axis system (also m_{T1}) and terminates at the location of the COM of the leg. Additionally, this vector is used to define the ACS with respect to the MCS. The MCS is assumed to be fixed with respect to the ACS. The calibration markers (m_{C1} and m_{T2}) are removed after this procedure. The procedure of defining ACS differs in the case of patients with cerebral palsy, owing to the difficulty in achieving anatomical position, the readers are referred to Ounpuu et al. (1996).

The transformation of the marker coordinate date from the GCS to ACS requires two successive matrix rotations, GCS to MCS ([G → M]) and MCS to ACS ([M → A]). The [G → M] is a 3 × 3 matrix, which depends on a particular instance of time since the MCS continuously changes relative to the GCS. Whereas, [M → A] is a 3 × 3 matrix, which does not depend on a particular instance of time. This matrix is obtained from the calibration procedure. The combination of these two rotation matrices results in the GCS to ACS matrix ([G → A]). The matrix is used to obtain the angular orientation of the anatomical axes with respect to the GCS. In order to track the coordinates of COM of a limb segment over time, a translational transformation is necessary. The $\boldsymbol{R_c}$ represents the location of COM, expressed as follows:

$$\boldsymbol{R_c} = \boldsymbol{R_m} + \boldsymbol{c}$$

where R_m is the GCS coordinates of the tracking marker, m_{T1}.

Rotation matrices: The rotation matrices facilitate the transformation of vectors from one coordinate system to another. The rotation matrix, involving rotation of θ about x-axis, in 3D is expressed as

$$[R_x] = \begin{bmatrix} 1 & 0 & 0 \\ 0 & \cos\theta_1 & \sin\theta_1 \\ 0 & -\sin\theta_1 & \cos\theta_1 \end{bmatrix}$$

Similarly, the rotations about y and z are expressed as

$$[R_y] = \begin{bmatrix} \cos\theta_2 & 0 & -\sin\theta_2 \\ 0 & 1 & 0 \\ \sin\theta_2 & 0 & \cos\theta_2 \end{bmatrix}$$

$$[R_z] = \begin{bmatrix} \cos\theta_3 & \sin\theta_3 & 0 \\ -\sin\theta_3 & \cos\theta_3 & 0 \\ 0 & 0 & 1 \end{bmatrix}$$

The product, $[R_z][R_y][R_x] = \begin{bmatrix} c_2c_3 & c_1s_3 + s_1s_2c_3 & s_1c_3 - c_1s_2c_3 \\ -c_2s_3 & c_1c_3 - s_1s_2s_3 & s_1c_3 + c_1s_2s_3 \\ s_2 & -s_1c_2 & c_1c_2 \end{bmatrix}$.

Here $c_i = \cos\theta_i$ and $s_i = \sin\theta_i$.

It should be noted that the matrix multiplication is not commutative, the order of transformation is such that transformation about x ($[R_x]$) occurs first and transformation about z ($[R_z]$) occurs last. This sequence of transformation is called as Cardan sequence. Let us now discuss a sample problem for a better understanding of the 3D kinematic and kinetic analysis.

Sample Problem 4.6

In the initial part of this sample problem, we will discuss the procedure for the calculation of various transformation matrices obtained from the calibration and tracking processes. Thereafter, using the rotation angles, we find out the variables required for the kinetics analysis.

Calibration: The calibration procedure is carried out using the position data of markers of a right leg segment (Fig. 4.17), in an anatomical position (Table 4.2). Marker position data defined in the GCS is given in Table 4.2. Following the procedure described earlier, the first step involves the determination of the long axis (y-axis of ACS) of the leg segment, using the following midpoints (obtained by averaging the position data):

$$\begin{aligned}&\text{Midpoint 1, considering } \mathrm{m_{T2}} \text{ and } \mathrm{m_{C1}}\\&\quad = (122.59, 6.51, 43.67)\\&\text{Midpoint 2, considering } \mathrm{m_{T3}} \text{ and } \mathrm{m_{C2}}\\&\quad = (127.94, 48.82, 45.67)\end{aligned}$$

The y-axis of the ACS is defined by the line drawn from Midpoint 1 to Midpoint 2, and expressed as a vector: $5.35i + 42.31j + 1.99k$. The coordinate axes are generally expressed as unit vectors, obtained by dividing the vector by its magnitude (|Midpoint 2 − Midpoint 1|). This unit vector representing the direction of the y-axis is calculated to be: $0.125\hat{i} + 0.991\hat{j} + 0.047\hat{k}$. Midpoint 1 and Midpoint 2 corresponds to the ankle centre and the knee centre, respectively. Using anthropometric data, the position of the COM is calculated as ($0.567 \times$ Midpoint 2 $+ 0.433 \times$ Midpoint 1), which is vector: $125.25i + 30.48j + 44.67k$.

Table 4.2 Marker coordinate data, defined with respect to GCS, during the calibration process

Marker	Location	X (cm)	Y (cm)	Z (cm)
$\mathrm{m_{T1}}$	Midleg	123.10	25.64	55.15
$\mathrm{m_{T2}}$	Lateral malleolus	122.37	6.61	48.76
$\mathrm{m_{T3}}$	Fibular head	128.13	49.11	50.81
$\mathrm{m_{C1}}$	Medial malleolus	122.81	6.41	38.58
$\mathrm{m_{C2}}$	Medial condyle	127.74	48.53	40.52

The next step involves the determination of x, y, and z axes. Let us assume, for instance, that the interim z-axis is the line joining the lateral malleolus to the medial malleolus. This interim z-axis denoted by z_i, is not the actual z-axis of the ACS. However, this interim axis simplifies the procedure of finding the z. The unit vector of the interim axis, z_i is expressed as

$$\begin{aligned} z_i &= \frac{\mathrm{m}_{\mathrm{T2}} - \mathrm{m}_{\mathrm{C1}}}{|\mathrm{m}_{\mathrm{C1}} - \mathrm{m}_{\mathrm{T2}}|} \\ &= \frac{-0.44i + 0.20j + 10.18k}{|0.44i + 0.20j + 10.18k|} \\ &= -0.043\hat{i} + 0.020\hat{j} - 0.999\hat{k} \end{aligned}$$

The x-axis is obtained from the cross-product of the z_i and the y axes, since x-axis is normal to the plane formed by these axes. Therefore, we can write, $x = y \times z_\mathrm{i}$ and is equal to:

$$x = 0.991\hat{i} - 0.127\hat{j} + 0.045\hat{k}$$

It should also be noted that the z_i axis is not necessarily perpendicular to the y-axis. In order to ensure the orthogonality of the anatomical axes, z is calculated as a cross-product $(x \times y)$ and is obtained as

$$z = -0.051\hat{i} - 0.041\hat{j} + 0.998\hat{k}$$

The transformation matrix from anatomical-to-global [A → G], is obtained as $[\{x\}\{y\}\{z\}]$:

$$[\mathrm{A} \rightarrow \mathrm{G}] = \begin{bmatrix} 0.991 & 0.125 & -0.051 \\ -0.127 & 0.991 & -0.041 \\ 0.045 & 0.047 & 0.998 \end{bmatrix}$$

We observe that the principal diagonal values are approximately equal to 1, indicating that the anatomical axes are almost aligned with the global axes. The transposition of the above matrix results in the global-to-anatomical matrix [G → A], which is expressed as

$$[\mathrm{G} \rightarrow \mathrm{A}] = \begin{bmatrix} 0.992 & -0.127 & 0.045 \\ 0.125 & 0.991 & 0.047 \\ -0.051 & -0.041 & 0.998 \end{bmatrix}$$

Determination of the local markers and joint centre vectors: Let us now define the coordinates of the markers and the joints (knee and ankle) relative to the origin of the ACS. The ankle joint vector (in ACS) is obtained by transforming the global ankle joint vector from GCS to ACS. The global ankle joint vector is determined from the position vectors of the ankle joint and the COM (origin of ACS) defined in the GCS.

$$\begin{aligned}&\text{Ankle vector (with respect to the COM)}\\&\quad= \text{position vector of the ankle joint}\\&\qquad- \text{position vector of COM}\\&\text{Ankle vector defined with respect to the COM}\\&\quad= (122.59i + 6.51j + 43.67k)\\&\qquad- (125.25i + 30.48j + 44.67k)\\&\quad= (-2.66i - 23.97j - 1.00k)\end{aligned}$$

This local ankle vector is transformed as shown below:

$$\begin{aligned}&\text{Local ankle vector}\\&\quad= [\mathrm{G} \rightarrow \mathrm{A}]\{\text{global ankle vector}\}\\&\text{Local ankle vector}\\&\quad= \begin{bmatrix} 0.992 & -0.127 & 0.045 \\ 0.125 & 0.991 & 0.047 \\ -0.051 & -0.041 & 0.998 \end{bmatrix}\begin{bmatrix} -2.66 \\ -23.97 \\ -1.00 \end{bmatrix}\\&\text{Local ankle vector}\\&\quad= \begin{bmatrix} 0.37 \\ -24.14 \\ 0.11 \end{bmatrix}\end{aligned}$$

Similar to the above procedure, other vectors corresponding to the knee joint and the markers are determined in the following:

$$\text{Local knee vector} \approx \begin{bmatrix} 0.37 \\ 18.56 \\ 0.11 \end{bmatrix}$$

$$\text{Local } \mathrm{m_{T1}} \text{ vector} \approx \begin{bmatrix} -1.04 \\ -4.58 \\ 10.76 \end{bmatrix};$$

$$\text{Local } \mathrm{m_{T2}} \text{ vector} \approx \begin{bmatrix} 0.37 \\ -23.83 \\ 5.20 \end{bmatrix};$$

$$\text{Local } \mathrm{m_{T3}} \text{ vector} \approx \begin{bmatrix} 0.76 \\ 19.11 \\ 5.22 \end{bmatrix}$$

The vector $\boldsymbol{c}$ in the ACS is equal and opposite to the location position vector corresponding to $\mathrm{m_{T1}}$.

$$\boldsymbol{c} = \begin{bmatrix} 1.04 \\ 4.58 \\ -10.76 \end{bmatrix}$$

This $\boldsymbol{c}$ vector is a constant vector, defined in the ACS, which is useful for the calculation of linear velocity and acceleration.

Determination of the constant marker-to-anatomical matrix: In this step, we calculate the constant marker-to-anatomical matrix. The $\mathrm{m_{T1}}$ is assumed as the origin of the marker plane. The line joining $\mathrm{m_{T1}}$ to $\mathrm{m_{T3}}$ is taken as the z-axis of the MCS, z_m. Additionally, the position vectors are defined with respect to the ACS, so that the [M → A] matrix can be determined, and are written as local position vectors. Now, we need to consider an interim vector denoted by $\boldsymbol{a}$, which is defined by a line joining $\mathrm{m_{T1}}$ to $\mathrm{m_{T2}}$. The y-axis of the MCS, y_m is perpendicular to the plane consisting $\boldsymbol{a}$ and z_m. The x-axis of the MCS, x_m is perpendicular to the plane consisting y_m and z_m. The following calculations are required to determine the [M → A] matrix.

$$\begin{aligned} z_m = {} & \text{local } \mathrm{m_{T3}} \text{ vector} \\ & - \text{local } \mathrm{m_{T1}} \text{ vector} \end{aligned}$$

$$z_m = \begin{bmatrix} 0.76 \\ 19.11 \\ 5.22 \end{bmatrix} - \begin{bmatrix} -1.04 \\ -4.58 \\ 10.76 \end{bmatrix} = \begin{bmatrix} 1.80 \\ 23.69 \\ -5.54 \end{bmatrix}$$

$$\begin{aligned} \boldsymbol{a} = {} & \text{local position vector of } \mathrm{m_{T2}} \\ & - \text{local position vector of } \mathrm{m_{T1}} \end{aligned}$$

$$\boldsymbol{a} = \begin{bmatrix} 0.37 \\ -23.83 \\ 5.20 \end{bmatrix} - \begin{bmatrix} -1.04 \\ -4.58 \\ 10.76 \end{bmatrix} = \begin{bmatrix} 1.41 \\ -19.25 \\ -5.57 \end{bmatrix}$$

$$y_m = (z_m \times \boldsymbol{a}) = \begin{bmatrix} 1.80 \\ 23.69 \\ -5.54 \end{bmatrix} \times \begin{bmatrix} 1.41 \\ -19.25 \\ -5.57 \end{bmatrix} = \begin{bmatrix} -238.49 \\ 2.20 \\ -68.03 \end{bmatrix}$$

$$x_m = (y_m \times z_m)$$
$$= \begin{bmatrix} -238.49 \\ 2.20 \\ -68.03 \end{bmatrix} \times \begin{bmatrix} 1.80 \\ 23.69 \\ -5.54 \end{bmatrix} = \begin{bmatrix} 1599.10 \\ -1443.60 \\ -5652.90 \end{bmatrix}$$

The constant [M → A] *matrix*: Since the skin movement is neglected, it is assumed that the MCS remains fixed with respect to the ACS. Upon normalising the above local vectors, the normalised [M → A] is expressed as $[\{x_m\}\{y_m\}\{z_m\}]$:

$$[\mathrm{M} \rightarrow \mathrm{A}] = \begin{bmatrix} 0.264 & -0.962 & 0.074 \\ -0.239 & 0.009 & 0.971 \\ -0.934 & -0.274 & -0.227 \end{bmatrix}$$

Tracking process: The [G → M] matrix is determined using the following procedure. As an example, we choose the time frame 10 to determine the position vectors of the marker axes, relative to the GCS, and thereby the [G → M] matrix (Table 4.3).

Following a similar procedure of determining [M → A] matrix and thereafter using the marker position data corresponding to the time frame 10, the position vectors are determined.

$$z'_m = \text{global position vector of } \mathrm{m_{T3}} - \text{global position vector of} \mathrm{m_{T1}}$$

$$z'_m = \begin{bmatrix} -15.13 \\ 48.57 \\ 49.17 \end{bmatrix} - \begin{bmatrix} -11.11 \\ 24.96 \\ 52.92 \end{bmatrix} = \begin{bmatrix} -4.02 \\ 23.61 \\ -3.75 \end{bmatrix}$$

$$\boldsymbol{a}' = \text{global position vector of } \mathrm{m_{T2}} - \text{global position vector of } \mathrm{m_{T1}}$$

Table 4.3 Marker coordinate data, defined with respect to GCS, during the tracking process

Frame	$\mathrm{m_{T1}}$			$\mathrm{m_{T2}}$			$\mathrm{m_{T3}}$		
	X	*Y*	*Z*	*X*	*Y*	*Z*	*X*	*Y*	*Z*
8	−11.53	25.01	53.03	−6.87	6.29	45.91	−15.72	48.57	49.20
9	−11.31	24.98	52.97	−6.83	6.24	45.84	−15.42	48.57	49.18
10	−11.11	24.96	52.92	−6.78	6.24	45.78	−15.13	48.57	49.17
11	−10.91	24.95	52.86	−6.70	6.25	45.72	−14.85	48.57	49.16
12	−10.72	24.95	52.81	−6.60	6.30	45.69	−14.57	48.58	49.14

$$a' = \begin{bmatrix} -6.78 \\ 6.24 \\ 45.78 \end{bmatrix} - \begin{bmatrix} -11.11 \\ 24.96 \\ 52.92 \end{bmatrix}$$

$$= \begin{bmatrix} 4.33 \\ -18.72 \\ -7.14 \end{bmatrix}$$

$$y'_m = (z'_m \times a')$$

$$= \begin{bmatrix} -4.02 \\ 23.61 \\ -3.75 \end{bmatrix} \times \begin{bmatrix} 4.33 \\ -18.72 \\ -7.14 \end{bmatrix} = \begin{bmatrix} -238.78 \\ -44.94 \\ -26.98 \end{bmatrix}$$

$$x'_m = (y'_m \times z'_m)$$

$$= \begin{bmatrix} -238.78 \\ -44.94 \\ -26.98 \end{bmatrix} \times \begin{bmatrix} -4.02 \\ 23.61 \\ -3.75 \end{bmatrix} = \begin{bmatrix} 805.45 \\ -786.96 \\ -5818.10 \end{bmatrix}$$

Upon normalising the above vectors, we can express the [M → G] matrix as

$$[\mathrm{M} \rightarrow \mathrm{G}] = \begin{bmatrix} -0.136 & -0.133 & -0.982 \\ -0.977 & -0.184 & -0.110 \\ -0.166 & 0.974 & -0.155 \end{bmatrix}$$

Therefore, the [G → M] matrix is expressed as

$$[\mathrm{G} \rightarrow \mathrm{M}] = \begin{bmatrix} 0.136 & -0.977 & -0.166 \\ -0.133 & -0.184 & 0.974 \\ -0.982 & -0.110 & -0.155 \end{bmatrix}$$

The [G → A] matrix is obtained from the matrix multiplication of [G → M] and [M → A]. The [M → A] matrix is a constant during the walking, obtained from the calibration process.

[G → A] = [M → A] × [G → M]

$$[\mathrm{G} \rightarrow \mathrm{A}] = \begin{bmatrix} 0.264 & -0.962 & 0.074 \\ -0.239 & 0.009 & 0.971 \\ -0.934 & -0.274 & -0.227 \end{bmatrix}$$

$$\begin{bmatrix} 0.136 & -0.977 & -0.166 \\ -0.133 & -0.184 & 0.974 \\ -0.982 & -0.110 & -0.155 \end{bmatrix}$$

$$= \begin{bmatrix} 0.963 & 0.214 & -0.165 \\ -0.202 & 0.976 & 0.083 \\ 0.179 & -0.047 & 0.983 \end{bmatrix}$$

The above matrix can be expressed in the form of the cosine and sine of the rotation angles (θ_1, θ_2, θ_3) required for coordinate transformation from GCS to ACS, as follows:

$$\begin{bmatrix} c_2c_3 & c_1s_3 + s_1s_2c_3 & s_1s_3 - c_1s_2c_3 \\ -c_2s_3 & c_1c_3 - s_1s_2s_3 & s_1c_3 + c_1s_2s_3 \\ s_2 & -s_1c_2 & c_1c_2 \end{bmatrix}$$

By equating the above two matrices, we get the rotation angles (θ_1, θ_2, θ_3)

$$\begin{aligned} \theta_1 &= 2.72^\circ \\ \theta_2 &= 10.29^\circ \\ \theta_3 &= 11.86^\circ \end{aligned}$$

Now we determine the $\boldsymbol{c}$, which was earlier defined in ACS, in the GCS is equal to

$$\begin{aligned} \{\boldsymbol{c}\} &= [\mathrm{A} \rightarrow \mathrm{G}]\{\boldsymbol{c}\} \\ \boldsymbol{c} &= \begin{bmatrix} 0.963 & -0.202 & 0.179 \\ 0.214 & 0.976 & -0.047 \\ -0.165 & 0.083 & 0.983 \end{bmatrix} \begin{bmatrix} 1.04 \\ 4.58 \\ -10.76 \end{bmatrix} \\ &= \begin{bmatrix} -1.84 \\ 5.19 \\ -10.37 \end{bmatrix} \end{aligned}$$

The $\boldsymbol{R_c}$ is determined using the R_m (the GCS coordinates of the tracking marker, $\mathrm{m_{T1}}$) and $\boldsymbol{c}$:

$$\boldsymbol{R_c} = \boldsymbol{R_m} + \boldsymbol{c} = \begin{bmatrix} -11.11 \\ 24.96 \\ 52.92 \end{bmatrix} + \begin{bmatrix} -1.84 \\ 5.19 \\ -10.37 \end{bmatrix} = \begin{bmatrix} -12.95 \\ 30.15 \\ 42.55 \end{bmatrix}$$

Determination of segment angular velocities and accelerations: The time derivative of the transformation angles helps to determine the segment angular velocities. The angular velocity, the first-time derivative of the angular data is expressed as:

$$\boldsymbol{\omega} = \frac{d\theta_1}{\mathrm{d}t} \cdot \hat{\boldsymbol{e}}_x + \frac{\mathrm{d}\theta_2}{\mathrm{d}t} \cdot \hat{\boldsymbol{e}}_{y'} + \frac{\mathrm{d}\theta_3}{\mathrm{d}t} \cdot \hat{\boldsymbol{e}}_{z''}$$

Here $\hat{\boldsymbol{e}}_x$, $\hat{\boldsymbol{e}}_{y'}$, and $\hat{\boldsymbol{e}}_{z''}$ denote the unit vectors of the three rotation axes x, y', and z''. These rotation axes are the axes about which the transformation takes place (from GCS to ACS). Consider angular velocity, ω', about the axis x when there are no rotations θ_2 or θ_3. This angular velocity is expressed as

$$\boldsymbol{\omega}' = \frac{d\theta_1}{\mathrm{d}t} \cdot \hat{\boldsymbol{e}}_x = \begin{bmatrix} \dot{\theta}_1 \\ 0 \\ 0 \end{bmatrix}$$

Now, after the rotation of the x-axis, the new axis system is x–y'–z' (Fig. 4.18). The rotation about the y'-axis results in the following. The second angular velocity $\boldsymbol{\omega}''$ is obtained as the sum of the term $\frac{\mathrm{d}\theta_2}{\mathrm{d}t} \cdot \hat{e}_{y'}$ (the time derivative of the angular data along the y' direction) and the contribution from ω', owing to the rotation about the y-axis.

$$\boldsymbol{\omega}'' = \frac{\mathrm{d}\theta_2}{\mathrm{d}t} \cdot \hat{\boldsymbol{e}}_{y'} + \begin{bmatrix} c_2 & 0 & -s_2 \\ 0 & 1 & 0 \\ s_2 & 0 & c_2 \end{bmatrix} \boldsymbol{\omega}'$$

$$= \begin{bmatrix} 0 \\ \dot{\theta}_2 \\ 0 \end{bmatrix} + \begin{bmatrix} c_2 & 0 & -s_2 \\ 0 & 1 & 0 \\ s_2 & 0 & c_2 \end{bmatrix} \begin{bmatrix} \dot{\theta}_1 \\ 0 \\ 0 \end{bmatrix} = \begin{bmatrix} c_2\dot{\theta}_1 \\ \dot{\theta}_2 \\ s_2\dot{\theta}_1 \end{bmatrix}$$

Similarly, after the rotation of the y'-axis, the new axis system is x'-y'-z''. The rotation about the z''-axis results in the following. The third term in the angular velocity, $\boldsymbol{\omega}'''$ is obtained as the sum of $\frac{\mathrm{d}\theta_3}{\mathrm{d}t} \cdot \hat{\boldsymbol{e}}_{z''}$ (the time derivative of the angular data along the z'' direction) and the contribution from ω'', owing to the rotation about the z''-axis.

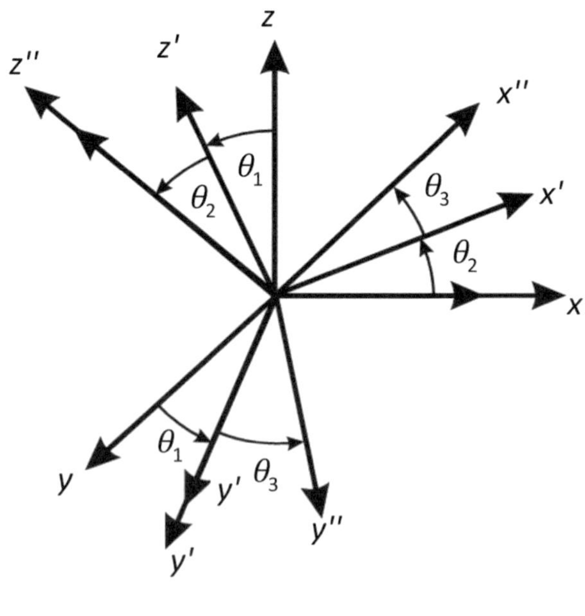

Fig. 4.18 The sequence of three rotations about various axes. The rotation, θ_1 about the x axis results in x, y', z'; the rotation, θ_2 about the y' axis results in x', y', z'', and the rotation, θ_3 about the z'' axis to get the final configuration x'', y'', z''

$$
\begin{aligned}
\boldsymbol{\omega}''' &= \frac{d\theta_3}{dt} \cdot \hat{\boldsymbol{e}}_{z''} + \begin{bmatrix} c_2 & 0 & -s_2 \\ 0 & 1 & 0 \\ s_2 & 0 & c_2 \end{bmatrix} \boldsymbol{\omega}'' \\
&= \begin{bmatrix} 0 \\ 0 \\ \dot{\theta}_3 \end{bmatrix} + \begin{bmatrix} c_3 & s_3 & 0 \\ -s_3 & c_3 & 0 \\ 0 & 0 & 1 \end{bmatrix} \begin{bmatrix} c_2\dot{\theta}_1 \\ \dot{\theta}_2 \\ s_2\dot{\theta}_1 \end{bmatrix} \\
&= \begin{bmatrix} c_3c_2\dot{\theta}_1 + s_3\dot{\theta}_2 \\ -s_3c_2\dot{\theta}_1 + c_3\dot{\theta}_2 \\ s_2\dot{\theta}_1 + \dot{\theta}_3 \end{bmatrix}
\end{aligned}
$$

The above expression is rewritten by decomposing into various components:

$$
\boldsymbol{\omega} = \begin{bmatrix} c_2c_3 & s_3 & 0 \\ -c_2s_3 & c_3 & 0 \\ s_2 & 0 & 1 \end{bmatrix} \begin{bmatrix} \dot{\theta}_1 \\ \dot{\theta}_2 \\ \dot{\theta}_3 \end{bmatrix} \tag{4.1}
$$

The time derivative quantity, $\dot{\theta}_i$, is calculated using the central difference method. In order to employ the central difference method, we require angular data of multiple time frames. For example, in order to find $\dot{\theta}_{10}$, we require the rotation angles corresponding to time frames 9 and 11. The rotation angles corresponding to the time frames 8, 9, 11, and 12, are calculated following a similar procedure used earlier for time frame 10. The rotation angles of various time frames are found to be

$$
\theta(t8) = \begin{bmatrix} 2.59^\circ \\ 9.31^\circ \\ 12.49^\circ \end{bmatrix}
$$

$$
\theta(t9) = \begin{bmatrix} 2.64^\circ \\ 9.91^\circ \\ 12.16^\circ \end{bmatrix}
$$

$$
\theta(t10) = \begin{bmatrix} 2.72^\circ \\ 10.29^\circ \\ 11.86^\circ \end{bmatrix}
$$

$$
\theta(t11) = \begin{bmatrix} 2.84^\circ \\ 10.55^\circ \\ 11.61^\circ \end{bmatrix}
$$

$$
\theta(t12) = \begin{bmatrix} 2.94^\circ \\ 10.56^\circ \\ 11.38^\circ \end{bmatrix}
$$

Using central difference method, the angular velocity at time frames 9 is calculated as (taking Δt as 0.002):

$$\begin{bmatrix} \dot{\theta}_1 \\ \dot{\theta}_2 \\ \dot{\theta}_3 \end{bmatrix}_{t9} = \frac{1}{2\Delta t}\{\theta(t10) - \theta(t8)\}$$

$$= \frac{1}{0.004}\left\{\begin{bmatrix} 2.72^\circ \\ 10.29^\circ \\ 11.86^\circ \end{bmatrix} - \begin{bmatrix} 2.59^\circ \\ 9.31^\circ \\ 12.49^\circ \end{bmatrix}\right\}$$

$$\begin{bmatrix} \dot{\theta}_1 \\ \dot{\theta}_2 \\ \dot{\theta}_3 \end{bmatrix}_{t9} = \begin{bmatrix} 32.48^\circ \\ 244.22^\circ \\ -158.15^\circ \end{bmatrix} \mathrm{s}^{-1} = \begin{bmatrix} 0.57 \\ 4.26 \\ -2.76 \end{bmatrix} \mathrm{rad\,s}^{-1}$$

Using a similar procedure, the angular velocity at the time frame 10 is

$$\begin{bmatrix} \dot{\theta}_1 \\ \dot{\theta}_2 \\ \dot{\theta}_3 \end{bmatrix}_{t10} = \frac{1}{2\Delta t}\{\theta(t11) - \theta(t9)\}$$

$$= \frac{1}{0.004}\left\{\begin{bmatrix} 2.84^\circ \\ 10.55^\circ \\ 11.61^\circ \end{bmatrix} - \begin{bmatrix} 2.64^\circ \\ 9.91^\circ \\ 12.16^\circ \end{bmatrix}\right\}$$

$$\begin{bmatrix} \dot{\theta}_1 \\ \dot{\theta}_2 \\ \dot{\theta}_3 \end{bmatrix}_{t10} = \begin{bmatrix} 49.83^\circ \\ 158.22^\circ \\ -137.86^\circ \end{bmatrix} \mathrm{s}^{-1} = \begin{bmatrix} 0.87 \\ 2.76 \\ -2.41 \end{bmatrix} \mathrm{rad\,s}^{-1}$$

The angular velocity at the time frame 11 is

$$\begin{bmatrix} \dot{\theta}_1 \\ \dot{\theta}_2 \\ \dot{\theta}_3 \end{bmatrix}_{t11} = \frac{1}{2\Delta t}\{\theta(t12) - \theta(t10)\}$$

$$= \frac{1}{0.004}\left\{\begin{bmatrix} 2.94^\circ \\ 10.56^\circ \\ 11.38^\circ \end{bmatrix} - \begin{bmatrix} 2.72^\circ \\ 10.29^\circ \\ 11.86^\circ \end{bmatrix}\right\}$$

$$\begin{bmatrix} \dot{\theta}_1 \\ \dot{\theta}_2 \\ \dot{\theta}_3 \end{bmatrix}_{t11} = \begin{bmatrix} 55.55^\circ \\ 69.32^\circ \\ -118.39^\circ \end{bmatrix} \mathrm{s}^{-1} = \begin{bmatrix} 0.97 \\ 1.21 \\ -2.07 \end{bmatrix} \mathrm{rad\,s}^{-1}$$

The angular acceleration of the segment at the time frame 10 is also calculated using central difference method on the segment angular velocities of the time frame 9 and 11. Using Eq. 4.1, the segment angular velocities of the time frames 9, 10, and 11 (in the ACS) are calculated.

$$\boldsymbol{\omega}(\mathrm{t9}) = \begin{bmatrix} c_2c_3 & s_3 & 0 \\ -c_2s_3 & c_3 & 0 \\ s_2 & 0 & 1 \end{bmatrix}_{t5} \begin{bmatrix} \dot{\theta}_1 \\ \dot{\theta}_2 \\ \dot{\theta}_3 \end{bmatrix}_{t9}$$

$$= \begin{bmatrix} 0.963 & 0.211 & 0 \\ -0.208 & 0.978 & 0 \\ 0.172 & 0 & 1 \end{bmatrix} \begin{bmatrix} 0.57 \\ 4.26 \\ -2.76 \end{bmatrix}$$

$$= \begin{bmatrix} 1.45 \\ 4.05 \\ -2.66 \end{bmatrix} \mathrm{rad\,s^{-1}}$$

$$\boldsymbol{\omega}(\mathrm{t10}) = \begin{bmatrix} c_2c_3 & s_3 & 0 \\ -c_2s_3 & c_3 & 0 \\ s_2 & 0 & 1 \end{bmatrix}_{t6} \begin{bmatrix} \dot{\theta}_1 \\ \dot{\theta}_2 \\ \dot{\theta}_3 \end{bmatrix}_{t10}$$

$$= \begin{bmatrix} 0.963 & 0.206 & 0 \\ -0.202 & 0.980 & 0 \\ 0.179 & 0 & 1 \end{bmatrix} \begin{bmatrix} 0.87 \\ 2.76 \\ -2.41 \end{bmatrix}$$

$$= \begin{bmatrix} 1.41 \\ 2.53 \\ -2.25 \end{bmatrix} \mathrm{rad\,s^{-1}}$$

$$\boldsymbol{\omega}(\mathrm{t11}) = \begin{bmatrix} c_2c_3 & s_3 & 0 \\ -c_2s_3 & c_3 & 0 \\ s_2 & 0 & 1 \end{bmatrix}_{t7} \begin{bmatrix} \dot{\theta}_1 \\ \dot{\theta}_2 \\ \dot{\theta}_3 \end{bmatrix}_{t11}$$

$$= \begin{bmatrix} 0.963 & 0.201 & 0 \\ -0.198 & 0.980 & 0 \\ 0.183 & 0 & 1 \end{bmatrix} \begin{bmatrix} 0.97 \\ 1.21 \\ -2.07 \end{bmatrix}$$

$$= \begin{bmatrix} 1.18 \\ 0.99 \\ -1.89 \end{bmatrix} \mathrm{rad\,s^{-1}}$$

The angular acceleration of the segment at the time frame 10 is

$$\boldsymbol{\alpha}(\mathrm{t10}) = \begin{bmatrix} \alpha_x \\ \alpha_y \\ \alpha_z \end{bmatrix}_{t10}$$

$$= \frac{1}{2\Delta t}\{\boldsymbol{\omega}(\mathrm{t11}) - \boldsymbol{\omega}(\mathrm{t9})\}$$

$$= \frac{1}{0.004}\left\{\begin{bmatrix} 1.18 \\ 0.99 \\ -1.89 \end{bmatrix} - \begin{bmatrix} 1.45 \\ 4.05 \\ -2.66 \end{bmatrix}\right\}$$

$$= \begin{bmatrix} -66.81 \\ -763.97 \\ 193.47 \end{bmatrix} \text{rad s}^{-1}$$

The $\boldsymbol{\alpha}$(t10) is defined with respect to the ACS.

Determination of linear velocity and acceleration: In order to find the linear velocity and acceleration using central-difference method, we need $\boldsymbol{R_c}$ corresponding to the time frames 8, 9, 10, 11, and 12. For the case of time frame 8, the $\boldsymbol{c}$ (in GCS) is obtained using the variable [A → G].

$$\begin{aligned}
\boldsymbol{c}|_{t8} &= [\mathrm{A} \rightarrow \mathrm{G}]_{t8} \begin{bmatrix} 1.04 \\ 4.58 \\ -10.76 \end{bmatrix} \\
&= \begin{bmatrix} c_2c_3 & c_1s_3 + s_1s_2c_3 & s_1s_3 - c_1s_2c_3 \\ -c_2s_3 & c_1c_3 - s_1s_2s_3 & s_1c_3 + c_1s_2s_3 \\ s_2 & -s_1c_2 & c_1c_2 \end{bmatrix}_{t8} \\
&\quad \begin{bmatrix} 1.04 \\ 4.58 \\ -10.76 \end{bmatrix} \\
&= \begin{bmatrix} 0.991 & -0.213 & 0.162 \\ 0.223 & 0.974 & -0.045 \\ -0.148 & 0.079 & 0.986 \end{bmatrix} \begin{bmatrix} 1.04 \\ 4.58 \\ -10.76 \end{bmatrix} \\
&= \begin{bmatrix} -1.71 \\ 5.17 \\ -10.40 \end{bmatrix} \\
\boldsymbol{R_c}|_{t8} &= \boldsymbol{R_m}|_{t8} + \boldsymbol{c}|_{t8} \\
&= \begin{bmatrix} -11.53 \\ 25.01 \\ 53.03 \end{bmatrix} + \begin{bmatrix} -1.71 \\ 5.17 \\ -10.40 \end{bmatrix} \\
&= \begin{bmatrix} -13.24 \\ 30.18 \\ 42.63 \end{bmatrix}
\end{aligned}$$

For the case of time frame 9: the $\boldsymbol{c}$ (in GCS) is calculated using the variable [A → G].

$$\boldsymbol{c}|_{t9} = [\mathrm{A} \rightarrow \mathrm{G}]_{t9} \begin{bmatrix} 1.04 \\ 4.58 \\ -10.76 \end{bmatrix}$$

$$= \begin{bmatrix} c_2c_3 & c_1s_3 + s_1s_2c_3 & s_1s_3 - c_1s_2c_3 \\ -c_2s_3 & c_1c_3 - s_1s_2s_3 & s_1c_3 + c_1s_2s_3 \\ s_2 & -s_1c_2 & c_1c_2 \end{bmatrix}_{t9} \begin{bmatrix} 1.04 \\ 4.58 \\ -10.76 \end{bmatrix}$$

$$= \begin{bmatrix} 0.963 & -0.207 & 0.172 \\ 0.218 & 0.975 & -0.045 \\ -0.158 & 0.081 & 0.984 \end{bmatrix} \begin{bmatrix} 1.04 \\ 4.58 \\ -10.76 \end{bmatrix}$$

$$= \begin{bmatrix} -1.80 \\ 5.18 \\ -10.39 \end{bmatrix}$$

$$\boldsymbol{R_c}|_{t9} = \boldsymbol{R_m}|_{t9} + \boldsymbol{c}|_{t9}$$

$$= \begin{bmatrix} -11.31 \\ 24.98 \\ 52.97 \end{bmatrix} + \begin{bmatrix} -1.80 \\ 5.18 \\ -10.39 \end{bmatrix}$$

$$= \begin{bmatrix} -13.11 \\ 30.16 \\ 42.58 \end{bmatrix}$$

For the case of time frame 10, the $\boldsymbol{c}$ (in GCS) is determined using the variable [A → G].

$$\boldsymbol{c}|_{t10} = [\mathrm{A} \rightarrow \mathrm{G}]_{t10} \begin{bmatrix} 1.04 \\ 4.58 \\ -10.76 \end{bmatrix}$$

$$= \begin{bmatrix} c_2c_3 & c_1s_3 + s_1s_2c_3 & s_1s_3 - c_1s_2c_3 \\ -c_2s_3 & c_1c_3 - s_1s_2s_3 & s_1c_3 + c_1s_2s_3 \\ s_2 & -s_1c_2 & c_1c_2 \end{bmatrix}_{t10} \begin{bmatrix} 1.04 \\ 4.58 \\ -10.76 \end{bmatrix}$$

$$= \begin{bmatrix} 0.963 & -0.202 & 0.179 \\ 0.214 & 0.976 & -0.047 \\ -0.165 & 0.083 & 0.983 \end{bmatrix} \begin{bmatrix} 1.04 \\ 4.58 \\ -10.76 \end{bmatrix}$$

$$= \begin{bmatrix} -1.84 \\ 5.19 \\ -10.37 \end{bmatrix}$$

$$\begin{aligned} \boldsymbol{R_c}|_{t10} &= \boldsymbol{R_m}|_{t10} + \boldsymbol{c}|_{t10} \\ &= \begin{bmatrix} -11.11 \\ 24.96 \\ 52.92 \end{bmatrix} + \begin{bmatrix} -1.84 \\ 5.19 \\ -10.37 \end{bmatrix} = \begin{bmatrix} -12.95 \\ 30.15 \\ 42.55 \end{bmatrix} \end{aligned}$$

For the case of time frame 11, the $\boldsymbol{c}$ (defined in the GCS) is calculated using the variable [A → G].

$$\begin{aligned} \boldsymbol{c}|_{t11} &= [A \rightarrow G]_{t11} \begin{bmatrix} 1.04 \\ 4.58 \\ -10.76 \end{bmatrix} \\ &= \begin{bmatrix} c_2c_3 & c_1s_3 + s_1s_2c_3 & s_1s_3 - c_1s_2c_3 \\ -c_2s_3 & c_1c_3 - s_1s_2s_3 & s_1c_3 + c_1s_2s_3 \\ s_2 & -s_1c_2 & c_1c_2 \end{bmatrix}_{t11} \\ &\quad \begin{bmatrix} 1.04 \\ 4.58 \\ -10.76 \end{bmatrix} \\ &= \begin{bmatrix} 0.963 & -0.198 & 0.183 \\ 0.210 & 0.977 & -0.049 \\ -0.169 & 0.085 & 0.982 \end{bmatrix} \begin{bmatrix} 1.04 \\ 4.58 \\ -10.76 \end{bmatrix} \\ &= \begin{bmatrix} -1.87 \\ 5.21 \\ -10.36 \end{bmatrix} \\ \boldsymbol{R_c}|_{t11} &= \boldsymbol{R_m}|_{t11} + c|_{t11} \\ &= \begin{bmatrix} -10.91 \\ 24.95 \\ 52.86 \end{bmatrix} + \begin{bmatrix} -1.87 \\ 5.21 \\ -10.36 \end{bmatrix} = \begin{bmatrix} -12.78 \\ 30.16 \\ 42.50 \end{bmatrix} \end{aligned}$$

For the case of time frame 12: the $\boldsymbol{c}$ (defined in the GCS) is determined using the variable [A → G].

$$\begin{aligned} \boldsymbol{c}|_{t12} &= [\mathrm{A} \rightarrow \mathrm{G}]_{t12} \begin{bmatrix} 1.04 \\ 4.58 \\ -10.76 \end{bmatrix} \\ &= \begin{bmatrix} c_2c_3 & c_1s_3 + s_1s_2c_3 & s_1s_3 - c_1s_2c_3 \\ -c_2s_3 & c_1c_3 - s_1s_2s_3 & s_1c_3 + c_1s_2s_3 \\ s_2 & -s_1c_2 & c_1c_2 \end{bmatrix}_{t12} \begin{bmatrix} 1.04 \\ 4.58 \\ -10.76 \end{bmatrix} \end{aligned}$$

$$= \begin{bmatrix} 0.964 & -0.194 & 0.183 \\ 0.206 & 0.977 & -0.051 \\ -0.169 & 0.087 & 0.982 \end{bmatrix} \begin{bmatrix} 1.04 \\ 4.58 \\ -10.76 \end{bmatrix} = \begin{bmatrix} -1.86 \\ 5.23 \\ -10.35 \end{bmatrix}$$

$$\boldsymbol{R_c}|_{t12} = \boldsymbol{R_m}|_{t12} + \boldsymbol{c}|_{t12}$$

$$= \begin{bmatrix} -10.72 \\ 24.95 \\ 52.81 \end{bmatrix} + \begin{bmatrix} -1.86 \\ 5.23 \\ -10.35 \end{bmatrix} = \begin{bmatrix} -12.58 \\ 30.18 \\ 42.46 \end{bmatrix}$$

Using the R_c corresponding to various time frames, we can now determine the velocities using the central difference method:

$$\boldsymbol{v}|_{t9} = \begin{bmatrix} v_X \\ v_Y \\ v_Z \end{bmatrix}_{t9} = \frac{1}{2\Delta t}\{\boldsymbol{R_c}|_{t10} - \boldsymbol{R_c}|_{t8}\}$$

$$= \frac{1}{0.004}\left\{\begin{bmatrix} -12.95 \\ 30.15 \\ 42.55 \end{bmatrix} - \begin{bmatrix} -13.24 \\ 30.18 \\ 42.63 \end{bmatrix}\right\}$$

$$= \begin{bmatrix} 72.52 \\ -7.01 \\ -19.20 \end{bmatrix} \text{cm s}^{-1}$$

$$\boldsymbol{v}|_{t11} = \begin{bmatrix} v_X \\ v_Y \\ v_Z \end{bmatrix}_{t11} = \frac{1}{2\Delta t}\{\boldsymbol{R_c}|_{t12} - \boldsymbol{R_c}|_{t10}\}$$

$$= \frac{1}{0.004}\left\{\begin{bmatrix} -12.58 \\ 30.18 \\ 42.46 \end{bmatrix} - \begin{bmatrix} -12.95 \\ 30.15 \\ 42.55 \end{bmatrix}\right\}$$

$$= \begin{bmatrix} 94.21 \\ 7.32 \\ -21.97 \end{bmatrix} \text{cm s}^{-1}$$

Using the above velocities, the acceleration at the time frame 10 is calculated as

$$\boldsymbol{a}|_{t10} = \begin{bmatrix} a_X \\ a_Y \\ a_Z \end{bmatrix}_{t10}$$

$$= \frac{1}{2\Delta t}\{\boldsymbol{v}|_{t11} - \boldsymbol{v}|_{t9}\}$$

$$= \frac{1}{0.004}\left\{\begin{bmatrix} 94.21 \\ 7.32 \\ -21.97 \end{bmatrix} - \begin{bmatrix} 72.52 \\ -7.01 \\ -19.20 \end{bmatrix}\right\}$$

$$= \begin{bmatrix} 5422.62 \\ 3581.25 \\ -694.06 \end{bmatrix} \text{cm s}^{-2}$$

This acceleration, calculated with respect to the GCS, will be useful to solve the force equilibrium equations in the kinetic analysis.

4.7.1 Kinetic Analysis

The reaction forces and moments acting at the proximal end of the right leg segment need to be calculated. Since the linear accelerations of the leg segment are available in the GCS from the kinematic analysis, it is easier to determine the segment joint reaction forces in the GCS. However, the segment moments are defined in the ACS, since the moment of inertia of the segment is known relative to the ACS. The kinetic analysis requires the three distal reaction forces, which are obtained from the analysis of the adjacent distal segment or from force plate data. The following are the steps involved in the calculation of joint forces and moments.

Step 1: Calculate the reaction forces at the proximal end of the segment in the GCS, using the force equilibrium equations.

$$\begin{aligned} F_X &= ma_X \text{ or } R_{XP} - R_{XD} - mg = ma_X \\ F_Y &= ma_Y \text{ or } R_{YP} - R_{YD} - mg = ma_Y \\ F_Z &= ma_Z \text{ or } R_{ZP} - R_{ZD} - mg = ma_Z \end{aligned}$$

where a_X, a_Y, a_Z are the segment COM accelerations in the *X*-, *Y*-, and *Z*-GCS directions, respectively. R_{XP}, R_{XD}, R_{YP}, R_{YD}, R_{ZP}, and R_{ZD} are the proximal and distal reaction forces in the *X*-, *Y*-, and *Z*-GCS directions, respectively.

Step 2: The proximal and distal reaction forces are transformed into the ACS using the [G → A] matrix transformation. The proximal and distal reaction forces in ACS are denoted by R_{xp}, R_{yp}, R_{zp}, R_{xd}, R_{yd}, and R_{zd}.

Step 3: The distal moments are transformed into the ACS, using the [G → A] matrix. The distal moments in ACS are denoted by M_{xd}, M_{yd}, and M_{zd}. These variables allow us to calculate the proximal moments in the anatomical axes.

In the following steps, we determine the proximal moments:

Step 4: The Euler's equations are written as follows:

$$\begin{aligned} & I_x\alpha_x + \left(I_z - I_y\right)\omega_y\omega_z \\ & \quad = \Sigma M_x = R_{zd}l_d + R_{zp}l_p + M_{xp} - M_{xd} \\ & I_y\alpha_y + (I_x - I_z)\omega_x\omega_z \\ & \quad = \Sigma M_y = M_{yp} - M_{yd} \\ & I_z\alpha_z + \left(I_y - I_x\right)\omega_x\omega_y \end{aligned}$$

$$= \Sigma M_z = -R_{xd} l_d - R_{xp} l_p + M_{zp} - M_{zd}$$

Here all quantities are defined with respect to ACS. The unknowns in the above equations are the moments (M_{xp}, M_{yp}, and M_{zp}) about the proximal x-, y-, z- axes. The moment of inertia, angular velocity, and angular acceleration are denoted by I, $\boldsymbol{\omega}$, and $\boldsymbol{\alpha}$, respectively.

Sample Problem 4.7

This sample problem involves calculation of the proximal forces and moments, for time frame 10. The anthropometric data used in the problem are as follows: mass of the segment (m) = 2.44 kg, moment of inertia: $I_x = 0.037$ kg m^2, $I_y = 0.0037$ kg m^2, $I_z = 0.037$ kg m^2; the distances from COM to the ends: $l_p = 22.06$ cm, $l_d = 16.84$ cm. The reaction forces at the distal end are denoted as R_{XD}, R_{YD}, and R_{ZD}, whereas the moments are denoted by M_{XD}, M_{YD}, and M_{ZD} (Table 4.4).

From the kinematic analysis, the linear accelerations were calculated as

$$\begin{aligned} a_X &= 54.22 \text{ m s}^{-2}, \\ a_Y &= 35.81 \text{ m s}^{-2}, \\ a_Z &= -6.94 \text{ m s}^{-2} \end{aligned}$$

Let us initiate the kinetic analysis by determining the proximal reaction forces, using the free body diagram shown in Fig. 4.19 and the equilibrium equations stated below:

$$\begin{aligned} R_{XP} - R_{XD} &= ma_X \\ R_{XP} &= ma_X + R_{XD} \\ &= 2.44 \times (54.22) + (-157.53) \\ &= -25.22 \text{ N} \end{aligned}$$

$$\begin{aligned} R_{YP} - R_{YD} - mg &= ma_Y \\ R_{YP} &= ma_Y + R_{YD} + mg \\ &= 2.44 \times (35.81) + (414.59) \\ &\quad + 2.44 \times 9.81 \\ &= 525.91 \text{ N} \end{aligned}$$

$$R_{ZP} - R_{ZD} = ma_Z$$

Table 4.4 Ankle reaction forces (in N) and moments (in Nm) defined in GCS

Frame	R_{XD}	R_{YD}	R_{ZD}	M_{XD}	M_{YD}	M_{ZD}
10	−157.53	414.59	−7.64	−2.23	1.64	0.09

$$
\begin{aligned}
R_{ZP} &= ma_Z + R_{ZD} \\
&= 2.44 \times (-6.94) + (-7.64) \\
&= -24.58\,\text{N}
\end{aligned}
$$

Using the above transformation matrix $[G \rightarrow A]_{t10}$, the distal reaction forces are transformed from the global to the ACS.

Fig. 4.19 Free body diagram of the right leg segment

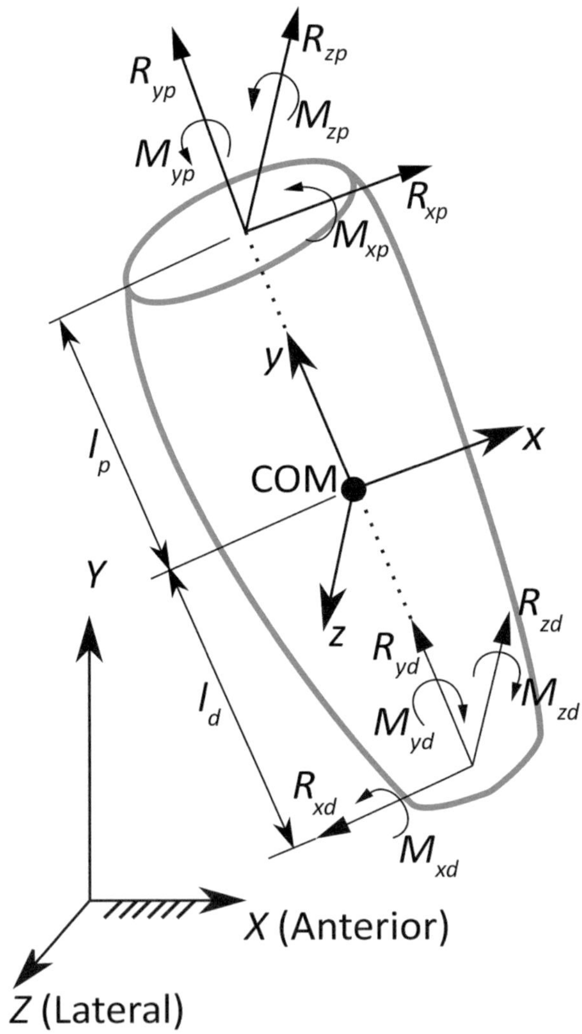

$$\begin{bmatrix} R_{xd} \\ R_{yd} \\ R_{zd} \end{bmatrix} = [\mathrm{G} \rightarrow \mathrm{A}]_{\mathrm{t}10} \begin{bmatrix} R_{XD} \\ R_{YD} \\ R_{ZD} \end{bmatrix}$$

$$= \begin{bmatrix} 0.963 & 0.214 & -0.165 \\ -0.202 & 0.976 & 0.083 \\ 0.179 & -0.047 & 0.983 \end{bmatrix} \begin{bmatrix} -157.53 \\ 414.59 \\ -7.64 \end{bmatrix}$$

$$= \begin{bmatrix} -61.91 \\ 435.78 \\ -55.00 \end{bmatrix} \mathrm{N}$$

Similarly, the proximal reaction forces and the ankle moments are transformed into the ACS:

$$\begin{bmatrix} R_{xp} \\ R_{yp} \\ R_{zp} \end{bmatrix} = [\mathrm{G} \rightarrow \mathrm{A}]_{\mathrm{t}10} \begin{bmatrix} R_{XP} \\ R_{YP} \\ R_{ZP} \end{bmatrix}$$

$$= \begin{bmatrix} 0.963 & 0.214 & -0.165 \\ -0.202 & 0.976 & 0.083 \\ 0.179 & -0.047 & 0.983 \end{bmatrix} \begin{bmatrix} -25.22 \\ 525.91 \\ -24.58 \end{bmatrix}$$

$$= \begin{bmatrix} 92.06 \\ 516.25 \\ -53.22 \end{bmatrix} \mathrm{N}$$

$$\begin{bmatrix} M_{xd} \\ M_{yd} \\ M_{zd} \end{bmatrix} = [\mathrm{G} \rightarrow \mathrm{A}]_{\mathrm{t}10} \begin{bmatrix} M_{XD} \\ M_{YD} \\ M_{ZD} \end{bmatrix}$$

$$= \begin{bmatrix} 0.963 & 0.214 & -0.165 \\ -0.202 & 0.976 & 0.083 \\ 0.179 & -0.047 & 0.983 \end{bmatrix} \begin{bmatrix} -2.23 \\ 1.64 \\ 0.09 \end{bmatrix}$$

$$= \begin{bmatrix} -1.81 \\ 2.06 \\ -0.39 \end{bmatrix} \mathrm{Nm}$$

The moment equilibrium equations can now be solved to obtain the proximal moments (M_{xp}, M_{yp}, M_{zp}). The angular velocities and accelerations, used in the following calculations, were obtained for time frame 10 during the kinematic analysis. Using the moment equilibrium about the x-axis (anatomical), we get

$$I_x\alpha_x + \left(I_z - I_y\right)\omega_y\omega_z = \Sigma M_x$$

$$I_x\alpha_x + \left(I_z - I_y\right)\omega_y\omega_z = -R_{zd}l_d - R_{zp}l_p + M_{xp} - M_{xd}$$

$$M_{xp} = I_x\alpha_x + \left(I_z - I_y\right)\omega_y\omega_z + R_{zd}l_d + R_{zp}l_p + M_{xd}$$

$$M_{xp} = 0.037 \times (-66.81) + (0.037 - 0.0037) \times 2.53 \times (-2.25) + (-55.00 \times 0.1684) + (-53.22 \times 0.2206) + (-1.81)$$

$$M_{xp} = -25.47\,\text{Nm}$$

Using the moment equilibrium about the y-axis (anatomical), we get

$$I_y\alpha_y + (I_x - I_z)\omega_x\omega_z = \Sigma M_y$$

$$I_y\alpha_y + (I_x - I_z)\omega_x\omega_z = M_{yp} - M_{yd}$$

$$M_{yp} = I_y\alpha_y + (I_x - I_z)\omega_x\omega_z + M_{yd}$$

$$M_{yp} = 0.0037 \times (-763.97) + (0.037 - 0.037) \times (1.41) \times (-2.25) + 2.06$$

$$M_{yp} = -0.77\,\text{Nm}$$

Using the moment equilibrium about the z-axis (anatomical), we get

$$I_z\alpha_z + \left(I_y - I_x\right)\omega_x\omega_y = \Sigma M_z$$

$$I_z\alpha_z + \left(I_y - I_x\right)\omega_x\omega_y = -R_{xd}l_d - R_{xp}l_p + M_{zp} - M_{zd}$$

$$M_{zp} = I_z\alpha_z + \left(I_y - I_x\right)\omega_x\omega_y + R_{xd}l_d + R_{xp}l_p + M_{zd}$$

$$M_{zp} = 0.037 \times (193.46) + (0.0037 - 0.037) \times (1.41) \times (2.53) + (-61.91 \times 0.1684) + (92.06 \times 0.2206) + (-0.39)$$

$$M_{zp} = 16.53\,\text{Nm}$$

The posture we have been discussing so far represents an instance after the heel strike and before the foot flat of the right foot, as shown in Fig. 4.20. The calculated proximal joint moments are acting on the right knee. The signs (positive or negative) of these joint moments are interpreted as follows. The negative sign of M_{xp} indicates a clockwise moment, which signifies the knee adduction moment during the initial stance phase of the gait. The M_{yp} represents the moment acting along the long axis of the leg. The negative sign (clockwise) of M_{yp} signifies the presence of hip

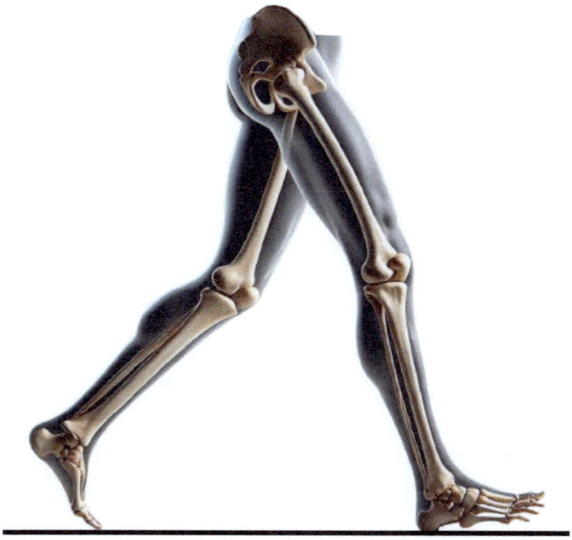

Fig. 4.20 The posture representing the time instance between heel strike and foot flat

external rotator moment, which usually causes the external rotation of hip during the initial stance phase. The positive sign of M_{zp} indicates a counter-clockwise moment, which signifies the presence of knee extensor moment which acts during the initial stance phase of walking. A MATLAB code that implements the steps involved in this problem is presented at the end of this chapter.

4.8 Use of Musculoskeletal Models for Force Estimation

Musculoskeletal models, in conjunction with static optimisation methods, are employed to investigate the distribution of muscle forces. These models are particularly useful when ethical concerns preclude in vivo measurements of joint contact forces and muscle forces. Subject-specific musculoskeletal models are increasingly utilised to achieve more precise estimations of joint loads during various activities. These models consider individual variations in anatomy and physiology. There is a range of both open-source and commercial software tools available for musculoskeletal modelling and simulation. Prominent open-source software includes OpenSim, while commercial software options include AnyBody, among others. These tools and techniques have led to a better understanding of the biomechanics of human movement, enabling researchers to gain insights into muscle function, joint forces, and the mechanics behind various activities.

An overview of the process for estimating joint forces using a musculoskeletal model is presented in Fig. 4.21. The following are the major steps:

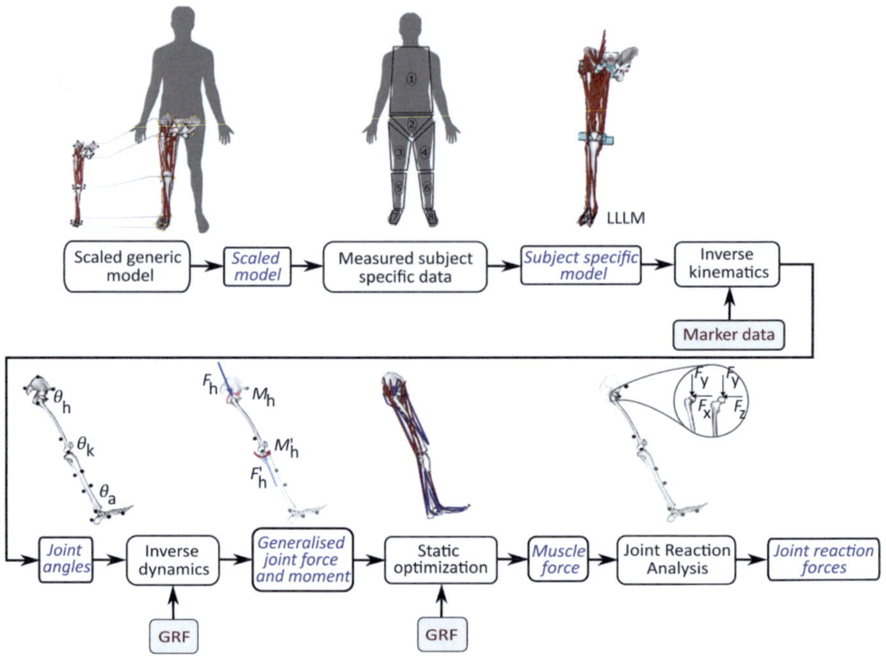

Fig. 4.21 Overview of joint force estimation indicating the important steps involved. Here θ denotes the joint angles, F and M represents the joint force and moments, respectively. LLLM stands for London Lower Limb Model, GRF denotes ground reaction force

1. Musculoskeletal Model Scaling: The process begins with a validated generic musculoskeletal model. The model is scaled to match the subject's specific dimensions and masses. This step ensures that the model is tailored to the individual.
2. Subject-Specific Parameters: The subject-specific parameters have to be incorporated into the model. These parameters may include the subject's anthropometric data, such as body segment lengths, joint locations, and mass distribution.
3. Kinematic data: Utilisation of inverse kinematics method to obtain kinematic information. This information is derived from the marker data and helps to define the movement of the subject.
4. Generalized Joint Forces and Moments: The kinematic information in combination with ground reaction force data and thereafter, using the method of inverse dynamics, one can calculate generalized joint forces and moments at various joints in the body.
5. Static Optimisation: Apply a static optimisation procedure to estimate the individual muscle forces. This step involves solving for the distribution of muscle forces that best matches the generalized joint forces and moments, while considering the biomechanical constraints of the musculoskeletal system.

6. Recalculation of the Joint Reaction Forces: After estimating the individual muscle forces, the joint reaction forces are recalculated. This step ensures that the joint forces are in equilibrium, considering the newly estimated muscle forces.

The joint force estimation of the London Lower Limb Model for various activities is presented in Fig. 4.22 (Mathai and Gupta 2019). This comparison demonstrated a close agreement between the estimates of the model and the measured hip joint forces in vivo (Bergmann et al. 2001). However, it is important to acknowledge the limitations of this method of force estimation, as discussed in the following.

(1) Accuracy of Generic Anatomy: The accuracy of the musculoskeletal model relies on how well it represents the generic anatomy of the human body. While efforts are made to match subject-specific parameters, there can still be variations that may affect the accuracy of the results.
(2) Objective Function Relationship: The objective function used in static optimisation techniques lacks a proven direct relationship with the actual distribution of muscle forces. It provides estimates based on minimising certain criteria, which may not precisely match the real distribution of forces.
(3) Limited Validation Data: Validating individual muscle forces is challenging. While electromyography (EMG) and in vivo joint reaction force data can be used for validation, data availability is often inadequate. Hence, it is difficult to validate the predicted results comprehensively.
(4) Complexity of Musculoskeletal Systems: The mechanical properties of musculoskeletal systems are inherently complex. Representing all aspects of this complexity in a model is challenging. Some simplifications and assumptions may be necessary, which may introduce uncertainties.

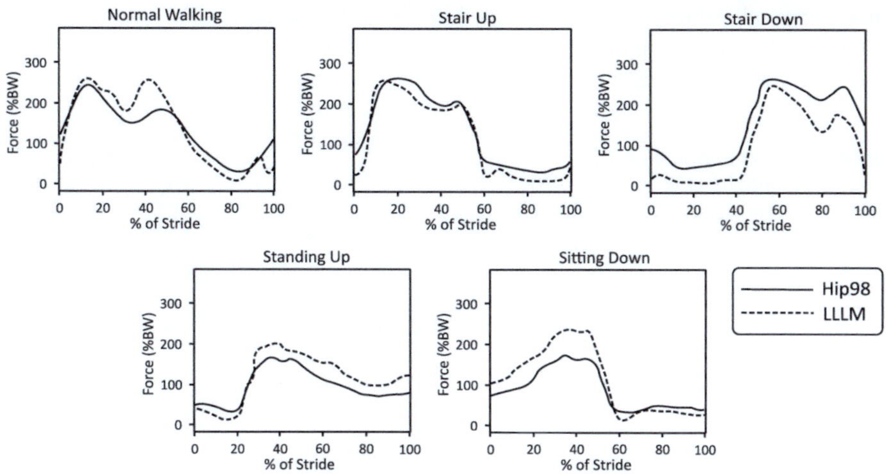

Fig. 4.22 Comparison of estimated hip contact force with Hip98 dataset (Bergmann et al. 2001)

In a summary, the musculoskeletal models and static optimisation techniques provide estimated values of muscle and joint forces during various activities. However, one should be aware of the above limitations and exercise caution when interpreting these estimated values.

4.9 Postural Stability

Postural stability is a fundamental aspect of human biomechanics, essential for maintaining balance and preventing falls during various activities. A key principle governing postural stability is the need for the body's centre of gravity (COG) to remain within the base of support, which is typically provided by the feet. The base of support serves as the foundation for providing stability. For example, a circular area is the base of support for a cone. Stability is achieved when the line of action of the gravitational force, passing through the COG of the entire body, remains within this base of support. Let us consider the case of a cone. Upon tilting the cone, as presented in Fig. 4.23, the size of the base of support is reduced and hence, stability is compromised. Critical angle (θ) determines the threshold for instability when the dimensions of the base of support (d) are reduced. The angle θ is given by

$$\theta = \tan^{-1}\frac{d}{h}$$

where h represents the height of COG of the body above ground (Fig. 4.23). Therefore, a reduction in the size of the base of support results in a smaller critical angle required to initiate instability. Additionally, the height of the COG profoundly influences postural stability. As demonstrated in Fig. 4.23, increasing the height of COG from h_1 to h_2 for a given base of support (d) reduces the angle θ necessary to trigger instability. Mathematically, this relationship is expressed as follows:

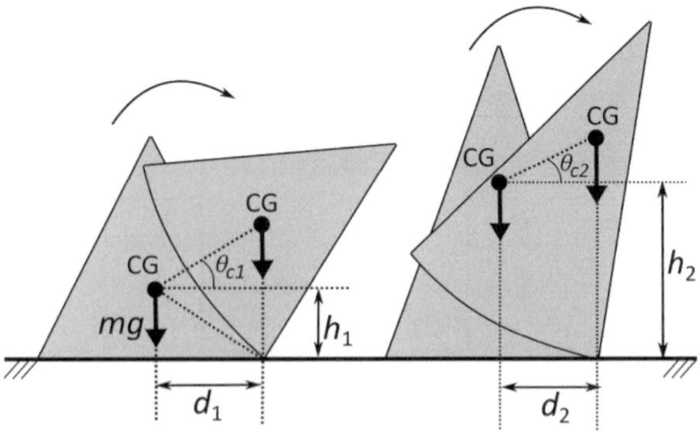

Fig. 4.23 Effect of height of the centre of gravity of the body on postural stability

$$\theta_{c1} = \tan^{-1}\left(\frac{d_1}{h_1}\right); \theta_{c2} = \tan^{-1}\left(\frac{d_2}{h_2}\right)$$

where θ_{c1} and θ_{c2} represent the critical angles corresponding to the heights h_1 and h_2, respectively. If h_2 is higher than h_1, then the θ_{c2} will become lesser than θ_{c1}, indicating that a smaller angle is needed to initiate instability, since the height of the COG is increased. This highlights the critical role played by both the base of support dimensions and the height of COG in determining postural stability, a crucial aspect of human biomechanics.

The positioning of the feet plays a vital role in determining the base of support and overall postural stability. Figure 4.24 depicts how variations in foot placement influence stability. When the two feet are close to each other, as shown in Fig. 4.24a, stability is compromised, exemplifying poor stability. Conversely, positioning the two feet apart, either in the medial–lateral direction or anterior–posterior direction, significantly enhances stability in the respective dimensions. Widening the distance between the feet provides excellent stability along the chosen direction. Moreover, the use of additional supporting structures, such as a single cane, crutch, or a pair of crutches, further increases the area of the base of support, consequently improving overall stability.

Understanding the relationship between the Centre of Mass (COM) and Centre of Pressure (COP) is crucial for joint kinematics and postural stability. The COM represents the net location of the body's mass in three-dimensional space and is

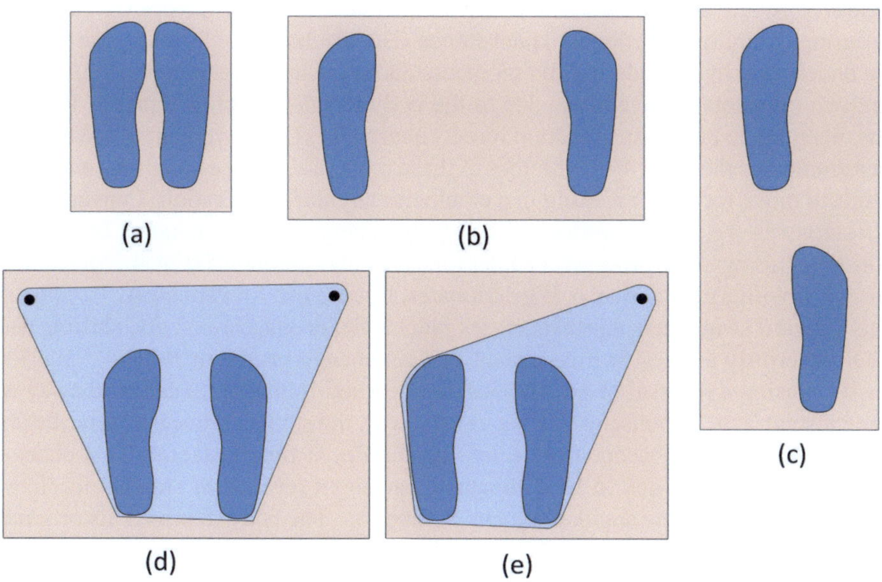

Fig. 4.24 Foot positioning; **a** poor stability, **b** good M-L stability, **c** good A-P stability, **d** pair of crutches, **e** single cane/crutch

calculated as a weighted average of the COM of each body segment. On the other hand, the COP signifies the location of the vertical ground reaction force vector, typically derived from a single force platform, assuming that all body contact points are on that platform. The location of COP is influenced by foot placement and the precise motor control of the ankle muscles, making it a key factor in maintaining postural stability.

4.10 Swaying During Quiet Stance

The phenomenon of swaying during a quiet stance provides valuable insights into the intricate relationship between the COP and the Centre of Gravity (COG) in the human body. In Fig. 4.25, we observe a subject swaying back and forth while standing on both feet atop a force platform. Even during seemingly motionless standing posture, the body's COG changes its position (Fig. 4.25). Let us discuss this change in COG in detail. Our attention is first drawn to the lines of action of two fundamental forces: the body weight (***W***) and the ground reaction force (***R***). The body weight acts through the COG, while ***R*** acts through the COP. Importantly, these lines of action are not aligned with the ankle joint but are offset by distances represented by lever arms, *p* and *g* corresponding to ***R*** and ***W***, respectively.

There are small horizontal fluctuations in COG, even during the quiet stance. Therefore, postural stability depends on the continual adjustment of the COP, achieved by the coordinated activity of the plantar flexor and dorsiflexor muscles, ensuring equilibrium during the quiet stance. The mechanics of swaying can further be understood by considering the moments about the ankle joint centre (C). There are two moments involved: one due to the body weight, which is equal to *Wg* and the other due to the ground reaction force, equal to *Rp*. The movement of the body is determined by the relative magnitudes of these moments. If *Wg* exceeds *Rp*, the body tends to move forward, resulting in a clockwise angular acceleration. Conversely, if *Rp* surpasses *Wg*, a counter-clockwise angular acceleration is generated. In order to illustrate this dynamic process, consider time instants 1 through 5 (Fig. 4.25). At time instant 1, with *g* greater than *p*, *Wg* dominates, leading to a forward sway. To counter this forward sway, the subject increases plantar flexor muscle activity, shifting the COP anteriorly as seen at time instant 2. This change results in ***R*** acting anterior to ***W***, causing a reversal of angular acceleration and initiating a backward sway at time instant 3. In response to this backward sway, the subject reduces plantar flexor muscle activation, subsequently decreasing the *Rp*, at time instant 4. This change reverts the moment back to ***W*** dominance, causing a reversal of angular acceleration and eventually, the angular velocity's direction. The body returns to its original state at time instant 5, with angular velocity and acceleration similar to those at time instant 1.

In a summary, the interplay between COP and COG resembles a cat-and-mouse game, where the continuous adjustments in COP of the body ensure postural stability during a quiet stance. The condition for stability is that the COP must remain within

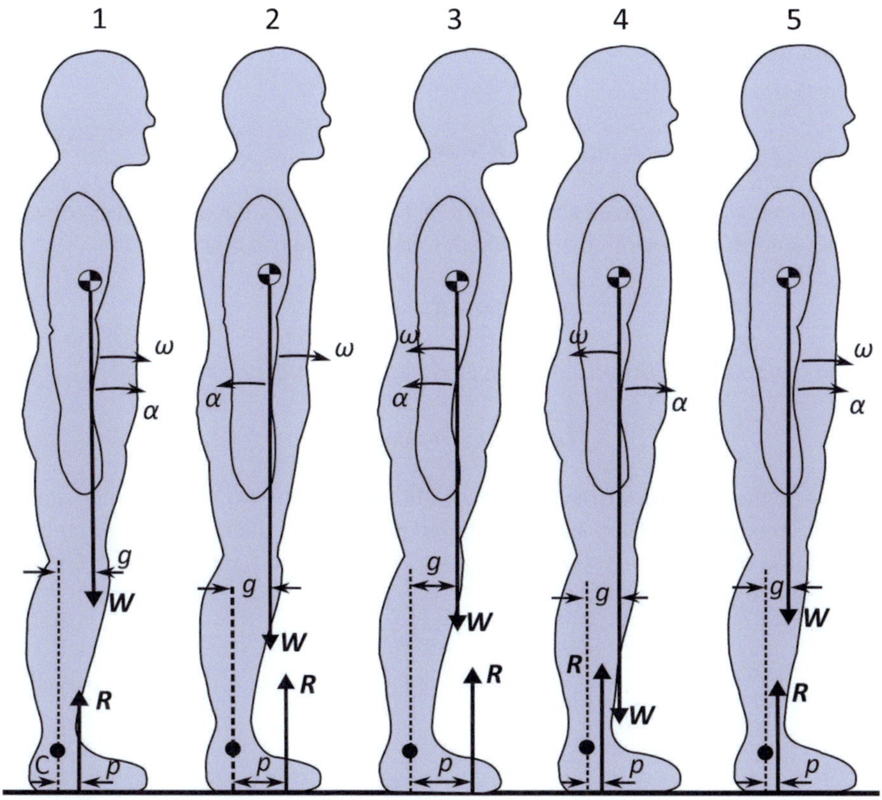

Fig. 4.25 Swaying back and forth of a subject during quiet stance. Five sequential time instants (1-5) are considered, indicating the locations of COM and COP as well as the associated angular velocities and accelerations of the body

the base of support. This definition of stability in standing balance is more accurate than one based solely on the location of the ground projection of the COM mentioned earlier. However, given the negligible velocity and acceleration of COM movements during quiet stance, criteria for stability based on both the COP and the ground projection of the COM are effectively equivalent.

4.11 Inverted Pendulum Model

The inverted pendulum model, relating the trajectories of the COP and COM, is a common approach to analyse balance dynamics (Winter et al. 1998). The sagittal plane model of the human body during quiet stance is presented in Fig. 4.26. The COP and COM are located at a distance, p and x anterior to the ankle joint centre (O). $\boldsymbol{R}$ and $\boldsymbol{R}_v$ denote the ankle reaction force and ground reaction force, respectively

(Fig. 4.27). Considering the foot alone (Fig. 4.27), the force equilibrium in the vertical direction is written as

$$\boldsymbol{R_v} + \boldsymbol{R} - m_f g = m_f a_v$$

where m_f and a_v are the mass and the vertical linear acceleration of foot, respectively. In quiet stance, a_v is equal to zero, which reduces the above equation to

$$\boldsymbol{R_v} = -\boldsymbol{R} + m_f g \tag{4.2}$$

The moment equilibrium about the ankle joint (O) is expressed as

$$\boldsymbol{M_a} + \boldsymbol{R_v} p - m_f g x_a = I_f \alpha_f$$

where I is the moment of inertia of foot about ankle joint (O). M_a is the sum of the moments generated by the ankle joint muscles. In quiet stance, the angular acceleration of the foot, α_f, is equal to zero.

The above equation is reduced as

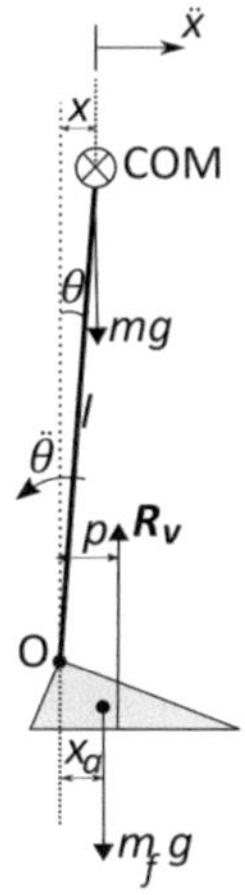

Fig. 4.26 Inverted pendulum model of the human body in the sagittal plane

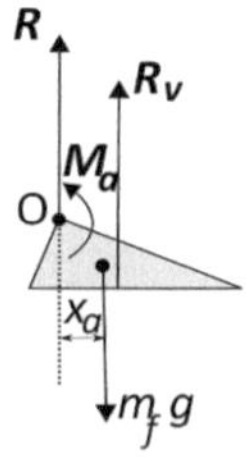

Fig. 4.27 Free body diagram of the foot

$$\boldsymbol{M_a} = -\boldsymbol{R_v}p + m_f g x_a \tag{4.3}$$

Combining Eqs. 4.2 and 4.3, we get

$$M_a = Rp + m_f g(x_a - p)$$

The value of $m_f g(x_a - p)$ is very less as compared to Rp and the above equation reduces to

$$M_a = Rp \tag{4.4}$$

Considering the free body diagram of the inverted pendulum (Fig. 4.28) and the moment equilibrium about the ankle joint (O)

$$I\ddot{\theta} = -\boldsymbol{M_a} - mgx \tag{4.5}$$

where I is the moment of inertia of the body (without the foot) about the ankle joint centre (O) in the sagittal plane and m is the mass of the body (without the feet). Combining Eqs. 4.4 and 4.5, we get

$$-\boldsymbol{R}p - mgx = I\ddot{\theta}$$

Since $\boldsymbol{R} = -mg$, from the force equilibrium along vertical direction (Fig. 4.28), the above equation reduces to

$$p - x = \frac{I\ddot{\theta}}{mg}$$

For small angles of sway (θ), the angular acceleration is

Fig. 4.28 Free body diagram of body without foot

$$\ddot{\theta} = -\frac{\ddot{x}}{h}$$

Substituting $\ddot{\theta}$, the relationship between the trajectory of the COP and COM is given by the following equation:

$$p - x = -\frac{I\ddot{x}}{mgh}$$

Similar correlation between the terms $(p - x)$ and $\ddot{x}$ was reported by Gage et al. (2004).

MATLAB Code for Solving the Numerical Problem Related to 3D Motion Analysis

```
%Calibration process
  %marker data of the stance is stored in the calibration.csv file
  calibration=importdata('calibration.csv')
  mt1=calibration(1,:);
  mt2=calibration(2,:);
  mt3=calibration(3,:);
  mc1=calibration(4,:);
  mc2=calibration(5,:);
  %the ankle and the knee centers are mid1 and mid2, respectively
  mid1=(mt2+mc1)/2;
  mid2=(mt3+mc2)/2;
  %using the anthroprometric relation, the center of mass of the leg segment
  COM_g=0.567*mid2+0.43*mid1;
  %the ACS (xa_g, ya_g, za_g) defined in the GRS.
  y_a=mid2-mid1;
  ya_g=y_a/norm(y_a);
  z_a_int=mt2-mc1;
  z_a_int=z_a_int/norm(z_a_int);
  x_a=cross(y_a,z_a_int);
  xa_g=x_a/norm(x_a);
  z_a=cross(x_a,y_a);
  za_g=z_a/norm(z_a);
  %the GRS to anatomical transformation matrix
  g2a=[xa_g' ya_g' za_g']';
  %%determination of marker to anatomical (constant): m2a matrix)
  mt1_g=mt1;
  mt2_g=mt2;
  mt3_g=mt3;
```

```
ankle_g=mid1-COM_g;
ankle_l=g2a*ankle_g';
knee_g=mid2-COM_g;
knee_l=g2a*knee_g';
% the following marker datas are now defined w.r.t to the ACS (COM)
mt1_l=g2a*(mt1_g-COM_g)';
mt2_l=g2a*(mt2_g-COM_g)';
mt3_l=g2a*(mt3_g-COM_g)';
c_vec=-mt1_l;
zm_a=mt3_l-mt1_l;
zm_a=zm_a/norm(zm_a);
a=mt2_l-mt1_l;
a=a/norm(a);
ym_a=cross(zm_a,a);
ym_a=ym_a/norm(ym_a);
xm_a=cross(ym_a,zm_a);
xm_a=xm_a/norm(xm_a);
%the marker to anatomical constant transformation matrix.
a2m=[xm_a ym_a zm_a]';
m2a=inv(a2m);
%%****** time series data follows****
%marker data of the tracking time steps is stored in the tracking.csv file
mt_tracker=importdata('tracking.csv')
mtt1_g4=mt_tracker(1,1:3);
mtt1_g5=mt_tracker(2,1:3);
mtt1_g6=mt_tracker(3,1:3);
mtt1_g7=mt_tracker(4,1:3);
mtt1_g8=mt_tracker(5,1:3);
mtt2_g4=mt_tracker(1,4:6);
mtt2_g5=mt_tracker(2,4:6);
mtt2_g6=mt_tracker(3,4:6);
mtt2_g7=mt_tracker(4,4:6);
mtt2_g8=mt_tracker(5,4:6);
mtt3_g4=mt_tracker(1,7:9);
mtt3_g5=mt_tracker(2,7:9);
mtt3_g6=mt_tracker(3,7:9);
mtt3_g7=mt_tracker(4,7:9);
mtt3_g8=mt_tracker(5,7:9);
mtt1=[mtt1_g4' mtt1_g5' mtt1_g6' mtt1_g7' mtt1_g8'];
mtt2=[mtt2_g4' mtt2_g5' mtt2_g6' mtt2_g7' mtt2_g8'];
mtt3=[mtt3_g4' mtt3_g5' mtt3_g6' mtt3_g7' mtt3_g8'];
%translation vector for velocity and acceleration calculation
Rc=zeros(5,3);
for j= 1: 5
mtt1_g=mtt1(:,j);
```

```
mtt2_g=mtt2(:,j);
mtt3_g=mtt3(:,j);
disp(['***********************']);
disp(['***********************']);
disp(['***********************']);
disp(['The calculations corresponding to the time frame: ' num2str(j+3) ' is being carried out \n']);
zm_g=mtt3_g-mtt1_g;
fprintf('The non-normalised marker z axis defined in global, z_m');
zm_g
a1=mtt2_g-mtt1_g;
fprintf('The non-normalised marker x axis (interim) defined in global, a=');
a1
ym_g=cross(zm_g,a1);
fprintf('The non-normalised marker y axis defined in global, y_m=');
ym_g
xm_g=cross(ym_g,zm_g);
fprintf('The non-normalised marker x axis defined in global, x_m=');
xm_g
zm_g=zm_g/norm(zm_g);
ym_g=ym_g/norm(ym_g);
xm_g=xm_g/norm(xm_g);
m2g=[xm_g'; ym_g'; zm_g'];
fprintf('The marker to global transformation matrix, M->G=');
m2g
g2m=m2g';
fprintf('The global to marker transformation matrix, G->M=');
g2m
g2m=[xm_g ym_g zm_g]';
g2a=m2a*g2m;
fprintf('The global to anatomical transformation matrix, G->A=');
g2a
theta2=asin(g2a(3,1));
theta(j,1)=rad2deg(acos(g2a(3,3)/cos(theta2)));
theta(j,3)=rad2deg(acos(g2a(1,1)/cos(theta2)));
theta(j,2)=rad2deg(theta2);
disp(['The joint angles corresponding to the time frame: ' num2str(j+3) ' are:']);
disp(['theta_1=' num2str(theta(j,1))]);
disp(['theta_2=' num2str(theta(j,2))]);
disp(['theta_3=' num2str(theta(j,3))]);
a2g=g2a';
fprintf('\nThe anatomical to global transformation matrix, A->G=');
a2g
c_g=a2g*c_vec;
fprintf('The c vector defined in the global coordinate system, c=');
```

```
c_g
Rm=mtt1_g;
Rc(j,:)=Rm+c_g;
fprintf('The R_c vector defined in the global coordinate system, c=');
Rc
if j==3
%record the G->A matrix of time frame 6 for kinetics calculations
G2A_t6=g2a;
end
end
%%the angular velocity calculations
omega=zeros(3,3);
j=1
for i=1:3
%the time interval is 0.002
disp(['************************']);
disp(['************************']);
disp(['************************']);
disp(['************************']);
disp(['************************']);
disp(['The angular velocity calculations corresponding to the time frame:' num2str(i+4)]);
%here the numbering is such a way that timeframe 4 is i=1, timeframe 5 is i=2
%and so on...
theta2= theta(i+1,2);
theta3= theta(i+1,3);
%angular velocity
theta_dot=(theta(i+2,:)-theta(i,:))/0.004;
disp(['The angular velocity, theta^dot, corresponding to the time frame:' num2str(i+4) ' =']);
theta_dot
disp(['\nThe angular velocity, theta^dot, corresponding to the time frame:' num2str(i+4) ' in radians =']);
theta_dot=deg2rad(theta_dot)
%transformation matrix for the angular velocity is dependent on the joint
%angle
ang_MT=[cos(deg2rad(theta2))*cos(deg2rad(theta3)) sin(deg2rad(theta3)) 0;-cos(deg2rad(theta2))*sin(deg2rad(theta3)) cos(deg2rad(theta3)) 0; sin(deg2rad(theta2)) 0 1];
%the transformed angular velocity
omega(j,:)=ang_MT*theta_dot';
disp(['The angular velocity, theta^dot, corresponding to the time frame: ' num2str(i+4) ' =']);
omega(j,:)
if i==3
```

```
%%the angular acceleration corresponding to time frame 6 calculations:
ang_acc=(omega(3,:)-omega(1,:))/0.004;
end
j=j+1;
end
fprintf('\nThe angular acceleration, alpha, corresponding to the time frame:6 =
');
ang_acc
%%the linear velocity calculations
velocity=zeros(3,3);
j=1;
for i=1:3
%the time interval is 0.002
disp(['*********************']);
disp(['*********************']);
disp(['*********************']);
disp(['*********************']);
disp(['*********************']);
disp(['The angular velocity calculations corresponding to the time frame: '
num2str(i+4) ' is being carried out \n']);
%angular velocity
velocity(j,:)=(Rc(i+2,:)-Rc(i,:))/0.004;
disp(['The velocity, v, corresponding to the time frame: ' num2str(i+4) ' =']);
velocity(j,:)
if i==3
%%the angular acceleration corresponding to time frame 6 calculations:
acc=(velocity(3,:)-velocity(1,:))/0.004;
end
j=j+1;
end
fprintf('\nThe acceleration, a, in cm/s^2 corresponding to the time frame:6 =');
acc
acc=acc*0.01; %acceleration in m/s
%%kinematic analysis
%distal reaction forces and moments as given in the textbook
R_XD=-157.53;
R_YD=414.59;
R_ZD=-7.64;
R_D=[R_XD;R_YD;R_ZD];
M_XD=-2.23;
M_YD=1.64;
M_ZD=0.09;
M_D=[M_XD;M_YD;M_ZD];
m=2.44;
%proximal reaction forces
```

```
R_XP=m*acc(1)+R_XD;
R_YP=m*acc(2)+R_YD+m*9.81;
R_ZP=m*acc(3)+R_ZD;
R_P=[R_XP;R_YP;R_ZP];
%the above reaction forces are w.r.t to the GRS
%The reaction forces w.r.t to the ACS are:
R_d=G2A_t6*R_D;
fprintf('\nThe distal reaction forces, corresponding to the time frame:6, w.r.t to the ACS: ');
R_d
R_p=G2A_t6*R_P;
fprintf('\nThe proximal reaction forces, corresponding to the time frame:6, w.r.t to the ACS: ');
R_p
M_d=G2A_t6*M_D;
fprintf('\nThe distal reaction moments, corresponding to the time frame:6, w.r.t to the ACS: ');
M_d
% solve for the proximal moments using the Euler equations:
I_x=0.037;
I_y=0.0037;
I_z=0.037;
length=0.389; % in meters
l_d=0.433*length;
l_p=length-l_d;
%the proximal moments using the Euler equations:
M_xp=I_x*ang_acc(1)+(I_z-I_y)*omega(2,2)*omega(2,3)+R_d(3)*l_d+R_p(3)*l_p+M_d(1);
M_yp=I_y*ang_acc(2)+(I_x-I_z)*omega(2,1)*omega(2,3)+M_d(2);
M_zp=I_z*ang_acc(3)+(I_y-I_x)*omega(2,1)*omega(2,2)+R_d(1)*l_d+R_p(1)*l_p+M_d(3);
```

Multiple Choice Questions (MCQ)

1. Which of the following variables cannot be obtained based on the co-ordinate data of markers?

   (A) Ground reaction force
   (B) Limb-segment and joint angles
   (C) Linear and angular velocities
   (D) Linear and angular acceleration

2. Which method is used to calculate the joint moments and forces, from the known information of kinetics?

(A) Newton's equation of motions
(B) Inverse dynamics
(C) Inverse kinematics
(D) Forward kinematics

3. Postural stability requires the COG to be located _____ the base of support.

(A) Within
(B) Outside
(C) On the boundary

4. Balance of the human body depends on the position of the center of gravity (COG) and _______.

(A) Center of mass (COM)
(B) Center of pressure (COP)
(C) Center of foot placement

5. Joint angle is defined as the angle between the axis of the proximal and distal segments of a joint. (True/False)

(A) True
(B) False

6. Anthropometric variables refer to data related to _____.

(A) Joint angle
(B) Joint forces
(C) Segment mass and inertial parameters

7. The contribution of each muscle force and distribution is solved using which numerical approach?

(A) Static optimisation
(B) Inverse dynamics
(C) Newton's equation of motion
(D) Forward dynamics

8. Which of the following transformation matrix is constant?

(A) Global to anatomical
(B) Global to marker
(C) Marker to anatomical

9. Which one of the following is not a kinematic variable?

(A) Angular velocity
(B) Angular acceleration
(C) Ground reaction force
(D) Displacement of limb segment

10. Which of the following is NOT an input parameter for inverse dynamics calculation using the link segment model?

    (A) Kinematic data
    (B) Kinetic data
    (C) Anthropometric data
    (D) Muscular force data

11. What is the angle between the long axis of the muscle and the fibre?

    (A) Physiologic cross-section angle (PCA)
    (B) Pennation angle
    (C) Angle of Inclination
    (D) Fiber angle

12. Muscles of the same group share the load, in proportion to their relative ____________.

    (A) Cross-sectional area
    (B) Length from the origin of insertion
    (C) Angle of pull

13. When modelling the musculoskeletal human body, a set of force or moment equilibrium equations, $Ax = b$, is formulated for the modelled structure. What does x represent here?

    (A) Unknown joint force in each segment
    (B) Unknown displacement of the segment
    (C) Unknown muscle force

14. ____________ is the unknown component in the inverse kinetics calculation of lower limb.

    (A) Reaction force at the proximal joint
    (B) Reaction force at the distal joint
    (C) Acceleration of segment
    (D) Net moment at distal joint

15. The physiological cross-section of muscle ________ with an increase in pennation angle.

    (A) Increases
    (B) Decreases
    (C) Remains unaltered

Answer Key to MCQs

1. A; 2. B; 3. A; 4. B; 5. A; 6. C; 7. A; 8. C; 9. C; 10. D; 11. B; 12. A; 13. C; 14. A; 15. B.

References

Awad A, Ali MS, Elassal MI (2021) Correlation between age, muscle architecture, and muscle strength in children with Erb's palsy. Bull Fac Phys Ther 26(1)

Bolsterlee B, D'Souza A, Gandevia SC, Herbert RD (2017) How does passive lengthening change the architecture of the human medial gastrocnemius muscle? J Appl Physiol 122(4):727–738

Bergmann G, Deuretzbacher G, Heller M, Graichen F, Rohlmann A, Strauss J, Duda GN (2001) Hip contact forces and gait patterns from routine activities. J Biomech 34(7):859–871

Brand RA, Pedersen DR, Friederich JA (1986) The sensitivity of muscle force predictions to changes in physiologic cross-sectional area. J Biomech 19(8):589–596

Chen Z, Franklin DW (2023) Musculotendon parameters in lower limb models: simplifications, uncertainties, and muscle force estimation sensitivity. Ann Biomed Eng 51(6):1147–1164

Crowninshield RD, Brand RA (1981) A physiologically based criterion of muscle force prediction in locomotion. J Biomech 14:793–801

Crowninshield RD, Johnston RC, Andrews JG, Brand RA (1978) A biomechanical investigation of the human hip. J Biomech 11:75–85

Davy DT, Audu ML (1987) A dynamic optimization technique for predicting muscle forces in the swing phase of gait. J Biomech 20:187–201

Dul J, Johnson GE, Shiavi R, Townsend MA (1984a) Muscle synergism-I: on criteria for load sharing between synergistic muscles. J Biomech 17:663–673

Dul J, Johnson GE, Shiavi R, Townsend MA (1984b) Muscle synergism-II: a minimum-fatigue criterion for load sharing between synergistic muscles. J Biomech 17:675–684

Erdemir A, McLean S, Herzog W, van den Bogert AJ (2007) Model-based estimation of muscle forces exerted during movements. Clin Biomech 22:131–154

Gage WH, Winter DA, Frank JS, Adkin AL (2004) Kinematic and kinetic validity of the inverted pendulum model in quiet standing. Gait Posture 19(2):124–132

Hardt DE (1978) Determining muscle forces in the leg during normal human walking: an application and evaluation of optimization methods. J Biomech Eng 100:72–78

Hu X, Charles JP, Akay T, Hutchinson JR, Blemker SS (2017) Are mice good models for human neuromuscular disease? Comparing muscle excursions in walking between mice and humans. Skeletal Muscle 7(1):26

Mathai B, Gupta S (2019) Numerical predictions of hip joint and muscle forces during daily activities: a comparison of musculoskeletal models. Proc Inst Mech Eng H 233(6):636–647

Pedotti A, Krishnan VV, Stark L (1978) Optimization of muscle-force sequencing in human locomotion. Math Biosci 38(1–2):57–76

Ounpuu S, Davis RB, DeLuca PA (1996) Joint kinetics: methods, interpretation, and treatment decision-making in children with cerebral palsy and myelomeningocele. Gait Posture 4:62–78

Seireg A, Arvikar RJ (1975) The prediction of muscular load sharing and joint forces in the lower extremities during walking. J Biomech 8(2):89–102

Seireg A, Arvikar RJ (1973) A mathematical model for evaluation of forces in lower extremities of the musculoskeletal system. J Biomech 6:313–326

Winter DA, Patla AE, Prince F, Ishac M, Gielo-Perczak K (1998) Stiffness control of balance in quiet standing. J Neurophysiol 80(3):1211–1221

Winter DA (2009) Biomechanics and motor control of human movement, 4th edn. Wiley, Hoboken

Chapter 5
Bone Tissue Mechanics

In the previous chapter, we considered the body segment to be non-deformable while analysing the joint dynamics. This chapter highlights the fundamentals of the mechanics of deformable body, pertinent to bone tissue mechanics. Subsequently, the mechanical properties of bone and their response to external forces will be discussed.

5.1 Fundamentals of Stress and Strain

In a spring-mass system, let us consider a mass M, hanging from the end of a spring (Fig. 5.1). When the system is in equilibrium, the spring remains undisturbed at a displacement of x. Notably, when the mass is doubled ($2M$), the elongation of the spring is proportionally doubled ($2x$). This observation leads to the understanding of the relationship between force and deformation.

Hooke's Law of Elasticity states that for relatively small deformations, the displacement 'x' is directly proportional to the deforming force 'F.'

$$F \propto x$$
$$F = kx$$

Here, the constant of proportionality (k) is the stiffness of the material, indicating its property to resist deformation.

5.1.1 Stress and Strain

Stress is defined as the internal resistive force per unit area that is generated within a body when it is acted upon by a set of external forces. When a specimen is loaded by

S. Gupta et al., *Biomechanics of Joints and Implants*,
https://doi.org/10.1007/978-981-96-0586-6_5

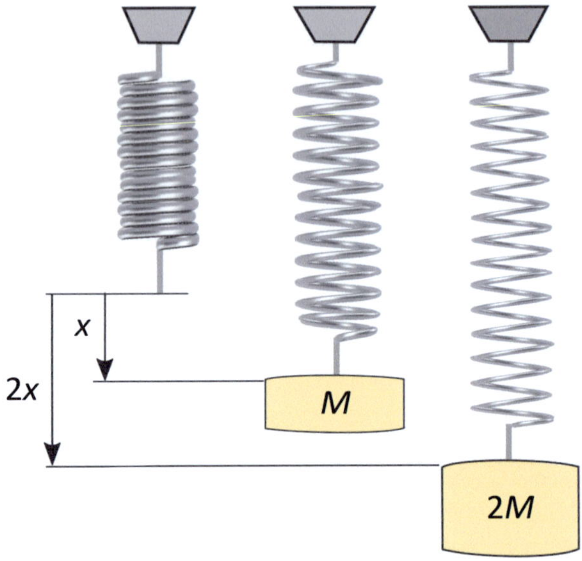

Fig. 5.1 Effect of mass on elongation of a spring

uniaxial force P (Fig. 5.2), the normal stress (σ) is defined as the ratio of the resisting force (equal to P) to the cross-sectional area of the specimen (A). The direction and magnitude of stress can vary depending on the applied forces and the specific area under consideration.

Strain is a measure of the deformation relative to the original size of a specimen. Linear strain (ε) is defined as the ratio of the change in length (Δl) to the original length (l) (Fig. 5.2). Unlike stress, strain is a dimensionless quantity, representing a fractional change in dimension. It is important to note that strain may occur as normal strain (change in size, elongation/contraction) and shear strain (change in shape, angular deformation).

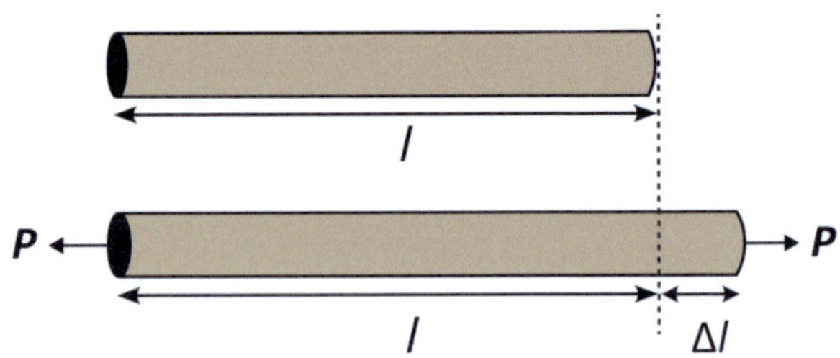

Fig. 5.2 Elongation of a specimen subjected to uniaxial force

5.1.2 Stress–Strain Curve

The stress–strain curve of a ductile material, is presented in Fig. 5.3. First, we encounter the proportionality limit, denoted as point 'a'. Until the proportionality limit, linear elastic behaviour prevails, showcasing a direct proportionality between the stress and the strain. This linear relationship is represented by the tangent of the angle formed by the linear portion of the curve with the strain axis. This proportionality constant, i.e. tangent of the angle, is called Young's modulus (E), a material property intrinsically defined by Hooke's law.

$$\sigma = E\varepsilon$$

The point 'b' denotes the elastic limit. Until this point, the specimen returns to its original configuration upon removal of the load. Beyond the elastic limit (point 'b'), the deformation remains irreversible and is known as plastic deformation, resulting in a permanent change in the shape of the specimen.

In case of a material exhibiting no distinct yield point, the yield strength is defined as the stress that corresponds to a permanent strain of 0.002 (oc′). The yield point is denoted by point 'c'. When the specimen is subjected to plastic deformation beyond the yield point, the strength of the specimen increases, indicated by an increase in load-carrying capacity, owing to strain hardening. The maximum limit of the stress is called ultimate strength (point 'd'). Beyond the point 'd', necking of the specimen sets in, resulting in a reduction in localised cross-sectional area. Since the stress is calculated in terms of the original area (undeformed cross-section), the stress value drops and eventually specimen ruptures at point 'e'.

The stiffness (k) of a specimen subjected to tensile force depends not only on Young's modulus (E) but also on the geometry of the specimen (length, l and area, A). The equation, $\sigma = E\varepsilon$, is rewritten as

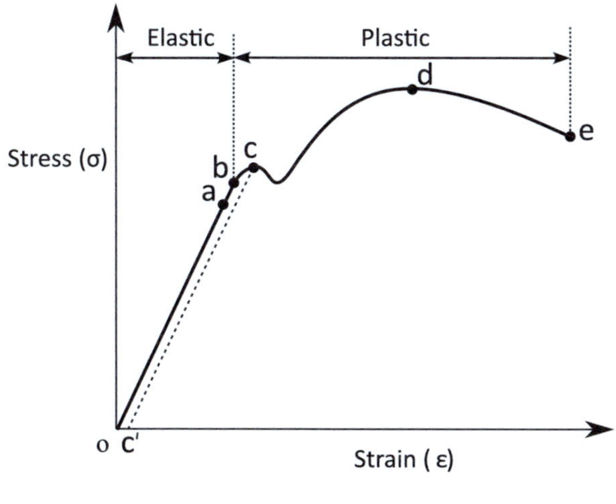

Fig. 5.3 A typical stress–strain curve of a ductile material, indicating the key points

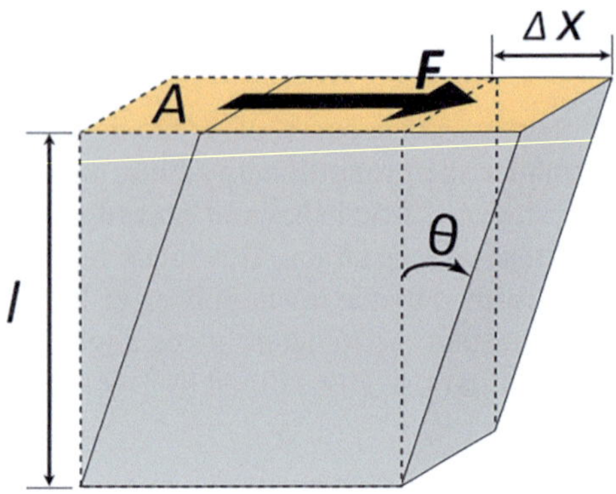

Fig. 5.4 A block subjected to a tangential force, F

$$\frac{F}{A} = E\left(\frac{\Delta l}{l}\right)$$

$$F = \left(\frac{EA}{l}\right)\Delta l = k\Delta l$$

Here, $k = \frac{EA}{l}$

In order to quantify the effect of a tangential force acting on a block shown in Fig. 5.4, we focus our attention on shear stress and shear strain. Similar to Young's modulus, the shear modulus or modulus of rigidity (G) is a material property. The shear modulus relates the shear stress (τ) with the shear strain (γ), according to

$$\tau = G\gamma$$

where shear stress and strain are expressed as follows:

$$\text{Shear stress,}\quad \tau = \frac{F}{A}$$

$$\text{Shear strain,}\quad \gamma = \tan(\theta) = \frac{\Delta x}{l}$$

5.1.3 *Shear Stress due to Torsion*

Let us consider a shaft subjected to torque (twisting moment) at one end, and fixed at the other end, as shown in Fig. 5.5. The torque generates shear stress within the shaft and the value of this shear stress varies linearly from the centre of the shaft to the periphery of the shaft. The shear stress experienced by any element at the centre of the shaft is equal to zero, whereas the shear stress is maximum at the periphery. The shear stress (τ) and strain (γ), due to torsion, are expressed as

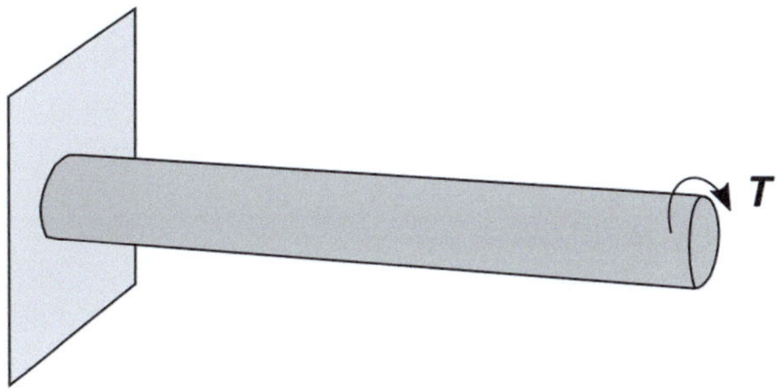

Fig. 5.5 A shaft subjected to torque, T

$$\tau = \frac{Tr}{J}$$

$$\gamma = r\frac{\phi}{L}$$

Here T is the external torque applied on the shaft, r is the radius of the shaft, J is the polar moment of inertia, ϕ is the angle of twist produced at the free end due to torsion, and L is the length of the shaft. The polar moment of inertia for a circular solid shaft is equal to

$$J = \pi\frac{r^4}{2}$$

5.1.4 Pure Bending

The Euler–Bernoulli beam theory, named after Leonhard Euler and Daniel Bernoulli, forms the basis of understanding the bending of slender beams. The Euler–Bernoulli beam theory makes several key assumptions to simplify the analysis of bending behaviour:

- The beam is slender, allowing shear deformations to be neglected.
- Deformations are small, ensuring that linear elastic material behaviour can be assumed.
- Cross-sections normal to the neutral axis remain plane even after deformation, i.e. plane sections remain plane, even after bending.

Let us consider a beam having a uniform cross-section, which was initially straight. The beam is subjected to bending moment (M) as given in Fig. 5.6.

Consider a small region ABCD (of the beam), as shown in Fig. 5.7a. After bending, the topmost layer AB is contracted to A′B′, whereas the bottom most layer CD is elongated to C′D′ as shown in Fig. 5.7b. The neutral layer, denoted by MN, does not exhibit any change in length. Let the angle subtended by A′C′ and B′D′, at point O,

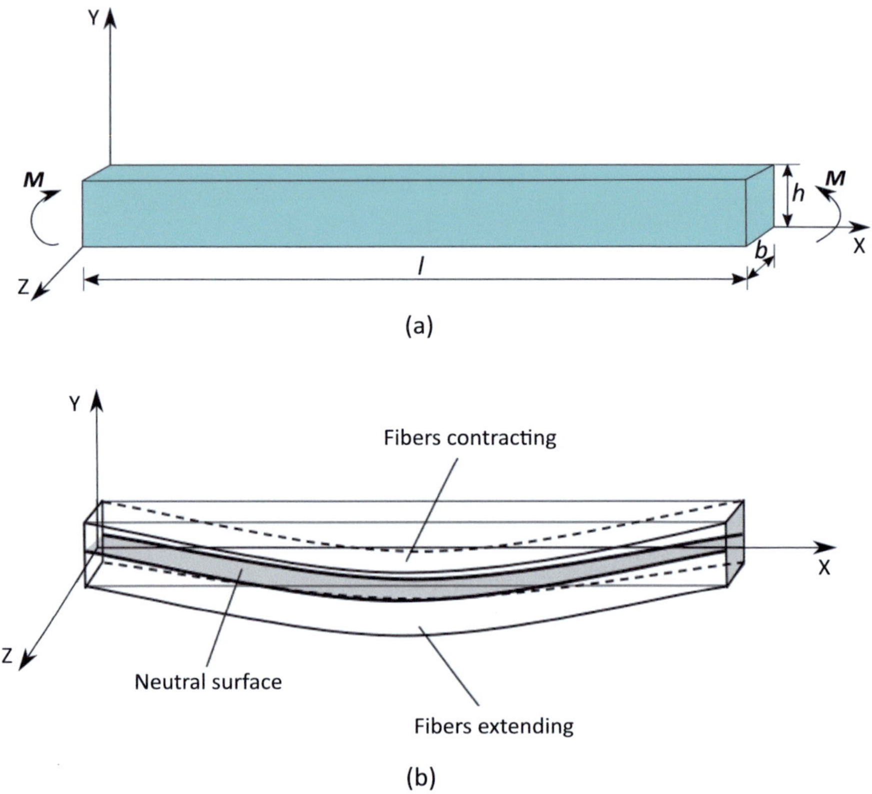

Fig. 5.6 Pure bending; **a** beam subjected to moment, M, at both ends, **b** the deformed beam

be equal to $d\theta$. Consider the layer PQ, which is contracted to P′Q′ at a distance of y from the neutral layer.

The change in the length of any fibre in the layer PQ,

$$\Delta \mathrm{PQ} = \mathrm{P'Q'} - \mathrm{PQ} = (R - y)d\theta - Rd\theta = -yd\theta$$

Strain exhibited by any fibre in layer, PQ is expressed as

$$\epsilon = \frac{\text{Change in length}}{\text{Original length}} = \frac{-yd\theta}{Rd\theta} = -\frac{y}{R}$$

Stress exhibited by any fibre in layer, PQ, $\sigma = E\epsilon = -\frac{Ey}{R}$
The above equation is rewritten as,

$$\frac{\sigma}{y} = -\frac{E}{R} \tag{5.1}$$

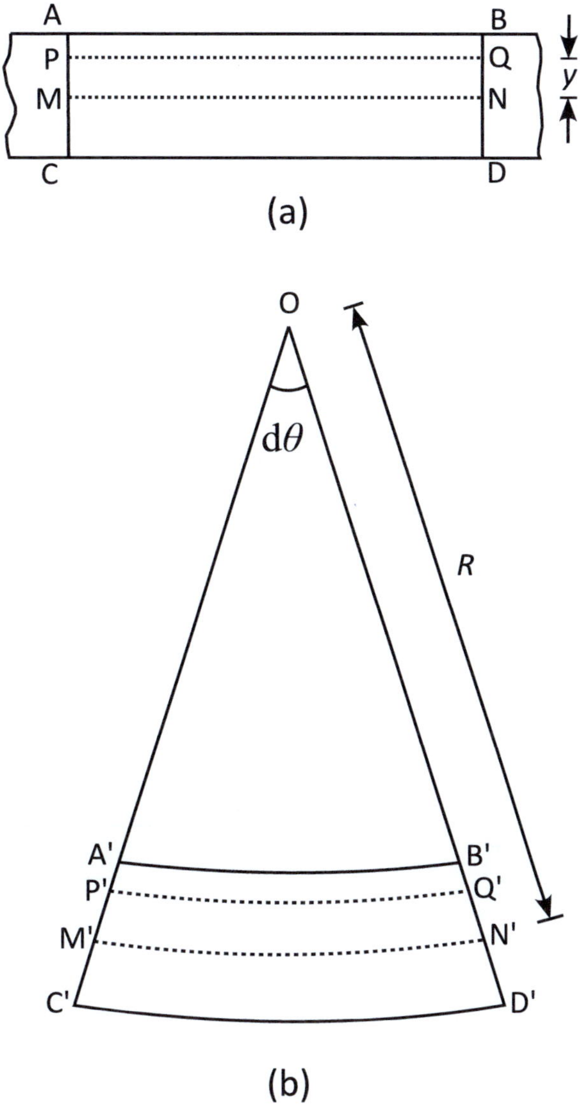

Fig. 5.7 Strain owing to bending. **a** Undeformed region, **b** deformed region

Hence the strain and stress owing to pure bending are linearly proportional to the distance of the fibre from the neutral layer.

The position of the neutral axis or layer is located as follows. An arbitrary cross-section of a beam is shown in Fig. 5.8. Consider a small area of breadth dy and the stress acting on it is equal to σ. The centroid of the strip is situated at a distance of y from the neutral axis (arbitrary). As we know there is no axial force involved in pure bending. According to the force equilibrium along the X-axis (Fig. 5.9),

$$\Sigma F_x = \int dF = \int_A \sigma dA = 0$$

The above equation is written as

$$\int_A \sigma dA = \frac{-E}{R} \int ydA = 0$$

The integral term, $\int ydA$ results in the first moment of area with respect to the neutral axis. The above equation is satisfied only if the neutral axis coincides with the centroidal axis. The moment equilibrium equation is written as (Fig. 5.9):

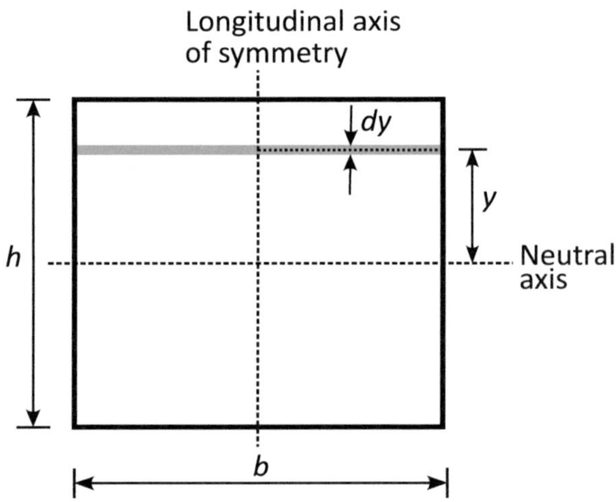

Fig. 5.8 Method of finding the location of neutral axis

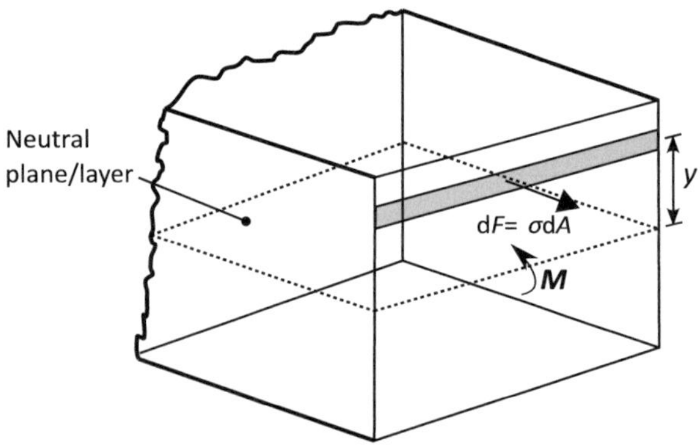

Fig. 5.9 Internal force associated with the strip at an arbitrary cross-section

$$\Sigma M_z = M + \int_A \sigma y dA = 0$$

$$M = -\int_A \sigma y dA$$

Using Eq. (5.1), the above equation is rewritten as follows:

$$M = -\int_A \frac{-Ey}{R} y dA$$

$$M = \int_A \frac{E}{R} y^2 dA = \frac{E}{R} \int_A y^2 dA = \frac{EI}{R}$$

$$\frac{M}{\mathrm{I}} = \frac{E}{R} \tag{5.2}$$

Here, I is equal to $\int_A y^2 dA$, the second moment of area (moment of inertia).
The radius of curvature (R) is expressed as

$$R = \frac{\left[1 + \left(\frac{dy}{dx}\right)^2\right]^{\frac{3}{2}}}{\frac{d^2y}{dx^2}}$$

However, for small deflections, the above equation is reduced to

$$R = \frac{1}{\frac{d^2y}{dx^2}} \tag{5.3}$$

Therefore, the differential equation governing pure bending is written as

$$M = EI\frac{d^2y}{dx^2}$$

Neglecting the negative sign in Eq. (5.1), and combining Eqs. (5.1), (5.2), and (5.3), the flexural equation is written as

$$\frac{M}{I} = \frac{E}{R} = \frac{\sigma}{y} \tag{5.4}$$

5.1.5 Generalized Hooke's Law

When we apply a uniaxial tensile force (F) on a body along the longitudinal direction, it exhibits a tendency to contract along the transverse direction (Fig. 5.10). This phenomenon is attributed to the Poisson's ratio (ν), which quantifies the transverse contraction associated with the longitudinal elongation. In 2D, the transverse strain ($\varepsilon_{\text{transverse}}$) owing to contraction is expressed in terms of Poisson's ratio and axial strain ($\varepsilon_{\text{axial}}$) as follows:

$$\varepsilon_{\text{transverse}} = -\nu\varepsilon_{\text{axial}}$$

These interrelated mechanical behaviours within an isotropic material in three dimensions are described by the constitutive equations as articulated by the generalized Hooke's law, stated below.

$$\varepsilon_x = \frac{\sigma_x}{E_x} - \frac{\nu\sigma_y}{E_y} - \frac{\nu\sigma_z}{E_z};\ \varepsilon_y = \frac{\sigma_y}{E_y} - \frac{\nu\sigma_x}{E_x} - \frac{\nu\sigma_z}{E_z};\ \varepsilon_z = \frac{\sigma_z}{E_z} - \frac{\nu\sigma_x}{E_x} - \frac{\nu\sigma_y}{E_y}$$

$$\gamma_{xz} = \frac{\tau_{xz}}{G_{xz}};\ \gamma_{xy} = \frac{\tau_{xy}}{G_{xy}};\ \gamma_{yz} = \frac{\tau_{yz}}{G_{yz}}$$

Here, σ_i denotes the normal stress along the ith direction. ε_i denotes the normal strains along the ith direction. γ_{jk} is shear strain experienced in the jth plane, along the kth direction. τ_{jk} is shear stress generated in the jth plane, along the kth direction. It should be noted that the strain along a particular direction is not solely related to the stress generated along that specific direction. It also accounts for the stresses acting along the other two orthogonal directions, as defined by the generalized Hooke's law.

Let us now focus our attention on a generalized scenario where a solid body is acted upon by a set of forces and constraints. The deformation response of the

Fig. 5.10 Lateral contraction associated with the longitudinal elongation owing to uniaxial tensile loading

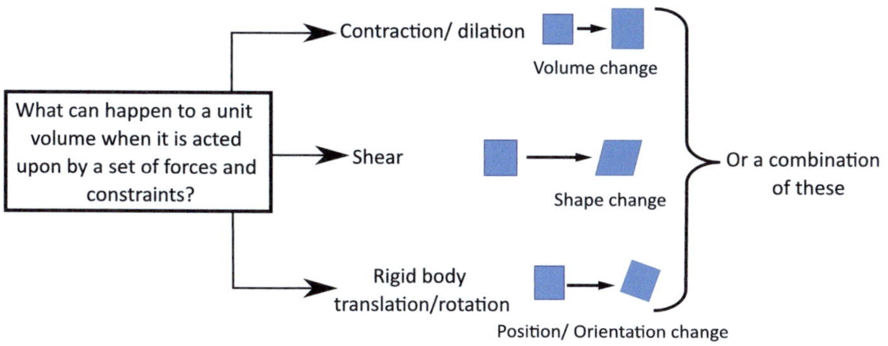

Fig. 5.11 Deformation of a solid due to external forces

solid body to these external forces and constraints can result in three primary cases (Fig. 5.11):

(1) It may undergo contraction or dilation, leading to a change in volume, due to normal force.
(2) It may undergo change in shape, due to shear force.
(3) It may translate or rotate, resulting in changes in position and orientation. It is noteworthy that a combination of these deformation modes may also occur in some complex loading scenarios.

5.1.6 Stress Tensor

Having explored the different modes of deformation, we now examine the concept of the stress tensor. The tensors are categorised into different orders. A zeroth-order tensor is a scalar quantity, a first-order tensor is a vector quantity, and the second-order tensor possesses both directional and orientation properties. Stress is a second-order tensor. For defining the state of stress, it is essential to specify the orientation of the area on which the stress is acting. In 2D, there are four components for the stress tensor (Fig. 5.12), whereas in 3D, there are nine components (Fig. 5.13). In 2D, the stress components in matrix form is expressed as

$$\sigma_{ij} = \begin{bmatrix} \sigma_{xx} & \tau_{xy} \\ \tau_{yx} & \sigma_{yy} \end{bmatrix}$$

In 3D, the stress components in matrix form is written as

$$\sigma_{ij} = \begin{bmatrix} \sigma_{xx} & \tau_{xy} & \tau_{xz} \\ \tau_{yx} & \sigma_{yy} & \tau_{yz} \\ \tau_{zx} & \tau_{zy} & \sigma_{zz} \end{bmatrix}$$

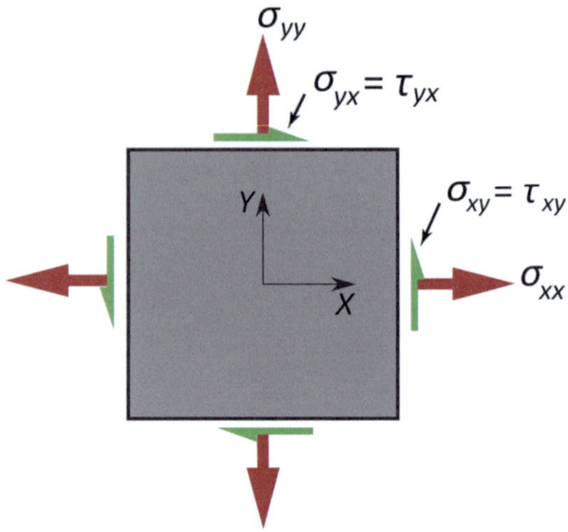

Fig. 5.12 State of stress in a two-dimensional element

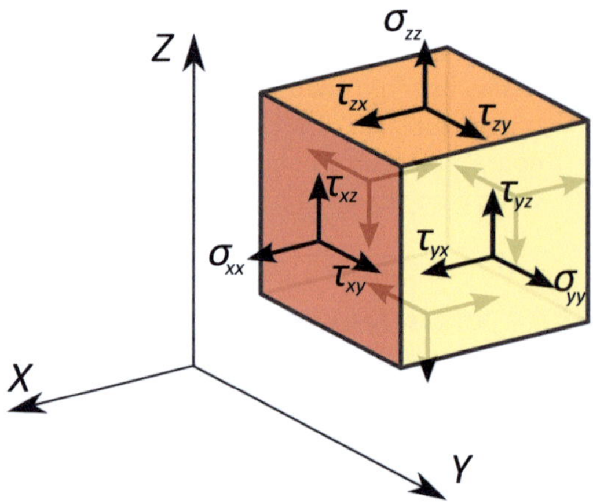

Fig. 5.13 State of stress in a three-dimensional element

The first index of the stress component represents the plane on which it acts and the second index represents the direction. Therefore, the stress component τ_{xy} corresponds to the stress acting on the x plane, along the y direction (Fig. 5.13). Conservation of angular momentum for a continuum requires the stress tensor to satisfy $\tau_{ij} = \tau_{ji}$ (here i and j can be x or y or z).

In general, any state of stress can be separated into two components: hydrostatic stress, which results in the change in volume, and deviatoric stress, which results in a change in shape (Fig. 5.14).

Hydrostatic stress refers to the uniform or isotropic state of stress within a material, wherein the stresses act equally along all directions (analogous to pressure). Hydrostatic stress results in a change in volume, but does not alter the shape of the material. It compresses or expands the material uniformly in all directions. Mathematically,

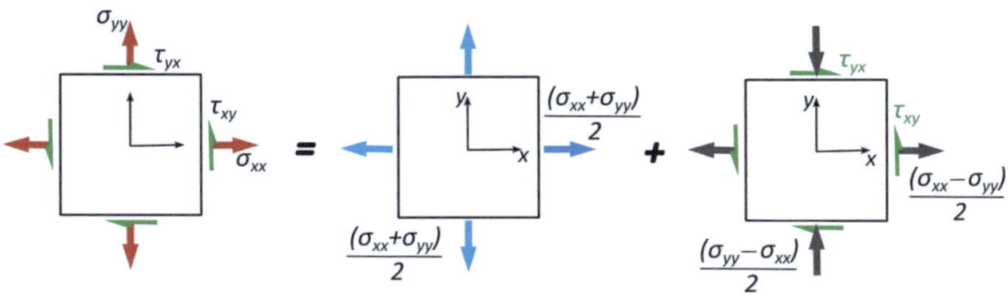

Fig. 5.14 Total stress is equal to the sum of hydrostatic and deviatoric component of stresses

hydrostatic stress tensor (σ_{hyd}) can be expressed as

$$\boldsymbol{\sigma}_{\boldsymbol{hyd}} = \boldsymbol{\sigma}_{\boldsymbol{m}} \boldsymbol{I}$$

Here, I is an identity matrix and σ_m is the average of the normal stresses; In 3D, $\sigma_m = \frac{\sigma_{xx}+\sigma_{yy}+\sigma_{zz}}{3}$.

Deviatoric stress represents the non-uniform or anisotropic state of stress within a material. Deviatoric stress results in shearing or distortion of the material without any change in volume. Mathematically, deviatoric stress tensor can be expressed as

$$\boldsymbol{\sigma}_{\boldsymbol{dev}} = \boldsymbol{\sigma} - \boldsymbol{\sigma}_{\boldsymbol{m}} \boldsymbol{I}$$

The stress tensor in 2D can be split into hydrostatic and deviatoric stress components as follows:

$$\sigma_{ij} = \begin{bmatrix} \sigma_m & o \\ o & \sigma_m \end{bmatrix} + \begin{bmatrix} \sigma_{xx} - \sigma_m & \tau_{xy} \\ \tau_{xy} & \sigma_{yy} - \sigma_m \end{bmatrix} = \begin{bmatrix} \frac{\sigma_{xx}+\sigma_{yy}}{2} & 0 \\ 0 & \frac{\sigma_{xx}+\sigma_{yy}}{2} \end{bmatrix} + \begin{bmatrix} \frac{\sigma_{xx}-\sigma_{yy}}{2} & \tau_{xy} \\ \tau_{xy} & \frac{\sigma_{yy}-\sigma_{xx}}{2} \end{bmatrix}$$

Similarly, the strain tensor in 2D can be split into hydrostatic and deviatoric strain components as follows:

$$\epsilon_{ij} = \begin{bmatrix} \epsilon_m & o \\ o & \epsilon_m \end{bmatrix} + \begin{bmatrix} \epsilon_{xx} - \epsilon_m & \epsilon_{xy} \\ \epsilon_{xy} & \epsilon_{yy} - \epsilon_m \end{bmatrix} = \begin{bmatrix} \frac{\epsilon_{xx}+\epsilon_{yy}}{2} & 0 \\ 0 & \frac{\epsilon_{xx}+\epsilon_{yy}}{2} \end{bmatrix} + \begin{bmatrix} \frac{\epsilon_{xx}-\epsilon_{yy}}{2} & \epsilon_{xy} \\ \epsilon_{xy} & \frac{\epsilon_{yy}-\epsilon_{xx}}{2} \end{bmatrix}$$

Let us discuss the concept of principal stress and von Mises stress that are crucial for stress analysis of biomechanical systems.

Principal stresses are the maximum and minimum normal stresses acting on planes where the shear stresses are absent. In three-dimensional stress analysis, there are typically three principal stresses, represented as σ_{11}, σ_{22}, and σ_{33}. The directions of these principal stresses are known as principal directions. The principal angles are the orientation of the principal planes (the planes on which the principal stresses act) with respect to the original coordinate axes.

The **von Mises stress**, also known as the equivalent stress, is a scalar value that combines the effect of both normal and shear stresses to estimate the overall state of stress in a material. The von Mises stress is often used to predict yielding or failure in ductile materials.

5.2 Stress Transformation

Stresses generated within a structures may act in different directions. Stress transformation helps us to find out the stress components acting on any arbitrary plane, which is oriented in a particular direction. Let us consider a body consisting of two materials and subjected to uniaxial tensile force P at both ends, as shown in Fig. 5.15. How do we know whether the interface can sustain the external force or not? For this, we need to determine the normal and shear stresses perpendicular and parallel to the interface, respectively. Therefore, we need to transform the coordinates (x, y) associated with the tensile force P to a different coordinate system (x', y') associated with the interface orientation (Fig. 5.16).

Using the free body diagram and equilibrium condition, we find out the set of equations relating the stresses $\sigma_{x'}$ and $\tau_{x'y'}$ to the stresses σ_x and τ_{xy}, acting on the area $\triangle$A (Fig. 5.17).

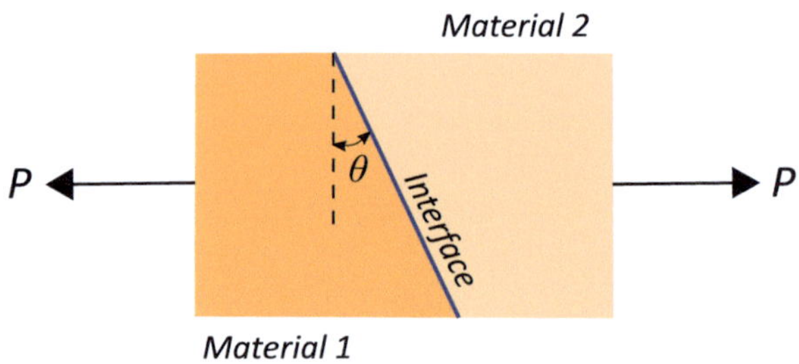

Fig. 5.15 Two materials glued at an interface making an angle θ with the direction of loading

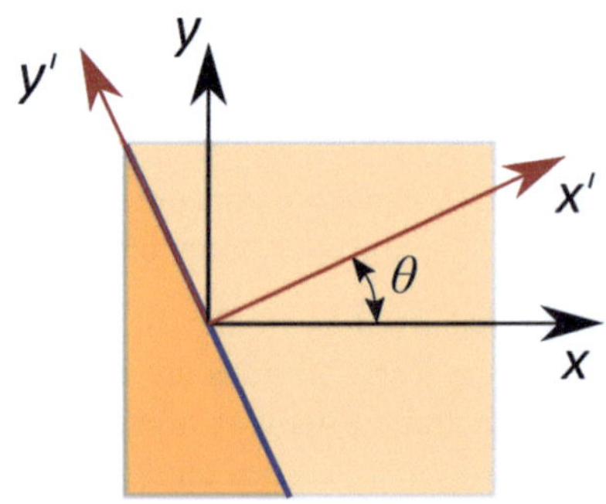

Fig. 5.16 Coordinate systems under consideration: reference (x, y) and new (x', y')

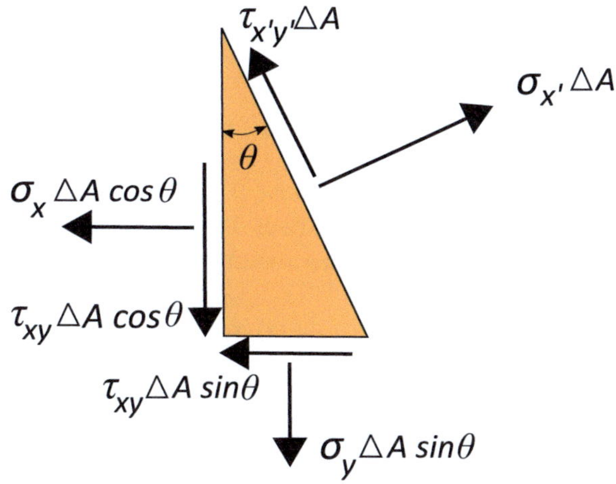

Fig. 5.17 Free body diagram consisting of the forces along the new and the old coordinate system

$$\sum F_{x'} = \sigma_{x'}\Delta A - (\tau_{xy}\Delta A\ \sin\theta)\cos\theta - (\tau_{xy}\Delta A\cos\theta)\sin\theta$$
$$- (\sigma_x\Delta A\cos\theta)\cos\theta - (\sigma_y\Delta A\sin\theta)\sin\theta = 0$$

$$\sigma_{x'} = \sigma_x\cos^2\theta + \sigma_y\sin^2\theta + 2\tau_{xy}\sin\theta\cos\theta \tag{5.5}$$

The normal stress component $\sigma_{y'}$, is determined by replacing θ by $90 + \theta$ in the above equation and is expressed as follows

$$\sigma_{y'} = \sigma_x\sin^2\theta + \sigma_y\cos^2\theta - 2\tau_{xy}\sin\theta\cos\ \theta \tag{5.6}$$

$$\sum F_{y'} = \tau_{x'y'}\Delta A + (\tau_{xy}\Delta A\sin\theta)\sin\theta - (\tau_{xy}\Delta A\cos\theta)\cos\theta$$
$$- (\sigma_y\Delta A\sin\theta)\cos\theta + (\sigma_x\Delta A\cos\theta)\sin\theta = 0$$

$$\tau_{x'y'} = (\sigma_y - \sigma_x)\sin\theta\cos\theta + \tau_{xy}(\cos^2\theta - \sin^2\theta) \tag{5.7}$$

Alternatively, the state of stress in the new coordinate system (x', y') can be determined by employing the transformation rule, obtained by rearranging the above set of Eqs. (5.5), (5.6), and (5.7), as follows:

$$[\sigma'] = [R][\sigma][R]^T$$

Here R is called as the rotational matrix, which in 2D is expressed as

$$[R] = \begin{bmatrix} \cos(\theta) & \sin(\theta) \\ -\sin(\theta) & \cos(\theta) \end{bmatrix}$$

The new 2D state of stress in terms of angle θ is expressed in matrix form:

$$[\sigma'] = \begin{bmatrix} \cos(\theta) & \sin(\theta) \\ -\sin(\theta) & \cos(\theta) \end{bmatrix} [\sigma] \begin{bmatrix} \cos(\theta) & -\sin(\theta) \\ \sin(\theta) & \cos(\theta) \end{bmatrix}$$

Sample Problem 5.1 The state of stress in the reference coordinate system (x, y) is given as the following. Find out the state of stress with respect to a new coordinate system that is obtained after rotating the reference coordinate system by 45°.

$$\begin{bmatrix} \sigma_x & \tau_{xy} \\ \tau_{yx} & \sigma_y \end{bmatrix} = \begin{bmatrix} 22 & 12 \\ 12 & -7 \end{bmatrix} \text{MPa}$$

The new state of stress ($[\sigma']$) is obtained using the transformation rule:

$$[\sigma'] = \begin{bmatrix} \cos(45^\circ) & \sin(45^\circ) \\ -\sin(45^\circ) & \cos(45^\circ) \end{bmatrix} \begin{bmatrix} 22 & 12 \\ 12 & -7 \end{bmatrix} \begin{bmatrix} \cos(45^\circ) & -\sin(45^\circ) \\ \sin(45^\circ) & \cos(45^\circ) \end{bmatrix}$$

$$[\sigma'] = \begin{bmatrix} 19.49 & -14.49 \\ -14.49 & -4.50 \end{bmatrix} \text{MPa}$$

Determination of Principal stresses and Maximum Shear stress: Let us determine the principal stresses and maximum shear stress, using stress transformation rule and after carrying out few trigonometric substitutions in Eqs. (5.5)–(5.7).

$$\sigma_{x'} = \frac{\sigma_x + \sigma_y}{2} + \frac{\sigma_x - \sigma_y}{2} \cos(2\theta) + \tau_{xy} \sin(2\theta) \tag{5.8}$$

$$\sigma_{y'} = \frac{\sigma_x + \sigma_y}{2} - \frac{\sigma_x - \sigma_y}{2} \cos(2\theta) - \tau_{xy} \sin(2\theta) \tag{5.9}$$

Upon differentiating the above equations with respect to the angle θ (the angle of the plane's orientation) and equating it to zero, we obtain the value of the angle θ_p (angle of inclination of the principal plane).

$$\tan(2\theta_p) = \frac{2\tau_{xy}}{(\sigma_x - \sigma_y)}$$

This can be further used to find out the principal stresses. Using the trigonometric relations from Fig. 5.18 and the transformation equations of the normal stress (Eqs. 5.8 and 5.9), the principal stresses are expressed as

$$\sigma_1 = \frac{\sigma_x + \sigma_y}{2} + \sqrt{\left(\frac{\sigma_x - \sigma_y}{2}\right)^2 + \tau_{xy}^2}$$

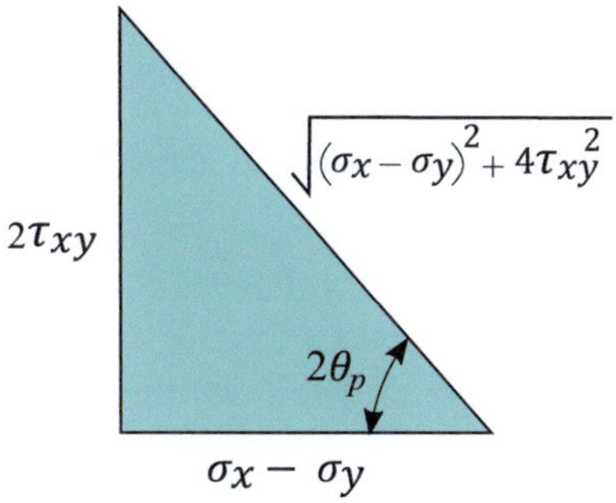

Fig. 5.18 Angle of inclination of the principal plane

$$\sigma_2 = \frac{\sigma_x + \sigma_y}{2} - \sqrt{\left(\frac{\sigma_x - \sigma_y}{2}\right)^2 + \tau_{xy}^2}$$

The transformation equation of the shear stress is written as

$$\tau_{x'y'} = \tau_{xy}\cos(2\theta) - \frac{\sigma_x - \sigma_y}{2}\sin(2\theta)$$

Upon differentiating with respect to θ and equating it to zero, we get the angle of inclination of planes of maximum shear.

$$\tan(2\theta_s) = \frac{\sigma_y - \sigma_x}{2\tau_{xy}}$$

Using the trigonometric relations from Fig. 5.19 and the transformation equation of the shear stress, the maximum shear stress is expressed as

$$\tau_{\max} = \sqrt{\left(\frac{\sigma_x - \sigma_y}{2}\right)^2 + \tau_{xy}^2}$$

We can express the above in terms of the principal stresses:

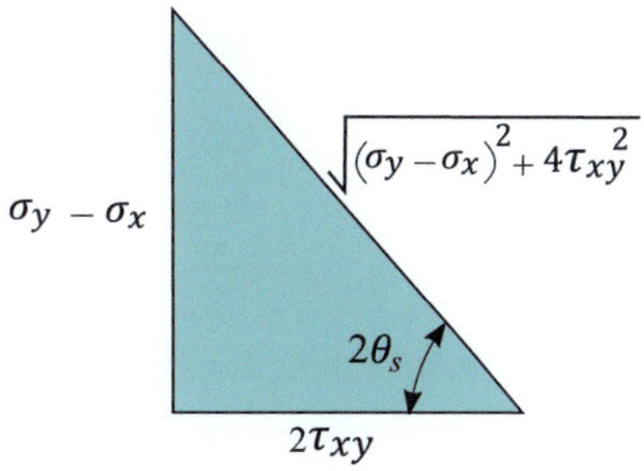

Fig. 5.19 Angle of inclination of the plane of maximum shear

$$\tau_{\max} = \frac{|\sigma_1 - \sigma_2|}{2}$$

The normal stresses acting in the plane of maximum shear stress are equal to

$$\sigma_{\text{avg}} = \frac{\sigma_{xx} + \sigma_{yy}}{2}$$

In order to determine the principal stresses, we can consider the stress tensor as a matrix 'A' in the eigenvalue problem, which involves finding the eigenvalues (principal stresses) and the associated eigenvectors (the orientations or directions of the principal planes).

$$Ax = \lambda x \tag{5.10}$$

The eigenvalue problem can be written in terms of stress as stated in the following:

$$\begin{bmatrix} \sigma_{xx} & \tau_{xy} & \tau_{xz} \\ \tau_{xy} & \sigma_{yy} & \tau_{yz} \\ \tau_{xz} & \tau_{yz} & \sigma_{zz} \end{bmatrix} \begin{bmatrix} n_1 \\ n_2 \\ n_3 \end{bmatrix} = \sigma \begin{bmatrix} n_1 \\ n_2 \\ n_3 \end{bmatrix}$$

The normal stress σ is an eigenvalue of the stress tensor whereas the unit vector $n = [n_1\ n_2\ n_3]^{\mathrm{T}}$ is the eigenvector. In 3D, there are three eigenvalues (principal stresses) and three corresponding eigenvectors (principal directions). Let us now discuss the non-trivial solution for this problem. The only non-trivial solution is obtained by setting the determinant of $(A - \lambda I)$ to zero (Eq. 5.10), which yields a cubic equation known as the characteristic equation as given in Eq. (5.11).

$$\sigma^3 - I_1\sigma^2 + I_2\sigma - I_3 = 0 \tag{5.11}$$

The coefficients of the cubic equation, I_1, I_2, and I_3 are known as the stress invariants. These coefficients do not vary with a change in the coordinate system. The solutions to the characteristic equation are the principal stresses $\sigma_1, \sigma_2,$ and σ_3.

$$I_1 = \sigma_{xx} + \sigma_{yy} + \sigma_{zz}$$

$$I_2 = \begin{vmatrix} \sigma_{xx} & \tau_{xy} \\ \tau_{xy} & \sigma_{yy} \end{vmatrix} + \begin{vmatrix} \sigma_{yy} & \tau_{yz} \\ \tau_{yz} & \sigma_{zz} \end{vmatrix} + \begin{vmatrix} \sigma_{xx} & \tau_{xz} \\ \tau_{xz} & \sigma_{zz} \end{vmatrix}$$

$$I_3 = \begin{vmatrix} \sigma_{xx} & \tau_{xy} & \tau_{xz} \\ \tau_{xy} & \sigma_{yy} & \tau_{yz} \\ \tau_{xz} & \tau_{yz} & \sigma_{zz} \end{vmatrix}$$

The principal directions (n) corresponding to each principal stress value are solved by substituting the principal stress into the Eigenvalue problem. Each of the principal directions, which is a unit vector, should satisfy the unit vector constraint (Eq. 5.12).

$$n_1^2 + n_2^2 + n_3^2 = 1 \tag{5.12}$$

In other words, the sum of squares of the components of each principal direction should be equal to 1.

Sample Problem 5.2 Let us now look into a sample problem solved using the concept of the eigenvalue problem. A 3D state of stress is given as below:

$$\sigma = \begin{bmatrix} \sigma_{xx} & \tau_{xy} & \tau_{xz} \\ \tau_{xy} & \sigma_{yy} & \tau_{yz} \\ \tau_{xz} & \tau_{yz} & \sigma_{zz} \end{bmatrix} = \begin{bmatrix} 22 & 12 & -15 \\ 12 & -7 & 3 \\ -15 & 3 & -17 \end{bmatrix} \text{MPa}$$

The principal stresses and directions are determined using the following steps. The Eigenvalue problem is written as

$$\begin{bmatrix} \sigma_{xx} & \tau_{xy} & \tau_{xz} \\ \tau_{xy} & \sigma_{yy} & \tau_{yz} \\ \tau_{xz} & \tau_{yz} & \sigma_{zz} \end{bmatrix} \begin{bmatrix} n_1 \\ n_2 \\ n_3 \end{bmatrix} = \sigma \begin{bmatrix} n_1 \\ n_2 \\ n_3 \end{bmatrix}$$

The invariants I_1, I_2 and I_3 are

$$I_1 = \sigma_{xx} + \sigma_{yy} + \sigma_{zz} = 22 + (-7) + (-17) = -2$$

$$I_2 = \begin{vmatrix} \sigma_{xx} & \tau_{xy} \\ \tau_{xy} & \sigma_{yy} \end{vmatrix} + \begin{vmatrix} \sigma_{yy} & \tau_{yz} \\ \tau_{yz} & \sigma_{zz} \end{vmatrix} + \begin{vmatrix} \sigma_{xx} & \tau_{xz} \\ \tau_{xz} & \sigma_{zz} \end{vmatrix} = \begin{vmatrix} 22 & 12 \\ 12 & -7 \end{vmatrix} + \begin{vmatrix} -7 & 3 \\ 3- & 17 \end{vmatrix} + \begin{vmatrix} 22 & 15 \\ 15 & -17 \end{vmatrix} = -787$$

$$I_3 = \begin{bmatrix} \sigma_{xx} & \tau_{xy} & \tau_{xz} \\ \tau_{xy} & \sigma_{yy} & \tau_{yz} \\ \tau_{xz} & \tau_{yz} & \sigma_{zz} \end{bmatrix} = \begin{bmatrix} 22 & 12 & -15 \\ 12 & -7 & 3 \\ -15 & 3 & -17 \end{bmatrix} = 5363$$

The characteristics equation is written as

$$\sigma^3 - I_1\sigma + I_2\sigma - I_3 = 0$$

Upon substituting the invariants in the above equation:

$$\sigma^3 + 2\sigma - 787\sigma - 5363 = 0$$

Upon solving the cubic equation, we obtain the principal stresses as

$$\sigma_1 = 30.08\,\text{MPa}$$
$$\sigma_2 = -7.15\,\text{MPa}$$
$$\sigma_3 = -24.94\,\text{MPa}$$

The principal directions corresponding to σ_1 are found out by substituting the stress value into the eigenvalue problem.

$$\begin{bmatrix} \sigma_{xx} & \tau_{xy} & \tau_{xz} \\ \tau_{xy} & \sigma_{yy} & \tau_{yz} \\ \tau_{xz} & \tau_{yz} & \sigma_{zz} \end{bmatrix} \begin{bmatrix} n_1 \\ n_2 \\ n_3 \end{bmatrix} = \sigma_1 \begin{bmatrix} n_1 \\ n_2 \\ n_3 \end{bmatrix}$$

Rearranging we get

$$\begin{bmatrix} \sigma_{xx} - \sigma_1 & \tau_{xy} & \tau_{xz} \\ \tau_{xy} & \sigma_{yy} - \sigma_1 & \tau_{yz} \\ \tau_{xz} & \tau_{yz} & \sigma_{zz} - \sigma_1 \end{bmatrix} \begin{bmatrix} n_1 \\ n_2 \\ n_3 \end{bmatrix} = 0$$

On substituting the values, we get

$$\begin{bmatrix} -8 & 12 & -15 \\ 12 & -37 & 3 \\ -15 & 3 & -47 \end{bmatrix} \begin{bmatrix} n_1 \\ n_2 \\ n_3 \end{bmatrix} = 0$$

Upon solving the above problem using the Gauss elimination method and employing the constraint of unit vector (Eq. 5.12), we obtain the principal direction corresponding to σ_1:

$$\begin{bmatrix} n_1 \\ n_2 \\ n_3 \end{bmatrix} = \begin{bmatrix} -0.92 \\ -0.28 \\ 0.28 \end{bmatrix}$$

Similarly, one can find the principal directions corresponding to other principal stresses. The principal direction corresponding to σ_2 is

$$\begin{bmatrix} n_1 \\ n_2 \\ n_3 \end{bmatrix} = \begin{bmatrix} -0.13 \\ 0.88 \\ 0.46 \end{bmatrix}$$

The principal direction corresponding to σ_3 is

$$\begin{bmatrix} n_1 \\ n_2 \\ n_3 \end{bmatrix} = \begin{bmatrix} -0.37 \\ 0.39 \\ -0.84 \end{bmatrix}$$

5.3 Bone Structure

Bone serves as a fundamental structural component of the human body, integral to the skeletal system. Structurally, bone provides protection to organs and soft tissues, transmits mechanical loads between different body parts during various activities, and helps maintain body shape. Bone is an adaptive material, that can alter its microstructure in response to changes in the mechanical environment. Moreover, the microstructure of bone is site-specific, exhibiting morphological variations based on local porosity.

From the molecular perspective, bone functions as a true composite material that consists of 65% mineral (Hydroxyapatite—HA), 35% organic matrix (mostly collagen fibres), water, cells and vessels. The mineral content is mostly impure HA in the form of tiny needle-shaped crystals, rods, and plates located within and between the organic matrix. The organic matrix is composed of 90% collagen and 10% of various non-collagenous proteins. Bone can be categorised into woven and lamellar types, based on collagen fibre architecture and arrangement. Woven bone (immature bone), a temporary and rapidly growing phase of bone, is typically observed in the foetus, young individuals, and during bone fracture healing. Its collagen fibres are randomly oriented and loosely packed as compared to lamellar bone. Lamellar bone (mature bone), characterised by more orderly arranged fibres, grows at a slower rate. Collagen and hydroxyapatite fibres in lamellar bone are organised into thin layers known as lamellae, and this type of bone generally has a lower mineral content than the woven bone. Both woven and lamellar bone contain small cavities (lacunae) connected by tubular canals (canaliculi) that house the bone cells (osteocytes).

Lamellar bone is further classified into cortical (or compact) and cancellous (or trabecular) bone, depending on the relative density or volume fraction of solids. The cortical bone is characterised by a dense outer shell surrounding the porous cancellous bone, with a volume fraction of solids greater than 70%. In contrast, cancellous bone is relatively porous, with a volume fraction of solids less than 70%. The mechanical properties of bone tissues are influenced by their organic and mineral composition. From an engineering standpoint, bone is a composite material consisting of a compliant organic matrix of collagen and non-collagenous proteins (NCP), along with a stiff mineral phase (Alijani and Vaughan 2024). Cancellous bone features a lattice-like network of interconnecting rods and plates, which is known as trabeculae. Whereas, cortical bone constitutes a solid mass with microscopic canals.

5.4 Cortical Bone

Cortical bone, or compact bone, is distinguished by its dense nature. The primary functional unit of compact bone is the osteon, or Haversian system, which plays a vital role in bone remodelling and maintenance. Osteons are approximately cylindrical, with diameters typically between 0.25 and 0.35 mm. They play a crucial role in the

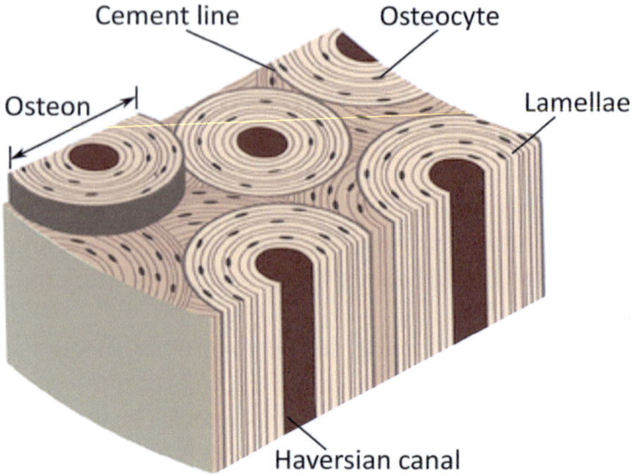

Fig. 5.20 Haversian system within cortical bone

repair and renewal of bone tissue, by adapting in response to mechanical stress and variations in mineral content (Thompson 1980). The Haversian system is organised into concentric layers of bone tissue called lamellae, which surround a central channel known as the Haversian canal. This canal contains blood vessels and nerves that are essential for the nourishment of the bone tissue (Fig. 5.20).

5.4.1 Mechanical Behaviour of Cortical Bone

Cortical bone comprises approximately 80% of the total bone mass in an adult human skeleton, despite representing only 20% of the total bone volume. It forms the diaphysis of long bones and constitutes the outer layer of the epiphyses and metaphyses. Ashman et al. (1984) employed a continuous wave ultrasound technique to measure the elastic moduli of cortical bone, estimating values of approximately 20–22 GPa along the axis of long bones and 12–14 GPa transversely. These measurements suggest that cortical bone exhibits transverse isotropy, being roughly 1.5–2 times stiffer and stronger along the longitudinal direction as compared to the radial or circumferential directions. An orthotropic or transversely isotropic constitutive relationship describes the cortical bone elastic properties fairly well (Guo 2008). A study by Carter et al. (1981) measured elastic moduli of cortical bone using mechanical testing and reported it to be in the range of 17.5 ± 1.9 GPa. However, these values are indicative and may vary based on factors, such as ethnicity, age, and sex of the individual.

The stress–strain behaviour of cortical bone, in the human femur, was reported by Özkaya and Nordin (1999). They observed three distinct regions in the stress–strain curve (Fig. 5.21). Initially, the bone exhibits a linear elastic region with an elastic modulus of around 17 GPa. This is followed by an intermediate region characterised by nonlinear elastoplastic behaviour, with bone yielding occurring at a yield strength

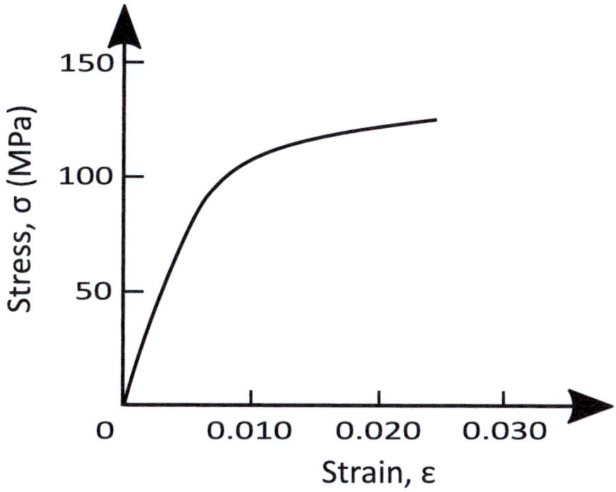

Fig. 5.21 Stress–strain diagram of human cortical bone, when loaded in the longitudinal direction (tensile) at a strain rate of 0.05 s^{-1} (Özkaya and Nordin 1999)

of approximately 110 MPa. The final region demonstrates linear plastic behaviour with a strain-hardening modulus of 0.9 GPa. The bone fractured when subjected to a tensile stress of 128 MPa and a corresponding tensile strain of 0.026 (Özkaya and Nordin 1999). Additionally, the study reported that the elastic moduli and strength of bone specimens varied with the strain rate, indicating the viscoelastic property of bone (Özkaya and Nordin 1999). Several Investigations on the material properties of the human femur under dynamic loading conditions were also carried out (Funk et al. 2004; Asgharpour et al. 2014). A more recent study by Mirzaali et al. (2016) reported that the yield strength of cortical bone is approximately 110 MPa in compression and 70 MPa in tension. Moreover, the strength and elastic moduli of cortical bone along the longitudinal direction (the direction aligned with the diaphyseal axis of femur) have been reported to be greater than those along the radial and circumferential directions (Morgan et al. 2018). Overall, the stress–strain behaviour of cortical bone resembles that of a brittle material, characterised by limited ductility and a tendency to fracture abruptly without substantial plastic deformation.

5.5 Cancellous Bone

The cancellous bone, alternatively referred to as trabecular bone, presents a porous network of rods and plates. Its volume fraction of solid material is less than 70%. Cancellous bone exhibits a spongy appearance due to the large number of interconnected trabeculae. While the trabeculae themselves lack blood vessels, adjacent tissues contain the necessary vasculature. Cancellous bone is primarily located in the inner core of bone structures, particularly in the epiphyseal regions of long bones. These trabeculae form either an open-cell structure when arranged in a network of rods, or a closed-cell structure when they form a network of plates. The mechanical behaviour of cancellous bone is similar to cellular material, such as polymeric foam.

The variations in density and structure of cancellous bone are observed upon examining the scanning electron micrograph images of cancellous bone (Fig. 5.22). These variations range from low-density cancellous bone with an asymmetric rod-like structure, typically extracted from the femoral head, to relatively higher density cancellous bone with a plate-like structure, also sourced from the femoral head. Additionally, the plate-like cancellous bone with a columnar structure has been found in specimens from the femoral condyle. These transformations, from rod-like to plate-like structures, represent transitions from open-cell to closed-cell type of structures.

Bone is a complex two-phase porous material. It is composed of a solid phase consisting of bone mineralised tissue and a fluid phase, comprising of blood vessels, blood, marrow, and interstitial fluid. Accurate measurement of the mineralised bone tissue is vital for understanding bone health and structure. The primary metrics used for this purpose are apparent density, ash density, and relative density.

Apparent Density (ρ), often referred to as structural density, is one of the most crucial measurements for bone mineralised tissue. It quantifies the mass of mineralised tissue per unit volume of the tissue. This measurement provides insights into the overall density of bone and its mineral content, making it valuable for assessing bone quality and strength.

Ash Density, is another metric used to measure bone mineralisation. It quantifies the ash weight (the residue left after the complete combustion of organic components) per unit tissue volume. Ash density is particularly useful in laboratory settings, where bone samples are subjected to controlled incineration to determine the mineral content precisely.

Relative Density (ρ_r), also known as the volume fraction of solid, is occasionally used to describe bone mineralised tissue. It is defined as the density of the cellular solid (the solid phase of bone) divided by the density of the solid cell wall material (ρ_s).

$$\rho_r = \frac{\rho}{\rho_s}$$

5.5.1 Mechanical Behaviour of Cancellous Bone

The mechanical behaviour of cancellous bone changes depending on the type of loading. The cancellous bone exhibits stress–strain characteristics similar to that of a ductile material. In the stress–strain curve of cancellous bone under compressive loading, three distinct phases can be identified (Fig. 5.23). The stress–strain curve of cancellous bone typically features an initial linear elastic phase, followed by a plateau

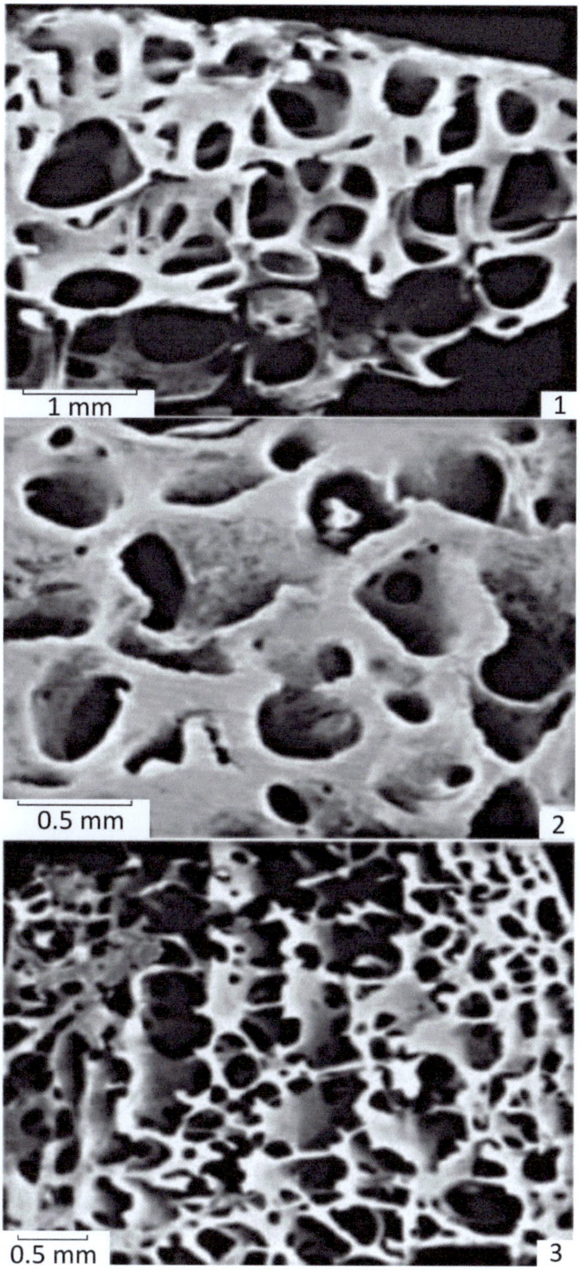

Fig. 5.22 Scanning electron micrographs of cancellous bone structure: **1**. Low density cancellous bone with an asymmetric rod-like structure; specimen taken from femoral head **2**. High density cancellous bone with an asymmetric plate-like structure; specimen taken from femoral head **3**. Plate-like cancellous bone with columnar structure; specimen taken from femoral condyle (Adapted from Gibson L. J. 1985)

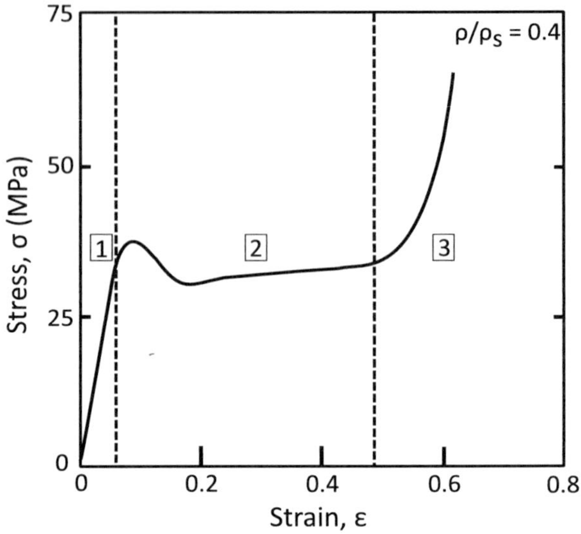

Fig. 5.23 Stress–strain behavior of cancellous bone under compressive loading

where the stress remains nearly constant despite increasing strain, and concludes with a steep upward slope leading to fracture (Gibson 1985; Özkaya and Nordin 1999).

Phase 1: In this phase, the bone tissue behaves like a linearly elastic material, with stress directly proportional to strain. During this initial phase, the behaviour is predominantly elastic, since the cell walls within the cancellous structure bend and compress axially under applied load.

Phase 2: As the applied load increases beyond a certain point, Phase 2 of "collapse and plateau" begins. In this phase, cells within the cancellous structure start to collapse. This collapse can occur through elastic buckling, plastic yielding, or brittle fracture of the cell walls. This phase progresses roughly at a constant load owing to almost zero reaction, until the cell walls meet and touch each other, resembling a plateau region. During this phase, the bone tissue undergoes appreciable deformation while the load remains relatively constant.

Phase 3: In the final phase, Phase 3, as the cell walls touch each other, the resistance to load increases steeply. This phase is marked by a rapid rise in the stress–strain curve. It culminates in fracture or failure of the bone tissue. After the collapse of the cell walls in Phase 2, the bone's resistance to further deformation increases significantly, resulting in the steep portion of the stress–strain curve observed in Phase 3.

5.5.2 *Density-Modulus Relationship of Cancellous Bone*

The material properties and the stress–strain characteristics of cancellous depend on the mode of loading and apparent density. Therefore, understanding the relationship between Young's modulus (E) and the apparent density of cancellous bone (ρ) is essential for understanding the mechanical behaviour of this porous bone tissue. Earlier studies have established mathematical models that describe this relationship, particularly through power-law models (Carter and Hayes 1977; Gibson 1985; Morgan et al. 2003; Helgason et al. 2008). Carter and Hayes (1977) reported a relationship between Elastic modulus (E), apparent density of cancellous bone (ρ) and strain rate ($\dot{\varepsilon}$), which was expressed as

$$E = 3790\dot{\varepsilon}^{0.06}\rho^3$$

Figure 5.24 presents the effect of apparent density on the compressive strength and elastic modulus.

Based on the experimental findings of Carter and Hayes (1977), the study by Gibson (1985) developed a theoretical model that provided further insights into the dependency of Young's modulus on bone apparent density. It was established both theoretically (Gibson 1985) and experimentally (Carter and Hayes 1977; Dalstra et al. 1993) that Young's modulus of cancellous bone is strongly dependent on bone apparent density. The power-law models, in comparison to linear models, allowed for a more accurate representation of the density-modulus relationship (Gibson 1985). At low relative densities or low apparent densities, cancellous bone exhibits an open cell structure characterised by interconnected rods (Fig. 5.25). Conversely, at higher densities, more material accumulates within the cell walls, resulting in a more closed network of plates (Fig. 5.25). The analysis of Gibson (1985) highlighted the transition in cancellous bone structure. Two power-law relationships between Young's modulus (E) and apparent density (ρ) were predicted:

1. $E \propto \rho^2$: This relationship corresponds to the open cell structure.
2. $E \propto \rho^3$: The cubic relationship corresponds to the closed-cell structure.

The experimental findings of Carter and Hayes (1977) corroborated well with the predictions of Gibson (1985). Their study provided empirical validation of the structural transition in cancellous bone based on apparent density. Specifically, the results suggested that this transition occurs at an apparent density of approximately 350 kg m^{-3} or a relative density of approximately 0.2.

The power-law relationship between bone density and elastic modulus was examined in a series of experimental studies, most notably by Morgan et al. (2003), and Morgan and Keaveny (2001). These studies deduced a variety of different equations, all based on the same basic relationship

$$E = C\rho^D$$

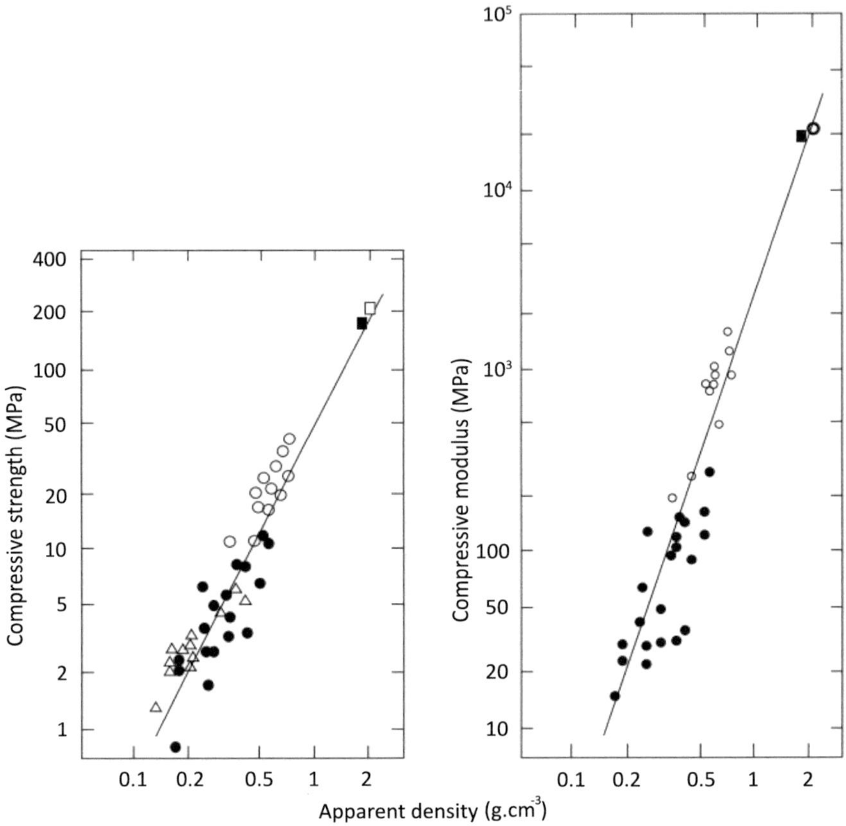

Fig. 5.24 The effect of apparent density on compressive strength and elastic modulus (Adapted from Carter and Hayes 1977). The open triangles represent fresh human compact bone. The open and filled squares represent fresh bovine and embalmed human compact bone, respectively. The filled and open circles represent human and bovine trabeculae bone, respectively

where *C* and *D* are constants. The *C* varied between 3,000 and 30,000, whereas the *D* ranged between 1.14 and 3.2 (Table 5.1). The local static strengths of cancellous bone were also related to the local bone density through power law relationships derived by Kaplan et al. (1985) and Stone et al. (1983).

Bone is a non-homogeneous anisotropic material. This is evident from the yield strength data for various anatomical locations and loading conditions (Table 5.2). However, there is relatively little difference between the yield strain results. Turner (1989) reported that the correlations of yield strain with (a) trabecular orientation, (b) structural density, and (c) bone density, hardly changed for various definitions of yield. This, in turn, is indicative of yield strain in cancellous bone being independent of the structural anisotropy. Subsequently, Turner et al. (1997) suggested

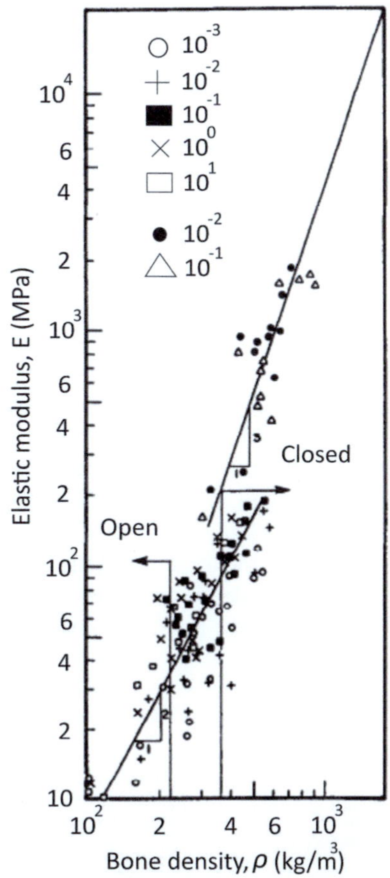

Fig. 5.25 Young's modulus plotted against apparent density of cancellous bone density. *Source* Gibson (1985). Here filled circle and empty triangle represents the bovine femoral condyle samples whereas other symbols represent the human tibial plateau samples. Respective strain rates are mentioned next to each symbols

Table 5.1 Range of constant C and D used in the power-law regression models between elastic modulus (E in MPa) and apparent density (ρ in g cm^{-3})

Anatomic site	Apparent density Mean (range)	$E = C\,\rho^D$	
		C	D
Vertebra (T10–L5)	0.18 (0.11–0.35)	4730	1.56
Proximal tibia	0.23 (0.09–0.41)	15,520	1.93
Greater trochanter	0.22 (0.14–0.28)	15,010	2.18
Femoral neck	0.56 (0.26–0.75)	6850	1.49
Pelvis	0.345 (0.109–0.959)	2017.3	2.46

Sources Morgan et al. (2003), Helgason et al. (2008)

that the apparent density of bovine cancellous bone adapts such that the continuum-level strains are uniform. The uniform strain criteria reproduced realistic density distributions in the proximal femur, which is also applicable to human bone.

Although load transfer and strength are most commonly expressed in terms of stress for engineering materials, studies indicate that strain-based criterion may be more mathematically simple and statistically powerful for human trabecular bone (Morgan and Keaveny 2001). Earlier experimental results indicated that yield strains are independent of density for bovine tibial bone (Keaveny et al. 1994; Turner 1989) and are isotropic (Chang et al. 1999). A study by Turner (1989) on bovine cancellous bone also suggested that the yield strain of cancellous bone is independent of its structural anisotropy. Within a single anatomic site, the yield strain is more or less uniform within the site providing a strong argument for the utility of strain-based description for describing failure of trabecular bone. In another experimental study, Morgan and Keaveny (2001) analysed the uniaxial compressive and tensile yield properties for cylindrical specimens of human trabecular bone from various anatomic sites (vertebra, femoral greater trochanter, proximal tibia, and femoral neck) and found that yield strains vary depending on the anatomical site. However, within a single anatomical site, yield strains were found to be relatively uniform, showing only a weak dependence on apparent density, despite notable variations in yield stress and elastic modulus. Reported inter-site variation of yield strain was much lower than

Table 5.2. Trabecular bone mechanical properties (mean ± standard deviation) by anatomic site and mode of loading

Anatomic site, Loading mode	Apparent density (g cm^{-3})	Modulus (MPa)	Yield strain (%)	Yield stress (MPa)
Vertebra				
Compression	0.18 ± 0.05	344 ± 148	0.77 ± 0.06	2.02 ± 0.92
Tension	0.19 ± 0.04	349 ± 133	0.70 ± 0.05	1.72 ± 0.64
p-Value	NS	NS	<0.001	NS
Proximal tibia				
Compression	0.23 ± 0.06	1091 ± 634	0.73 ± 0.06	5.83 ± 3.42
Tension	0.23 ± 0.10	1068 ± 840	0.65 ± 0.05	4.50 ± 3.14
p-Value	NS	NS	<0.001	NS
Greater trochanter				
Compression	0.22 ± 0.05	622 ± 302	0.70 ± 0.05	3.21 ± 1.83
Tension	0.22 ± 0.04	597 ± 330	0.61 ± 0.05	2.44 ± 1.26
p-Value	NS	NS	<0.001	NS
Femoral neck				
Compression	0.58 ± 0.11	3230 ± 936	0.85 ± 0.10	17.45 ± 6.15
Tension	0.54 ± 0.12	2700 ± 772	0.61 ± 0.03	10.93 ± 3.08
p-Value	NS	NS	<0.001	0.003

NS indicates no significant differences ($p > 0.05$). *Source* Morgan and Keaveny (2001)

that of the yield stress (1.2 versus 8 times), as presented in Table 5.2. Importantly, the small coefficients of variation in yield strain (less than 12%) observed at each site indicated uniformity of the yield strain. This supports the effectiveness of strain-based descriptions in understanding bone failure. This understanding is essential in fields like orthopaedics, biomechanics, and bone tissue engineering, where knowledge of the mechanical properties of cancellous bone at specific anatomical sites is critical for accurate diagnosis, treatment, and design of implant.

5.5.3 Loading and Deformation of Bone Structure

Bone loading and the resulting deformation are influenced by various mechanical scenarios that bones can encounter in daily activities, accidents, and specific conditions. The main types of loading scenarios for bones are as follows (Fig. 5.26).

Compression: In this case a compressive force is applied on the bone along its longitudinal axis, causing shortening and widening of the bone structure (Fig. 5.26). Bones are generally strong under compression as they are designed to support the body weight. In load-bearing bones such as femur, compressive forces are predominant during standing and walking. Compression can lead to oblique fractures if the load exceeds the bone strength.

Tension: In this case a tensile (pulling) force is applied on the bone along its longitudinal axis, leading to lengthening and narrowing of the bone structure (Fig. 5.26). Bones are weaker under tension as compared to compression. The tensile loading induces tensile stresses and if excessive, results in a transverse fracture.

Shear: In this case, the forces are applied parallel to the surface of a bone, causing adjacent layers to slide past each other (Fig. 5.26). Bones are weakest under shear forces. The shear loads are common in spine, especially during rotational movements or lifting heavy objects improperly. Shear forces can lead to oblique fractures.

Torsion: In this case a bone is subject to twisting moment (or torque) about its longitudinal axis (Fig. 5.26), resulting in a helical deformation. Bones resist torsion through their cylindrical shape and the presence of the cortical (compact) bone. Torsion usually occurs in bones like the tibia during activities involving sudden changes in direction. Torsional loads can cause spiral fractures.

Bending: In this case a bone is subjected to lateral forces, causing the bone structure to bend (Fig. 5.26). Long bones are particularly designed to withstand bending forces. Bending is common in long bones, such as the radius and ulna. Bending can result in transverse or oblique fractures.

Combined loading: In this case, two or more types of loading act on the bone simultaneously, such as compression and bending, or tension and torsion. This combined state of loading is expected to result in complex deformation patterns. For instance, a football tackle can involve compression, bending, and torsion, leading to compound

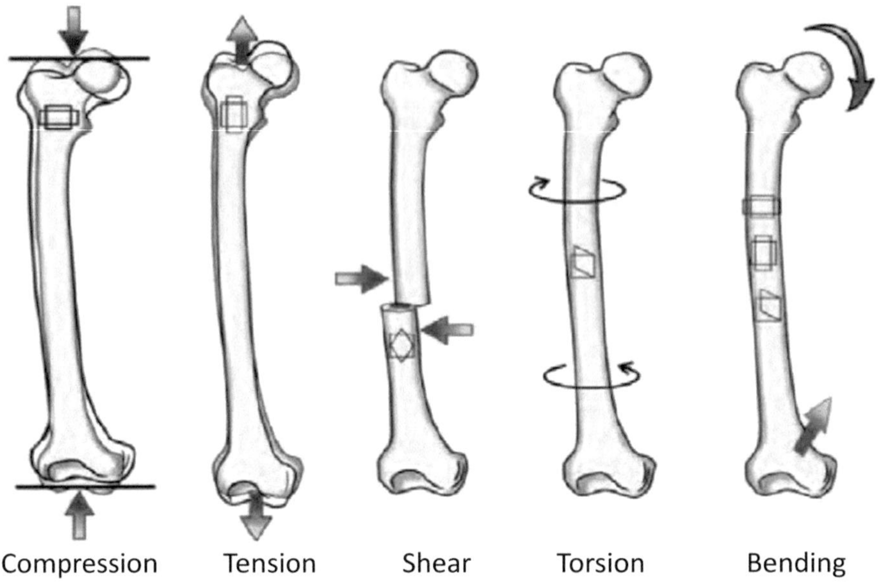

Fig. 5.26 Different scenarios of loading and the associated deformation in bone

fractures. Different types of fractures and modes of fracture healing will be discussed in Chap. 10 (Fig. 5.26).

5.5.4 Viscoelastic Property of Bone

Bone is a dynamic and adaptive tissue. Bone exhibits viscoelastic properties, which means its mechanical behaviour is dependent on the rate of loading. In order to gain insight into the viscoelastic properties of bone, one can examine how bone responds to constant loads at varying rates. This section is focussed on the viscoelastic properties of bone tissue, its measurement, and its influence on bone properties. Viscoelasticity in bone tissue can be quantified through techniques such as nanoindentation. This method involves applying a constant load to the bone surface and measuring the resulting creep behaviour, which is the gradual deformation over time.

Several key mechanical properties of bone tissue, including ultimate strength, energy to failure, and fracture toughness, are influenced by the rate of deformation or strain rate. An increase in strain rate leads to a corresponding increase in bone stiffness and strength (Fig. 5.27). The relationship is such that faster stretching requires higher stress levels to induce deformation, making bone more resistant to mechanical forces at higher strain rates. Additionally, it suggests that the rate at which bone is subjected to mechanical stress significantly impacts its mechanical behaviour and resilience. The fatigue life of bone, or its ability to endure repeated mechanical loading, is

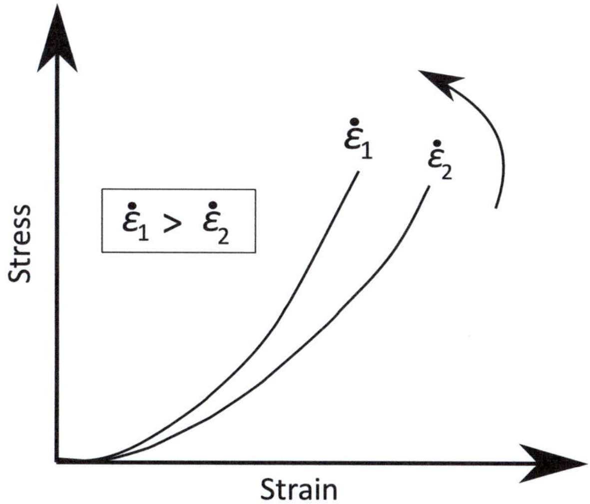

Fig. 5.27 Stress–strain curve of cancellous bone for varying strain rate

also influenced by the frequency of loading. Moreover, the growth of cracks within bone tissue is both cycle-dependent (related to the number of loading cycles) and frequency-dependent (related to the rate at which these cycles occur) at high-stress intensity levels. This highlights the complex interplay of viscoelasticity, fatigue, and damage in bone.

The viscoelastic properties of bone display various characteristics as outlined below:

1. **Sensitivity to loading rate**: Material properties of bone change in response to the rate of applied load.
2. **Stress relaxation and creep**: Bone exhibits stress relaxation, where stress decreases over time at a constant strain, and creep, where strain increases over time at a constant stress.
3. **Hysteresis**: The energy loss observed during loading and unloading cycles. During loading, the energy is stored in a material, whereas the energy is recovered during unloading. There is a lag or difference in the stress–strain curves during loading and unloading, which indicates energy loss or dissipation.

Let us now examine the experimental findings from Carter and Hayes (1977), specifically the influence of marrow on the mechanical behaviour of bone (Fig. 5.28). The study used specimens taken from the human tibial plateau and bovine femoral condyles, which had different geometries. These specimens were subjected to a range of strain rates, varying from 0.001 to 10 s^{-1}. The goal of their experiments was to investigate how the mechanical properties of cancellous bone were influenced by factors, such as the strain rate and the presence of marrow (Carter and Hayes 1977). They found that the viscous flow of marrow had a considerable impact on the mechanical properties of trabecular bone, but primarily at high strain rates. This suggested that the presence of marrow within cancellous bone becomes particularly

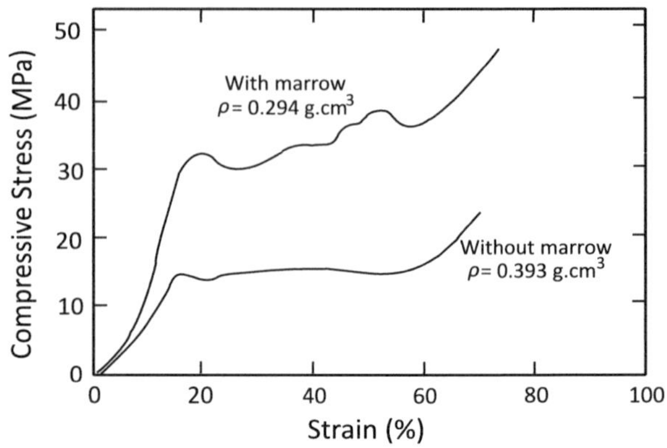

Fig. 5.28 Compressive stress–strain curve of cancellous bone, with and without marrow

influential during severe traumatic compressive loading scenarios, such as impact loads or accidents in vivo. The marrow can absorb a substantial amount of energy under these conditions. The stress–strain behaviour of bone, both with and without marrow, is presented in Fig. 5.28. It is worth noting that the strain rate applied during these mechanical tests was 10 s^{-1}. Moreover, the study reported that the viscous flow of marrow influenced the mechanical properties of trabecular bone primarily at high strain rates (10 s^{-1}). This observation implies that during scenarios involving severe traumatic and compressive loading, the presence of marrow may play a significant role in absorbing a substantial amount of energy.

The investigation of Carter and Hayes (1977) reported that both the compressive strength and modulus of both compact and trabecular bone exhibit a relationship with the strain rate, with an approximate proportionality to the strain rate raised to the power of 0.06 (Fig. 5.29). However, it is important to note that for bone, apparent density plays a much more significant role than the strain rate. In other words, while strain rate does have an effect, the density of the bone appears to be a more influential factor in determining its mechanical behaviour.

5.6 Bone Remodelling

Bone is an adaptive material that undergoes continuous processes of formation, reinforcement, and decay in response to changes in mechanical and biochemical environment. One of its most remarkable features is its ability to adapt both its external geometry and internal structure to variations in mechanical loading through bone apposition (gain) and bone resorption (loss). Bone apposition is facilitated by osteoblasts, while bone resorption is carried out by osteoclasts. This adaptive process is known as bone remodelling. Remodelling is categorised into internal remodelling, which involves changes in bone density and trabecular architecture, and external remodelling, which affects the periosteal geometry (Fig. 5.30) (Huiskes et al. 1987). Although both processes of bone remodelling occur concurrently, changes in

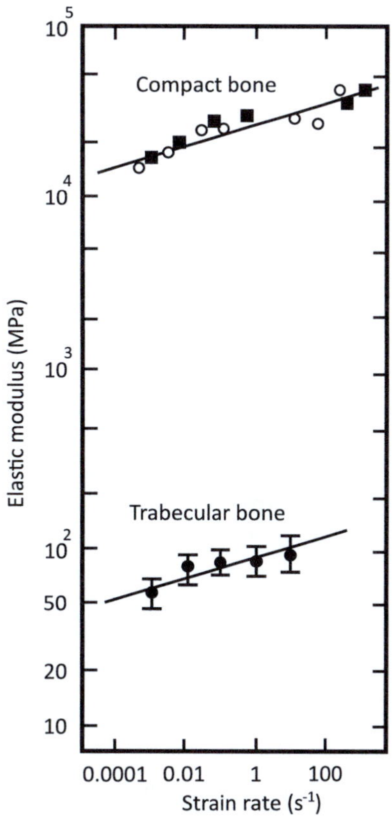

Fig. 5.29 Effect of strain rate on the elastic modulus (Adapted from Carter and Hayes 1977)

bone geometry are relatively minimal in adults as compared to internal adaptation. This is because cancellous bone generally exhibits a higher metabolic activity and responds more rapidly to changes in mechanical loads than the cortical bone (García et al. 2002). Consequently, internal remodelling has been more extensively studied. Nevertheless, several studies have explored the combined effects of internal and external remodelling models, simultaneously (García et al. 2001; Huiskes et al. 1987; Weinans et al. 1993).

In a healthy bone, the rates of bone resorption and bone apposition typically remain in equilibrium, resulting in no net changes to bone morphology (Frost 1964; Weinans 1991). When the mechanical loading environment within the bone is altered, this equilibrium is disturbed. Consequently, bone strives to adapt to a new equilibrium state. Surgical reconstruction, using a prosthesis, significantly alters the mechanical environment, since the prosthesis carries a major part of the load that was previously carried solely by the bone.

As early as the seventeenth century, Galileo observed a relationship between mechanical forces (such as body weight) and bone morphology. This observation sparked considerable scientific interest in understanding how bone structure and

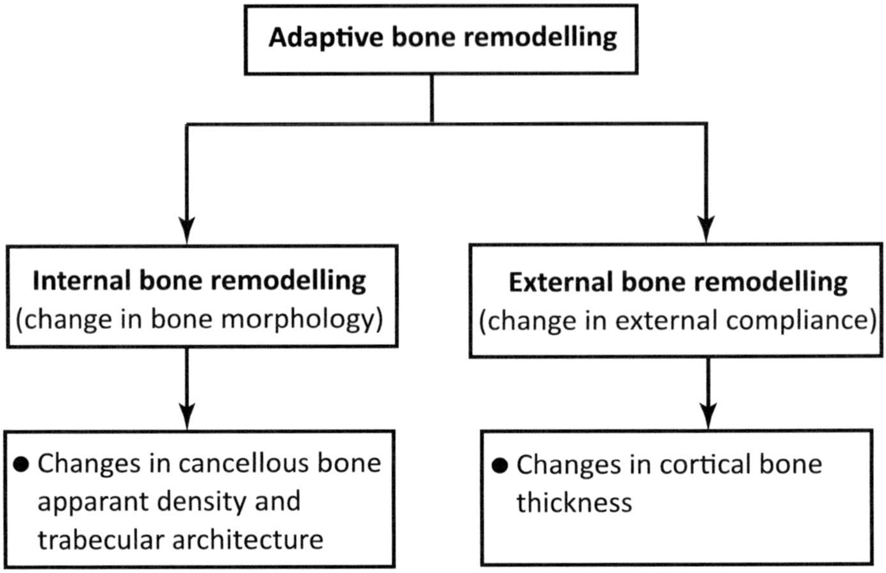

Fig. 5.30 Classifications of adaptive bone remodelling

function are related. Wolff (1892) made a significant contribution to this understanding, building on Roux's theory of functional adaptation (Roesler 1981; Roux 1881). Roux (1881) suggested that bone apposition and resorption are biologically controlled processes dependent on local stress conditions. Wolff hypothesised that every change in the form and the function of a bone is followed by certain definite changes in its internal architecture and equally definite secondary alterations in its external geometry, in accordance with mathematical laws (Wolff 1892). Later, the proposed 'law of bone transformation' by Wolff was referred to as 'Wolff's law'. Wolff's law has been the subject of numerous attempts to develop mathematical models to quantify bone remodelling (Huiskes et al. 1987; Doblaré and García 2001; Geraldes and Philips 2014).

In general, the proposed mathematical formulations assumed that bone possesses a sensing capability to detect changes in the local mechanical environment (stimulus) within it and respond accordingly, in conjunction with other biological factors, through the activity of osteoblasts and osteoclasts. Most models utilise apparent density (ρ) as a variable to represent the remodelling state, although the specific mechanical stimulus varies across different models. The following are the mechanical stimuli used in different formulations of bone remodelling.

- Strain
- Strain energy density
- Elastic strain energy per unit of bone mass
- Principal stress
- Combined strain and microdamage

- Combined strain and orthotropy

While several mechanical stimuli have been suggested, the elastic strain energy per unit of bone mass has been widely used as the mechanical stimulus for the numerical prediction of bone remodelling. Mathematical formulations and computational frameworks of adaptive bone remodelling simulations will be discussed in Chap. 9.

5.7 Trabecular Architecture

Trabecular bone, a spongy and interconnected structure within bones, plays a pivotal role in bone strength and load-bearing capacity. The orientation and arrangement of trabeculae are critical factors in withstanding mechanical stresses experienced by the skeletal system during physiological activities. The discussion in this subsection explores the alignment of trabecular structures in relation to stress directions and their significance in bone biomechanics. Julius Wolf compared the alignment of trabecular structures in the proximal femur with the principal directions in a simplified Fairbairn crane model, as presented in Fig. 5.31 (Wolff 1892). It was found that the principal directions, owing to the vertically applied load in the crane model, roughly corresponded to those in the proximal femur (Fig. 5.31). It has been established that both the shape of the femoral shaft and the orientation of internal trabeculae are adaptations that resist the stresses imposed by the body weight during various physiological activities. These adaptations contribute to the bone's ability to withstand mechanical loads. The notable study by Koch (1917) investigated the stress trajectories and distribution within a proximal femur. His mathematical analysis confirmed that the trabeculae in the femur are predominantly arranged in an orthogonal manner. Subsequent studies further revealed that in regions experiencing complex loading, trabeculae may intersect at acute angles to effectively resist shear forces (Pidaparti and Turner 1997; Skedros and Baucom 2007).

Trabeculae in the proximal femur can be classified into five main groups based on their functions and orientations (Fig. 5.32):

1. Primary compression group
2. Primary tension group
3. Secondary compression group
4. Secondary tension group
5. Greater trochanter group

These trabecular groups often intersect orthogonally. However, in certain regions subjected to intricate loads, they intersect non-orthogonally. This structural adaptability allows the femur to distribute and manage mechanical stresses efficiently. The morphology and anisotropy of bone, particularly in the proximal femur, are profoundly influenced by mechanical loading. Trabecular orientation, in particular, plays a crucial role in shaping the bone structure. The following discussion examines bone morphology in different regions of the proximal femur. Figure 5.32 displays

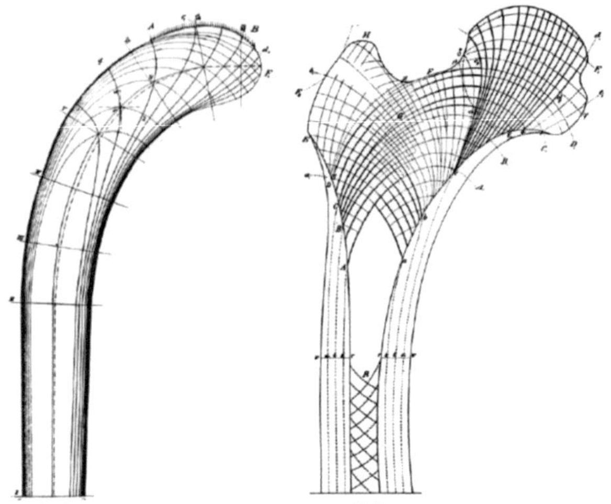

Fig. 5.31 Comparison of **a** Fairbairn crane with **b** proximal femur. Adapted from Wolff (1892)

bone samples from different areas of the proximal femur, each exhibiting distinct morphology and bone density.

1. Samples 1 and 2: Both samples, corresponding to the principal compression group, exhibit similar bone morphology having higher bone density.
2. Samples 3 and 4: Both samples, corresponding to the principal tension group, exhibit similar bone morphology. These samples have lower bone density as compared to samples 1 and 2.
3. Samples 5 and 6: These samples share similar bone morphology with even lower bone density, representing the secondary tension and compression groups.
4. Sample 7: This region, known as Ward's triangle, is partly dense and lies between the compressive and tensile groups.

5.8 Anisotropic Properties of Bone

Bone is structurally anisotropic due to its porous and trabecular architecture. In the proximal femur, anisotropy primarily stems from the orientation of trabeculae within the tensile and compressive groups. The study by Julius Wolf suggested that trabecular orientation might align with the direction of principal stresses (Wolff 1892). Experimental measurements of bone mechanical properties have led to the assumption of orthotropy as the closest approximation to bone anisotropy (Ashman et al. 1984; Turner and Burr 1993; Cuppone et al. 2004). Orthotropic materials have different properties along three mutually perpendicular axes. Stress-induced changes in bone microstructure can result in alterations in bone apparent density and trabecular orientation, contributing to its anisotropic behaviour.

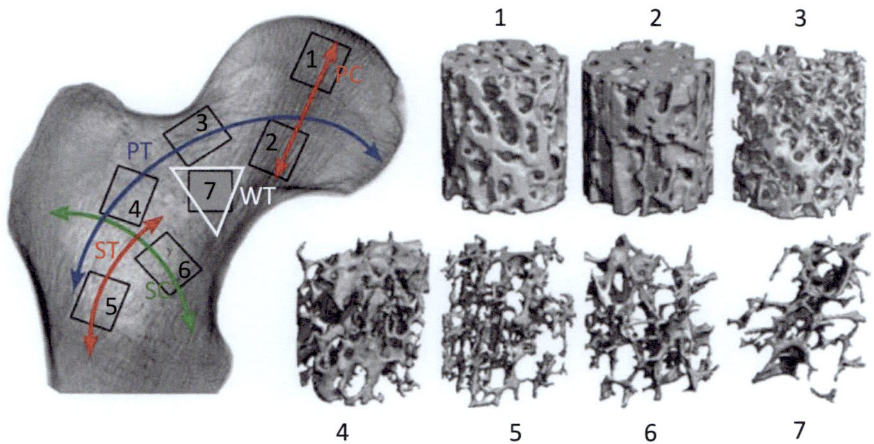

Fig. 5.32 Trabecular groups in proximal femur. PT: Primary tensile groups, PC: Primary compressive groups, ST: Secondary tensile groups, SC: Secondary compressive groups, WT: Ward's triangle (Adapted from Wang et al. 2013)

In several in silico studies, bone tissue properties were traditionally assumed to be isotropic in order to simplify the material model. However, recognizing the importance of bone anisotropy, researchers have sought different ways to include this property in the numerical models. Various techniques for implementing bone anisotropy in numerical simulations are discussed in the following.

- **Micro-Finite Element (FE) models**: Micro-FE models are highly detailed representations of bone at the microstructural level which addresses the structural anisotropy. These models account for the trabecular orientation. Micro-FE simulations provide a realistic depiction of bone behaviour under various loading conditions.
- **Fabric tensor method**: The fabric tensor method involves the calculation of a fabric tensor that characterises the orientation of trabeculae within the bone. This tensor is used to incorporate anisotropic properties into the FE models. The method offers a quantitative description of trabecular orientation and its effect on bone behaviour.
- **Strain-based algorithm**: Strain-based algorithms focus on adapting numerical models to consider the varying mechanical properties of bone along different directions. These algorithms adjust the material properties of the FE model based on the local strains and orientations. They enable the simulation of bone anisotropy in response to complex loading scenarios.
- **Principles of continuum damage mechanics**: Continuum damage mechanics principles are used to model the degradation of material properties in response to damage accumulation (Doblare and Garcia 2001). These principles can be applied to represent the progressive alteration of bone microstructure under load.

5.8.1 Micro-FE Models

Micro-FE models have been used to analyse the structural anisotropy in the cancellous bone, owing to its intrinsic trabecular architecture. These models are based on high-resolution micro-CT images. Figure 5.33 represents the reconstructed micro-CT model of a bone segment (Verhulp et al. 2006). However, the micro-CT-based FE simulations are computationally expensive, since it involves a large number of elements. In a detailed study by Verhulp et al. (2006), two types of micro-FE models were considered: one representing healthy bone and the other osteoporotic bone. The element size of the microscale model was equal to 0.08 mm. A constant material property was assigned to the elements of the microscale model, based on the presence of trabecular and cortical bone. The microscale model was coarsened to produce the continuum-based model of element sizes 0.64 and 3.04 mm voxels. The element-wise material property in the continuum-based model was governed by the elastic modulus-density relationship. The apparent density of each element in the continuum model was determined based on the percentage of the cortical and cancellous bone within the element. The results of this study indicated that continuum models are more appropriate for describing the mechanical properties of healthy femurs rather than

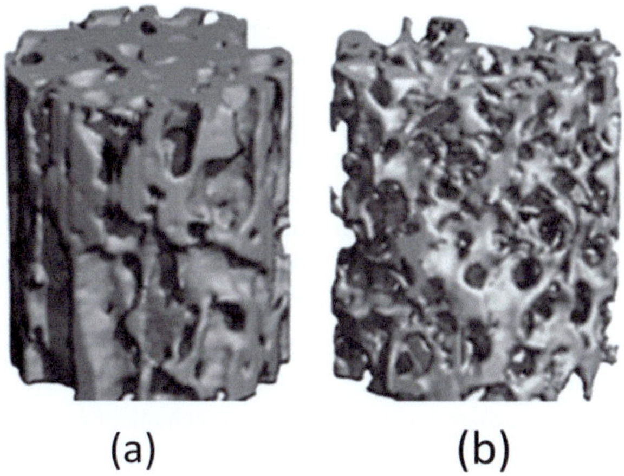

Fig. 5.33 Reconstructed micro-CT images of bone segments from proximal femur: **a** healthy femur, **b** osteoporotic femur (Adapted from Verhulp et al. 2006)

those affected by osteoporosis, since bone anisotropy intensifies with the progression of the osteoporosis (Verhulp et al. 2006).

5.8.2 *Fabric Tensor Method*

The fabric tensor method is an effective approach used to describe the orientation and anisotropy of trabecular bone. The fabric tensor is a second-rank tensor that characterises the geometric arrangement of the microstructure. Cowin (1985) introduced the term "fabric tensor" along with the model equations to evaluate the material constants. Boundary-based methods, such as the mean intercept length tensor (Whitehouse 1974) and the gradient structure tensor (Tabor and Rokita 2007) are a few popular methods used to calculate the fabric tensor.

5.8.3 *Strain-Based Algorithm*

Let us now consider the strain-based orthotropic bone adaptation. Orthotropic bone adaptation refers to the process of bone remodelling, which considers the orthotropic orientation of the trabecular structure (Geraldes and Philips 2014; Mathai et al. 2021). A summary of the steps involved in this algorithm is outlined as follows:

(a) Trabecular orientation and principal directions: In orthotropic bone adaptation, the orientation of the trabeculae within the bone is determined based on the principal directions.
(b) Stress and strain tensors: During each iteration, stress and strain tensors are extracted for each bone element.

(c) Eigen analysis of stress tensors: The Eigen analysis of the stress tensors provides the local principal directions and their corresponding principal stress values.
(d) Implementation of bone remodelling rule: This rule governs how the bone adapts its microstructure in response to mechanical stimuli. The bone's elastic constants are updated based on the strain along the principal directions (Geraldes and Philips 2014; Mathai et al. 2021).
(e) Resulting trabecular orientation: The outcome of the orthotropic bone remodelling process is the redistribution of material properties based on strain levels along the principal directions.

It is worth noting that orthotropic bone remodelling involves complex computational simulations and modelling techniques. The process aims to mimic how bone naturally adapts to changes in the mechanical environment.

Multiple Choice Questions (MCQ)

1. What is strain?

   (A) The ratio of change in force to the original length
   (B) The ratio of change in length to the original length
   (C) The ratio of tensile force to the change in length
   (D) The ratio of change in length to the tensile force applied.

2. The materials that have the ____ elastic properties in _____ directions are called isotropic.

   (A) Different, all
   (B) Same, all
   (C) Different, specific

3. Which of the following property characterises bone most accurately?

   (A) Isotropic
   (B) Homogeneous
   (C) Anisotropic

4. What effect does deviatoric stress have?

   (A) Changes in volume only
   (B) Changes in shape only
   (C) Changes in both volume and shape
   (D) No change in volume or shape

5. How many numbers of independent elastic constants are required to describe an orthotropic linear elastic material?

   (A) 2
   (B) 3

(C) 9
(D) 12

6. Which of the following is not a property of any material?

(A) Young's Modulus
(B) Bulk Modulus
(C) Stress
(D) Compressibility

7. Stress is defined as

(A) External force per unit area
(B) Internal resistive force per unit area
(C) Internal resistive force per change in length

8. Based on the given stress tensor, which is the first invariant?

$$\begin{bmatrix} \mathbf{1} & \mathbf{3} & \mathbf{5} \\ \mathbf{3} & \mathbf{8} & \mathbf{-5} \\ \mathbf{5} & \mathbf{-5} & \mathbf{1} \end{bmatrix}$$

(A) 1
(B) 3
(C) 5
(D) 10

9. Apparent density is the __________ per total volume of the tissue sample.

(A) Mineralised tissue mass
(B) Tissue mass
(C) Unmineralised tissue mass

10. Endurance limit is the stress level below which an infinite number of loading cycles can be applied to a material without causing ________.

(A) Flexural failure
(B) Tensile failure
(C) Fatigue failure

11. Based on the given stress tensor, which one of the following options is a principal stress?

$$\begin{bmatrix} \mathbf{7} & \mathbf{0} & \mathbf{0} \\ \mathbf{0} & \mathbf{6} & \mathbf{2} \\ \mathbf{0} & \mathbf{3} & \mathbf{5} \end{bmatrix}$$

(A) 2
(B) 7

(C) 1
(D) 9

12. ______ is the tendency of a solid material to deform permanently under the influence of persistent mechanical stress.

(A) Creep
(B) Hysteresis
(C) Stress relaxation

13. Stress is a _____ order tensor.

(A) 1st
(B) 4th
(C) 6th
(D) 2nd

14. Modulus of rigidity is the ratio of ____ to shear strain.

(A) shear stress
(B) tensile stress
(C) flexural stress
(D) compressive stress

15. Due to the viscoelastic property, the stiffness and strength of bone ____ with ____ in strain rate.

(A) Increases, decrease
(B) Increases, increase
(C) Decreases, increase

Answer Key to MCQs

1. B; 2. B; 3. C; 4. B; 5. C; 6. C; 7. B; 8. D; 9. A; 10. C; 11. B; 12. A; 13. D; 14. A; 15. B

Practice Problems

5.1. The femur is loaded with an effective hip joint reaction force (F_{HJR}) of 2000 N as shown in Fig. 5.34. The femoral neck width is given as 30 mm. The section *X–X* is normal to the axis of the femoral neck. Take the area of the cross-section as 800 mm^2 and the second moment of area as 40,000 mm^4. Find the maximum normal stress acting at the section *X–X*.

5.2. The state of the stress and strain at a given point in the trabeculae bone is given below. Find the directions in which the orthotropic-strain-based adaptation occurs.

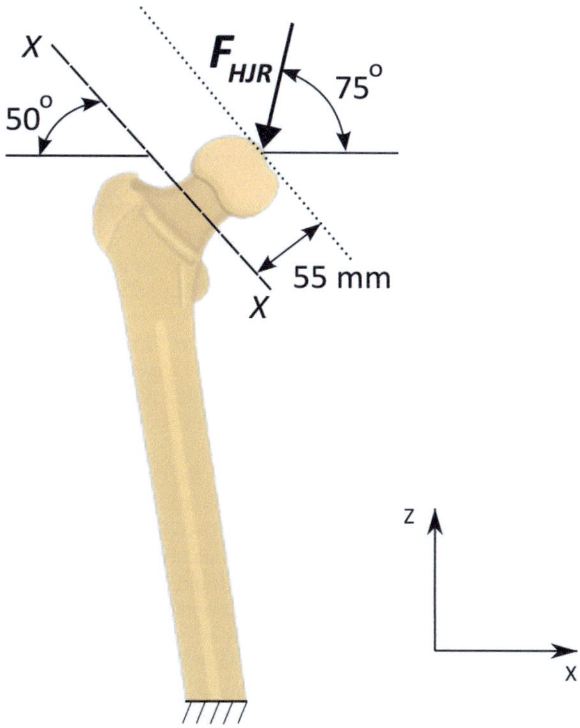

Fig. 5.34 A femur subjected to hip joint reaction force (F_{HJR}) and fixed distally

$$\sigma = \begin{bmatrix} 5 & 1 & 0 \\ 1 & 3 & 2 \\ 0 & 2 & 6 \end{bmatrix} \text{kPa}; \epsilon = \begin{bmatrix} 0.001 & 0.002 & 0 \\ 0.002 & 0.004 & 0.001 \\ 0 & 0.001 & 0.006 \end{bmatrix}$$

References

Alijani H, Vaughan TJ (2024) Exploring the hierarchical structure of lamellar bone and its impact on fracture behaviour: a computational study using a phase field damage model. J Mech Behav Biomed Mater 153:106471

Asgharpour Z, Zioupos P, Graw M, Peldschus S (2014) Development of a strain rate dependent material model of human cortical bone for computer-aided reconstruction of injury mechanisms. Forensic Sci Int 236:109–116

Ashman RB, Cowin SC, Van Buskirk WC, Rice JC (1984) A continuous wave technique for the measurement of the elastic properties of cortical bone. J Biomech 17(5):349–361

Carter DR, Hayes WC (1977) The compressive behavior of bone as a two-phase porous structure. J Bone Joint Surg Am 59(7):954–962

Carter DR, Caler WE, Spengler DM, Frankel VH (1981) Uniaxial fatigue of human cortical bone: the influence of tissue physical characteristics. J Biomech 14(7):461–470

Chang WC, Christensen TM, Pinilla TP, Keaveny TM (1999) Uniaxial yield strains for bovine trabecular bone are isotropic and asymmetric. J Orthop Res 17(4):582–585

Cowin SC (1985) The relationship between the elasticity tensor and the fabric tensor. Mech Mater 4(2):137–147

Cuppone M, Seedhom B, Berry E, Ostell A (2004) The longitudinal Young's modulus of cortical bone in the midshaft of human femur and its correlation with CT scanning data. Calcif Tissue Int 74:302–309

Dalstra M, Huiskes R, Odgaard A, van Erning L (1993) Mechanical and textural properties of pelvic trabecular bone. J Biomech 26:523–535

Doblaré M, García JM (2001) Application of an anisotropic bone-remodelling model based on a damage-repair theory to the analysis of the proximal femur before and after total hip replacement. J Biomech 34(9):1157–1170

Frost HM (1964) The laws of bone structure. Charles C Thomas Publisher

Funk JR, Kerrigan JR, Crandall JR (2004) Dynamic bending tolerance and elastic-plastic material properties of the human femur. Annu Proc Assoc Adv Automot Med 48:215–233

García JM, Doblaré M, Cegoñino J (2002) Bone remodelling simulation: a tool for implant design. Comput Mater Sci 25:100–114

García JM, Martínez MA, Doblaré M (2001) An anisotropic internal-external bone adaptation model based on a combination of CAO and continuum damage mechanics technologies. Comput Methods Biomech Biomed Eng 4:355–377

Geraldes DM, Phillips AT (2014) A comparative study of orthotropic and isotropic bone adaptation in the femur. Int J Numer Methods Biomed Eng 30(9):873–889

Gibson LJ (1985) The mechanical behaviour of cancellous bone. J Biomech 18(5):317–328

Guo XE (2008) Mechanical properties on cortical and cancellous bone tissue. In: Cowin SC (ed) Bone mechanics handbook, 2nd edn. Informa Healthcare, pp 10.1–10.23

Helgason B, Perilli E, Schileo E, Taddei F, Brynjólfsson S, Viceconti M (2008) Mathematical relationships between bone density and mechanical properties: a literature review. Clin Biomech 23(2):135–146

Huiskes R, Weinans H, Grootenboer HJ, Dalstra M, Fudala B, Sloof TJ (1987) Adaptive bone remodelling theory applied to prosthetic design analysis. J Biomech 20:1135–1150

Kaplan SJ, Hayes WC, Stone JL, Beaupré GS (1985) Tensile strength of bovine trabecular bone. J Biomech 18(9):723–727

Keaveny TM, Wachtel EF, Ford CM, Hayes WC (1994) Differences between the tensile and compressive strengths of bovine tibial trabecular bone depend on modulus. J Biomech 27(9):1137–1146

Koch JC (1917) The laws of bone architecture. Am J Anat 21(2):177–298

Mathai B, Dhara S, Gupta S (2021) Orthotropic bone remodelling around uncemented femoral implant: a comparison with isotropic formulation. Biomech Model Mechanobiol 20(3):1115–1134

Mirzaali MJ, Schwiedrzik JJ, Thaiwichai S, Best JP, Michler J, Zysset PK, Wolfram U (2016) Mechanical properties of cortical bone and their relationships with age, gender, composition and microindentation properties in the elderly. Bone 93:196–211

Morgan EF, Keaveny TM (2001) Dependence of yield strain of human trabecular bone on anatomic site. J Biomech 34(5):569–577

Morgan EF, Bayraktar HH, Keaveny TM (2003) Trabecular bone modulus-density relationships depend on anatomic site. J Biomech 36:897–904

Morgan EF, Unnikrisnan GU, Hussein AI (2018) Bone mechanical properties in healthy and diseased states. Annu Rev Biomed Eng 20:119–143

Pidaparti RMV, Turner CH (1997) Cancellous bone architecture: advantages of nonorthogonal trabecular alignment under multidirectional loading. J Biomech 30:979–983

Özkaya N, Nordin M (1999) Fundamentals of biomechanics: equilibrium, motion, and deformation, 2nd edn. Springer, New York

Roesler H (1981) Some historical remarks on the theory of cancellous bone structure (Wolff's law). In: Cowin SC (ed) The mechanical properties of bone. The American Society of Mechanical Engineers, New York, pp 27–42

Roux W (1881) Der ziichtende kampf der teile, oder die 'tei/auslese' imorganismus (theorie der 'finktionellen unpassung'). Wilhelm Engelmann

Skedros JG, Baucom SL (2007) Mathematical analysis of trabecular 'trajectories' in apparent trajectorial structures: the unfortunate historical emphasis on the human proximal femur. J Theor Biol 244(1):15–45

Stone JL, Beaupré GS, Hayes WC (1983) Multiaxial strength characteristics of trabecular bone. J Biomech 16(9):743–752

Tabor Z, Rokita E (2007) Quantifying anisotropy of trabecular bone from gray-level images. Bone 40(4):966–972

Thompson DD (1980) Age changes in bone mineralization, cortical thickness, and Haversian canal area. Calcif Tissue Int 31(1):5–11

Turner CH (1989) Yield behavior of bovine cancellous bone. J Biomech Eng 111(3):256–260

Turner CH, Burr DB (1993) Basic biomechanical measurements of bone: a tutorial. Bone 14:595–608

Turner CH, Anne V, Pidaparti RMV (1997) A uniform strain criterion for trabecular bone adaptation: do continuum-level strain gradients drive adaptation? J Biomech 30(6):555–563

Verhulp E, van Rietbergen B, Huiskes R (2006) Comparison of micro-level and continuum-level voxel models of the proximal femur. J Biomech 39(16):2951–2957

Wang J, Zhou B, Guo X (2013) Trabecular microstructure differs greatly between trabecular groups in proximal femurs of postmenopausal women. ORS 2013 annual meeting

Weinans H (1991) Mechanically induced bone adaptations around orthopaedic implants. Doctoral dissertation, University of Nijmegen, The Netherlands

Weinans H, Huiskes R, van Rietbergen B, Sumner DR, Turner TM, Galante JO (1993) Adaptive bone remodelling around bonded noncemented total hip arthroplasty: a comparison between animal experiments and computer simulation. J Orthop Res 11:500–513

Whitehouse WJ (1974) The quantitative morphology of anisotropic trabecular bone. J Microsc 101(Pt 2):153–168

Wolff J (1892) Das Gesetz der Transformation der Knochen. Springer, Berlin, Germany

Chapter 6
Biomechanics of Orthopaedic Implants Part I: Hip and Knee

This chapter discusses the classifications of orthopaedic implants and the common failure mechanisms associated with different implants. The concept of composite beams and their application in implant-bone structures has also been discussed. The biomechanics of hip and knee replacements are also presented in this chapter.

6.1 Joint Replacement

Osteoporosis and osteoarthritis are common musculoskeletal disorders that affect the human bones and major weight-bearing joints, respectively. These chronic disorders prevail more in the aging population. Decreased strength and mass of bone are observed in osteoporosis, causing a considerably higher risk of bone fractures owing to falls (Ridzwan et al. 2018). In osteoarthritis, the cartilages at the end of the bones degenerate, exposing the bone surfaces to come into contact. This leads to pain, discomfort and limited mobility. Moreover, gait abnormalities and limited range of motion can cause gradual muscle weakness. The number of patients affected by osteoporosis is increasing worldwide. In end-stage osteoarthritis, the patients suffer from debilitating pain and have very limited mobility.

Joint replacement or arthroplasty is considered to be an effective treatment for end-stage osteoarthritis and certain types of bone fracture, depending on the severity and location of the fracture. Joint replacement offers pain relief and restoration of normal movements. It may also be warranted for treating injuries of bones and joints due to severe trauma, including sports injuries and road traffic accidents. The trauma includes fractures in bones or dislocation of joints, disturbing the natural functioning of the joint. Post-traumatic arthritis is very similar to osteoarthritis in symptoms and appearance.

In joint replacement, the affected joint is surgically reconstructed by removing the damaged part and replacing it with an artificial joint known as a prosthesis

S. Gupta et al., *Biomechanics of Joints and Implants*,
https://doi.org/10.1007/978-981-96-0586-6_6

or implant. Implants are designed to mimic the operations of the natural joint closely. Orthopaedic implants have been used effectively in clinical practice over the last several decades, with their design and performance evolving over time. However, implantation of the joints often leads to alteration of the biomechanics of the natural joints. Implant design, implant positioning, implant materials, fixation techniques, and implant-bone interface characteristics are the major factors that influence the biomechanics of the implanted joints. In this chapter, we will discuss the biomechanics of hip and knee implants.

6.1.1 Classification of Implants: Fixation Techniques

There are two main fixation techniques by which an implant is fixed with the surrounding host bone: cemented and cementless (Fig. 6.1). In the cemented fixation, acrylic bone cement (polymethyl methacrylate—PMMA) is used to grout the implant within the surrounding bone (Pal et al. 2013). Acrylic cement consists of two components: a liquid containing monomethylmethacrylate (MMA) monomer and a powder containing MMA-styrene co-polymer. These two components are mixed outside the body to produce PMMA through a polymerization process. A lump of PMMA dough is injected within a reamed (prepared) bone (e.g. the femoral canal for hip replacement) using a cement gun. Subsequently, the implant is introduced and held in the desired position for a few minutes. The acrylic cement penetrates the porous bone and hardens, thereby fixing the implant to the host bone. The technique was introduced in the 1960s by Sir John Charnley in cemented THA, as a material to fill the gap between the hip stem and the bone. Ever since its introduction, remarkable improvement in hip arthroplasty has been achieved in the last few decades. Although modern cementing techniques have considerably reduced the failure rates of cemented hip arthroplasty, these techniques have not entirely eliminated failure. The surgical technique adopted for cementation, the thickness of the cement mantle, and implant design can influence the outcome of the cemented fixation. The outcome of cemented implants can be improved by using distal plugs and cement pressurizing devices.

In the case of cementless fixation, the initial fixation is obtained by inserting an implant into a slightly smaller cavity. A press-fit is thus obtained and compression hoop stresses are generated. The initial stability is provided by a press fit or frictional contact between the implant and the adjacent bone. This is often coupled with geometric (lattice porous structure) or surface (hydroxyapatite or plasma-sprayed porous coating) features that encourage bone growth into or onto the implant. Aseptic loosening of cementless implants is primarily attributed to the deficiency of biological bonding between the implant and the bone. Successful bone growth determines the implant stability for this type of implant. The stability is categorised into primary stability (mechanical) and secondary stability (biological). Under physiological loading conditions, the magnitude of implant-bone relative micromotion during the early postoperative period determines the implant's primary stability. In

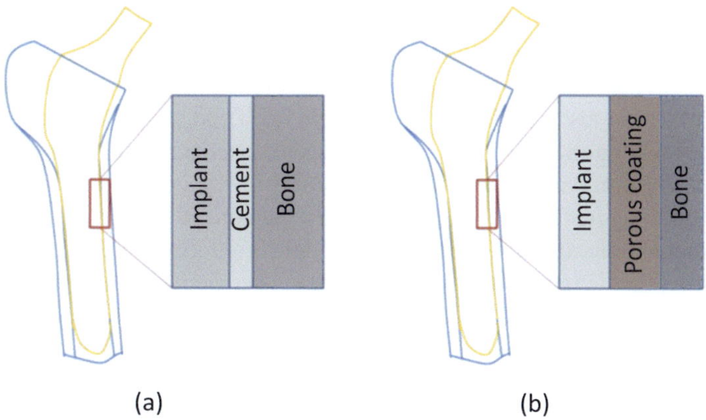

Fig. 6.1 Schematic representation of a typical **a** cemented fixation and **b** uncemented fixation

this period, the bone growth process is incomplete. In contrast, the magnitude of micromotion observed under physiological loading when the bone growth process is complete indicates secondary stability. Bone growth into/onto the implant is dependent on the magnitude of implant-bone micromotion under physiological activity. Generally, bone growth is enhanced by small micromotions (20-40 μm), and inhibited by large micromotions (> 150 μm), forming a fibrous tissue layer at the implant-bone interface. Cell activity and implant design can also influence the bone growth process.

The combination of these two fixation techniques is called the hybrid fixation technique. Here, one or more components are cemented, with the rest being cementless. It may also be possible that part of a component is fixed using acrylic bone cement, but the remaining part of the component is not cemented. The choice of the method of fixation is multifactorial. It is dependent on implant design, patient's age, available bone stock and bone quality, and surgeon's experience. Satisfactory performances have been reported for both of these fixation methods in specific patient groups. Each method, however, demonstrates a unique set of advantages and disadvantages.

Let us now discuss the major advantages and disadvantages of both fixation methods. The cemented technique offers fixation of implants in patients with poor (porous) bone quality, in particular, patients suffering from osteoporosis. The acrylic bone cement solidifies within 10–15 min of application; hence almost immediate fixation of the implant is achieved. This technique is surgically less challenging with regard to the precision of the required bone cut. Diverse bone morphology and bone qualities are easily catered by this method. Antibiotics may be added to the bone cement to reduce infection at the surgical sites. Reduced end-of-stem pain, low intra- and post-operative fractures and easier component positioning are a few other advantages of the cemented fixation method. However, the polymerization process of preparing the bone cement is exothermic, which causes thermal injury and necrosis in the surrounding bone. Bone fractures in the implanted femur have been associated with bone necrosis, possibly resulting from the heat (thermal injury) generated by cement polymerization. Acrylic bone cement has a low tensile strength (static: 20–25 MPa) and is susceptible to failure within itself and also at the implant-cement and cement-bone interfaces. Since cement is weak in tension and shear, bone cement degrades with time, causing a gradual reduction of the interfacial stability. Under repetitive loading, cracks are generated within the cement mantle leading to failure of the implanted bone structure. Cement wear debris abraded from the cement mantle can cause particulate reactions by macrophages, osteolysis, and eventual loosening. Intra-operatively, fat and marrow embolisms have been reported along with respiratory distresses for patients undergoing cemented arthroplasty. This is attributed to the increased intramedullary pressure generated by introducing the implant into cement. Such respiratory and hemodynamic complications are higher for patients with lower baseline pulmonary capacity (Issack et al. 2009; Donaldson et al. 2009).

In contrast, the cementless fixation technique offers reduced operation time, hence reduced blood loss and associated risks and costs. The technique is dependent on the biological attachment with bone and potentially offers a long-term stable fixation. The uncemented implant eliminates the long- and short-term problems of acrylic bone cement. However, a higher risk of peri-prosthetic fracture, hip stem subsidence, and end-of-stem pain is observed (Cinotti et al. 2013; Su and Barrack 2013; Ulvi et al. 2013). Such complications may be reduced by establishing an appropriate fit and an initial stable implant of appropriate size for bone ingrowth. Press-fitted implants require healthy bones; hence, this type of implant may not be suitable for patients with low bone quality owing to osteoporosis. Since the cementless technique is dependent on peri-prosthetic bone ingrowth, a longer rehabilitation period as compared to cemented implants is required. Moreover, the cementless implants are more expensive than the cemented ones.

6.1.2 Implant Failure Mechanisms

Despite the overall success of joint replacement surgeries, failures of the implant-bone structure have been reported. The cause of the failure of implanted bone is multifactorials. Although eventual failures of implants are believed to be due to biological causes, the failure process may be initiated by mechanical events. According to the national joint registry data of various countries, the major reasons for the failure of joint replacements are aseptic loosening, dislocation/subluxation, periprosthetic fracture, infection, and adverse tissue reaction to wear particulate debris (The New Zealand Joint Registry 2020; National Joint Registry for England and Wales 2023). Failure rates have been significantly affected by implant design-related factors. Some of the important factors are excessive stresses and strains within the implant-bone structure and at the interfaces, stress-shielding-induced adverse bone remodelling, generation of wear particle debris, and poor fixation of the implant with bone (osseointegration) owing to reduced bone ingrowth.

Two common failure scenarios for cemented hip implants, identified by Huiskes (1993), can also be used to analyse the failure mechanisms of other joint replacements. These failure scenarios are the following:

- Accumulated damage failure scenario,
- Particulate reaction failure scenario.

According to the accumulated damage failure scenario, the likelihood of mechanical failure depends on the stress generated in a material versus the strength of that material. The implants are subjected to dynamic load over a considerable period of time. The repetitive natures of the external loads generate stresses in the materials and at the interfaces. The strengths of the implant materials or the implant-cement-bone interfaces may not be enough to withstand the long-term effects of dynamic loading. The cement layer is subjected to mechanical damage in the form of micro-cracks, reducing the strength of the cement layer and the implant-cement-bone interfaces. Subsequently, interface debonding might lead to mechanical loosening of a weak interface.

The interface debonding is dependent not only on the multi-axial state of stresses at the interface but also on the gradually changing bone density and bone strength adjacent to the interface. In the case of uncemented prostheses, loosening may occur due to implant-bone interface debonding. Hoffman failure criterion has been employed to predict the implant-bone interface failure, using normal (s) and shear stresses (τ) acting at the interface. Let us consider S_t, S_c, and S_s as the uniaxial tensile, compressive, and interfacial shear strength of bone, respectively, at a given location. The Hoffman number (FL) at that location is calculated using the following expression (Hoffman, 1967):

$$FL = \frac{\sigma_n^2}{S_t S_c} + \left(\frac{1}{S_t} - \frac{1}{S_c}\right)\sigma_n + \frac{\tau^2}{S_s^2}$$

Interface failure is likely to happen if FL is greater than 1.

Let us consider a sample problem on interfacial debonding. The state of stress at a specific location on the implant-bone interface is expressed as

$$\sigma = \begin{bmatrix} 50 & 10 \\ 10 & 30 \end{bmatrix} \text{MPa}$$

The implant-bone interface is inclined at 45°, as shown in Fig. 6.2. The density of the nearest bone element is equal to 1 g cm^{-3}. The local bone strengths (in MPa) is calculated using the following equations (Kaplan et al. 1985).

$$S_t = 14.5\rho^{1.71};\ S_c = 32.4\rho^{1.85};\ S_s = 21.6\rho^{1.65};$$

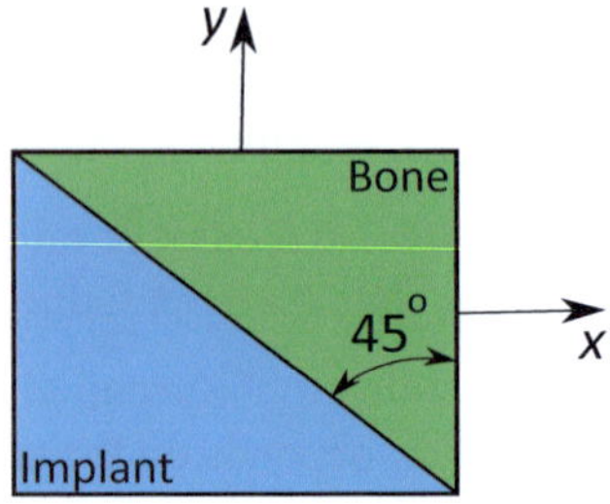

Fig. 6.2 Simplified 2D representation of an implant-bone interface

As the first step, we need to find the normal and tangential stress acting at the interface using the stress transformation matrix discussed in Sect. 5.2.

$$[\sigma'] = \begin{bmatrix} \cos(45°) & \sin(45°) \\ -\sin(45°) & \cos(45°) \end{bmatrix} \begin{bmatrix} 50 & 10 \\ 10 & 30 \end{bmatrix} \begin{bmatrix} \cos(45°) & -\sin(45°) \\ \sin(45°) & \cos(45°) \end{bmatrix}$$

$$[\sigma'] = \frac{1}{2}\begin{bmatrix} 1 & 1 \\ -1 & 1 \end{bmatrix} \begin{bmatrix} 50 & 10 \\ 10 & 30 \end{bmatrix} \begin{bmatrix} 1 & -1 \\ 1 & 1 \end{bmatrix} = \begin{bmatrix} 50 & -10 \\ -10 & 30 \end{bmatrix} \text{MPa}$$

From $[\sigma']$, it is clear that $\sigma_n = 50$ MPa and $\tau = -10$ MPa.
The strength of the bone depends on the apparent density as expressed below:

$$S_t = 14.5\rho^{1.71};\ S_c = 32.4\rho^{1.85};\ S_s = 21.6\rho^{1.65}$$

Using the above expressions, the strength values are equal to

$$S_t = 14.5\,\text{MPa};\ S_c = 32.4\,\text{MPa};\ S_s = 21.6\,\text{MPa}$$

Using the strength and stress values, the *FL* number is found as follows.

$$\begin{aligned} FL &= \frac{\sigma_n^2}{S_t S_c} + \left(\frac{1}{S_t} - \frac{1}{S_c}\right)\sigma_n + \frac{\tau^2}{S_s^2} \\ &= \frac{50^2}{14.5 \times 32.4} + \left(\frac{1}{14.5} - \frac{1}{32.4}\right) \times 50 + \frac{(-10)^2}{21.6^2} \\ FL &= 7.44 \end{aligned}$$

Since the *FL* number is higher than 1, the interface is likely to fail at that specific location.

As per the particulate-reaction failure scenario, the cement–bone interface gradually disintegrates owing to the migration of wear particle debris. Cracks in the cement layer, due to excessive stresses, generate tiny particles of cement debris. The metal debris generated from the implant, polyethylene wear particles generated at the articulating surface, or the cement particle debris can migrate in the reconstructed joint space. In response to these foreign tiny particles, the immune system of the body can produce an inflammatory action at the joint space. This inflammatory response causes localised bone loss around the implant, a phenomenon known as osteolysis that can lead to soft-tissue interposition and implant loosening for both cemented and cementless devices. Polyethylene wear debris, in particular, is a major threat to any joint replacements (Harris 2001; Holding et al. 2006).

Other aseptic failure mechanisms of the implanted joints include the effect of stress shielding and bone remodelling, dislocation, and migration.

Stress-shielding and bone remodelling: The stiffness of a prosthesis material, in most cases, is very high compared to the host bone. As a result, after implantation, the bulk of the load is carried by the implant, which was formerly carried by the bone itself in the preoperative situation. This shields the bone from carrying the musculoskeletal loads and produces abrupt changes in the mechanical environment within the bone structure. This phenomenon is commonly known as 'stress/strain shielding'. Consequently, the bone is subject to reduced levels of stress and strain, which leads to adverse bone remodelling. The bone mineral content is reduced significantly, leading to periprosthetic bone resorption and eventual loosening of the implant (Huiskes 1993).

Dislocation: The stable structure of the normal joint is disturbed and weakened by surgical intervention, enhancing the risk of dislocation. For example: the small bearing size at the articulating surface of the hip joint may also lead to dislocation (Zijlstra et al. 2017). Impaction allografting, a technique to treat cavitary bone defects, has also been associated with dislocation (Dattani and Blunn 2007).

Migration: Implants can migrate to some extent in bone. However, it is more common for cementless implants owing to the lack of initial fixation. The implant subsides due to micro-cracking of the supporting trabecular bone (Burr et al. 1985). This is typically observed around the tip of the implant where high stresses are generated. Quick early migration is attributed to premature implant loosening (Glyn-Jones et al. 2004).

In a summary, the multi-axial stresses (tensile, compressive, and shear) induced at the material interfaces should not be too high to cause interface debonding. Moreover, implant-induced adverse changes in stress–strain distribution should be low enough to achieve a physiological load transfer (closer to natural bone). Moreover, the initial migration of the implant should be minimal.

6.2 Structural Analysis of Implant-Bone Structures

Pre-clinical analysis of the implant-bone structure has commonly been carried out using the Finite Element (FE) Method, a popular numerical technique used to test and validate clinical hypothesis that requires a computational facility. However, in regions within the structure, where the simple theory of elasticity methods is applicable, the use of these methods ascertains the FE analysis results. The structure of the implant-bone may be analysed with the help of classical beam theory, which was discussed earlier in Chap. 5. In this section, we focus our discussion on the concepts of composite beam and how it can be applied to the implant-bone structure subjected to axial loading and bending.

6.2.1 Composite Bar Subjected to Axial Loading

Let us consider a composite bar, subjected to an axial load (***F***) (Fig. 6.3). The area of the cross-section of the inner material and outer materials are A_1 and A_2, respectively (Fig. 6.4). The elastic modulus of the inner material and outer materials are E_1 and E_2, respectively. Let us first write down the force equilibrium equation along the x direction, for an arbitrary section (Fig. 6.4).

$$\begin{aligned} &\Sigma F_x = 0 \\ &-\int_{A_1} \sigma_1 \mathrm{d}A_1 - \int_{A_2} \sigma_2 \mathrm{d}A_2 + F = 0 \end{aligned} \tag{6.1}$$

Here σ_1 and σ_2 are the normal stresses generated in materials 1 and 2, respectively. According to the Hooke's law, Eq. 6.1 is rewritten as follows:

$$-\int_{A_1} E_1 \varepsilon_1 \mathrm{d}A_1 - \int_{A_2} E_2 \varepsilon_2 \mathrm{d}A_2 + F = 0$$

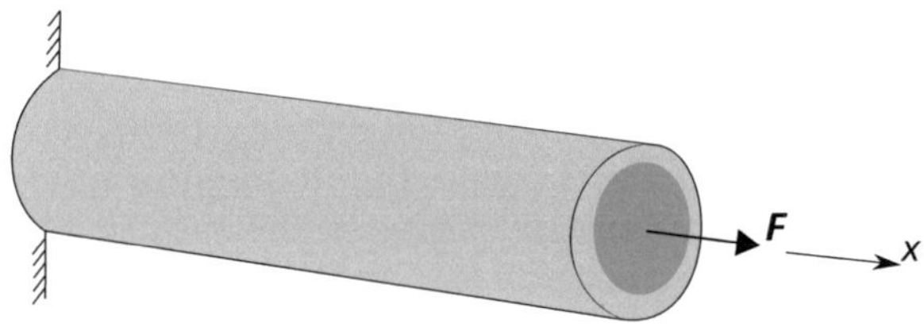

Fig. 6.3 A composite bar subjected to axial load

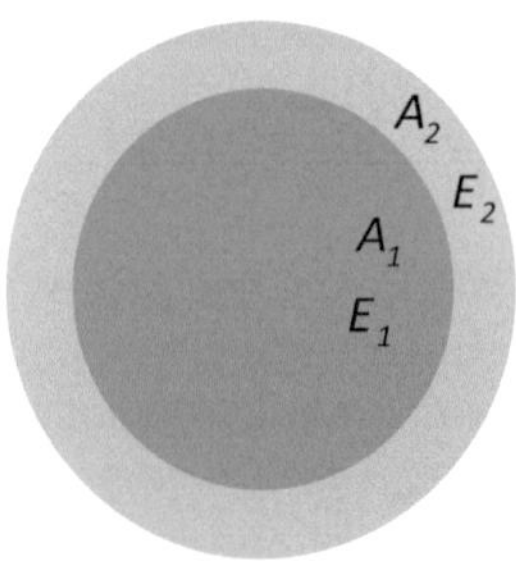

Fig. 6.4 Cross section of the composite bar shown in Fig. 6.3

Since the interface between two materials is fully bonded, i.e. the strains $\varepsilon_1 = \varepsilon_2 = \varepsilon$, the above equation is simplified as

$$F = \varepsilon\left(\int_{A_1} E_1 dA_1 + \int_{A_2} E_2 dA_2\right)$$

Since the materials are homogenous throughout, we obtain

$$F = \varepsilon(E_1A_1 + E_2A_2)$$
$$\varepsilon = \frac{F}{(E_1A_1 + E_2A_2)}$$

The stresses induced in the material 1 and 2 are expressed as

$$\sigma_1 = E_1\varepsilon_1 = E_1\varepsilon = \frac{FE_1}{(E_1A_1 + E_2A_2)}$$
$$\sigma_2 = E_2\varepsilon_2 = E_2\varepsilon = \frac{FE_2}{(E_1A_1 + E_2A_2)}$$

From the above equations, we can infer that the stress generated by the material 'i' is proportional to the quantity, $\frac{E_i}{\Sigma E_i A_i}$.

Let us now discuss the stress/strain shielding phenomenon with regard to the axial loading of an implant-bone structure. A simplified representation of the proximal femur implanted with a cementless hip implant is presented in Fig. 6.5. A hip-joint load (F) of 300 N is assumed to be applied on the implant head. Following the concept of load transfer in a composite structure, the load carried by the cementless stem (F_s) and the femur (F_b) are expressed as follows:

$$F_s = \sigma_s A_s = \frac{FE_sA_s}{(E_sA_s + E_bA_b)} \tag{6.2}$$

$$F_b = \sigma_b A_b = \frac{FE_bA_b}{(E_sA_s + E_bA_b)} \tag{6.3}$$

As evident from Eqs. 6.2 and 6.3, the load carried by each component is dependent on its respective elastic modulus and cross-sectional area. The cross-sectional areas of the stem and femur bone may be arbitrarily taken as 200 mm^2 and 140 mm^2, respectively. The elastic modulus of the stem and femur bone may be assumed as 110,000 MPa and 15,000 MPa, respectively. Using these values, the load shared by the stem and the femur is calculated as

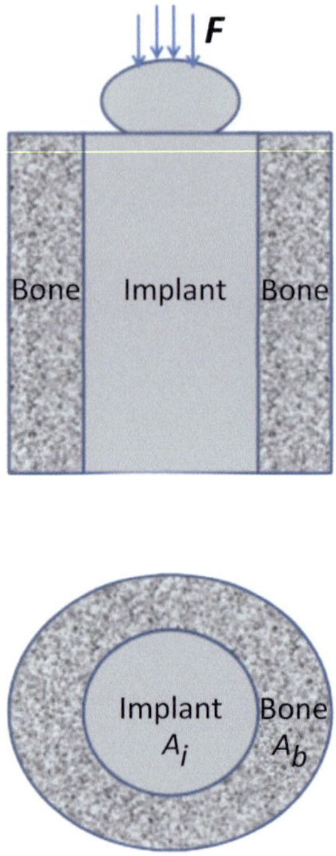

Fig. 6.5 Schematic of simplified section of a proximal femur implanted with a cementless hip stem

$$F_s = \frac{FE_sA_s}{(E_sA_s + E_bA_b)} = \frac{300 \times 110{,}000 \times 200}{(110{,}000 \times 200 + 15{,}000 \times 140)} = 273.86\,\text{N}$$
$$F_b = \frac{FE_bA_b}{(E_sA_s + E_bA_b)} = \frac{300 \times 15{,}000 \times 140}{(110{,}000 \times 200 + 15{,}000 \times 140)} = 26.14\,\text{N}$$

The load shielded by the implant can be expressed in terms of percentage as $\left(1 - \frac{F_b}{F}\right) \times 100 = 91.29\%$. The effect of shielding of load by the implant with higher stiffness material is evident from the above sample problem. The implant transfers the load to the bone through interfacial stresses at the implant-bone interface. Therefore, higher implant stiffness results in lower load transfer to the host bone and thereby, lower interfacial stresses (Huiskes 1993).

6.2.2 Composite Beam Subjected to Pure Bending

A fractured femur treated with an intramedullary nail, post-osseointegration, can be represented as a multi-material beam. In the case of cemented implant fixation, three materials such as bone, cement, and implant can be considered in the multi-material beam idealization. Beam theory, when applied within its constraints, offers insights into the overall behaviour of bone-implant systems. The application of beam theory to multi-material beams provides a framework for understanding the mechanical behaviour of implant-bone systems (Engh and Bobyn 1988; Silva et al. 1999).

According to the beam theory, the reduction in stress generated within bone immediately after implantation is proportional to the mechanical and cross-sectional parameters of the bone and implant (Engh and Bobyn 1988; Silva et al. 1999). The cross-sectional parameter for axial loading is the cross-sectional area; for bending, it is the moment of inertia; and for torsion, it is the polar moment of inertia. The mechanical parameter relevant to axial and bending loads is the elastic modulus, whereas it is the shear modulus for the torsional load.

Using the theory of classical beam discussed in Chap. 5, we will evaluate the load sharing in composite beam subjected to pure bending. Let us consider a three-material beam having a cross-section as presented in Fig. 6.6. The centroids C_i ($i = 1$–3) are at a distance of y_i from the reference coordinate (Y–Z). Assuming that the three-material beam is subjected to pure bending (moment, M), the moment equilibrium equation (Fig. 6.6) is written as follows:

$$\Sigma M_z = M + \int_{A_1} \sigma y dA + \int_{A_2} \sigma y dA + \int_{A_3} \sigma y dA = 0$$

$$M = -\left(\int_{A_1} \sigma y dA + \int_{A_2} \sigma y dA + \int_{A_3} \sigma y dA\right)$$

$$M = -\left(\int_{A_1} \frac{-E_1 y}{R} y dA + \int_{A_2} \frac{-E_2 y}{R} y dA + \int_{A_3} \frac{-E_3 y}{R} y dA\right)$$

However, in the above equation, the y is measured with respect to the neutral axis which will be discussed later. Since the radius of curvature is constant throughout the beam, it is taken outside the integral operator.

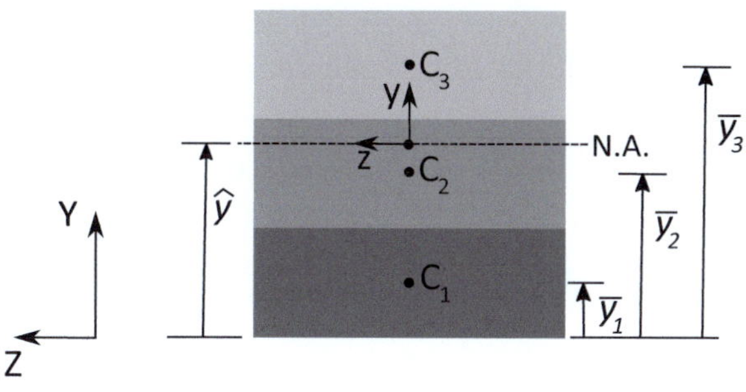

Fig. 6.6 Cross-section of a three-material beam

$$M = -\left(\frac{-E_1}{R} \int_{A_1} y^2 dA + \frac{-E_2}{R} \int_{A_2} y^2 dA + \frac{-E_3}{R} \int_{A_3} y^2 dA \right)$$

The bending moment is expressed in terms of the elastic modulus (E_i), moment of inertia (I_i) and radius of curvature (R) as given below.

$$M = \frac{E_1 I_1 + E_2 I_2 + E_3 I_3}{R}$$

Here, the moment of inertia (I_i) is measured about the neutral axis of respective materials. In order to find the individual stresses generated in different materials of the beam, we substitute ($1/R$) in the above equation using the following relation:

$$-\frac{E_i}{R} = \frac{\sigma_i}{y}$$

Upon substituting the term ($1/R$), we obtain

$$M = \frac{\sigma_i (E_1 I_1 + E_2 I_2 + E_3 I_3)}{E_i y}$$

The stresses generated by different materials are expressed as follows:

$$\sigma_i = \frac{E_i M y}{E_1 I_1 + E_2 I_2 + E_3 I_3} \tag{6.4}$$

Here y is the distance measured from the neural axis which can be determined following the procedure described in the following paragraph. Equation 6.4 indicates how the stress differs for different materials within a composite beam. We can, therefore, infer that the flexural stress generated by the material 'i' is proportional to the quantity, $\frac{E_i}{\Sigma E_i I_i}$. The concept of stress-shielding, which will be also discussed in Chap. 8, can be easily comprehended with the help of a composite beam.

Let us write the force equilibrium equation as

$$\Sigma F_x = \int dF = \int_{A_1} \sigma dA + \int_{A_2} \sigma \, dA + \int_{A_3} \sigma dA = 0$$

Using the flexural equation, the above equation is rewritten as

$$\int_{A_1} \frac{-E_1}{R} y dA + \int_{A_2} \frac{-E_2}{R} y dA + \int_{A_3} \frac{-E_3}{R} y dA = 0$$

$$\frac{-E_1}{R}\int_{A_1} y\mathrm{d}A + \frac{-E_2}{R}\int_{A_2} y\mathrm{d}A + \frac{-E_3}{R}\int_{A_3} y\mathrm{d}A = 0$$

Here, y is the distance measured from the neutral axis. Let us consider the distance to be measured from the reference coordinate (Y–Z) as shown in Fig. 6.6. Therefore, we can write $y = Y - \hat{y}$.

Now, the above equation is rewritten as follows.

$$E_1\int_{A_1}(Y-\hat{y})\mathrm{d}A + E_2\int_{A_2}(Y-\hat{y})\mathrm{d}A + E_3\int_{A_3}(Y-\hat{y})\mathrm{d}A = 0$$

$$E_1\int_{A_1} Y\mathrm{d}A + E_2\int_{A_2} Y\mathrm{d}A + E_3\int_{A_3} Y\mathrm{d}A = E_1\int_{A_1}\hat{y}\mathrm{d}A + E_2\int_{A_2}\hat{y}\mathrm{d}A + E_3\int_{A_3}\hat{y}\mathrm{d}A$$

Using the equation, $\int_{A_i} Y\mathrm{d}A = \overline{y_i}A_i$, the above equation is rewritten as

$$E_1\overline{y_1}A_1 + E_2\overline{y_2}A_2 + E_3\overline{y_3}A_3 = \hat{y}(E_1A_1 + E_2A_2 + E_3A_3)$$

Here, $\overline{y_i}$ is the centroid of the cross-sectional area A_i.

The location of the neutral axis is expressed as

$$\hat{y} = \frac{E_1\overline{y_1}A_1 + E_2\overline{y_2}A_2 + E_3\overline{y_3}A_3}{E_1A_1 + E_2A_2 + E_3A_3} \tag{6.5}$$

The implant-bone structure of a cemented hip stem is idealized as a three-material beam as shown in Fig. 6.7. So far, we have dealt with a case in which the beam had an axis of symmetry and the moment was acting on the plane of symmetry. However, in reality, the cross-sections of the bone-implant structures are not symmetric.

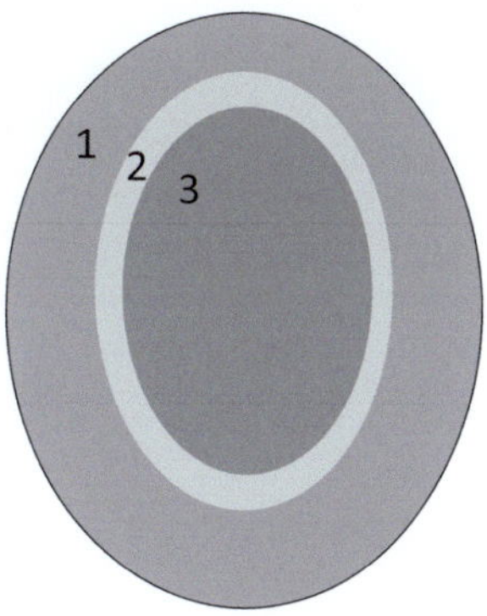

Fig. 6.7 Simplified representation of cross-section of a cemented femoral implant-bone structure as a three-material beam

6.2.3 Bending of Unsymmetrical Beam

Let us now discuss the pure bending of unsymmetrical beams, considering a convenient set of centroidal axes (y, z) about which the moments of inertia (I_{zz} and I_{yy}) and product of inertia (I_{yz}) are calculated. A bending moment, M, is applied on the unsymmetrical beam, as shown in Fig. 6.8. This moment is resolved into components along y- and z-axis as M_y and M_z, respectively. Let us assume the neutral axis to be BB′. The strain at any point in the cross-section is expressed as

$$\varepsilon = k'y'$$

Therefore, the stress is expressed as

$$\boldsymbol{\sigma} = Ek'y' = ky'$$

where k and k' are non-zero constants. For force equilibrium, the following conditions need to be satisfied:

$$\Sigma F_x = \int \mathrm{d}F = \int_A \sigma \,\mathrm{d}A = 0$$

$$k \int_A y' \mathrm{d}A = 0$$

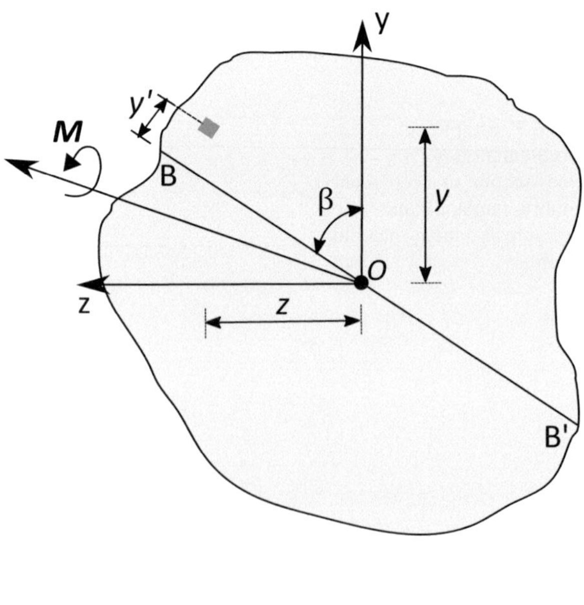

Fig. 6.8 Pure bending of an unsymmetrical beam having centroid at O

This indicates that the neutral axis passes through the centroid or the modulus-weighted centroid in the case of a multi-material beam. Let us assume β to be the angle between BB′ and the y-axis.

The y' is expressed as, $y' = y\sin\beta - z\cos\beta$.

The moment equilibrium conditions result in the following:

$$M_z = -\int_A \sigma y \mathrm{d}A = -\int_A k y' y \mathrm{d}A$$

$$M_y = \int_A \sigma z \mathrm{d}A = \int_A k y' z \mathrm{d}A$$

Upon substituting for y' in the above equations:

$$\begin{aligned} M_z &= -\int_A k(y\sin\beta - z\cos\beta) y \mathrm{d}A \\ &= -\int_A k\left(y^2\sin\beta - yz\cos\beta\right)\mathrm{d}A \\ &= -kI_{zz}\sin\beta + kI_{yz}\cos\beta \\ M_y &= \int_A k(y\sin\beta - z\cos\beta) z \mathrm{d}A \\ &= \int_A k\left(yz\sin\beta - z^2\cos\beta\right)\mathrm{d}A \\ &= kI_{yz}\sin\beta - kI_{yy}\cos\beta \end{aligned} \tag{6.6}$$

From the above two equations, we get

$$\frac{M_z}{M_y} = \frac{-I_{zz}\sin\beta + I_{yz}\cos\beta}{I_{yz}\sin\beta - I_{yy}\cos\beta} = -\frac{\left(I_{zz}\tan\beta - I_{yz}\right)}{I_{yz}\tan\beta - I_{yy}}.$$

After rearranging the above expression, we can write

$$\tan\beta = \frac{I_{yy}M_z + I_{yz}M_y}{I_{yz}M_z + I_{zz}M_y} \tag{6.7}$$

The location of the neutral axis BB′ with respect to the axes (y, z) is found out with the help of Eq. 6.7. Using Eq. 6.6, the stress is expressed in terms of the y-component of the moment (M_y) as

$$\sigma = ky' = \frac{M_y y'}{I_{yz}\sin\beta - I_{yy}\cos\beta} = \frac{M_y(y\sin\beta - z\cos\beta)}{I_{yz}\sin\beta - I_{yy}\cos\beta}$$

The above equation is simplified as

$$\sigma = \left\{\frac{M_y(y\tan\beta - z)}{I_{yz}\tan\beta - I_{yy}}\right\}$$

The term $\tan\beta$ is substituted using Eq. 6.7 and the above equation is rewritten as

$$\sigma = \frac{M_z(yI_{yy} - zI_{yz}) + M_y(yI_{yz} - zI_{zz})}{I_{yz}^2 - I_{yy}I_{zz}} \tag{6.8}$$

If the chosen y- and z-axis are the principal axes (axes of symmetry), then the product of inertia, I_{yz}, is equal to zero. In the case of a symmetric beam, the above equation reduces to

$$\sigma = \frac{M_z y I_{yy} - M_y z I_{zz}}{-I_{yy}I_{zz}} = -\frac{M_z y}{I_{zz}} + \frac{M_y z}{I_{yy}}$$

Alternative method: An alternative method for finding bending stress in the unsymmetrical beam is discussed in the following (Fig. 6.9). The strain at any point is a linear function of the coordinates (s, t), according to the classical beam theory. Therefore, the strain (ε) at any point (s, t) in the cross-section is expressed as

$$\varepsilon(s, t) = C_1 + C_2 t + C_3 s$$

The stress at any point is expressed as

$$\sigma = E(s, t)\varepsilon(s, t) = E(s, t)(C_1 + C_2 s + C_3 t) \tag{6.9}$$

For force equilibrium, the following conditions need to be satisfied:

$$\begin{aligned}\Sigma F_t = \int \mathrm{d}F &= \int_A \sigma \mathrm{d}A \\ &= \int_A E(s, t)(C_1 + C_2 s + C_3 t)\mathrm{d}A \\ &= 0\end{aligned}$$

The above equation results in the following conditions:

$$C_1 \int_A E(s, t)\mathrm{d}A = 0 \tag{6.10}$$

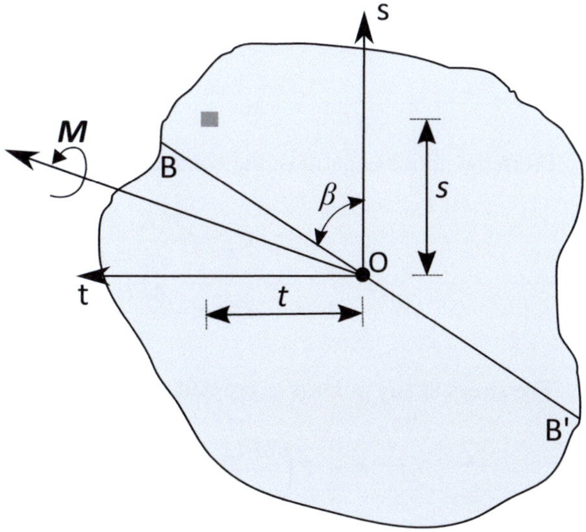

Fig. 6.9 Method 2, pure bending of an unsymmetrical beam

$$C_2 \int_A E(s,t) s \mathrm{d}A = 0 \tag{6.11}$$

$$C_3 \int_A E(s,t) t \mathrm{d}A = 0 \tag{6.12}$$

Equation 6.10 results in $C_1 = 0$, which means that the stress is zero at the origin (Eq. 6.9). Equations 6.11 and 6.12 indicate that the neutral axis passes through the origin of the axes *s*-*t*. The moment equilibrium condition results in the following:

$$M_s = \int_A \sigma t \mathrm{d}A = \int_A E(s,t)(C_2 s + C_3 t) t \mathrm{d}A \tag{6.13}$$

$$M_t = -\int_A \sigma s \mathrm{d}A = \int_A E(s,t)(C_2 s + C_3 t) s \mathrm{d}A \tag{6.14}$$

where M_s and M_t are the component of M about s and t directions. For the time being, let us assume $E(s, t) = E$, a constant (single material beam).

$$M_s = \int_A E(s,t)(C_2 s + C_3 t) t \mathrm{d}A = E \int_A \left(C_2 st + C_3 t^2\right) \mathrm{d}A = EC_2 I_{st} A + EC_3 I_{ss} A$$

$$M_t = -\int_A E(s,t)(C_2 s + C_3 t) s \, \mathrm{d}A = -E \int_A \left(C_2 s^2 + C_3 st\right) \mathrm{d}A$$
$$= -[EC_2 I_{tt} A + EC_3 I_{st} A]$$

Upon using the above two equations,

$$\frac{M_s}{M_t} = -\frac{[C_2 I_{st} + C_3 I_{ss}]}{C_2 I_{tt} + C_3 I_{st}}$$

From the above equations, the constants C_2 and C_3 is expressed as

$$C_2 = \frac{M_t I_{ss} + M_s I_{st}}{I_{st}^2 - I_{ss} I_{tt}}$$
$$C_3 = -\frac{M_t I_{st} + M_s I_{tt}}{I_{st}^2 - I_{ss} I_{tt}}$$

The stress at any point is expressed as

$$\sigma = E\left(\frac{M_t I_{ss} + M_s I_{st}}{I_{st}^2 - I_s I_t} s - \frac{M_t I_{st} + M_s I_{tt}}{I_{st}^2 - I_s I_t} t\right) \tag{6.15}$$

The location of the neutral axis is obtained by equating the above to zero.

$$E\left(\frac{M_t I_{ss} + M_s I_{st}}{I_{st}^2 - I_{ss} I_{tt}} s - \frac{M_t I_{st} + M_s I_{tt}}{I_{st}^2 - I_{ss} I_{tt}} t\right) = 0$$
$$\tan(\beta) = \frac{t}{s} = \frac{M_t I_{ss} + M_s I_{st}}{M_t I_{st} + M_s I_{tt}} \tag{6.16}$$

where β is the angle made by the neutral axis with the y-axis.

6.2.4 Bending of Unsymmetrical Composite Beam

Now, let us consider the case of the composite beam (Fig. 6.10). Upon substituting $s = y - \hat{y}$ and $t = z - \hat{z}$ in Eqs. 6.13 and 6.14 and according to the sign convention followed so far, we get

$$M_s = \int_A E(s,t)(C_2(y - \hat{y}) + C_3(z - \hat{z}))(z - \hat{z})\mathrm{d}A \tag{6.17}$$

$$M_t = -\int_A E(s,t)(C_2(y - \hat{y}) + C_3(z - \hat{z}))(y - \hat{y})\mathrm{d}A \tag{6.18}$$

In order to account for three materials in the composite beam, let us use the summation operator as given below in the equations.

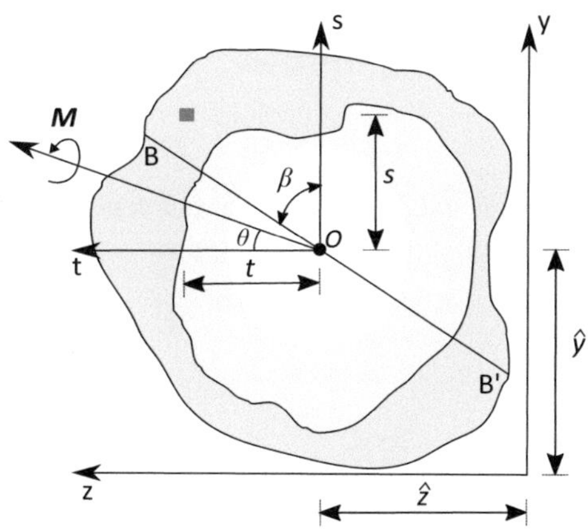

Fig. 6.10 Bending of unsymmetrical composite beam

$$M_s = \sum_{i=1}^{3} \int_{A_i} E_i(C_2(y - \hat{y}) + C_3(z - \hat{z}))(z - \hat{z})\mathrm{d}A$$

$$M_t = -\sum_{i=1}^{3} \int_{A_i} E_i(C_2(y - \hat{y}) + C_3(z - \hat{z}))(y - \hat{y})\mathrm{d}A$$

Let us closely look into the integrands in the above equations:

$$\begin{aligned} &E_i(C_2(y - \hat{y}) + C_3(z - \hat{z}))(z - \hat{z}) \\ &\quad = E_i\left[C_2(yz - y\hat{z} - \hat{y}z + \hat{y}\hat{z}) + C_3\left(z^2 - 2z\hat{z} + \hat{z}^2\right)\right] \\ &E_i(C_2(y - \hat{y}) + C_3(z - \hat{z}))(y - \hat{y}) \\ &\quad = E_i\left[C_3(yz - y\hat{z} - \hat{y}z + \hat{y}\hat{z}) + C_2\left(y^2 - 2y\hat{y} + \hat{y}^2\right)\right] \end{aligned}$$

Using the above expression, Eqs. 6.17 and 6.18 are rewritten as

$$\begin{aligned} M_s &= \sum_{i=1}^{3} \int_{A_i} E_i\left[C_2(yz - y\hat{z} - \hat{y}z + \hat{y}\hat{z}) + C_3\left(z^2 - 2z\hat{z} + \hat{z}^2\right)\right]\mathrm{d}A \\ &= \sum_{i=1}^{3} E_i\left[C_2\left(I_{yz_i} - \hat{z}\overline{y}_i A_i - \hat{y}\overline{z}_i A_i + \hat{y}\hat{z}A_i\right) + C_3\left(I_{yy_i} - 2\hat{z}\overline{z}_i A_i + \hat{z}^2 A_i\right)\right] \end{aligned} \quad (6.19)$$

$$M_t = -\sum_{i=1}^{3} \int_{A_i} E_i\left[C_3(yz - y\hat{z} - \hat{y}z + \hat{y}\hat{z}) + C_2\left(y^2 - 2y\hat{y} + \hat{y}^2\right)\right]\mathrm{d}A$$

$$= -\sum_{i=1}^{3} E_i\left[C_3\left(I_{yz_i} - \hat{z}\overline{y}_i A_i - \hat{y}\overline{z}_i A_i + \hat{y}\hat{z}A_i\right) + C_2\left(I_{zz_i} - 2\hat{y}\overline{y}_i A_i + \hat{y}^2 A_i\right)\right] \tag{6.20}$$

The $\overline{y}_i$ and $\overline{z}_i$ are the different centroid positions with respect to the axes *y*–*z*. The above equations are expressed as

$$M_s = C_2 I^*_{yz} + C_3 I^*_{yy}$$
$$M_t = -\left(C_2 I^*_{zz} + C_3 I^*_{yz}\right)$$

Here,

$$I^*_{yz} = \sum_{i=1}^{3} E_i\left(I_{yz_i} - \hat{z}\overline{y}_i A_i - \hat{y}\overline{z}_i A_i + \hat{y}\hat{z}A_i\right)$$
$$I^*_{yy} = \sum_{i=1}^{3} E_i\left(I_{yy_i} - 2\hat{z}\overline{z}_i A_i + \hat{z}^2 A_i\right)$$
$$I^*_{zz} = \sum_{i=1}^{3} E_i\left(I_{zz_i} - 2\hat{y}\overline{y}_i A_i + \hat{y}^2 A_i\right)$$

Using the above equations,

$$\frac{M_s}{M_t} = -\frac{\left[C_2 I^*_{yz} + C_3 I^*_{yy}\right]}{C_2 I^*_{zz} + C_3 I^*_{yz}}$$

From the above equation, the constants C_2 and C_3 are expressed as

$$C_2 = \frac{M_t I^*_{yy} + M_s I^*_{yz}}{I^{*2}_{yz} - I^*_{yy} I^*_{zz}}$$
$$C_3 = -\frac{M_t I^*_{yz} + M_s I^*_{zz}}{I^{*2}_{yz} - I^*_{yy} I^*_{zz}}$$

Similar to Eq. 6.9, the stress at any point is expressed as

$$\sigma = E\left(\frac{M_t I^*_{yy} + M_s I^*_{yz}}{I^{*2}_{yz} - I^*_{yy} I^*_{zz}} s - \frac{M_t I^*_{yz} + M_s I^*_{zz}}{I^{*2}_{yz} - I^*_{yy} I^*_{zz}} t\right)$$

The location of the neutral axis is obtained by equating the above equation to zero.

$$E\left(\frac{M_tI^*_{yy}+M_sI^*_{yz}}{I^{*2}_{yz}-I^*_{yy}I^*_{zz}}s-\frac{M_tI^*_{yz}+M_sI^*_{zz}}{I^{*2}_{yz}-I^*_{yy}I^*_{zz}}t\right)=0$$

$$\frac{M_tI^*_{yy}+M_sI^*_{yz}}{I^{*2}_{yz}-I^*_{yy}I^*_{zz}}s-\frac{M_tI^*_{yz}+M_sI^*_{zz}}{I^{*2}_{yz}-I^*_{yy}I^*_{zz}}t=0$$

$$\tan(\beta)=\frac{t}{s}=\frac{M_tI^*_{yy}+M_sI^*_{yz}}{M_tI^*_{yz}+M_sI^*_{zz}}$$

Here $M_s = M \sin\theta$ and $M_t = M \cos\theta$ (Fig. 6.10).

According to the classical beam theory, the equations in Sect. 6.2 hold true for pure bending of the implant-bone structure which is considered as an unsymmetrical composite beam. These equations are useful for stress analysis of implant-bone composite structure and to determine the location of failure owing to high stresses when subjected to pure bending.

6.3 Biomechanics of Hip Replacement

Based on the implant geometry, hip replacements are classified as stemmed hip replacement and hip resurfacing (Fig. 6.11). Stemmed hip replacements are further categorised into Total Hip Replacement (THR) and hemi hip replacement, depending on whether both the bones (femur and pelvis) of the joint are replaced or the femur only. In the following subsections, the mechanics of THRs and hip resurfacing will be discussed.

6.3.1 Total Hip Replacement

In Total Hip Replacement (THR) or Total Hip Arthroplasty (THA), both the femoral and acetabular sides of the joint are replaced. An acetabular component (hemispherical cup) is used to replace the damaged acetabular cavity. This component is usually made of a special grade of polyethylene (ultra-high molecular weight polyethylene (UHMWPE), often used with a metal backing. The damaged femoral head and neck are usually removed and a stem-like structure is inserted into the femoral medullary canal. The components are fixed with the bone using the cemented or uncemented fixation technique. The femoral head, which is made of metal (Co-Cr alloy) or ceramic (alumina or zirconia), articulates with the polyethylene (UHMWPE) liner of the acetabular component (Fig. 6.12). The femoral stem is made of metallic biomaterials, such as Titanium (Ti) alloy and Cobalt-Chromium (Co-Cr) alloy.

Based on their construction, the hip implants can be classified into monoblock and modular (Park and Bronzino 2003). A monoblock implant consists of a single part and is usually less expensive. If two or more parts are assembled together during

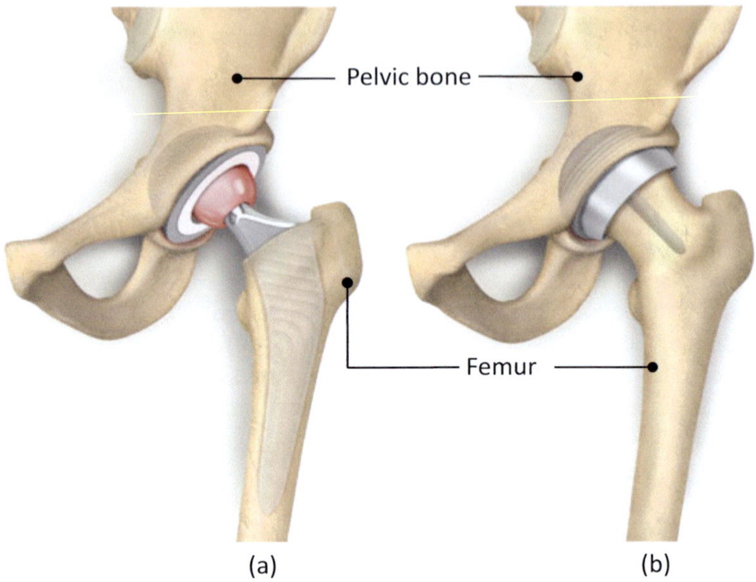

Fig. 6.11 Classification of the hip replacement surgery; **a** stemmed hip replacement, **b** hip resurfacing. *Source* Palazzuolo et al. (2023)

Fig. 6.12 Various components involved in total hip arthroplasty; **a** assembled view, **b** uncemented hip stem, **c** cemented hip stem

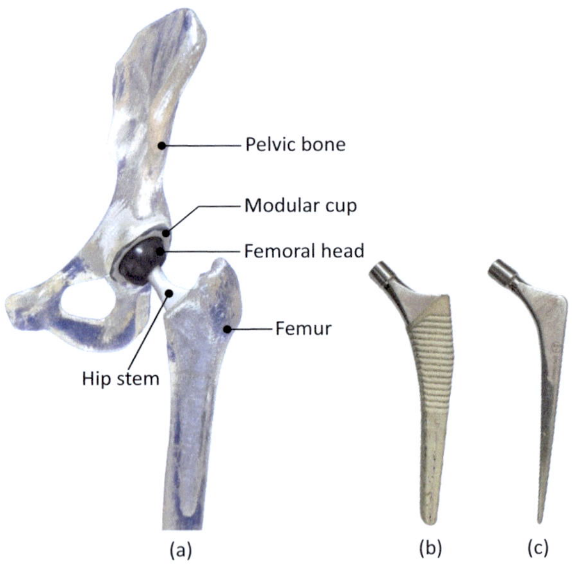

the implantation, it is called a modular implant. In such implants, the femoral head is fitted to the femoral neck with a Morse taper. Such design allows changes in head material, head size, and neck length, providing provisions for customization of the implant components during surgery. However, such implants are more prone to corrosion than the monoblock components. A monoblock acetabular cup is made of UHMWPE, while a modular cup consists of a metallic shell (metal-backing) and a congruent UHMWPE liner inserted into the shell. The metal backing provides stiffness to the implant and thereby reduces the deformation of the UHMWPE liner. The outer porous surface of the metallic shell facilitates the biological fixation of the cup within the host bone. The modular cup allows easier revision involving liner exchange.

The implant components are designed to provide a wide range of motion and support the demanding musculoskeletal loading of the joint. However, dislocation of the implanted joint should be avoided, and the implant's impingement on the rim of the acetabular cup should be prevented. The femoral neck length, restoration of the centre of motion and femoral offset can influence the biomechanics of the implanted hip joint. Appropriate selection and restoration of such parameters decrease the bending stress on the implant-bone interface.

Polyethylene wear debris is generated at the articulating surface of the implanted joint due to load bearing and movements of the implant head within the acetabular cup. Micromotion between the UHMWPE insert and the acetabular cup shell can also generate polyethylene debris. Cement debris is generated at the implant-cement and cement-bone interfaces owing to relative movements (micromotion). Small fragments of porous coating can also be produced at the implant-bone interfaces. Such tiny wear debris can easily spread around the joint space with the body fluids. The immune system of the body can identify such foreign tiny particles and initiate an inflammatory response in the joint space. In the process, rapid localised bone loss may occur around the implant. Such bone loss is called osteolysis, which may lead to implant loosening and/or fracture of the bone. In order to reduce the risk of osteolysis, the wear properties of UHMWPE have been improved by the use of crosslinked polyethylene. Due to their low wear properties, the metal-on-metal and the ceramic-on-ceramic bearing surfaces have also been used in the implanted hip joint.

6.3.1.1 Effect of Stem Length

Hip implants have been in clinical use for several decades. Several designs have evolved over time with improvement in performance. Implantation with a hip stem leads to alterations in the load transfer within an implanted femur. Usually, a femoral stem transfers the load more distally, leaving the proximal femur shielded from the load. A part of the load is transferred through the tip of the stem. Alteration of the load transfer in the femur triggers the bone remodelling process, leading to changes in bone density. Stress-shielding leads to bone resorption (reduction in bone density) in the proximal part of the implanted femur. However, bone apposition (increase in bone density) has been observed around the distal tip of the stem. A shorter stem can lead to more favourable load transfer in the implanted femur by reducing stress-shielding in the proximal femur.

The short-stem femoral implant offers the advantage of less bone removal from the medullary canal, owing to reduced stem length. Such stem designs usually rely on metaphyseal fixation with the bone and are commonly used in younger and active patients. Short hip-stems can preserve the pattern of natural load transfer more closely since the load is transferred more proximally to the bone than for a long-stem design. Conventional long hip-stems, which have diaphyseal fixation, cause stress-shielding in the proximal femur. The metaphyseal engagement of the stem with the bone allows smooth transfer of loads to the proximal femur for the short stem designs. Changes in load transfer, owing to the shortening of the stem, lead to a reduction in stress-shielding and adverse bone remodelling (bone resorption). Despite the advantages of reduced stress-shielding and adverse bone remodelling, a short hip-stem can have reduced anchoring strength of the implant than that of a long-stem design.

6.3.1.2 Effect of Interface Characteristics

In an implanted-bone structure, the implant-bone interface (cementless fixation) and the implant-cement and cement-bone interfaces (cemented fixation) are often the weakest links of the structure. The strength of the cement-bone and implant-cement interfaces can degrade over time. Depending on the degree of osseointegration, the implant-bone interface might not be perfectly bonded. Load transfer in an implanted bone structure is influenced by the interface characteristics. Finite element analysis of the implanted bone structure revealed that the load is transferred through the stem tip to the distal femur in the case of a bonded interface. Consequently, the proximal part of the femur is shielded from carrying load to a variable extent. In contrast, an imperfectly bonded interface allows micromovements of the components, causing partial load transfer due to friction and shear at the interfaces. The remaining part of the load is transferred through the distal tip of the stem. Assuming a frictional contact interface, therefore, leads to relatively lower stress-shielding in the proximal femur as compared to the fully bonded interface.

6.3.1.3 Effect of Porous Lattice Structure

The Ti-6Al-4V alloy is more commonly used as a hip-stem material owing to its high strength-to-weight ratio, superior corrosion resistance, bio-compatibility, excellent osseointegration character, and relatively lower elastic modulus ($\approx$110 GPa). Despite having lower elasticity as compared to other biomaterials, the elastic modulus of Ti-6Al-4V is significantly higher in comparison to human cortical bones ($\approx$18 GPa). A stiffer implant alters the load transfer pattern in the adjacent bone since a part of the bone is partially shielded from the load. Consequently, bone remodels itself based on the altered mechanical stimulus. The shielded part of the bone gradually becomes less dense and weaker. This reduction in bone density leads to aseptic loosening and eventual failure of the implant. In order to reduce the stiffness of implants and minimise the effect of stress-shielding, the concept of porous implants has evolved. Fabrication of such porous structures is nowadays possible using additive manufacturing techniques. Moreover, embedding porosity into the implants is expected to enhance bone growth at the implant-bone interface and thereby improve fixation and osseointegration.

Both regular structures and irregular bionic architectures are commonly used as porous structures for an implant (Chowdhury et al. 2021). The geometries in the regular structure have repetitive features, whereas the irregular structures have random features. Various unit cell designs have been investigated to establish their suitability for applications in porous hip implants. Simple unit-cell designs, such as cubic, body-centred-cubic, face-centred-cubic type cells, and more complex designs like a tetrahedron, octet, and hexagonal unit cells have been extensively explored. Various triply periodic minimal surface-based porous structures have also drawn the attention of the researchers. The regular structures have been employed in several orthopaedic and dental implant designs, owing to their enhanced controllability of the structural parameters. The mechanical properties of such structures can be tailored to meet the demand of a specific application by changing the porosity. Therefore, evaluation of the mechanical properties of such structures is required to establish the ideal porosity ranges for a given application. FE-based simulation and mechanical characterisation of 3D-printed porous structures are commonly performed to optimise porous implant design (Chowdhury et al. 2021; Chaudhuri et al. 2024).

6.3.1.4 Effect of Implant Alignment

It is debatable whether the neutral, valgus, and varus orientation of the implant can influence the load transfer in the implanted bone or not. Some studies suggest that varus/valgus mal-positioning of the hip stem can have adverse outcomes of hip surgery (Teh et al. 2024). Load transfer pattern may be abnormal for a varus-cemented hip stem. Implant subsidence, loosening and even periprosthetic fractures are reported for such hip stems. Femoral head offset and limb length may be affected as well. A higher risk of revision has been reported for varus uncemented hip-stems. Likewise, valgus uncemented stems may have similar detrimental effects. It induces more stress-shielding in the proximal femur and increases wear in the long term. However, other studies suggest that the alignment of a tapered-wedge cementless femoral component did not affect the results of hip surgery, at least at a mean of 5 years post-surgery (Min et al. 2008; Tsujino et al. 2024). Stem alignment in 98 consecutive primary THR (neutral—62 hips; valgus—20 hips; and varus—16 hips) with a cementless hip-stem was investigated up to 7.7 years post-surgery (Min et al. 2008). No significant differences in clinical or radiographic results between groups were observed. Harris hip scores and incidence of thigh pain were also similar between groups. Predicted changes in bone density due to bone remodelling were found to be similar in all three groups.

6.3.2 Common Classification of Stem Designs

Cemented femoral stem designs can be categorised according to their pattern of load transfer to the femur; shape-closed design and force-closed design (Shen 1998; Chulsomlee et al. 2023). "Shape-closed" stem design has a rough surface finish and possesses a mild taper or no taper (Figs. 6.13a and 6.14a) and debonding at the implant–cement interface. This design is also called composite beam design since the stem-cement-bone construct functions mechanically as a "composite beam," and it transfers load principally at the level of the femoral diaphysis.

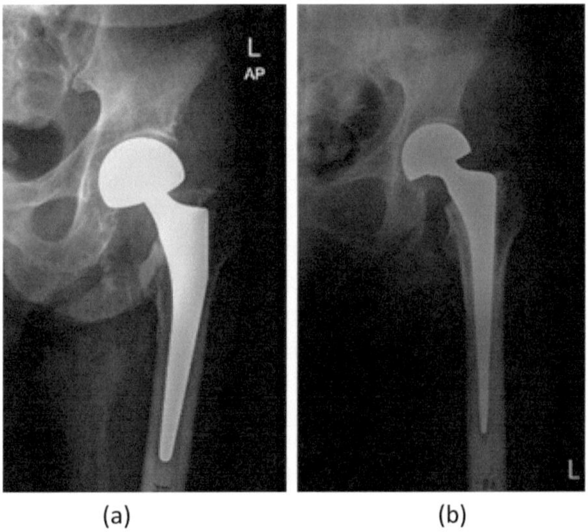

Fig. 6.13 Radiograph showing cemented femoral stem design. **a** Shape-closed stem as cemented Corail stem (DePuy International, Leeds), **b** force-closed stem as CPT stem (Zimmer Biomet, Warsaw, IN). *Source* Chulsomlee et al. (2023)

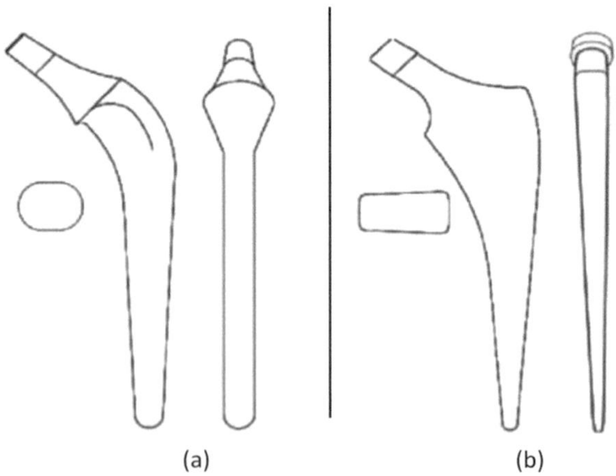

Fig. 6.14 The stem designs along with the cross-section details. **a** Rounded cross-section of shape-closed stem, **b** tapered cross-section of force-closed stem

These design features can enhance rotational stability and reduce stem subsidence. In simpler words, shape-closed stems achieve stability by closely matching the shape of the femur, relying on the fit within the femoral canal for stability. Such a design is known to exhibit minimal postoperative subsidence.

The Force-closed design, also known as taper-slip design, has a polished stem with double or triple-taper (Figs. 6.13b and 6.14b). These stems are designed to subside within the cement mantle. This subsidence causes hoop stress in the proximal cement layer and transfers loads directly to the surrounding bone. This is why the force-closed design is expected to transfer load more proximally than the composite beam design. A third taper can be included to further enhance load transfer to the proximal medial femoral cortex, the region prone to high periprosthetic bone resorption.

6.3.3 Hip Resurfacing

In a conventional THA, the entire femoral head and neck are removed to allow insertion of the hip stem within the resected femur. Bone stock preserved in the femoral shaft is sometimes inadequate for revision surgery due to stress-shielding-induced bone resorption. Hip Resurfacing Arthroplasty (HRA) has emerged as an alternative to the conventional THA. It is a type of hip replacement where only the articulating surfaces of the femoral head and the acetabulum are replaced with prostheses (Fig. 6.15). More bone is preserved in the femoral head, neck and shaft than the conventional THA. The damaged femoral head is resurfaced with a cup-like device with a small stem inserted into the femoral head and neck. Although the conventional THA is a well-established and highly successful reconstructive procedure for most hip joint disorders, it has been found to be less effective for young and active patients than for aged patients with a subsequent need for revision surgery. HRA has evoked considerable interest, particularly for young and active arthritic patients with expectations of restoring lifestyle in addition to improved mobility and pain relief.

The advantages of HRA over THR include preservation of bone stock for subsequent revision surgery, easier revision, more precise biomechanical restoration of the joint, normal femoral loading, and enhanced stability due to the larger diameter of articulation. Unlike the THR, the femoral medullary canal is not violated in the resurfacing procedure; only the articulating surface of the femoral head is replaced by an implant. Preservation of bone in the femoral head, neck, and shaft ensures more physiological load transfer within the implanted femur and reduces the effect of stress-shielding. Large diameter femoral component offers the inherent stability of the reconstructed joint and optimal range of movements. As such, the femoral resurfacing retains the physiological aspect of the hip. The prevalence of the dislocation was markedly reduced as compared to the THA. Moreover, the chances of infections due to the insertion of foreign materials are low as the implant is smaller in size. It is less likely to be susceptible to osteonecrosis (loss of blood supply), due to retaining of the vascularised femoral head and reduced surgical intervention.

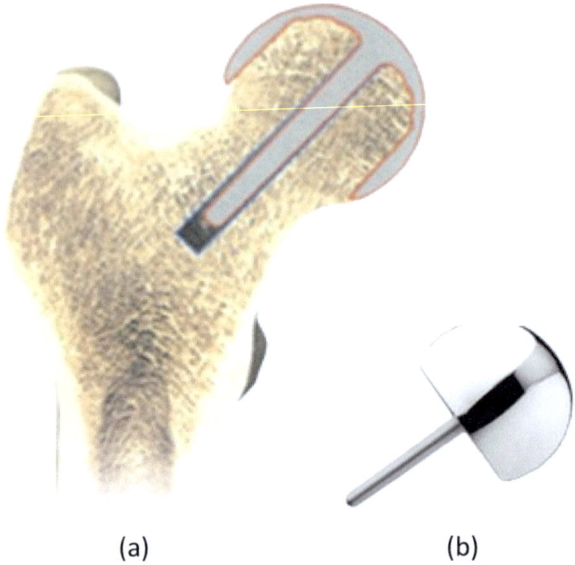

Fig. 6.15 Hip resurfacing; **a** resurfaced proximal femur after HRA, **b** hip resurfacing component

Despite its advantages, short-term and long-term failures of HRA have been reported. In the short-term, femoral neck fracture is a concern (Fig. 6.16), particularly within a few months after surgery (Shimmin et al. 2008). Post-operatively, HRA generates higher strains in the proximal femoral neck-component junction, irrespective of the variation in interface conditions, indicating a potential risk of femoral neck fracture (Pal et al. 2009; Gupta et al. 2006). Strain-shielding has been observed in the femoral head post-implantation, leading to periprosthetic bone resorption and eventual implant loosening in the long term. The proximal femoral shaft around the trochanteric region is also subjected to bone resorption; however, bone resorption was observed to be lower than the femoral head. Subsequent to bone remodelling, this strain concentration in the cancellous bone around the proximal femoral neck-component junction and the distal tip of the stem was considerably reduced, lowering the risk of neck fractures (Pal et al. 2009; Gupta et al. 2006). Although HRA preserves bone stock during primary surgery, strain-shielding and periprosthetic bone resorption may cause eventual gradual loosening.

6.3.3.1 Effect of Stem Length

The metaphyseal stem plays a crucial role in load transfer within the resurfaced femur. Ideally, the stem is used as a visual guide for placement of the femoral implant during surgery. An over-reamed hole is used to insert the stem within the femoral head and neck to avoid direct contact between the stem and bone. However, bone can gradually grow in the gap between the stem and the bone. Once stem-bone contact is established, part of the load is transferred distally through the stem tip.

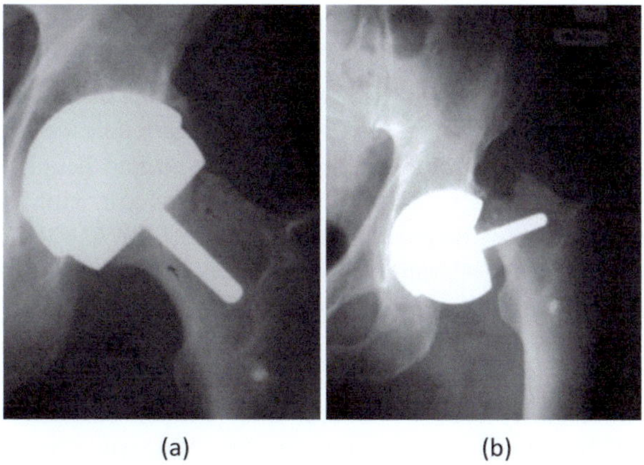

Fig. 6.16 A radiograph of hip resurfacing arthroplasty showing propagation of fracture line (indicated by black arrows) from superior femoral neck leading to eventual femoral neck fracture. *Source* Shimmin et al. (2008)

This causes more strain-shielding and, subsequently, adverse bone remodelling in the femoral head underlying the implant cup. In order to reduce the risk of implant loosening due to adverse bone remodelling, the stem length may be shortened to an extent to ensure physiological load transfer. The effect of short stem HRA (50% of the standard length) on load transfer and bone remodelling within the resurfaced femur was investigated by Pal et al. (2010a). The effect of shortening the stem was predominantly evident in strain-shielding and bone remodelling results (Fig. 6.17). Post-operatively, strain-shielding was observed in a smaller (approximately half) bone volume underlying the cement than that of the long-stem design. This effect lessened considerably towards the distal part of the head. In contrast to the long-stem design, the short-stem model did not experience much elevated strains in the bone around the stem tip (Fig. 6.17).

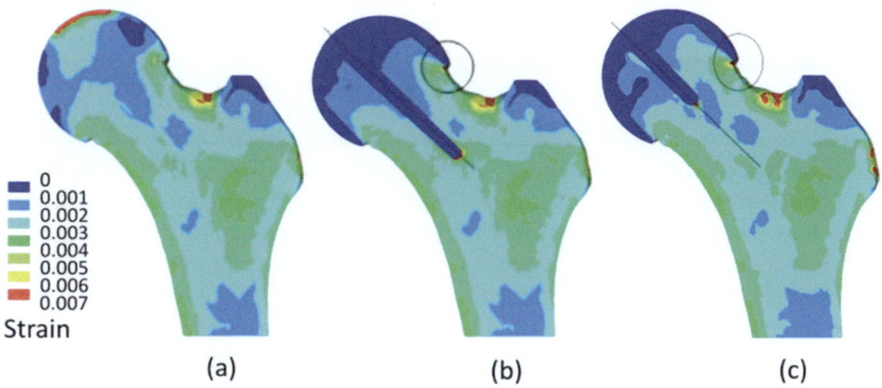

Fig. 6.17 Equivalent strain distributions during stair climbing, mid-frontal sectional view, **a** intact femur; **b** resurfaced femur with conventional long stem; **c** short-stem (50% reduction) resurfaced femur. Area of failure initiation (high strain) is encircled

6.3.3.2 Effect of Alternative Materials

Metal-on-metal (MoM) bearing surface has been commonly reported for hip resurfacing arthroplasty. However, this bearing surface is associated with the release of metal ions, which are detrimental to human health. Adverse tissue reactions and systemic toxicity are common concerns for MoM bearings. Alternative bearing surfaces have been explored to replace the MoM bearing surface. The Ceramic-on-metal and Ceramic-on-ceramic bearing surfaces have been used alternatively. The ceramic-on-ceramic surface is reported to have a lower wear rate compared to the MoM surface. Moreover, it does not have the risk associated with metal ion release. The ceramic femoral resurfacing component (shorter stem) has been successfully used in clinical trials (https://classic.clinicaltrials.gov/ct2/show/NCT03326804). However, there may be concerns regarding the use of ceramic as an alternative resurfacing component with standard stem length. An increased risk of femoral neck fracture was reported with a ceramic resurfacing component of standard stem length (Pal et al. 2010b). A higher volume of bone is exposed to elevated strain in the proximal femoral neck region and around the distal tip of the stem. Moreover, the ceramic component was subjected to higher tensile stresses. Ceramics are brittle in nature and may be susceptible to fracture under excessive tensile stresses. Moreover, a squeaking sound is generally associated with such bearing surfaces several years after the surgery. The exact reasons for squeaking are not known fully yet, but micro-separation, impingement, and edge loading causing stripe wear have been linked as possible reasons. For ceramic-on-metal bearing surfaces, the amount of metal ion release is halved as metal is used only for one component. Similarly, the squeaking problem associated with the ceramic-on-ceramic surfaces is also reduced. Despite the perceived advantages, ceramic-on-metal surfaces are not frequently used, perhaps due to an increased awareness of the risk of metal ion release.

6.4 Biomechanics of Knee Replacement

In total knee replacements (TKR), the damaged femoral condyles and the tibial plateaus are replaced with a femoral and tibial component, respectively (Fig. 6.18). The femoral component is commonly made of Co-Cr alloy. Some designs are made of ceramics or ceramic/metal mixtures. The tibial component is made of either polyethylene or metallic alloys (Ti-alloy or Co-Cr alloy). Typically, a tibial insert made of polyethylene (UHMWPE) is used on top of the metallic tibial plate. This polyethylene insert acts as the bearing component between the femoral component and the tibial plate. The tibial and femoral components have stems and/or pegs that are inserted into the proximal tibia and distal femur. A patellar component is sometimes used to replace the patellofemoral joint.

The geometry and kinematics of the knee joint is much more complicated than that of the hip joint. In a normal knee joint, the associated ligaments provide stability to the joint. Based on the extent of stability obtained by the ligaments, knee implants are

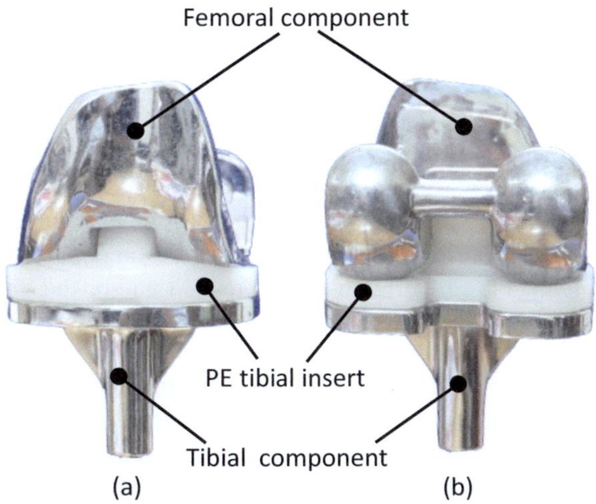

Fig. 6.18 Typical knee implant showing different components

classified as constrained, partially constrained/semi-constrained and unconstrained (Park and Bronzino 2003). Constrained implants are used in the absence of any ligaments and rely on a hinge articulation with a fixed axis of rotation. Hinged knee implants fall under this category. The highest level of constraint is provided by such implants, including rotational stability. Over time, the first-generation hinged knee implants evolved into the Rotating Hinge Knee (RHK) implant. The RHK implant is suitable for grossly unstable knee joints owing to the absence of natural functioning ligaments, poor soft tissue envelope, severe bone loss, and severe varus-valgus and flexion contracture (Das et al. 2017).

Partially-constrained or semi-constrained implants partly depend on the stability provided by the remaining ligaments and joint capsule. Used more commonly in patients with severe angular deformities or for revision cases with moderate ligamentous instability, these implants can control the posterior displacement of the tibia on the femur and medial–lateral rotation of the knee. Posterior stabilised (PS) knee implants fall under this category. It sacrifices the deficient posterior cruciate ligament (PCL). Posterior stability is obtained by the engagement of a tibial polyethylene post in the intercondylar cam of the femoral component. A slot on the distal femur is created by removing bone to accommodate the cam-post mechanism. The post of the conventional PS implant is slightly rounded. Varus-valgus and rotational stability cannot be obtained with such implants. Similar to the PS implants, the constrained condylar knee (CCK) implants also rely on a post-cam mechanism. However, a larger post and a deeper femoral box with rectangular flat lateral and medial surfaces are used to reduce internal–external rotation and improve varus-valgus stability. Non-constrained implants depend on the stability provided by the ligaments of the joint. Such implants usually do not provide any constraint and are typically used for patients with minimum or no ligamentous instability. Cruciate retaining (CR) knee implants fall within this category. The natural functioning posterior cruciate ligament (PCL) is retained during the surgery. Such implants preserve nearly normal knee movement. The presence of other knee stabilising ligaments, good bone quality, and proper soft tissue balancing are required with this type of implant. With the increasing degree of constraints, the length of the femoral and tibial stem extensions is higher. This helps in diverting the load originally borne by the ligament to the stem-bone interfaces.

Knee replacements are either cemented or uncemented or hybrid fixation type. In the cemented fixation, bone cement is used to grout the implant with the bone. On the other hand, the uncemented fixation depends on the biological fixation. The implants are fitted with slight press-fit and/or the implant surface is coated with a porous coating. The Co-Cr alloy is commonly used for the femoral component. The tibial counterpart has two components. A tibial polyethylene insert is assembled with a titanium alloy tibial tray. The wear characteristics of the tibial polyethylene insert are different from those of the acetabular liner. Delamination and fatigue wear of the UHMWPE insert are observed due to sliding motion at the articulating surface (Bhattacharya et al. 2022). The wear particle size is bigger than that observed in the hip joint. Osteolysis around the knee joint is less prevalent due to the larger wear particle size.

6.4.1 *Effect of Implant Alignment*

Rotational malalignment of the knee implants has been associated with poor outcomes of knee joint replacement (Zhang et al. 2016). Malalignment of the tibial component can lead to abnormal distribution of the joint force in the medial and lateral sides of the tibia. This may lead to overloading of the implant-bone interface, promoting interface failure. Research studies have attempted to evaluate the effect of the variation of medial–lateral force distributions, due to the alignment positions on the performance of both metal backing (MB) and all-polyethylene (AP) tibial components. Misaligned MB component, particularly a posteriorly misaligned tibial plate, reduces stress-shielding in the tibia. Tibial plates inserted in a varus position can produce lower stress-shielding than a plate positioned in the neutral position (Zhang et al. 2016). However, increased strain in the medial condyle in such implant orientation is a matter of concern. In general, the influence of the malorientation of an all-polyethylene tibial plate on stress shielding is small. However, in the varus position, increased strain observed in the medial condyle owing to offloading of the lateral condyle may indicate a potential risk of failure (Perillo-Marcone et al. 2000; Perillo-Marcone and Taylor 2007; Berend et al. 2010).

6.4.2 *Effect of Implant Materials*

Implant materials and design also influence the stress shielding in bone after knee joint replacement. Research studies suggest that the effect of tibial plate material is more than that of implant design on stress shielding and subsequent bone loss in the proximal tibia and distal femur (Zhang et al. 2016; Galas et al. 2023). A significant amount of bone loss is observed in total knee replacement patients down the years after surgery, particularly for MB plate. Changing the material from Ti alloy to Co-Cr alloy further aggravates the tibia bone loss in patients with MB tibial plate (Zhang et al. 2016). Similarly, lower stress-shielding in the distal femur is observed with the titanium alloy femoral component than with the Co-Cr alloy. In other words, higher bone stresses are induced in the femur with a Ti-alloy component than the Co-Cr component, mostly in the medial compartment at higher knee flexion angles (Galas et al. 2023). Implant manufacturers and clinicians should consider the effect of implant material on the changes in bone quality after surgery for future design of implants and/or surgical decisions. As mentioned earlier, the femoral component is usually made of Co-Cr alloy, since the material has improved scratch/wear resistance properties. This has a proven track record of excellent long-term success. Titanium (Ti-6AL-4V) alloy also has low wear resistance to be used as the femoral component. Other wear-resistant coating materials, such as oxidized zirconium (OxZr), titanium nitride (TiN), titanium niobium nitride (TiNbN), and zirconium nitride (ZrN) are available for the femoral component (Skjöldebrand et al. 2022).

6.4.3 Implant Stability: Role of Geometric Features

Certain geometric features of implants have a considerable effect on the stability of the implant. Small pegs and/or stems are used in knee implants to achieve fixation and stability. A short intramedullary stem is used in the tibial components to provide stability against tilting and lift-off of the tibial tray. The stem causes distal load transfer. However, the cancellous bone in the distal tibia is weaker than the proximal tibia. This may be more prone to failure under load. Alternatively, some designs of the tibial plates come with pegs or wide delta-shaped keels (Fig. 6.19). Tibial trays with pegs may help with uniform distribution of force across the cement–bone interface, thus lowering the risk of aseptic loosening. Revision of a pegged tibial tray is easier than a stemmed tibial tray.

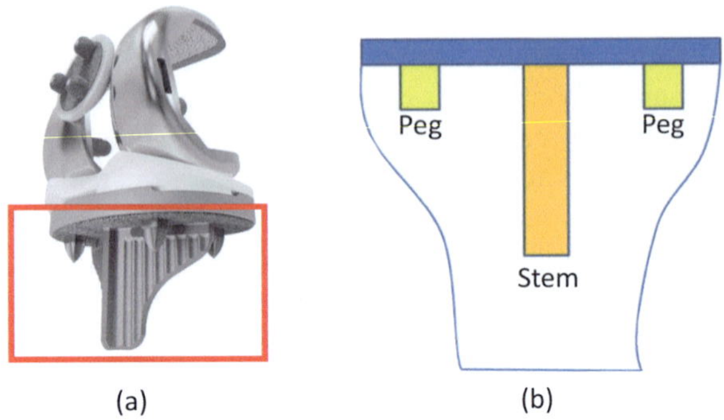

Fig. 6.19 A typical knee implant: **a** the tibial component, **b** enlarged view of the tibial component with pegs and stem

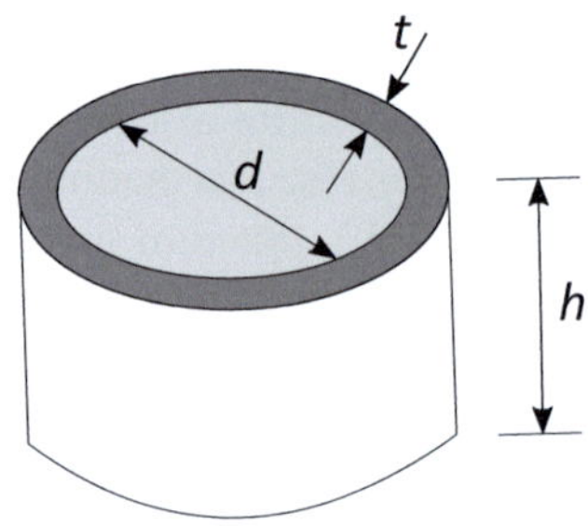

Fig. 6.20 A vertebral body, consisting of cortical and cancellous bone

Practice Problems

1. The vertebral body, consisting of cortical and cancellous bone, has a height of h, as shown in Fig. 6.20. The top of the vertebral body is loaded with a central compressive force ($\boldsymbol{P}$) via a thin rigid plane (not shown in figure). The diameter of the cancellous region is d and the thickness of the outer cortical region is t, as depicted in Fig. 6.20. The bottom of the vertebral body is assumed to be fixed. The Young's modulus and area of cross-section of cancellous region are denoted by E_{canc} and A_{canc}, respectively. The Young's modulus and area of the cross-section of the cortical region are denoted by E_{cortical} and A_{cortical}, respectively. Find the load shared by the cortical and the cancellous bone.

2. The state of stress at a specific location on the implant-bone interface is expressed as

$$\sigma = \begin{bmatrix} 45 & -15 \\ -15 & 25 \end{bmatrix} \text{MPa}$$

The implant-bone interface is inclined at 45° as shown in Fig. 6.21. The bone density of the nearest bone element is equal to 1 g cm^{-3}. Find the Hoffman failure index.

Fig. 6.21 Simplified 2-D representation of implant-bone interface

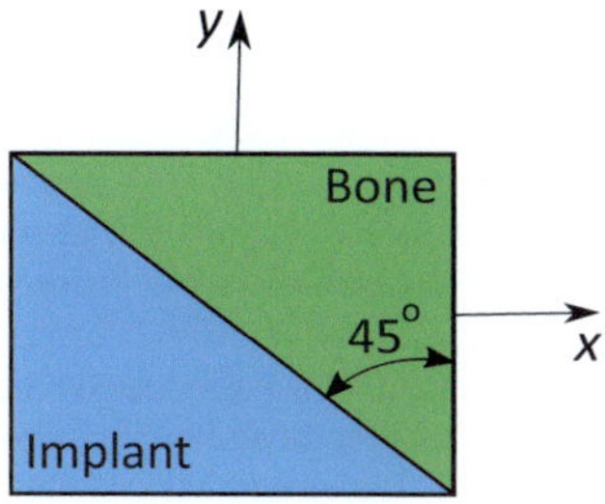

References

Berend ME, Small SR, Ritter MA, Buckley CA (2010) The effects of bone resection depth and malalignment on strain in the proximal tibia after total knee arthroplasty. J Arthroplasty 25(3):314–318

Bhattacharya R, Mukherjee K, Pal B (2022) Polyethylene in orthopedic implants: recent trends and limitations. In: Hashmi MSJ (ed) Encyclopedia of materials: plastics and polymers, vol 4, pp 777–794

Burr DB, Martin RB, Schaffler MB, Radin EL (1985) Bone remodeling in response to in vivo fatigue microdamage. J Biomech 18(3):189–200

Chaudhuri A, Mahato PK, Pal B (2024) Evaluation of the mechanical characteristics of Ti64 cubic and body-centered-cubic porous structures: a finite element study validated with physical tests. Mech Adv Mater Struct 1–14

Chowdhury S, Anand A, Singh A, Pal B (2021) Evaluation of mechanical properties of Ti-25Nb BCC porous cell structure and their association with structure porosity: a combined finite element analysis and analytical approach for orthopedic application. Proc Inst Mech Eng H 235(7):827–837

Chulsomlee K, Prukviwat S, Tuntiyatorn P, Vasaruchapong S, Kulachote N, Sirisreetreerux N, Tanphiriyakun T, Chanplakorn P, Sa-Ngasoongsong P (2023) Correlation between shape-closed femoral stem design and bone cement implantation syndrome in osteoporotic elderly femoral neck fracture undergoing cemented hip arthroplasty: a retrospective case-control study in 128 patients. Orthop Traumatol Surg Res 109(1):103450

Cinotti G, Della Rocca A, Sessa P, Ripani FR, Giannicola G (2013) Thigh pain, subsidence and survival using a short cementless femoral stem with pure metaphyseal fixation at minimum 9-year follow-up. Orthop Traumatol Surg Res 99(1):30–36

Das S, Arkesh A, Selvanayagam R (2017) Constraints in total knee arthroplasty: current status and review of literature. Int J Res Med Sci 5(2):374–378

Dattani R, Blunn G (2007) Revision of the femoral prosthesis with impaction allografting. Acta Orthop Belg 73(5):558–565

Donaldson AJ, Thomson HE, Harper NJ, Kenny NW (2009) Bone cement implantation syndrome. Br J Anaesth 102(1):12–22

Engh CA, Bobyn JD (1988) The influence of stem size and extent of porous coating on femoral bone resorption after primary cementless hip arthroplasty. Clin Orthop Relat Res 231:7–28

Galas A, Banci L, Innocenti B (2023) The effects of different femoral component materials on bone and implant response in total knee arthroplasty: a finite element analysis. Materials 16(16):5605

Glyn-Jones S, Gill HS, McLardy-Smith P, Murray DW (2004) Roentgen stereophotogrammetric analysis of the Birmingham hip resurfacing arthroplasty. J Bone Joint Surg Br 86(2):172–176

Gupta S, New AMR, Taylor M (2006) Bone remodelling inside a cemented resurfaced femoral head. Clin Biomech 21(6):594–602

Harris WH (2001) Wear and peri-prosthetic osteolysis: the problem. Clin Orthop Relat Res 393:66–70

Hoffman O (1967) The brittle strength of orthotropic material. J Compos Mater 1(2):200–206

Holding CA, Findlay DM, Stamenkov R, Neale SD, Lucas H, Dharmapatni AS, Callary SA, Shrestha KR, Atkins GJ, Howie DW, Haynes DR (2006) The correlation of RANK, RANKL, TNF-α expression with bone loss volume and polyethylene wear debris around hip implants. Biomaterials 27(30):5212–5219

Huiskes R (1993) Failed innovation in total hip replacement: diagnosis and proposals for a cure. Acta Orthop Scand 64(6):699–716

Issack PS, Lauerman MH, Helfet DL, Sculco TP, Lane JM (2009) Fat embolism and respiratory distress associated with cemented femoral arthroplasty. Am J Orthop 38(2):72–76

Kaplan SJ, Hayes WC, Stone JL, Beaupré GS (1985) Tensile strength of bovine trabecular bone. J Biomech 18(9):723–727

Min BW, Song KS, Bae KC, Cho CH, Kang CH, Kim SY (2008) The effect of stem alignment on results of total hip arthroplasty with a cementless tapered-wedge femoral component. J Arthroplasty 23(3):418–423

National Joint Registry for England and Wales (2023) 20th annual report. Hemel Hempstead, UK. ISSN 2054-183X

New Zealand Joint Registry (2020) Sixteen-year report January 1999 to December 2019. Wellington, New Zealand

Pal B, Gupta S, New AM (2009) A numerical study of failure mechanisms in the cemented resurfaced femur: effects of interface characteristics and bone remodelling. Proc Inst Mech Eng H 223(4):471–484

Pal B, Gupta S, New AMR (2010a) Influence of the change in stem length on load transfer and bone remodelling for a cemented resurfaced femur. J Biomech 43(15):2908–2914

Pal B, Gupta S, New AMR (2010b) Design considerations of ceramic resurfaced femoral head: effect of interface characteristics on failure mechanism. Comput Methods Biomech Biomed Eng 13(2):143–155

Pal B, Puthumanapully P, Amis AA (2013) Biomechanics of implant fixation. Orthop Trauma 27(2):76–84

Palazzuolo M, Bensa A, Bauer S, Blakeney WG, Filardo G, Riegger M (2023) Resurfacing hip arthroplasty is a safe and effective alternative to total hip arthroplasty in young patients: a systematic review and meta-analysis. J Clin Med 12(6):2093

Park JB, Bronzino JD (2003) Biomaterials: principles and applications. CRC Press

Perillo-Marcone A, Taylor M (2007) Effect of varus/valgus malalignment on bone strains in the proximal tibia after TKR: an explicit finite element study. J Biomech Eng 129(1):1–11

Perillo-Marcone A, Barrett DS, Taylor M (2000) The importance of tibial alignment: finite element analysis of tibial malalignment. J Arthroplasty 15(8):1020–1027

Ridzwan MIZ, Sukjamsri C, Pal B, van Arkel RJ, Bell A, Khanna M, Baskaradas A, Abel R, Boughton O, Cobb J, Hansen UN (2018) Femoral fracture type can be predicted from femoral structure: a finite element study validated by digital volume correlation experiments. J Orthop Res 36(3):993–1001

Shen G (1998) Femoral stem fixation: an engineering interpretation of the long-term outcome of Charnley and Exeter stems. J Bone Joint Surg Br 80:754–756

Shimmin A, Beaule PE, Campbell P (2008) Metal-on-metal hip resurfacing arthroplasty. J Bone Joint Surg Am 90(3):637–654

Silva MJ, Reed KL, Robertson DD, Bragdon C, Harris WH, Maloney WJ (1999) Reduced bone stress as predicted by composite beam theory correlates with cortical bone loss following cemented total hip arthroplasty. J Orthop Res 17(4):525–531

Skjöldebrand C, Tipper JL, Hatto P, Bryant M, Hall RM, Persson C (2022) Current status and future potential of wear-resistant coatings and articulating surfaces for hip and knee implants. Mater Today Bio 15:100270

Su EP, Barrack RL (2013) Cementless femoral fixation: not all stems are created equally. Bone Joint J 95-B(1):53–56

Teh HL, Abounouh M, Haibock P, Selvaratnam V, Ganapathy SS, Graichen H (2024) A comprehensive analysis on contributing factors for varus or valgus malposition of femoral stems in uncemented total hip arthroplasty via DAA. J Orthop 50:42–48

Tsujino M, Iwakiri K, Ohta Y, Minoda Y, Kobayashi A, Nakamura H (2024) Stem alignment with the tapered wedge cementless stem "Accolade II" does not affect bone mineral density over 5 years after total hip arthroplasty. Arthroplasty Today 27:101433

Zhang QH, Cossey A, Tong J (2016) Stress shielding in periprosthetic bone following a total knee replacement: effects of implant material, design, and alignment. Med Eng Phys 38(12):1481–1488

Zijlstra WP, De Hartog B, Van Steenbergen LN, Scheurs BW, Nelissen RGHH (2017) Effect of femoral head size and surgical approach on risk of revision for dislocation after total hip arthroplasty. Acta Orthop 88(4):395–401

Chapter 7
Biomechanics of Orthopaedic Implants Part II: Shoulder and Spine

In continuation to Chap. 6, this chapter discusses the biomechanics of shoulder arthroplasty and spine surgery. We will be discussing various techniques and implants used in these surgical interventions. As we know, the implant material plays a vital role in the clinical outcome of surgical interventions. An overview of the common biomaterials used for orthopaedic implants has also been presented.

7.1 Biomechanics of Shoulder Arthroplasty

Total Shoulder Arthroplasty involves surgical reconstruction of the glenohumeral (GH) joint when it is affected by osteoarthritis, rheumatoid arthritis, and traumatic injuries. The GH joint is reconstructed by inserting a humeral and a glenoid component in the humerus and the scapula, respectively. In 1893, Jules Emile Pean (1830–1898), a French surgeon, performed a shoulder replacement in a patient with arthritis of the shoulder for the first time. He used a prosthesis made of platinum and rubber, which lasted for approximately two years after implantation and was removed because of uncontrollable inflammation. Despite this complication, the short-term benefits were encouraging and the advent of shoulder arthroplasty was realised. Shoulder arthroplasty can be classified into two types: hemi-arthroplasty and Total Shoulder Arthroplasty (TSA). In the hemi-arthroplasty, a humeral component is inserted within the humerus while the scapula remains intact, causing direct metal-bone contact. In the TSA, in addition to the humeral component, a glenoid component is inserted into the glenoid part of the scapula. Hemi-arthroplasty has provided excellent clinical outcomes for patients suffering from glenohumeral arthritis along with irreparable rotator cuff tears (Wirth and Rockwood 1996). It has the advantage of avoiding a relatively complicated procedure of the insertion of glenoid prosthesis, which requires adequate technical skill of the surgeon. However, because of the imperfect fit of the humeral head in the glenoid cavity, TSA is more widely used. The

S. Gupta et al., *Biomechanics of Joints and Implants*,
https://doi.org/10.1007/978-981-96-0586-6_7

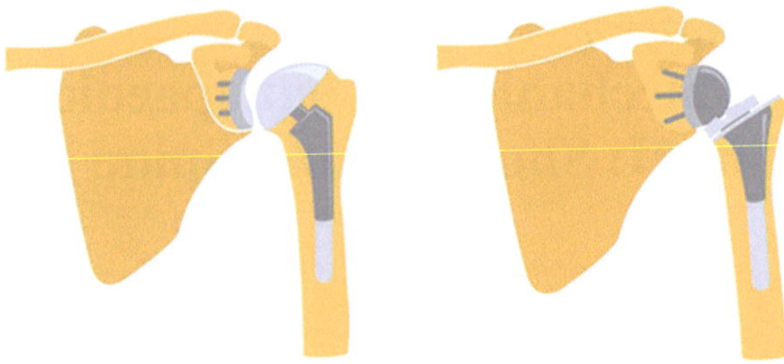

Fig. 7.1 Shoulder replacement techniques: **a** conventional TSA, **b** reverse TSA

TSA was born with the introduction of a new total polyethylene glenoid component by Neer (1974). The Neer I prosthesis has been used extensively over the decades. In 1984, C. S. Neer introduced the metal-backed polyethylene component as the Neer II prosthesis.

Shoulder joint replacement is not so common as compared to hip and knee joint replacements. The shoulder replacement techniques are classified into conventional (or anatomical) TSA and reverse TSA (Fig. 7.1). A conventional TSA consists of a humeral head articulating on a polyethylene glenoid component (Fig. 7.1a). The ball-shaped humeral head is usually made of metal or ceramic. Higher contact stresses are observed in this configuration. Moreover, there is a concern about edge loading. The point of application, as well as the magnitude and direction of the dominant GH-joint reaction force, varies with the change in the abduction angle of the arm. Berms (1993) observed that the resultant GH-joint reaction force (F) moves from inferior to superior and then to the centre of the glenoid cavity, which has been termed the "rocking horse effect" and may be responsible for the high rate of glenoid lucent lines in the TSA. In contrast, when the relative positions of the components are reversed, it is known as reverse TSA (Fig. 7.1b). This reverse configuration is more conforming and can provide more stability to the joint while limiting the range of motion.

7.1.1 Conventional Total Shoulder Arthroplasty

Conventional or anatomical TSA is performed for patients suffering from pain owing to osteoarthritis, inflammatory arthritis, post-traumatic degenerative joint disease, and osteonecrosis in the glenoid. In TSA, the diseased or damaged spherical head of the humerus and glenoid cavity are replaced with a prosthesis (Fig. 7.2). Either a pegged or keeled ultra-high-molecular-weight polyethylene cemented implant is used to replace the glenoid cavity (Fig. 7.2b, c). A metal-backed polyethylene glenoid component is sometimes used to increase the durability and fixation of the component, particularly for shoulder joints with poor bone quality. The use of metal backing, however, is still debatable, since this may adversely affect the load distribution in the

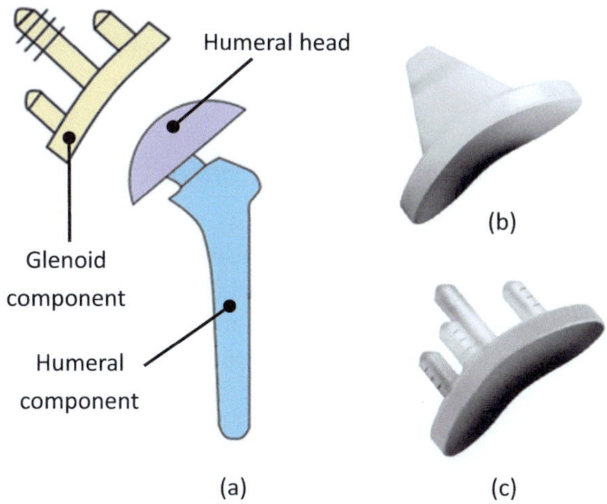

Fig. 7.2 Various components used in conventional TSA; **a** Glenoid and humeral components; **b** Keeled glenoid component, **c** Pegged glenoid component

implanted glenoid. The humeral component has a metallic ball attached to a metaphyseal stem (Fig. 7.2a). The stem can be fixed with the humerus by two means, cemented fixation (fully or proximally) and press-fit fixation. The stability of the fully cemented humeral component does not differ from the proximally cemented component. Similar to surface replacement of the hip joint, the shoulder joint may also be resurfaced. Only the damaged surface of the humeral head may be replaced by the humeral component without the stem. The outcomes of resurfaced shoulder joint and standard stemmed shoulder replacement have been reported to be comparable.

Aseptic glenoid component loosening is a frequently occurring complication threatening the TSA (Cofield 1984; Barrett et al. 1989; Berms 1993; Torchia et al. 1997; Wallace et al. 1999). Loosening is defined as the progression of radiolucent lines exceeding 1.5 mm in width or any definite change in the component position (Torchia et al. 1997; Wallace et al. 1999). It is worsened by conforming implants, incorrect implant positioning, capsular contractures, and rotator cuff tears. Loosening may occur shortly after surgery, probably due to excessive stresses in the cement or at the interfaces of cement-bone, implant-cement, and implant-bone. Medium- (3–10 years) and long-term (10 years or more) follow-up observations revealed radiographic changes that indicated loosening (Berms 1993; Torchia et al. 1997; Wallace et al. 1999). This may be due to bone remodelling, which is caused by TSA-induced changes in the stress distributions. A follow-up study using radiographic evaluation of cemented Neer glenoid components observed that the percentage of cement-bone radiolucent lines around the glenoid component, excluding the area around the keel, remained constant (84%) (Torchia et al. 1997). However, the percentage of radiolucent lines around the keel was observed to increase from 42 to 76%. A similar trend was evident in the percentage of components that changed their position from 10 to 38% (Torchia et al. 1997). The chances of initial interface failure (primary stability) are dependent on the surgical precision, the peri-prosthetic bone quality, and the cement penetration within bone for the cemented implants or the degree of osseointegration for the uncemented implants.

7.1.2 Implant Stability: Role of Geometric Features

In the case of conventional TSA, the glenoid component designs often come with pegs or keels to provide an adjunct function (Fig. 7.2b, c). Research studies indicated that pegged components provide more osseointegration, hence better fixation of the implant. In the case of total polyethylene design, the chances of cement–bone interface debonding were found to be higher near the tip of the keel as compared to the superior edge of the prosthesis (Gupta et al. 2004). Whereas for the metal-backed design, this interface near the tip of the keel was observed to be more susceptible to failure as compared to the non-backed design. Apart from these, the conformity of the components, the design and size of the keels, or the pegs also influence the implant stability. Pegs provide a greater area for implant fixation, resulting in better component stability, particularly against the shear forces. However, there seems to potential risk of strain concentration around the tips of the pegs. Similarly, the number, size, and aspect ratio of the pegs are important design parameters to be considered for achieving a stable implant fixation.

7.1.3 Reverse Total Shoulder Replacement

Reverse total shoulder arthroplasty (RTSA) is an efficient surgical procedure for the treatment of glenohumeral joint diseases in patients suffering from severe rotator cuff deficiency. The implant socket and the ball are reversed in the RTSA as compared to the conventional TSA. RTSA is a better alternative for patients with large rotator cuff tears and shoulder arthritis since the rotator cuff muscles cease to function. The RTSA depends on the deltoid muscle, rather than rotator cuff muscles, to position and actuate the arm. It includes a polished metal ball that replaces the glenoid cavity and a stemmed-humeral component containing a PE socket replacing the humeral head (Fig. 7.3). The RTSA was first introduced by Charles Neer in the year 1974 and has gained popularity over the past five decades. RTSA has the advantage of treating patients suffering not only rotator cuff deficiency but also glenohumeral arthrosis. However, major limitations of the first design of Charles Neer included glenoid component loosening and implant failure, owing to the lateralization of the centre of rotation of the shoulder joint, resulting in large forces being generated at the implant-bone glenoid surface (Berliner et al. 2015). In the year 1985, a novel "ball-and-socket" design was introduced by Paul Grammont, based on four key principles as stated below (Frank et al. 2022; Berliner et al. 2015; Kontaxis and Johnson 2009):

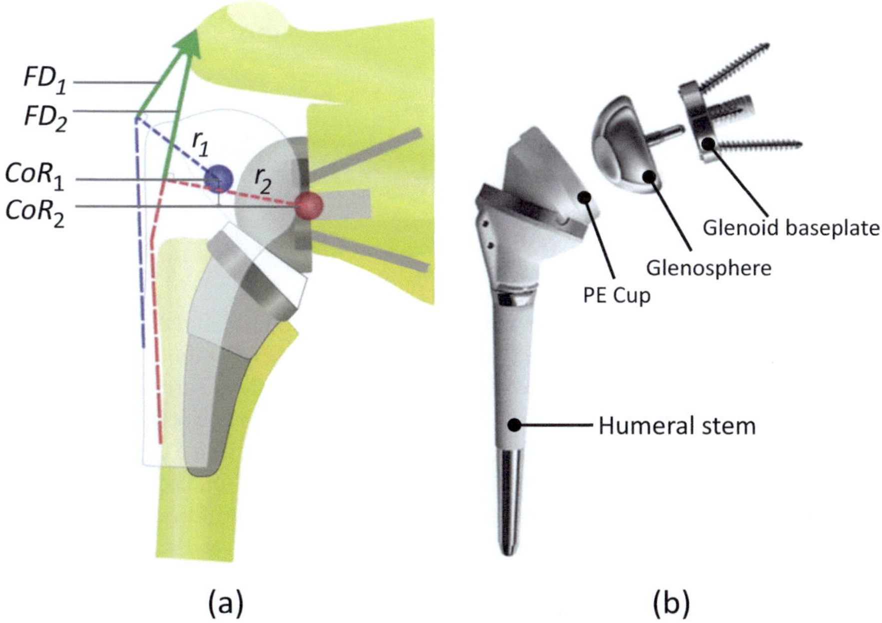

Fig. 7.3 The reverse total shoulder arthroplasty; **a** A simplified 2D representation of the biomechanics of RTSA, The centre of rotation (*CoR*) is shifted medially and inferiorly (from CoR_1 to CoR_2), thus lengthening the moment arm (r) from (r_1) to (r_2) and increasing the deltoid tensile force (*FD*) from FD_1 to FD_2, **b** The Grammont Delta III prosthesis showing various components: the glenoid baseplate with diverging polar screws and two equatorial screws, the glenosphere shaped as a hemisphere, the humeral stem or stem-less components including a modular metaphyseal implant, and the articulating polyethylene inlay. Adapted from Frank et al. (2022) and Boileau et al. (2005)

(1) the centre of rotation (CoR) should be fixed, located distally and medially with regard to the level of the glenoid surface.
(2) the prosthesis should be inherently stable. Shifting the CoR medially and distally resulted in a reduction of the moment at the glenoid component, thus reducing the potential risk of glenoid loosening and leading to a stable implant configuration.
(3) the moment arm of the deltoid should be effective from the initiation of the movement. Medialization of the CoR also facilitated a larger moment arm for the muscle deltoid, permitting the muscle to be effective enough to initiate humeral abduction. Moreover, lowering the humerus to apply tension in the muscle deltoid eventually increases the muscle fibre recruitment of the anterior and posterior deltoid to compensate for the deficiency in the rotator cuff.
(4) a large glenosphere and a small humeral cup small is necessary to create a semi-constrained articulation. The initial design of the glenosphere was 2/3rd of a sphere and the CoR was located lateral to the glenoid surface, resulting in high destabilising forces at the implant-bone surface and eventual failure (Fig. 7.4a). After the review of the first set of eight cases, the second-generation

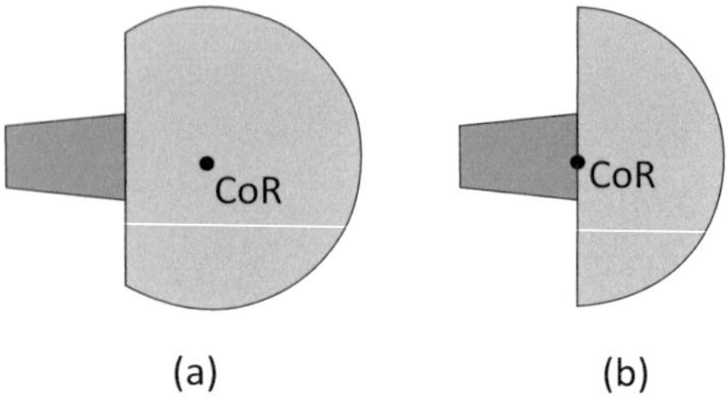

Fig. 7.4 Different glenosphere designs, **a** Two-thirds of a sphere, **b** Hemisphere

design (Delta III) of Paul Grammont's RTSA was considered a hemispherical glenosphere. In this revised design, the CoR was located at the glenoid surface (Fig. 7.4b). A medialised CoR leads to a reduction in shear forces. Moreover, a central peg and two divergent screws attached to the baseplate of the glenosphere resisted the destabilising forces at the implant-bone surface. This improved design led to the reduction of shear forces and moment on the implants, thus ensuring a strong fixation. In the initial Delta III design, a threaded screw was used to attach the glenosphere to the baseplate (Sabharwal and Bale 2021). This attachment was further improved using a Morse taper and a counter-sunk screw. The humeral component featured a small cup, covering less than half of the glenosphere, oriented almost horizontally with a non-anatomic neck-shaft angle of 155°. Thereafter over time, several changes in implant designs were undertaken to reduce complications and improve the clinical scope of RTSA.

Although RTSA has been performed in the past 30 years, complication rates ranging from 0 to 69% have been reported. The medialization of the CoR increases the potential risk of scapular notching. This primary complication is characterised by the impingement of the supero-medial humeral component against the inferior scapular neck during humeral adduction (Berliner et al. 2015; Mollon et al. 2017). Implant design and surgical techniques are known to affect the postoperative function and minimise complications, such as scapular notching, instability, and component failure (Kontaxis and Johnson 2009; Berliner et al. 2015). Several techniques have been suggested to reduce scapular notching including increasing the lateral offset of CoR, positioning the glenoid baseplate with inferior inclination, eccentric positioning of the glenoid, and decreasing the neck-shaft angle of the humeral component (Berliner et al. 2015). Hence, research and development towards improved RTSA implants continued in order to maximise functional outcomes and extend the durability of the implant. Even though variable biomechanical inferences of different implant designs with subtle modifications evolved with time, the basic design of RTSA remains unchanged (Fig. 7.3). It is important for orthopaedic surgeons to comprehend the biomechanical principles behind these design refinements to justify the implant selection and optimise the patient outcomes (Sabharwal and Bale 2021).

7.2 Biomechanics of Spinal Surgery

The occurrence of low back pain is quite common in adult humans. This is primarily attributed to degenerative disc diseases (DDD) of the spine, as discussed in Chap. 1. In the case of severe DDDs, there are two different surgical treatments available: fusion and non-fusion surgeries, as shown in Fig. 7.5. Spinal fusion surgery is used to join two or more vertebras together to prevent motion in the destabilised segments of the spine. It is considered to be the gold standard and the most effective surgical treatment for disc degeneration. Although fusion surgery is effective in relieving back pain, it affects the biomechanics of the lumbar spine. Adjacent segment degeneration (ASD) is commonly observed in patients who underwent fusion surgery. This phenomenon occurs when the segment adjacent to a fused region experiences accelerated degenerative changes over time owing to increased mechanical load. The ASD may lead to revision surgery and call for a further fusion of adjacent segments. In contrast to fusion surgery, non-fusion surgery preserves the motion of the spine at the disc level (Fig. 7.5b). Non-fusion methods have been suggested as an alternative to fusion for a variety of spinal disorders, including discogenic low-back pain. However, the existing designs of non-fusion type of implants suffer from concerns, such as subsidence, difficulty in implantation, and the requirement of revision surgeries. Over the years, researchers have investigated the biomechanical behaviour of the spine implanted with prostheses following various surgical approaches. Computational models have proven to be effective in predicting the outcomes of lumbar spine implantation.

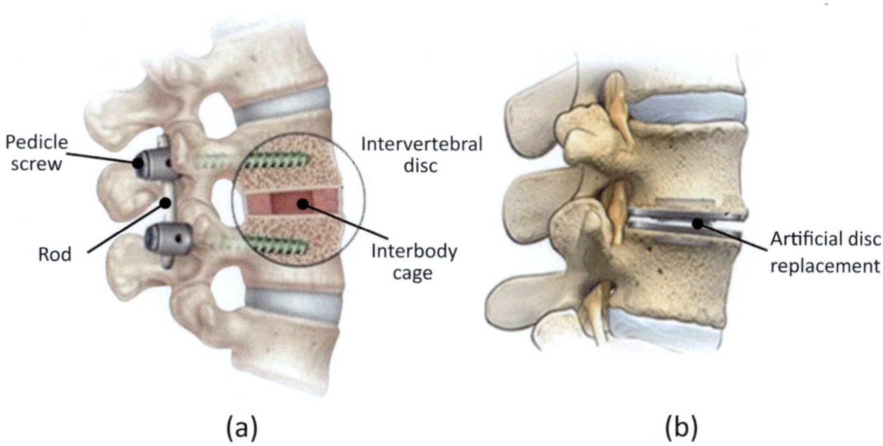

Fig. 7.5 Surgical procedure for treating disk degenerative disease in spine: **a** Fusion technique, **b** Non-fusion technique

7.2.1 Spinal Fusion

Spinal fusion was first introduced in 1911 by Albee and Hibbs, who utilised autologous bone for posterior spinal stabilisation. Since then, spinal fusion has become one of the most important and frequently employed operations of the spine. A spinal fusion removes pathology by obliterating painful movements with subsequent stabilisation of the affected spine segments. The biomechanical goals of the fusion are to restore intervertebral disc height and promote fusion by providing immediate stability (the ability to restrict the motion of a segment under physiological load). Interbody fusion cage plays an important role in the biomechanical outcome of spinal fusion by preserving the disc height as bone graft incorporation occurs (Faizan et al. 2014).

Various surgical approaches have been established for lumbar interbody fusion (LIF) and insertion of implants, namely, anterior lumbar interbody fusion (ALIF), transforaminal lumbar interbody fusion (TLIF), posterior lumbar interbody fusion (PLIF), lateral lumbar interbody fusion (LLIF) and oblique lumbar interbody fusion (OLIF), as shown in Fig. 7.6. The choice of the surgical approach depends on the location of degeneration and the patient's condition. One of the original approaches for spinal fusion surgery is the PLIF technique, where the surgical access to the intervertebral disc is gained from the posterior direction (Fig. 7.6). Another approach from the posterior side for fusion surgery is the TLIF, which involves direct unilateral access to the intervertebral foraminal space, thereby reducing the chances of damage to nerve roots and ligamentum flavum. The ALIF allows adequate access to the ventral surface of the intervertebral disc. However, this approach is limited to L4/L5 and L5/S1 due to vascular anatomy. The LLIF technique involves accessing the disc space via a lateral retroperitoneal trans psoas corridor (Fig. 7.6). This surgery is not performed on the L5/S1 level since the lateral access is obstructed by the location of the iliac crest. The OLIF approach involves minimally invasive surgery access to the disc space via a corridor between the psoas muscles and the peritoneum.

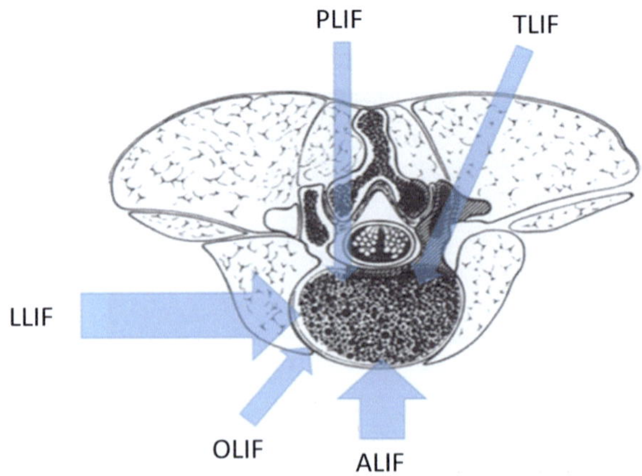

Fig. 7.6 Types of spinal fusion surgery: anterior lumbar interbody fusion (ALIF), transforaminal lumbar interbody fusion (TLIF), posterior lumbar interbody fusion (PLIF), lateral lumbar interbody fusion (LLIF) and oblique lumbar interbody fusion (OLIF). Adapted from Reid et al. (2019)

The surgical approaches have distinct challenges. The thecal sac needs to be moved sideways in PLIF to access the IVD. This procedure has a potential risk since any damage to the spinal cord will induce serious injuries. Several abdominal muscles are needed to be incised to reach the IVD in ALIF and LLIF. In comparison, the TLIF has been reported to be the preferred approach having minimal risks. The parts of facet joints are removed in TLIF, avoiding the risk of injury to the thecal sac and abdominal muscles. Each surgical approach has a unique set of advantages and limitations. The selection of surgical approach is multifactorial, which depends on clinical indication, surgeons' experience, and patient-related factors.

7.2.1.1 Interbody Cage

Fusion restores functionality by immobilising the joint, thereby preventing mechanical instability and offering pain relief. Interbody cage plays a major role in the biomechanical outcome of fusion surgery, as they restore disc height and relieve nerve root pressure. The cages are filled with bone graft material and implanted between the vertebrae. The bone graft then begins to grow into and around the cages, eventually forming a solid bond or fusion between the two vertebrae. These implants are inserted anteriorly, laterally, or posteriorly and are often supplemented with posterior instrumentations like pedicle rods and screw systems to aid the fusion process.

Bagby first introduced the interbody cage in 1988, where a basket cage made of stainless steel was used to hold the bone graft for cervical spine fusion in a horse (Chong et al. 2015). Since then, different designs and types of cages continuously evolved with different material properties. Titanium (Ti), polyetheretherketone (PEEK), and carbon-fibre-reinforced (CFR) polymers have been primarily used for the manufacturing of cage implants (Fig. 7.7). Ti-alloys have excellent corrosion-resistant properties with sufficiently high strength under physiological loading and their surfaces can be modified to promote osseointegration and cell adhesion. The clinical results of the patients were favourable with the elimination of pain and a successful bony fusion (Yamagata et al. 2012). However, due to a mismatch of elastic modulus between the bone and the implant, local bone resorption occurred adjacent to the implant leading to cage subsidence and implant migration. Another limitation of titanium is its radiopacity, which limits the post-operative monitoring of spinal fusion.

In the recent past, PEEK cages were introduced as an alternative to titanium cages with the advantages of lower elastic modulus close to cortical bone, and radiolucency that aids the clinician in monitoring fusion over time. However, the limitations of PEEK cages lie in their hydrophobic or poor osteoconductive nature, which can inhibit bone growth into the cages from vertebral endplates, eventually resulting in non-fusion. Carbon-fibre reinforced polymers cages achieved good clinical outcomes. However, the rate of subsidence cases was high and was replaced by CFR-PEEK cages owing to its superior elastic modulus (Li et al. 2015; Kurtz

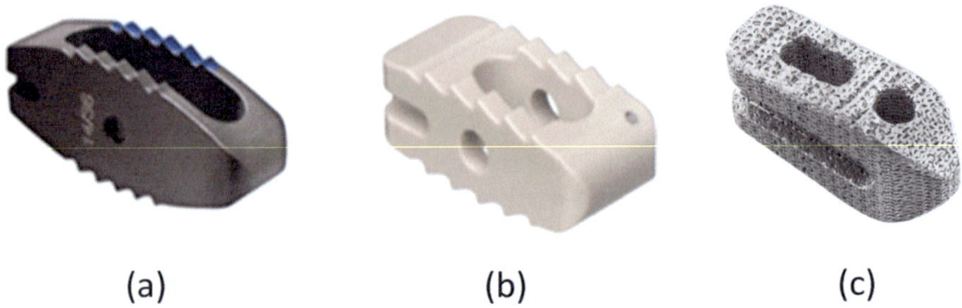

Fig. 7.7 Different types of interbody cage: **a** Titanium (Ti) alloy cage, **b** Polyether ether ketone (PEEK) cage, **c** Porous titanium alloy cage

2019). Few studies have used titanium-coated PEEK implants to increase osteogenesis and the outcome was a high rate of arthrodesis with low implant subsidence (Walsh et al. 2015; Sclafani et al. 2017). Although solid cages made of titanium and PEEK have favourable clinical outcomes and achieve spinal fusion, their stiffness is high as compared to the natural intact model, which affects the loading mechanism in a spine (Lee et al. 2016; Zhang et al. 2018).

Porous cages have been recently used to facilitate bone tissue ingrowth into the holes of the cages, thereby forming a strong fusion between the bone and the cage and reducing post-operative complications like stress shielding and cage subsidence. Porous cages made of titanium alloy or metallic foams can not only reduce the bone-implant stiffness mismatch but also can achieve favourable osteoblast adhesion and differentiation than PEEK cages (Fig. 7.7c). Moreover, porous cage acts like scaffolds and maintain space within it, so that bone cells can migrate and form new bone tissue. Trabecular metal or porous tantalum is another biomaterial that evolved as an alternative to porous titanium cages, which mimics the trabecular structure of the cancellous bone. It has a low elastic modulus close to that of cancellous bone and the biomechanical properties are sufficient to withstand physiological loading conditions. Studies reported that trabecular metal proved to be very efficient in fusion and can be used as a stand-alone cage without the aid of posterior fixation (Van de Kelft and Van Goethem 2015; Hanc et al. 2016). However, porous tantalum comes at a high manufacturing cost that eventually increases the overall operation cost. Cost reduction and customization might be the next step for the application of trabecular metal.

7.2.2 Non-fusion Surgery

Non-fusion surgery involves surgical procedures that focus on treating spinal disorders without fusing the vertebrae together. Non-fusion techniques aim to maintain or restore motion and flexibility in the spine, in contrast to traditional spinal fusion, where adjacent vertebrae are fused together to eliminate motion at a specific spinal

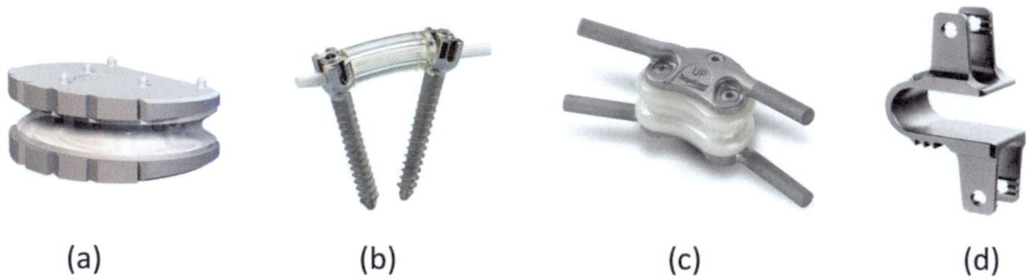

Fig. 7.8 Types of non-fusion implants: **a** Artificial intervertebral disc, **b** Dynamic stabilisation system, **c** Facet replacement system, **d** Interspinous processes spacer

segment. These procedures are designed to preserve the biomechanics of the lumbar spine, shorten the recovery period, and decrease the likelihood of adjacent segment degeneration.

Artificial disc replacement (ADR) is a common type of non-fusion surgery. In this procedure, an artificial implant is used to replace the damaged intervertebral disc (Fig. 7.8a). The implant is designed to mimic the natural movement of the spine, thus preserving motion at the treated segment. Another non-fusion approach involves the use of dynamic stabilisation systems (Fig. 7.8b). These systems provide spinal support, while enabling precise movement, using flexible rods that have the ability to bend and flex. Motion preservation devices, such as facet replacement systems and annular repair devices, are designed to restore the natural movement of the spine while treating specific spinal conditions. Facet replacement systems replace the facet joints to address degenerative conditions affecting the facet joints to preserve motion and to reduce pain while annular repair devices reinforce the annulus fibrosis to prevent disc herniation (Fig. 7.8c). Interspinous process spacers are specialised devices utilised in spinal surgery to address conditions, such as spinal stenosis, a condition characterised by the narrowing of spaces within the spine that may result in nerve compression (Fig. 7.8d). These spacers are placed between the spinous processes, which are the bony protrusions on the back of the vertebrae, in order to create more space and relieve pressure on the spinal nerves. Non-fusion spine surgery encompasses a variety of techniques and devices aimed at treating spinal conditions while preserving the natural movements and biomechanics of the spine. Each technique offers a unique set of benefits and the choice is based on the specific needs and conditions of a patient.

7.3 Biomaterials

According to Williams (2009), “a biomaterial is a substance that has been engineered to take a form which, alone or as part of a complex system, is used to direct, by control of interactions with components of living systems, the course of any therapeutic or diagnostic procedure, in human or veterinary medicine”. In simple words, a substance that has been engineered to interact with biological systems for a medical purpose is called biomaterials. These are synthetic or natural materials used to replace parts of a biological system to treat diseases or injuries. Such materials remain in contact with the living tissue when implanted inside the body. They should be physiologically accepted by the tissue and be safe, reliable, and economical. Various requirements of biomaterials are discussed in the following sections.

7.3.1 Primary Requirements

The primary requirements of biomaterials are compatibility with the biological system, suitable mechanical properties as required for a given application, and corrosion-resistant properties. Biomaterials must not cause adverse biological reactions in the host body. Such materials generally show resistance to degradation and are chemically stable and bioactive. Mechanical and physical properties should not alter post-implantation. Since orthopaedic applications generally involve weight bearing, the mechanical properties, such as elastic modulus, ductility, hardness, wear properties, and strength, must be within the acceptable limits for a particular application. The composition and microstructure of the materials should be such that they do not corrode inside the body under the influence of biological fluids. The implants are generally subjected to demanding biological and mechanical environments. Hence, the characterisation of the compatibility of the materials with the host tissues is of vital importance.

The term "biocompatibility" is often used to describe implantable biomaterials. "Biocompatibility" means that the implant material can produce favourable biological and chemical responses inside the host tissues. It is the ability of a biomaterial to perform its intended function when implanted inside or on the surface of the body, without causing any undesirable effects in the body. According to Williams (2008), biomaterials should generate the most appropriate cellular or tissue response, and optimise the clinical performance. Biocompatibility has two facets: surface compatibility and structural compatibility. Surface compatibility of a material indicates that the host tissue can tolerate the physical, biological, and chemical behaviours of the material. Whereas, by structural compatibility, it is indicated whether the mechanical properties of the host tissue can be matched. Material composition, elastic properties, and surface characteristics are important considerations for biomaterials. Specific cell types can tolerate specific materials. It may be noted that a biomaterial that demonstrates good biocompatibility for a given application may not be well-suited to another application.

In the context of cell culture in vitro, another term, "Cytocompatibility," qualitatively describes the behaviour of living cells with a synthetic material substrate. It is the ability of biomaterials that produce no adverse effect in vitro when in contact with proliferating cells. A biomaterial should be pro-healing rather than pro-inflammatory. Cytotoxicity is the assessment of cytocompatibility to device or implant-related materials and their leached-out components. An implant is supposed to be in contact with blood. Therefore, haemocompatibility is also considered important. It is defined as an in vitro assessment of the blood compatibility of any material, surface, and their leached-out components to change the clotting factors. Additionally, the osseointegration property is critical for implant stability. Biological fixation of an implant achieved by direct bone-to-implant contact is known as osseointegration, which is a time-dependent healing process. The process of bone healing, as in fractures, is primarily dependent on osteoinduction, the process by which osteogenesis is induced. Osteogenesis is the ability of a material to create bone intrinsically. Osteoconduction is the process by which bone grows onto a surface or into a porous structure. An osteoconductive surface permits bone growth on its surface or down into pores, channels, or pipes. This feature is required for cementless implants in joint replacements.

In a summary, the primary requirements of a biomaterial are the following:

- Biocompatibility (Bio-inert and non-toxic—must not react with the body)
- Adequate mechanical strength in static and fatigue loading—tension, compression, and shear
- High fracture toughness
- Sufficient stiffness and hardness
- Reasonable ductility
- Resistance to corrosion—in the presence of body fluids, singly or in combination with other materials
- Wear resistant materials with regard to articulating parts
- Long-term stability—both dimensional and chemical

7.3.2 Types of Biomaterials and Applications

Various grades of materials demonstrate biocompatibility to a great extent. Such materials can be grouped into metals, polymers, ceramics, and their composites. Common metallic alloys used as biomaterials are stainless steel (surgical grade—AISI 316L), titanium alloy (Ti–6Al–4V), cobalt–chromium–molybdenum (Co–Cr–Mo) alloy, tantalum (Ta), which are known to possess high strength and high fracture toughness, corrosion resistance, and biocompatible. However, the major concerns of these metallic biomaterials are the high elastic modulus as compared to host bone leading to stress shielding and the toxic effects caused by metal-ion release from the implants. Among the ceramics, alumina, zirconia, titania, bioactive glasses, and hydroxyapatite (HA) are commonly used biomaterials. Specific advantages of these materials are excellent wear properties, high hardness, non-corrosive over time, bio-inert, and stability. However, these materials have high elastic modulus and are brittle in nature.

Polymers have been extensively used in a wide variety of biomedical applications. Polymers are macromolecules consisting of a long chain of smaller repeating units (monomers). The specific advantages of polymers are elastic modulus almost equal to the cancellous bone, lightweight and the properties can be modified easily according to molecular weight. However, the disadvantages are poor wear properties and less favourable for osseointegration. Polymers that are often used as biomaterials include polyethylene (PE), polyurethane (PU), polyacetal (PA), polysulfone (PS), silicone rubber (SR), polyethylene terephthalate (PET), polylactic acid (PLA), poly (glycolic acid) (PGA), polytetrafluoroethylene (PTFE), polymethacrylate (PMMA), polyetheretherketone (PEEK). A polymer composite is a multi-phase material, in which reinforcing fillers are integrated with a polymer matrix, resulting in an improved mechanical property that cannot be obtained from the individual component. Polymer composite materials for biomedical applications include HA/PE, silica/SR, carbon fibre/ultra-high molecular weight polyethylene (CF/UHMWPE), carbon fiber/epoxy (CF/epoxy), and CF/PEEK (Chowdhury et al. 2022). Advantages of polymer composites include high mechanical strength, high fatigue resistance, low volumetric wear rate, and the Elastic modulus being closer to cortical bone, which eliminates the stress shielding effect. These materials can transmit X-rays, CT, or MRI scans without artefacts, hence making it easier to monitor bone growth and the healing process. However, carbon-fibre-reinforced PEEK composites have a lower surface energy owing to their relatively hydrophobic surface, which limits cell adhesion.

Biomaterials find their application in solving various clinical problems. Some of the major application areas include replacing diseased or damaged body parts, assisting in the healing of body parts, improving the function of malfunctioning organs, and correcting abnormalities (Park and Bronzino 2003). Biomaterials also help in the diagnosis and treatment of many diseases and are widely used for treating various skeletal problems. Artificial joint replacement (orthopaedic implant), used for treating damaged or diseased joints, is one of the common applications. Other than joint replacement, orthopaedic implants are also used for fracture fixation and dynamic stabilisation. These include reconstructive implants, fracture management products (bone plate, intramedullary rod, and screw), and spinal products. Some common biomaterials used in joint replacement are presented in Table 7.1. The following are some of the common application areas of biomaterials:

- Orthopaedics: for example, hip joint replacements, knee joint replacements, intervertebral discs, fracture fixation, and bone grafts
- Surgical: staples, sutures, scalpels
- Cardiovascular: heart valves, pacemakers, catheters, grafts, stents
- Soft tissue: wound healing, reconstruction and augmentation, ocular
- Dental: fillings, prosthetics, orthodontics

Table. 7.1 Selected biomaterials used in joint replacements

Biomaterials	Major properties	Applications
Pure titanium	Good osseointegration	Porous coating on the implant surface
Titanium alloy	Excellent strength, low modulus of elasticity, high corrosion resistance, low wear resistance, heat treatable	Hip stem, knee stem, porous coating, metal backing of the acetabular cup
Cobalt–chrome alloy	Good mechanical properties, hard, stiff, high wear resistance	Hip stem, knee stem, femoral implant head acetabular cup, porous coating, metal backing of the acetabular cup
UHMWPE	Low friction wear resistance, low creep resistance	Acetabular cup and tibial insert of the knee implant
Alumina	Lightweight, hard, brittle, high wear resistance	Femoral implant ball, acetabular cup
Zirconia	High toughness and high wear resistance	Femoral implant ball
PMMA	Brittle, weak in tension, instability, low fatigue strength	Bone cement fixation
Polyether Ether Ketone (PEEK)	Excellent biological, mechanical, and chemical properties, high toughness and rigidity, easy to process, and self-lubricating	Femoral implant ball and acetabular cup in hip replacement, femoral component in knee replacement, and shoulder and knee and arthroscopy, cage for spinal fusion surgery
Carbon Fibre Reinforced (CFR)-PEEK	Additional stiffness and weight-bearing capabilities over natural PEEK	Acetabular cup in hip replacement, femoral component in knee replacement
Hydroxyapatite (HA)-PEEK	Higher osseointegration efficiency	Bioactive scaffolds/implants
Tricalcium phosphate (TCP)-PEEK	Excellent mechanical properties, bio-active and biodegradability	Anchoring device for anterior cruciate ligament reconstruction

Multiple Choice Questions (MCQ) for Chaps. 6 and 7

1. The choice of implant fixation technique depends on

   (A) Patient's age
   (B) Bone quality
   (C) Surgeon's experience
   (D) All of the above

2. The primary stability of a cementless implant is obtained by a close fit and frictional contact. This statement is

(A) True
(B) False

3. The failure scenario arising from wear particle debris from the articulating surfaces is known as _______.

(A) Accumulated damage failure
(B) Particulate reaction failure
(C) Interface debonding
(D) Necrosis

4. Osteolysis may be attributed to which of the following factors

(A) Stress shielding
(B) Polyethylene wear particles generated at the articulating surface
(C) Migration
(D) None of the above

5. Which of the following is the most prevailing failure scenario of cemented hip stems?

(A) Accumulated damage
(B) Failed ingrowth
(C) Stress shielding
(D) Breakage of implant

6. In total hip arthroplasty, only the femoral side of the joint is replaced. This statement is _____.

(A) True
(B) False

7. A short hip stem can reduce the effect of stress shielding in the proximal femur compared to a conventional long stem. This statement is

(A) True
(B) False

8. Possible sources of generation of wear particle debris in a cemented hip implant are

(A) Wear of bearing surfaces
(B) Abrasion of cement particles
(C) Fretting between metal parts in modular prosthesis.
(D) All the above

9. The primary reason for stress shielding in an implanted bone is_____.

(A) Fixation technique
(B) Stiffness mismatch between implant and bone
(C) Bone quality
(D) Surgical inaccuracy

10. Which of the following is a short-term failure mechanism in surface replacement of the hip joint?

    (A) Neck fracture
    (B) Aseptic loosening
    (C) Osteolysis
    (D) None of the above

11. Which of the following is linked with cellular porous implants

    (A) Increased implant strength
    (B) Increased elastic modulus
    (C) Enhanced bone growth
    (D) None of the above

12. Stress shielding in an implanted bone increases with an increase in _____.

    (A) Implant-bone interface stress
    (B) Stiffness of the implant
    (C) Bone density
    (D) None of these

13. Squeaking is associated with which of the following bearing surfaces?

    (A) Metal-on-metal
    (B) Metal-on-polyethylene
    (C) Ceramic-on-ceramic
    (D) None of the above

14. A cam post-mechanism is used in posterior stabilised knee implants. Which of the following is the primary function of this mechanism?

    (A) To replace the function of the posterior cruciate ligament
    (B) To replace the function of the medial collateral ligament
    (C) To replace the function of the lateral collateral ligament
    (D) None of the above

15. The functioning of the reverse total shoulder replacement is dependent on

    (A) Rotator cuff muscle
    (B) Deltoid muscle
    (C) Both the rotator cuff and deltoid muscles
    (D) Neither the rotator cuff nor the deltoid muscle

16. Which surgery joins two or more vertebrae to prevent motion in the affected spine segments?

    (A) Fusion surgery
    (B) Non-fusion surgery
    (C) Laminectomy
    (D) Discectomy

17. Which of the following types of artificial disc replacements are associated with excessive load on the facet joints?

    (A) Constrained
    (B) Semi-constrained
    (C) Unconstrained
    (D) None of the above

18. Commonly used biomaterial for hip stems is

    (A) Ti alloy
    (B) Co–Cr alloy
    (C) UHMWPWE
    (D) Ceramic

19. In a metal-backed acetabular cup, the acetabular liner is usually made of

    (A) Ti alloy
    (B) Co–Cr alloy
    (C) UHMWPWE
    (D) Ceramic

20. Which of the following properties is critical to consider for a biocompatible material?

    (A) Physical
    (B) Mechanical
    (C) Chemical
    (D) All of the above

21. What is the main consequence of 'stress-shielding' in bone surrounding a prosthetic implant?

    (A) Increased bone mineral content leading to stronger bone
    (B) Enhanced joint stability and reduced risk of dislocation
    (C) Reduced bone density leading to aseptic loosening
    (D) Decreased migration of the implant due to improved fixation

22. What is the primary function of the porous lattice structure in hip implants?

    (A) To increase the stiffness of the implant for better load transfer
    (B) To reduce the elastic modulus and improve osseointegration
    (C) To prevent the migration of the implant within the bone
    (D) To enhance the load transfer to the distal femur and reduce proximal bone stress

23. Which stem design category relies on a polished surface and multiple tapers to achieve load transfer and stability?

    (A) Force-closed design
    (B) Shape-closed design

(C) Composite beam design
(D) Taper-slip design

24. What is one major advantage of Hip Resurfacing Arthroplasty (HRA) over traditional Total Hip Arthroplasty (THA)?

    (A) Increased risk of femoral neck fractures
    (B) Better preservation of bone stock for future revision surgeries
    (C) Higher likelihood of stress shielding in the proximal femur
    (D) Greater susceptibility to osteonecrosis due to reduced vascularisation

25. What is the primary advantage of using a rotating hinge knee (RHK) implant in total knee replacement?

    (A) It provides minimal rotational stability and relies on natural ligaments
    (B) It improves varus-valgus stability with a cam-post mechanism
    (C) It offers partial constraint and relies on residual ligament support
    (D) It is suitable for patients with grossly unstable knee joints due to severe ligament loss

Answer Key to MCQ

1. D, 2. A, 3. B, 4. B, 5. A, 6. B, 7. A, 8. D, 9. B, 10. A, 11. C, 12. B, 13. C, 14. A, 15. B, 16. A, 17. C, 18. A, 19. C, 20. D; 21 C; 22 B; 23 A; 24. B; 25 D.

References

Barrett WP, Thornhill TS, Thomas WH, Gebhart EM, Sledge CB (1989) Non-constrained total shoulder arthroplasty for patients with polyarticular rheumatoid arthritis. J Arthroplasty 4(1):91–96

Berliner JL, Regalado-Magdos A, Ma CB, Feeley BT (2015) Biomechanics of reverse total shoulder arthroplasty. J Shoulder Elbow Surg 24(1):150–160

Berms J (1993) The glenoid component in total shoulder arthroplasty. J Shoulder Elbow Surg 2:47–54

Boileau P, Watkinson DJ, Hatzidakis AM, Balg F (2005) Grammont reverse prosthesis: design, rationale, and biomechanics. J Shoulder Elbow Surg 14(Suppl S):147S-161S

Chong E, Pelletier MH, Mobbs RJ, Walsh WR (2015) The design evolution of interbody cages in anterior cervical discectomy and fusion: a systematic review. BMC Musculoskelet Disord 16:99

Chowdhury S, Datta P, Pal B (2022) Composite biomaterials in bone grafting and other biomedical applications. In: Wahab MA (ed) Encyclopedia of materials: plastics and polymers. Elsevier, pp 1–15

Cofield RH (1984) Total shoulder arthroplasty with Neer prosthesis. J Bone Joint Surg 66(6):899–906

Faizan A, Kiapour A, Kiapour AM, Goel VK (2014) Biomechanical analysis of various footprints of transforaminal lumbar interbody fusion devices. J Spinal Disord Tech 27(4):E118–E127

Frank JK, Siegert P, Plachel F, Heuberer PR, Huber S, Schanda JE (2022) The evolution of reverse total shoulder arthroplasty: from the first steps to novel implant designs and surgical techniques. J Clin Med 11:1512

Gupta S, van der Helm FCT, van Keulen F (2004) Stress analysis of cemented glenoid prostheses in total shoulder arthroplasty. J Biomech 37(11):1777–1786
Hanc M, Fokter SK, Vogrin M, Molicnik A, Recnik G (2016) Porous tantalum in spinal surgery: an overview. Eur Spine J 26(1):1–7
Kontaxis A, Johnson GR (2009) The biomechanics of reverse anatomy shoulder replacement: a modeling study. Clin Biomech 24:254–260
Kurtz SM (2019) Development and clinical performance of PEEK intervertebral cages. In: PEEK biomaterials handbook, 2nd edn. Elsevier, pp 263–280
Lee YH, Chung CJ, Wang CW, Peng YT, Chang CH, Chen CH, Chen YN, Li CT (2016) Computational comparison of three posterior lumbar interbody fusion techniques using porous titanium interbody cages with 50% porosity. Comput Biol Med 71:35–45
Li CS, Vannabouathong C, Sprague S, Bhandari M (2015) The use of carbon fiber-reinforced (CFR) PEEK material in orthopedic implants: a systematic review. Clin Med Insights Arthritis Musculoskelet Disord 8:33–45
Mollon B, Mahure SA, Roche CP, Zuckerman JD (2017) Impact of scapular notching on clinical outcomes after reverse total shoulder arthroplasty: an analysis of 476 shoulders. J Shoulder Elbow Surg 26(7):1253–1261
Neer CS II (1974) Replacement arthroplasty for glenohumeral osteoarthritis. J Bone Joint Surg 56-A:1–13
Park JB, Bronzino JD (2003) Biomaterials: principles and applications. CRC Press
Reid PC, Morr S, Kaiser MG (2019) State of the union: a review of lumbar fusion indications and techniques for degenerative spine disease. J Neurosurg Spine 31(1):1–14
Sabharwal S, Bale S (2021) The biomechanics of reverse shoulder arthroplasty. J Arthrosc Joint Surg 8:7–12
Sclafani JA, Bergen SR, Staples M, Liang K, Raiszadeh R (2017) Arthrodesis rate and patient-reported outcomes after anterior lumbar interbody fusion utilizing a plasma-sprayed titanium-coated PEEK interbody implant: a retrospective, observational analysis. Int J Spine Surg 11(1):4
Torchia ME, Cofield RH, Settergren CR (1997) Total shoulder arthroplasty with Neer prosthesis: long-term results. J Shoulder Elbow Surg 6(6):495–505
Van de Kelft E, Van Goethem J (2015) Trabecular metal spacers as standalone or with pedicle screw augmentation in posterior lumbar interbody fusion: a prospective, randomized controlled trial. Eur Spine J 24(11):2597–2606
Wallace AL, Phillips RL, MacDougal GA, Walsh WR, Sonnabend DH (1999) Resurfacing of the glenoid in total shoulder arthroplasty: a comparison, at a mean of five years, of prostheses inserted with and without cement. J Bone Joint Surg 81-A(4):510–518
Walsh WR, Bertollo N, Christou C, Schaffner D, Mobbs RJ (2015) Plasma sprayed titanium coating to polyetheretherketone improves the bone-implant interface. J Spine 15(5):1041–1049
Williams DF (2008) On the mechanisms of biocompatibility. Biomaterials 29(20):2941–2953
Williams DF (2009) On the nature of biomaterials. Biomaterials 30(30):5897–5909
Wirth MA, Rockwood CA (1996) Current concepts review: complications of total shoulder-replacement arthroplasty. J Bone Joint Surg 78-A:603–616
Yamagata T, Takami T, Uda T, Ikeda H, Nagata T, Sakamoto S, Tsuyuguchi N, Ohata K (2012) Outcomes of contemporary use of rectangular titanium stand-alone cages in anterior cervical discectomy and fusion: cage subsidence and cervical alignment. J Clin Neurosci 19(12):1673–1678
Zhang Z, Li H, Fogel GR, Xiang D, Liao Z, Liu W (2018) Finite element model predicts the biomechanical performance of transforaminal lumbar interbody fusion with various porous additive manufactured cages. Comput Biol Med 95:167–174

Chapter 8
Biomechanical Modelling, Analysis and Design of Orthopaedic Implants

Biomechanical modelling and analysis, using the finite element (FE) method, has become an essential tool in orthopaedic research, particularly for pre-clinical evaluations of implant designs. Hence, the development of realistic FE models of intact and implanted bone is of vital importance. In this chapter, we shall learn about the major steps of FE analysis of implant-bone structures. The accuracy of the numerical (FE) predictions, however, depends on the quality of the FE model. Validation of the FE model is therefore necessary before it can be further used for clinically important research. In this chapter, methods of experimental validation have also been discussed. An overview of the steps followed for the design and development of orthopaedic implants has also been presented.

8.1 The Need for Finite Element Modelling for Biomechanical Analysis

Stress analysis is required for evaluations of the mechanical behaviours of implant-bone structures. It helps to analyse implant design variables, predict the risk of failure, and suggest measures for improvements of the design. The FE method has emerged as an important tool in orthopaedic biomechanics to determine deformations, stresses, and strains in bone and implants and to study the effect of implant-induced stress/strain shielding. It is a fast and effective pre-clinical testing method that is used to test and validate clinical hypotheses, as compared to the clinical follow-up procedure. This numerical approach provides full-field data. Using this method, the influence of several parameters can be evaluated in a short time with minimum expenses, compared to clinical investigations and mechanical testing. Evaluation of bone remodelling around implants has been studied extensively, using FE analysis of implanted bone structures and simulation of the implant-induced bone remodelling process. The degree of stress shielding and bone remodelling depends on

S. Gupta et al., *Biomechanics of Joints and Implants*,
https://doi.org/10.1007/978-981-96-0586-6_8

several factors, such as implant design variables (size, geometry, material), and bone variables (bone density and geometry). The FE models of implant-bone structures provide the opportunity to investigate the effects of such variables on its biomechanical behaviour, both in the short-term and long-term. Detailed 3-D FE models of implanted and intact (natural) bone are, therefore, required to understand the changes in load transfer in the implanted bone and to investigate stress-strain related failure mechanisms. The major steps of FE modelling and analysis of implant-bone structures have been described in the following sections.

8.2 Development of FE Model

Finite element analysis of bones and implant-bone structures has effectively been used as a pre-clinical technique to investigate the mechanical consequences of implantation. FE models based on patient-specific bone geometry and material properties have commonly been used. The geometry and material properties of such FE models are based on medical image data, such as computed tomography (CT-) and magnetic resonance imaging (MRI-) scan data. A FE modeller carefully decides the extent to which the details of the structure need to be included in the model. The abilities of the modelling technique and the sensitivity of the model to the input parameters are to be examined to address any clinically relevant questions (Taylor and Prendergast 2015). The major steps for the development of a subject-specific FE model of intact and implanted bones are shown in Fig. 8.1. In the following subsections, let us learn the major steps of the development of an FE model of an implant-bone structure.

8.2.1 *Image Processing and Geometrical Reconstruction of Bone*

Accurate identification of the bone geometry from the acquired image data set is essential to create a subject-specific FE model of bone, precisely. The bone geometry should be separated from the adjacent soft-tissue structures. This is performed using a process called segmentation. Manual segmentation is generally laborious, time-consuming, and dependent on the skill of the operators. This calls for some form of automation of the process. A commonly used segmentation method is a region-based algorithm. In this method, segmentation is initiated in a small region and propagated to the neighbouring pixels with similar features. It is possible that this segmentation can extend to two or more neighbouring bones. Several commercial software and freeware are available nowadays for easier segmentation and further processing of medical images. Figure 8.2 presents a CT-scan slice of a human hip joint in the transverse plane.

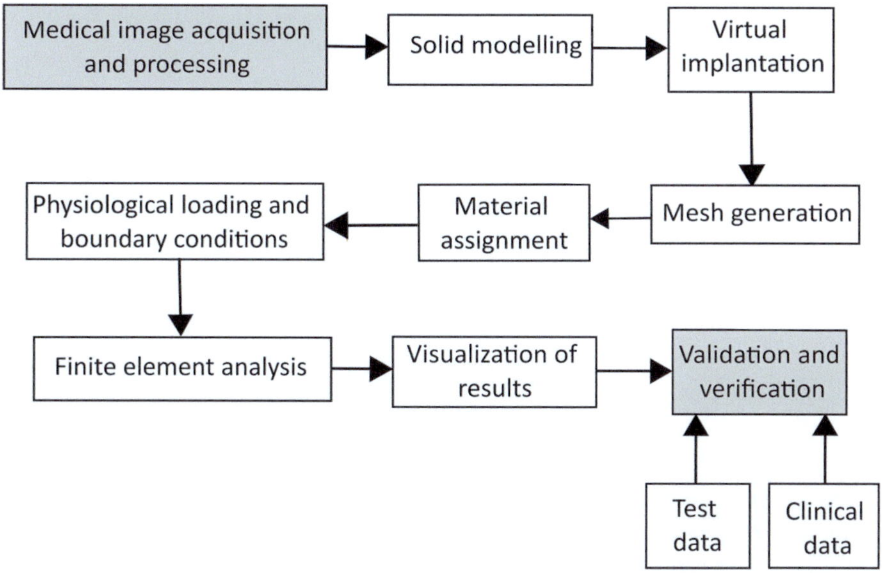

Fig. 8.1 A representative flowchart showing the major steps in FE analysis of intact and implanted bone

Fig. 8.2 A representative CT-scan slice of a hip joint in the transverse plane

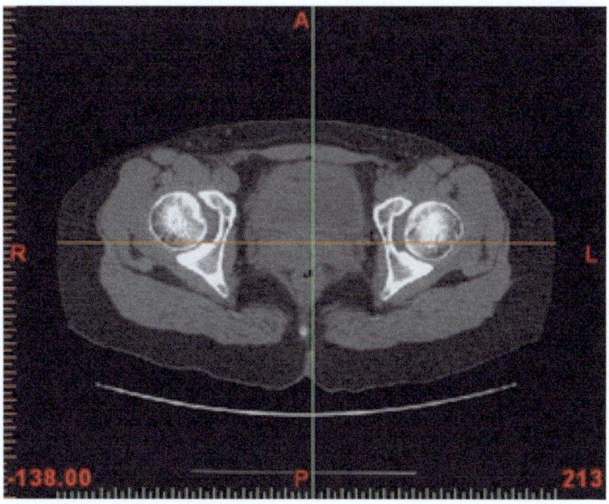

Within a 2D CT-scan slice, image segmentation is usually performed to identify the bone from surrounding soft tissue, employing a suitable range of thresholds of the image's gray-scale values (Hounsfield Units). Semi-automatic and manual segmentation methods, or a combination of both are commonly used (Byrne et al. 2016). The segmented hard tissue from the adjacent soft tissue is shown in Fig. 8.3.

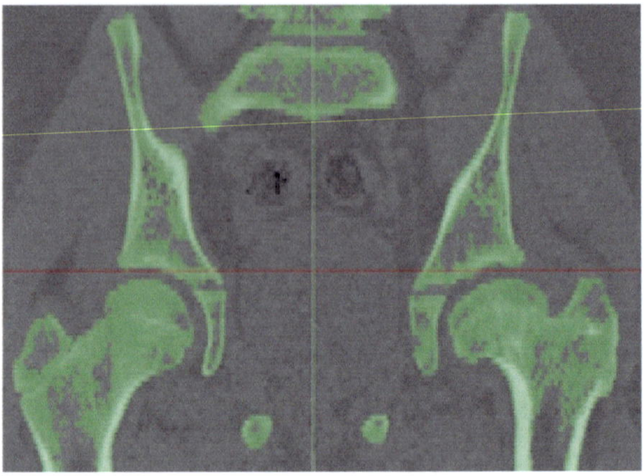

Fig. 8.3 Segmentation of the hard tissue from the adjacent soft tissues around the hip joint

The next step is region growing. In this algorithm, the initial selection of a small region gets propagated if the adjoining pixels have similar characteristics (Poelert et al. 2012). This semiautomatic method is capable of splitting the segmented images into separate objects, which is valuable for separating bone and soft-tissue structures. The selection of the most effective algorithms for a particular study needs careful combinations of components. A pelvic bone model separated from the surrounding bones using region-growing operation is presented in Fig. 8.4.

Further image processing steps, such as smoothing, cavity fill, and filters for removing islands and noise, are helpful for achieving better quality segmentation. A surface mesh model of the segmented geometry can be obtained after thresholding and region-growing operations. This mesh may need subsequent cleaning, optimisation, and correction for any remaining defects. The cleaned and corrected surface mesh can then be converted to a volumetric FE mesh. A 3D model of a pelvic bone reconstructed from a CT-scan dataset is shown in Fig. 8.5.

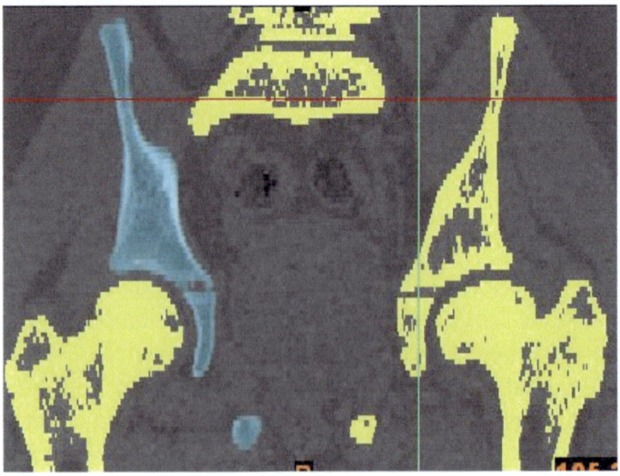

Fig. 8.4 Region growing operation to separate the right pelvic bone from the adjacent bones

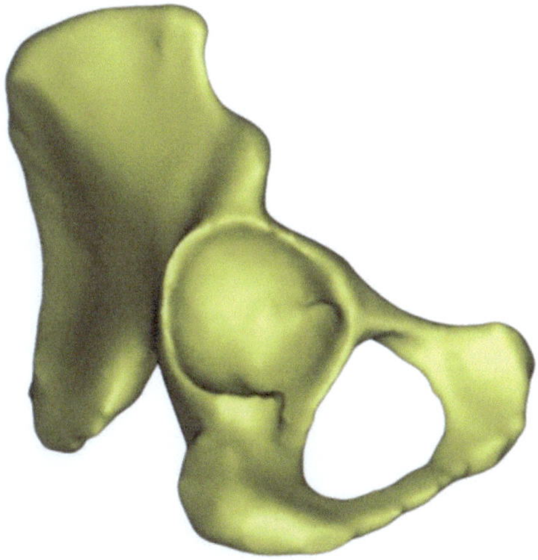

Fig. 8.5 A 3D solid model of a pelvic bone developed from CT-scan dataset

The quality of the segmented parts may depend on pixel resolution, slice thickness, offset between slices, and other imaging parameters.

8.2.2 FE Modelling of Implanted Bone

Once the segmentation of bone images is complete, the bone geometry can be developed. The geometry of the implant can be modelled using computer-aided design (CAD) solid modelling tools. Both the bone and the implant models can then be imported into a computer-aided engineering (CAE) platform, where the implant can be placed within the bone in a desired position and orientation, as recommended by the surgical guidelines. Virtual implantation of the bone model can then be performed using Boolean subtraction and other operations. A virtually operated implanted pelvic bone model is presented in Fig. 8.6. This model can be used for generating volumetric FE mesh.

Usually, voxel-based (automatic) or geometry-based (semi-automatic) techniques are used for mesh generation. The voxel-based meshing technique (Keyak et al. 1990) was often employed since it is a rapid, geometry-independent, and fully automated process. In the voxel-based meshing technique, elements occupy a predefined number of voxels from the CT scan (Poelert et al. 2012), and the element faces are kept parallel to the CT-scanner coordinate axes (Lengsfeld et al. 1998). The mesh is obtained by positioning the nodes in a cubic-lattice form and converting a voxel or a set of neighbouring voxels into a brick element (Fig. 8.7a). No surface or solid bodies are required in this method. However, the unsmoothed geometry of the model might cause numerical issues with convergence and accuracy (Lengsfeld et al. 1998;

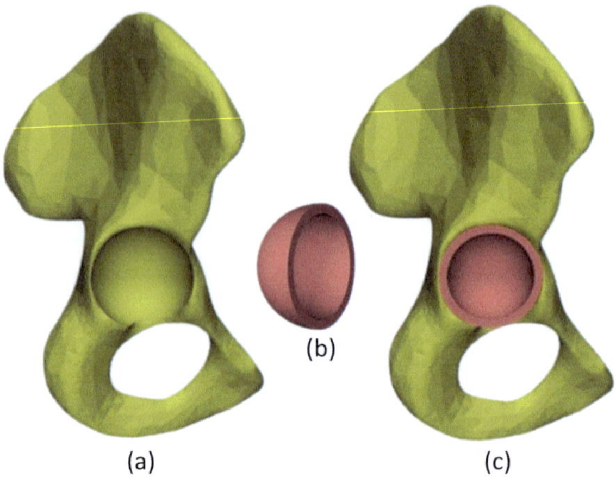

Fig. 8.6 Virtual implantation of acetabular component in the pelvic bone; **a** reamed acetabulum, **b** hemispherical acetabular component, **c** implanted acetabulum

Viceconti et al. 1998). In contrast, some topological information about the geometry is required to generate an appropriate mesh in the semi-automatic process. This meshing technique, however, is more common and advanced (Poelert et al. 2012; Lengsfeld et al. 1998), owing to the ability to generate finer mesh in the desired areas and element faces parallel to bone surfaces. In the FE modelling of implant-bone structures, geometry-based unstructured solid mesh (linear or quadratic tetrahedral elements) is often employed due to their convenience of mesh adaptivity in complicated domains. Voxel-based and geometry-based FE meshes of a femur bone are shown below (Fig. 8.7). Elements with abnormal aspect ratios and distorted and excessive curvature should preferably be avoided in the mesh generation procedure.

The element sizes should be wisely selected to capture the geometry adequately and to assign bone material properties in the FE models precisely. A small uncertainty in the segmentation may predict unrealistic results, losing its clinical relevance.

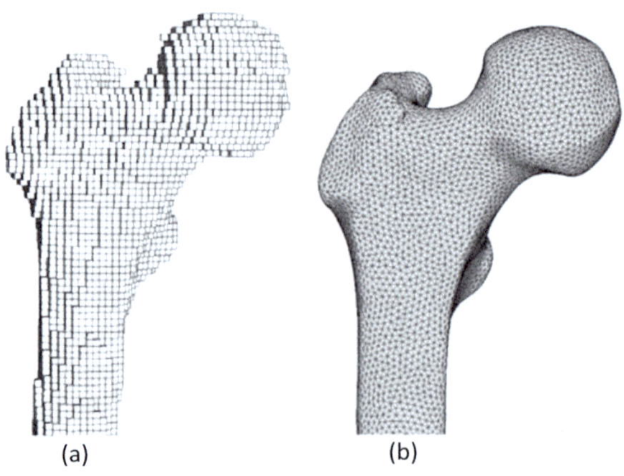

Fig. 8.7 FE meshes of a proximal femur; **a** voxel-based, **b** geometry-based (adapted from Sas et al. 2020)

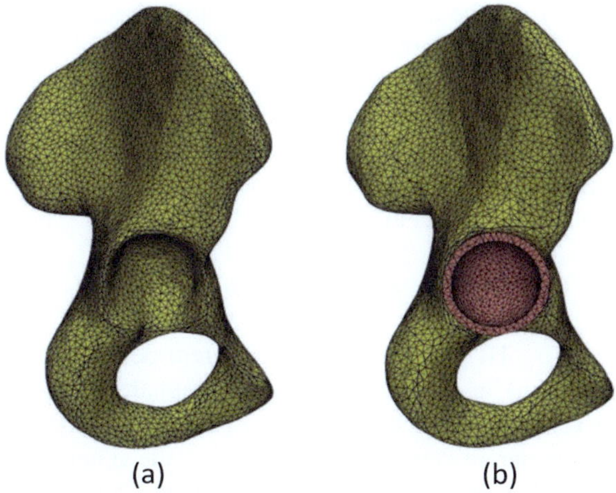

Fig. 8.8 Finite element models of **a** an intact and **b** an implanted pelvic bone

Therefore, the segmentation and subsequent mesh generation process should be automatic, accurate, robust, and general. The FE models of an intact and implanted pelvic bone are shown in Fig. 8.8.

Although the segmentation and subsequent development of FE models for long bones are comparatively accurate, scapula and pelvic bone, owing to their intricate geometries and excessively curved boundaries, need more attention during segmentation and meshing. It may be noted that the FE-predicted strain depends on the assigned cortical bone thickness. A CT-based FE model may actually overestimate or underestimate the cortical bone thickness, leading to errors in FE predictions (Anderson et al. 2005).

8.2.3 Assignment of Bone Material Property

The state-of-the-art technique for the assignment of heterogenous cancellous bone material properties to the subject-specific FE model is based on the pixel gray value of the CT-scan dataset. In this method, an element-specific (heterogeneous) distribution of bone material properties is obtained, assuming linear elastic and isotropic properties of bone (Taddei et al. 2004; Zannoni et al. 1999). CT numbers in the Hounsfield Unit (HU) are used to calculate the apparent density (ρ in g cm^{-3}) for individual bone elements (Taddei et al. 2004). The CT numbers are linearly related to the apparent bone density over a wide range (Hvid et al. 1989), as presented below in Eq. (8.1).

$$\rho_n = c + \times \text{HU}_n \tag{8.1}$$

where ρ_n is the uniform apparent density assigned to the nth element of the model, HU_n is the uniform (averaged) CT number of the nth element, and c and d are the calibration coefficients. These coefficients can be calculated from two points specified by the bone apparent density and CT numbers, (ρ_1, HU_1) and (ρ_2, HU_2), measured at two locations of the CT-scan dataset (no bone condition and outer cortical bone). The ρ_n of an element (n) is determined by interpolating between the said two points (Zannoni et al. 1999), using Eq. (8.2) as stated below.

$$\rho_n = \rho_1 + [(\rho_2 - \rho_1)/(HU_2 - HU_1)] \times [HU_n - HU_1] \tag{8.2}$$

Here (ρ_1, HU_1) represents the apparent density and CT number of water (no bone condition), and (ρ_2, HU_2) is the bone apparent density in the outer cortical bone and the CT number for cortical bone. The Young's modulus (E in MPa) is related to the bone apparent density (ρ in g cm^{-3}) by the general power law empirical relationship (Eq. 8.3) as stated below (as discussed in Chap. 5):

$$E = C\rho^D \tag{8.3}$$

The values of constant C vary from 3000 to 30,000, whereas values for constant, D, range from 1.14 to 3.2, depending on the anatomic site (Morgan et al. 2003). Such power law relationships were used in subject-specific bone FE models, with a strong agreement between the FE-predicted results and experimentally measured data (Schileo et al. 2007). CT-based heterogeneous material property assignment can be performed using the custom written code, freeware BONEMAT, and commercial medical image processing software, e.g. MIMICS. The BONEMAT freeware connects the FE mesh with the CT-scan dataset by mapping a bone element of the FE model into the CT-scan dataset of bone (Taddei et al. 2004; Zannoni et al. 1999). In this approach, each element is assigned a single value of density, thus a single value of elastic modulus. Variation of material properties within a bone element is not accounted for. The pixel CT-gray values, contained within an element, are averaged and eventually assigned to the element. The BONEMAT allows the calibration of theoretical or empirical relationships, which is useful to account for subject-specific variability of the heterogeneous bone material properties in FE models based on CT-scan datasets. The heterogeneous bone density distribution based on CT-gray value in a natural femur bone is shown in Fig. 8.9.

Several power-law relationships between the elastic modulus and apparent density, substantially different from each other, have been reported. The relationships are site-specific, which means a single relationship for a particular anatomic site (say femur) may not be suited for another anatomic site (say pelvic bone). The relationships between CT number (HU) and apparent density and between the apparent density and elastic modulus are independent and thus may act as a potential source of error (Poelert et al. 2012). Appropriate calibration of the relationship between the CT number (HU) and the apparent bone density is important in assigning material properties of bone. Yosibash et al. (2007) investigated the sensitivity of the FE

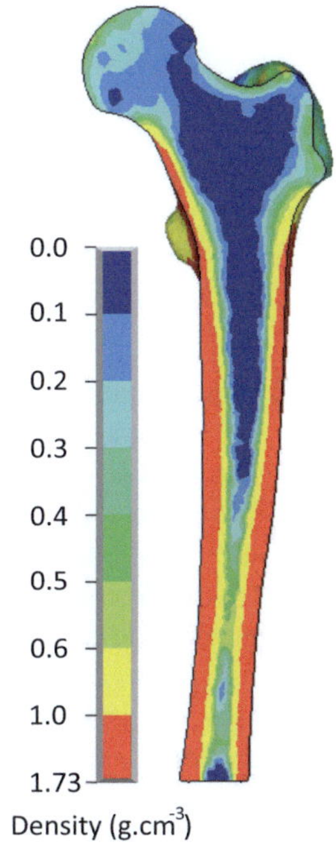

Fig. 8.9 Heterogeneous bone density (g cm^{-3}) distribution in a coronal plane within a femur

model to such relationships by comparing five different density–elasticity relationships. The overall accuracy of the FE predictions was remarkably influenced by the density-elasticity relationship, emphasising the need for due attention to the relationships used. Such relationships may vary with the subjects as well. Laz et al. (2007) concluded that stresses and fracture risk of the femur varied significantly with the stochastic variation of such material relations.

A few words must be discussed about the errors and approximations involved in the development of FE models of intact and implanted bone structures using CT-scan data. While locating the position of a bone element in the CT image during the material property assignment, some elements or a portion of an element may actually be located outside the bone contour of the CT image. The pixel gray values of such elements are generally approximated to the nearest element located within the contour. Moreover, a finite element may be mapped (overlap) on pixels belonging to cancellous and cortical bone. Taking an average of the voxel values on both sides of the boundary should ideally be avoided. Otherwise, this can over- or underestimate the material properties of the element. This is known as the 'Partial Volume Effect', defined as the loss of apparent property in a region due to the limited resolution of the imaging system. A suggested solution is image segmentation at the boundary of cortical and cancellous bone, using a single value thresholding. In this regard, Ghosh et al. (2015) used a thresholding method based on a density value at the endocortical boundary to segment the cortical bone and cancellous bone boundaries in a pelvic bone. In the case of modelling soft and hard tissues, surface geometry and the calculations of materials properties (density and elastic modulus) may also be affected by the partial volume effect, owing to the inclusion of fat/body fluid/soft tissue as well as bone in a single voxel.

8.2.4 Musculoskeletal Loading and Boundary Conditions for FE Models

Physiological loading and boundary conditions are essential to the FE analysis of intact and implant-bone structures, as they play a major role in the load transfer within a reconstructed joint. Boundary conditions prescribe the restrictions in degrees of freedom of selected nodes, mimicking the physiological constraints. The components and magnitudes of musculoskeletal loading on the joint and/or bone need to be applied at the desired locations (area). Non-invasive measurements of muscle forces and joint reaction forces are challenging. Researchers have measured joint reaction forces for various daily activities, using instrumented implants and telemetry (Heller et al. 2001; Bergmann et al. 2001). However, the measurement of muscle forces is a very complex and difficult task, which might not always be possible due to technical and ethical limitations. Instead, muscle forces are often predicted using musculoskeletal models (Mathai and Gupta 2019). Such models are simulated for movements corresponding to specific activities. Recorded video data of such movements are required for this purpose. The corresponding muscle forces that produce such movements can be predicted using an inverse dynamic approach, as discussed earlier in Chap. 4. However, such a solution may not be unique. Static optimisation of the solution can aid in such a situation.

Heller et al. (2001) and Bergmann et al. (2001) presented extensive data on joint loading and muscle loading during normal walking and stair climbing for a number of patients. Bergmann and his research group have reported an extensive amount of work on the telemetric measurement of the joint loads for various human joints (https://orthoload.com/). The joint load and muscle forces are usually distributed uniformly on selected nodes over a specified area of the FE model. For example, Pal et al. (2009) distributed the hip joint reaction force uniformly for the intact femur model over a roughly circular area of diameter 30 mm. A comparable area of the hip contact force was observed in an in vitro investigation of contact stress patterns in an intact human hip (Brown and Shaw 1983). The locations of the muscle attachment in bone are usually available from relevant literature. The distribution of the muscle forces is uniform over a patched area, centred on the attachment point of the respective muscles in the bone. For example, Duda et al. (1996) reported the estimated muscle attachment site in the femur for a number of subjects. Although there is an inter-individual variation of the musculoskeletal system, data on the area of muscle insertions are usually taken from the literature. Muscle attachment sites for the pelvic bone are shown in Fig. 8.10. Applications of hip contact force, muscle

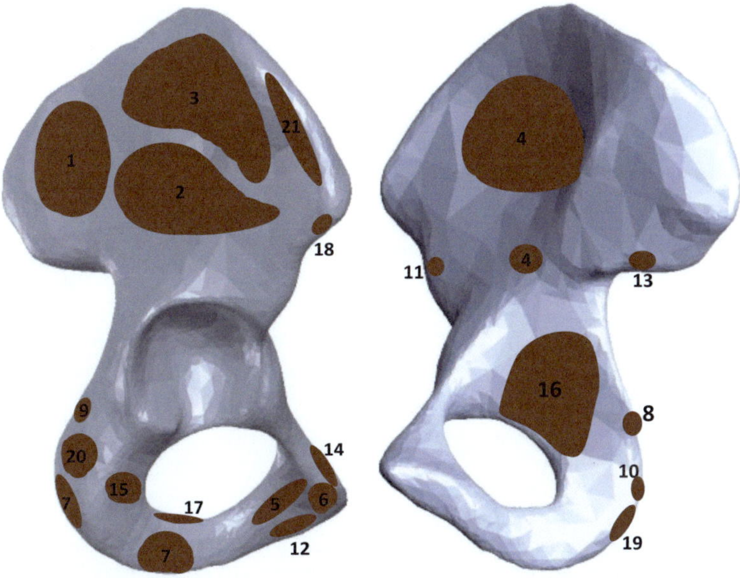

Fig. 8.10 Muscle attachment sites in a pelvic bone. (1) Gluteus maximus, (2) Gluteus minimus, (3) Gluteus medius, (4) Iliopsoas, (5) Adductor brevis, (6) Adductor longus, (7) Adductor magnus, (8) Biceps femoris, (9) Gemellus superior, (10) Gemellus inferior, (11) Rectusfemoris, (12) Gracilis, (13) Piriformis, (14) Pectineus, (15) Quadratus femoris, (16) Obturator femoris, (17) Obturator externus, (18) Sartorius, (19) Semitendinosus, (20) Semimembranosus, (21) Tensor fasciae latae

forces, and boundary conditions on the femur and pelvic bone model are illustrated in Fig. 8.11.

It is a challenge to mimic the *in vivo* loading conditions, owing to the lack of information on the exact position of many muscle attachment sites. The boundary conditions may also change from time to time during the dynamic activities. In the FE model, the prescribed boundary conditions should most closely match the physiological constraints. Unrealistic stresses around the constraints may be generated for any artificial constraints.

The method of application of the predominant joint reaction force in a subject-specific FE model of bone affects the load transfer in the implanted joint. The FE study by Ghosh et al. (2015) observed that the inclusion of the cartilage layer between the anatomical femoral head and the acetabulum leads to a more accurate stress-strain distribution within the acetabulum (Fig. 8.11c). Application of the hip-joint reaction force at the centre of an anatomical femoral head covered by a cartilage layer as compared to a spherical femoral head was observed to be more suitable than a spherical femoral head, resulting in a more physiological load transfer. High compressive and low tensile stresses were generated within the acetabulum, primarily due to the transfer of the hip joint force. This detailed FE model of the pelvic bone, used in combination with musculoskeletal loading conditions in a normal walking

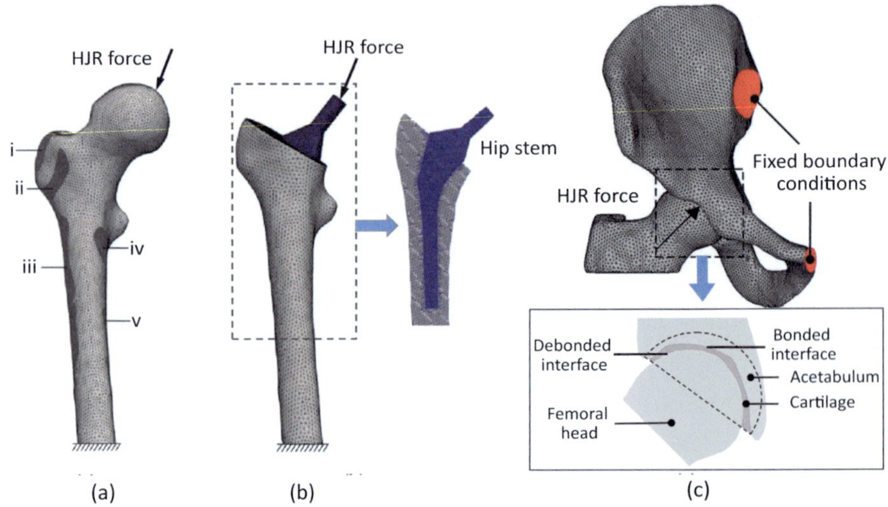

Fig. 8.11 Loading and boundary conditions in FE models; **a** femur with attachment sites of various muscles: (i) gluteus, (ii) vastus lateralis, (iii) vastus intermedius, (iv) vastus medialis and (v) adductors, **b** implanted proximal femur with sectional view, **c** intact pelvic bone model including femoral head and cartilage, the interfacial conditions are also presented here

cycle, served as a reference solution for investigating deviations in load transfer post-implantation (Ghosh et al. 2015).

In order to investigate stress-strain-related failure mechanisms within an implanted bone structure, it is often necessary to concentrate more on the region surrounding the implant-bone interfaces. This calls for developing a submodel of an implanted bone, with a finer element size around the implant (Fig. 8.12). The effect of forces acting on different bony regions in the proximity of the implanted bone area cannot be neglected, since the bone (e.g. scapula, pelvic bone, lumber spine) might be subjected to bending in different planes (Gupta et al. 2004a; Ghosh et al. 2013a; Talukdar et al. 2021). As a first step towards implementing the submodelling technique, a reference solution of a full model of the implanted bone is required. This reference solution corresponds to a set of musculoskeletal loading conditions of joint reaction and muscle forces during an activity. The submodel and the full model need to be coupled for the exchange of appropriate information. Nodal displacements of the implanted full model are transferred to the submodel at the cut boundaries, located adequately distant from the implanted region, as shown in Fig. 8.12 (Ghosh et al. 2013a). Any other form of approximate boundary conditions, applying artificial constraints at locations far away from the region of interest to restrain rigid body motion, is considered to be inappropriate and is expected to lead to doubtful results. A finely meshed full FE model of the implant-bone structure is an alternative. However, the computational requirements for solving such a finer full FE model are much too large.

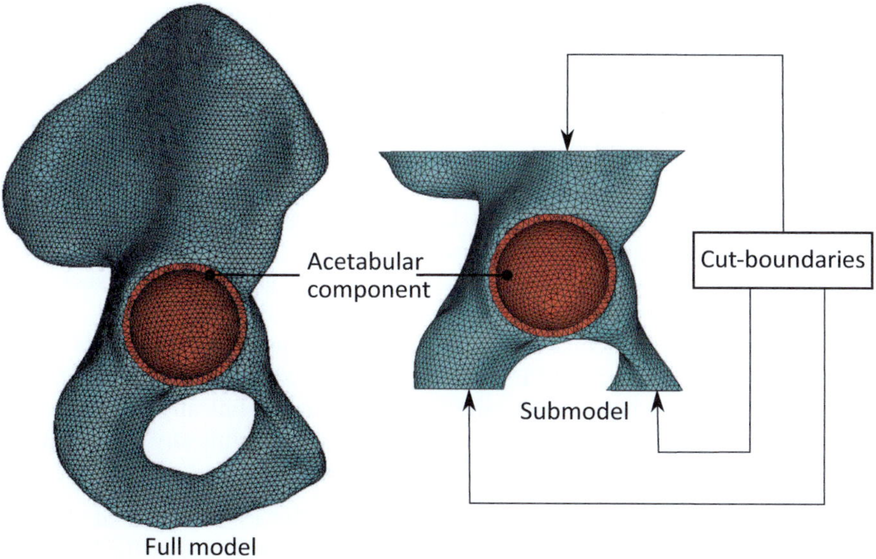

Fig. 8.12 Illustration of the submodelling technique used for numerical evaluations on acetabular components. The nodal displacements are transferred from the full model to the cut-boundaries of submodel

8.2.5 Population-Based Model

Whilst evaluating the performance of orthopaedic implants, a majority of the numerical studies have been based on subject-specific FE models. Although clinically relevant, these results cannot be ideally generalised owing to variability in bone geometry and material properties within a population. A few studies have considered modelling multiple bones from different subjects, accounting for inter-subject variability (Radcliffe et al. 2007a, b). Another widely used approach is population-based models. In this approach, the geometry and material properties of the bones of a population are statistically represented (Taylor et al. 2013; Laz and Browne 2010; Bryan et al. 2010). Employing a limited number of training datasets, this method can produce several artificial models by incorporating the variation in bone geometry and material properties comparable to that observed in the population. Such models can be used to predict the mechanical behaviour and risk of failures of implant–bone structures across a population (Galloway et al. 2013; Bryan et al. 2010).

8.3 Verification and Validation of FE Models

A numerical model comprises a large number of mathematical equations and depends on computational techniques for finding an approximate solution to the underlying physical phenomenon. In biomechanics and allied research, the major challenge is the uncertainty in extracting data on complex geometry (micro and macro) and material properties of the biological tissues considered in an analysis. Numerical models are capable of representing complex systems and simulating a large variety of problems in biomechanics, including structural analysis. The tools and techniques, such as FE modelling and analysis, are effective only when used with due caution.

The accuracy of FE-predicted results (e.g. stresses and strains) depends largely on the quality of reconstructed FE models. Hence, the soundness of the FE predictions must be assessed before the model is used to address clinically relevant issues. Viceconti et al. (2005) reported the requirements of mining clinically important data from FE analysis of bone and implant-bone structures.

- "Verification is about solving the equations correctly" (Viceconti et al. 2005)—the behaviour of the theoretical base model should be predicted by the FE model. A model can be verified by measuring its numerical accuracy (e.g. convergence test).
- "Validation is about solving the correct equations" (Viceconti et al. 2005)—the behaviour of the allied physical phenomenon should be predicted by the FE model. Validation is assessing the accuracy with which the FE model predicts the reality.

More explicitly, verification is the process of establishing that a model precisely embodies the conceptual description and solution to the model (Anderson et al. 2007). Conversely, validation is a process by which the numerical predictions are compared to the experimentally measured data to assess the modelling errors (if any) and, eventually, the validity of the FE model.

Three levels of requirements are prescribed for the FE studies to consider addressing clinically relevant questions. As a minimum requirement, the studies partially or fully based on FE analysis should pay attention to multiple points, such as model selection, verification, appropriate parameter identification, sensitivity analysis, and inter-subject variability, to make any clinical recommendation (Viceconti et al. 2005). The first three points (model selection, verification, appropriate parameter identification) may be considered sufficient for studies making only theoretical speculation and deductive reasoning (Viceconti et al. 2005). The second level of requirements includes validation against controlled in vitro studies to be satisfied by the models used in biomechanics research. The third level of requirements, such as the need for risk-benefit analysis and prospective and retrospective studies, applies to research that is spaced between biomechanics and clinical research. Also, Anderson et al. (2007) presented the methodical and multi-step approach for verification and validation. A researcher should ensure the following three important aspects:

- Whether the governing mathematical equations of the model are correct?
- Whether the model accurately represent the underlying physics of the problem?

- Whether the error and uncertainty in the model predictions are assessed?

A computational modelling and analysis should include precise strategies for both verification and validation. FE analyses on the implant-bone structures suggest that biomechanical factors play a role in the failure mechanisms of the reconstructed joints. Model validation is necessary to give confidence in the FE-predicted results and is achieved by comparing the numerical predictions with the experimental measurements (Gupta et al. 2004b; Pal et al. 2010; Ghosh et al. 2012; Chanda et al. 2015). Published subject-specific FE models of bones and implant-bone structures may not always be appropriately verified and adequately validated. However, a one-to-one experimental validation of a FE model is not possible, if the model is based on CT scan data of a living subject.

8.3.1 Validation Using Strain Gauge Method

Experimental validation is a laborious but useful tool to judge the accuracy of FE predictions. Quite often, investigations have focused on relating FE strains with experimentally measured strains, using strain gauges or rosettes, at different locations on the bone surface, including composite bone and cadaver bone (Gupta et al. 2004b; Pal et al. 2010; Ghosh et al. 2013b; Chanda et al. 2015). Strain gauge measurements provide local average strain occurring underneath the gauge mounted at discrete locations. Several factors are known to affect strain measurements, including local surface preparation, associated instrumentation, and the ambient temperature (Dally and Railey 1991). A strong bond between the strain gauge and the test specimen should be ensured. The local bone surfaces are generally cleaned with alcohol. Repetitive abrading with grit emery paper and degreasing with alcohol and acetone should follow. A set of general-purpose strain gauges or strain rosettes with appropriate gauge factor and resistance can be used at the desired locations and orientations. The gauges are generally mounted on the bone surface using a single-component cyanocrylate adhesive. Strain gauges are mounted on similar locations and orientations on the intact and implanted bone for comparing pre- and post-implantation measurements.

The experimental validation study by Gupta et al. (2004b) was carried out on a fresh cadaveric scapula with eighteen strain gauges attached at various locations and orientations. Load was applied to the scapula in a mechanical testing machine. Constraints, interconnected by ball joints, were applied at three locations, as shown in Fig. 8.13. These constraints were used to unambiguously inflict bending and deflection of the tested scapula along all directions, as related to applied loads at different locations. The equivalent FE model of the tested scapula was based on a CT-scan dataset and a combination of solid and shell elements, with material properties and thickness taken from CT-scan data. The validity of the FE model predictions was assessed using linear regression analysis; reasonably high correlation coefficients (0.89–0.97) and low standard errors (29–105 $\mu\varepsilon$) and percentage errors

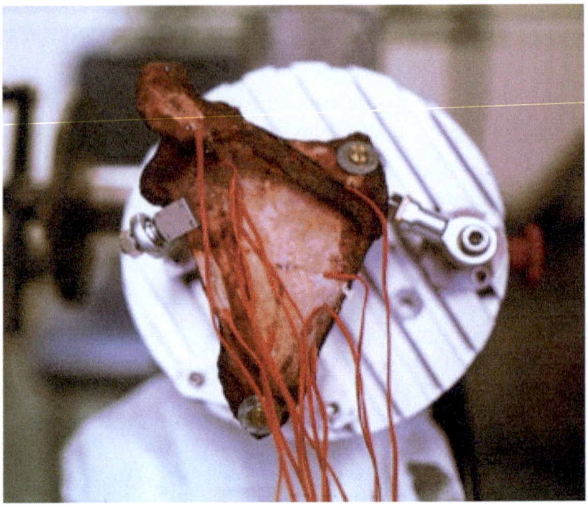

Fig. 8.13 Mechanical testing set-up for strain gauge measurements on fresh scapula (adapted from Gupta et al. 2004b)

in regression slope (9–23%) were observed between the experimentally measured and FE-predicted strains (Gupta et al. 2004b).

An experimental study by Pal et al. (2010) investigated the effect of mechanical parameters on the risk of failure of femoral resurfacing, using strain gauges at different locations and orientations on 4th generation composite femur model (Sawbones) (Fig. 8.14). Using a mechanical testing machine, the specimens were tested under loading vertically applied on the femoral head (Fig. 8.15). The measured strains were then related to those predicted by the equivalent FE models of the bone and implant-bone structure. A strong agreement between the measured strains and the FE predicted strains was observed, with a high correlation coefficient (0.92–0.95), a low standard error of the estimate (170–379 $\mu\varepsilon$), and a low percentage error in regression slope (12–17%). Similar experimental validation of intact and implanted pelvic bones, using strain gauges, was carried out on composite hemi-pelvises (Ghosh et al. 2013b). Although a comparison of localised surface strains at different locations of the intact and implanted bone structures might be useful for validating the equivalent FE models, a full-field strain distribution is required to capture the sharp gradient in strain fields within a structure.

Strain gauges can provide local strain data only at selected locations of fixation. The irregular geometry of a bone and associated heterogeneous material property causes large strain gradients in the structure, thus requiring a large number of gauges for measurement. However, the flat surface area available for fixing a sufficient number of gauges is limited. Moreover, a sharp gradient in the strain field might not be captured well using measurements at discrete locations. Full-field strain data, as predicted by FE analysis, is highly useful when dealing with irregular structures with non-homogeneous material property distribution, such as bones. Researchers have attempted to use the photoelastic method to obtain bone surface strain distributions upon loading. However, this technique is not well-suited for excessively curved bone

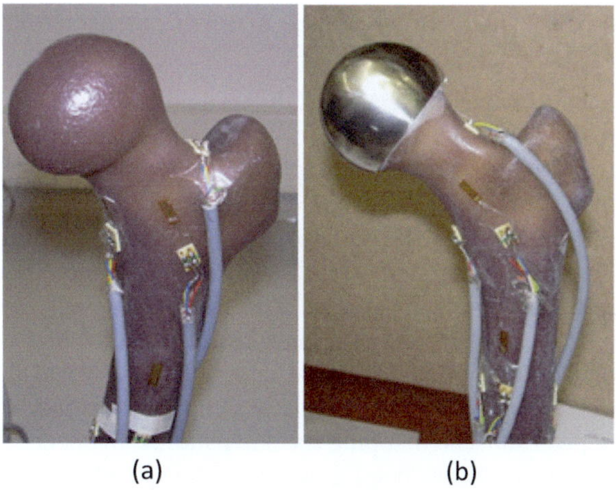

Fig. 8.14 Strain gauges mounted on composite femur specimens, **a** intact and **b** implanted

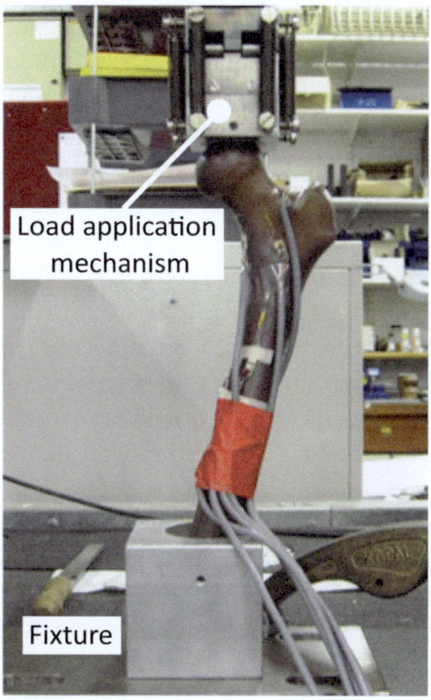

Fig. 8.15 Mechanical testing of an intact femur using strain gauge technique

structures. Non-contact full-field strain measurement techniques, such as Digital Image Correlation (DIC), can deal with the above limitations.

8.3.2 *Validation Using Digital Image Correlation Technique*

Digital Image Correlation (DIC) is a non-contact optical metrology to measure full-field surface deformations of any material subjected to forces. In this method, the surface of the specimen is sprayed with a randomised speckle pattern of contrasting paints, usually a black speckle pattern on the white-painted surface. The displacements of the speckle patterns, owing to load application, can be recorded using two digital cameras. The use of two cameras in stereoscopic configuration allows the capture of the deformation, both in the plane normal to the camera and out of the plane. Images of the specimen are captured before and after load application. The main concept behind DIC is to compare the set of final (deformed) and the initial (undeformed) images. Displacements are calculated by correlating the coordinates of pixel subsets (mapping speckle patterns) in the undeformed and deformed images. Subsequently, the displacements of the speckle patterns are processed by a software to calculate the strains in a region. The DIC measurements are highly dependent on the quality of the generated speckle pattern. It is required to generate non-repetitive, isotropic, and high-contrast speckle patterns against a white light source.

In order to validate the FE-predicted full-field strain distributions in the intact and implanted hemi-pelvises, the DIC strain measurement technique was employed by Ghosh et al. (2012) (Fig. 8.16). Additionally, the deviations in load transfer due to implantation were investigated. Strain rosettes were used to validate the DIC measurements. An excellent linear relationship between the measured and FE-predicted strains was found, with a high correlation coefficient ($R = 0.96$ for intact, 0.94 for implanted) and a low standard error of the estimate ($SE = 69.54$ $\mu\varepsilon$ for intact, 75.09 $\mu\varepsilon$ for implanted). Moreover, the strain rosette and DIC measurements closely match each other, validating the DIC technique. Implantation led to an increase in cortex strain in the superior part of the acetabulum, indicating increased load transfer through the cortical bone at several locations around the acetabulum. The FE model served as a valid predictor of the actually measured strain distribution in the intact and implanted pelvis models.

Chanda et al. (2015) performed a rigorous experimental validation of FE models of intact and implanted composite femurs using the DIC technique and a custom-made experimental setup (Fig. 8.17). Full-field strain distributions for both specimens were in close agreement with the FE-predicted strain patterns, with a high correlation coefficient ($R = 0.92$–0.94), a low standard error of the estimate ($SE = 26$–38 $\mu\varepsilon$), and a low percentage error in linear regression slope ($PE = 2$–4%). These results were found to be similar to the findings of similar studies (Dickinson et al. 2011; Grassi et al. 2014). However, the deviations in strains observed in some curved areas are due to the subtended acute angle in the viewing direction of the cameras. In order to improve the accuracy of measurement, a finer speckle pattern and closely positioned cameras for data acquisition (preferable view direction normal to the surface) are recommended. This study also observed that the implant-induced changes in surface strains (DIC measurements) closely matched the clinically observed bone density changes around an uncemented femoral stem.

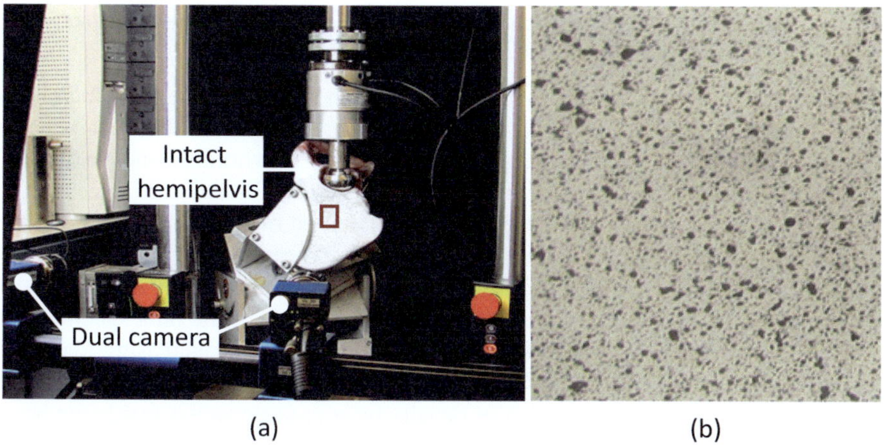

Fig. 8.16 An experimental setup for DIC measurement, **a** load application on an intact hemipelvis (composite), **b** speckle pattern corresponding to the rectangular portion/specimen

Pronounced reduction in proximal bone density corresponded to a 50–64% reduction in surface strain, and a trivial increase in distal bone density corresponded to a 4–14% increase in strain, postoperatively. These results confirm that the FE model generation procedure was good enough for further numerical investigations on failure mechanisms and implant design evaluations.

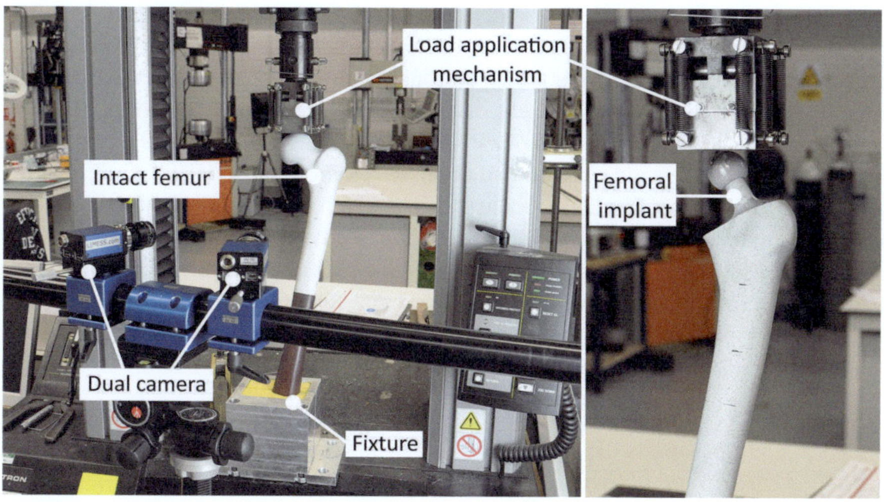

Fig. 8.17 The experimental set-up for DIC measurement of intact (left) and implanted (right) femur specimens, consisting of fixtures, load application planer bearing mechanism and cameras

8.3.3 Validation Using Digital Volume Correlation Technique

Three-dimensional (3D) Digital Volume Correlation (DVC) technique, which is an extension of the DIC technique to three spatial dimensions, has also been used to compute 3D full-field displacement and strain in bone. The working principle of DVC is very similar to that of DIC. The major difference is that 3D volume images of the specimen (undeformed and deformed), usually obtained from CT/MRI systems, are used in DVC instead of 2D images captured by the cameras in DIC. Unlike the DIC technique, which can measure deformation and strain only on the surface of the specimen, the DVC method can calculate the deformation of the bulk specimen. It tracks a set of digital images of the specimen, obtained under increasing load. For tracking, it is necessary to have a random feature in the images. This random feature may be represented by the variation of density or gray-scale values inside the bone specimen. The displacement of patterns of voxel gray-scale values in two sequential images is used to determine the deformation (Ridzwan et al. 2018). The total specimen volume is subdivided into multiple sub-volumes for easier identification of the displacement pattern and subsequent calculation of deformation. Subsequently, the strains can be computed from the displacements. While many DVC studies employed micro-CT images, Ridzwan et al (2018) used clinical CT images in a study of hip fracture. Fourteen femora (cadavers) were CT-scanned and mechanically tested to fracture in a sideways falling configuration. Strain patterns inside the femur were measured using the DVC technique. The FE-predicted strain pattern was compared to the DVC-measured pattern, and the fracture was observed experimentally in the femora. The DVC-measured strain inside the femora corroborated fairly well with the FE-predicted strain patterns. The study concluded that the prediction of FE models agreed well with the experimental observations (86% of the cases), and were able to predict the clinical hip fracture types. A considerable amount of work has already been undertaken in the measurement of bone strains using DVC (Madi et al. 2013; Dall'Ara et al. 2017; Peña Fernández et al. 2018; Rapagna et al. 2019). Employing high-resolution peripheral quantitative computed tomography (HR-pQCT) imaging and DVC, Rapagna et al. (2019) measured bone deformation due to press-fit knee joint implantation in six cadaver femora. Their study concluded that the methods can be used to compare the performance of different implants and the influence of different press-fit magnitudes. The method can also provide information on the bone-implant interface that can be used for experimental validation of the FE models (Sukjamsri et al. 2015).

8.3.4 Validation Using Linear Displacement Sensor

Measurement of other quantities within the implant-bone structure, such as implant-bone micromotion, can also be performed. Implant-bone micromotions were measured using a Linear Displacement Sensor (LDS) in a resurfaced femur and an implanted hemi-pelvis, as shown in Fig. 8.18 (Pal et al. 2010; Ghosh et al. 2013b). Special purpose holders need to be developed to mount an LDS at the desired location and orientation on an implant-bone structure. Experimental studies conducted on the resurfaced femur and implanted acetabulum involved a precise measurement of implant-bone relative displacements (Pal et al. 2010; Ghosh et al. 2013b). The study by Pal et al. (2010) attempted to measure the stem-bone micromotion and implant cup-bone relative displacements in a resurfaced femur using LDSs. Stem-bone micromotion and implant cup-bone relative displacements, both experimentally measured (0–13.7 μm) and FE-predicted (0–25 μm), were small. In the case of the implanted hemi-pelvis, the measured implant-bone relative displacements were low (less than 30 μm), and agreed with the trends predicted (less than 41 μm) by the equivalent FE model (Ghosh et al. 2013b). Despite these quantitative deviations in the measured and predicted relative displacement results, the FE models successfully projected the overall experimental measurements. Experimentally measured micromotion may slightly differ from the FE-predicted results, depending on bone deformation due to bone bending. It is suggested that modelling the linkages, used to hold the displacement sensors in the experimental setup, can account for the bone deformation during the measurement of micromotion.

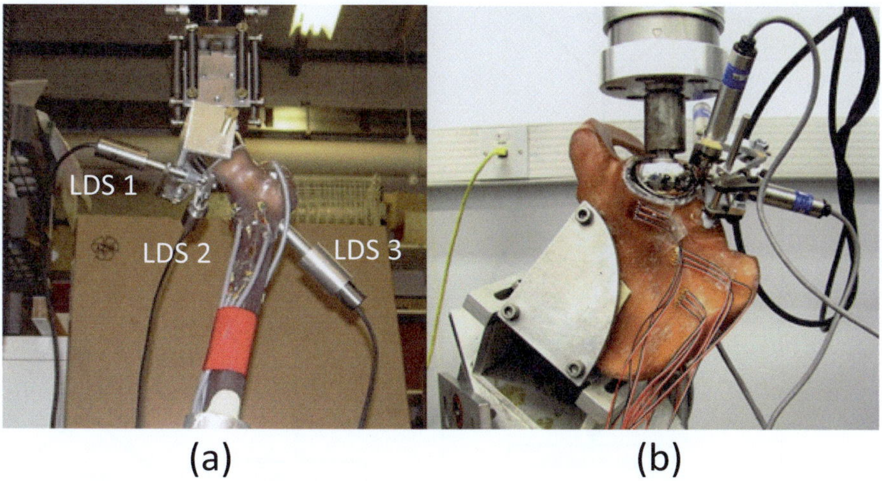

Fig. 8.18 Measurements of implant-bone micromotions using Linear Displacement Sensors (LDS) in, **a** resurfaced femur, **b** implanted hemi-pelvis

8.4 Steps in the Design of Orthopaedic Implants

Orthopaedic implant design is a multi-step process. The major steps are design feasibility, concept and detailed design, design verification, manufacture, validation, and design transfer, as shown in Fig. 8.19. An overview of these steps is discussed in the following sub-sections.

8.4.1 Design Feasibility

Clinical problems associated with the existing implants give rise to the requirements of an improved or a new design. Usually, the orthopaedic surgeons and the engineers, working collaboratively, come up with a novel or improved design idea to overcome the existing problems. A market survey for the new design idea is required to recognise the potential demands. A product with a limited demand may not be

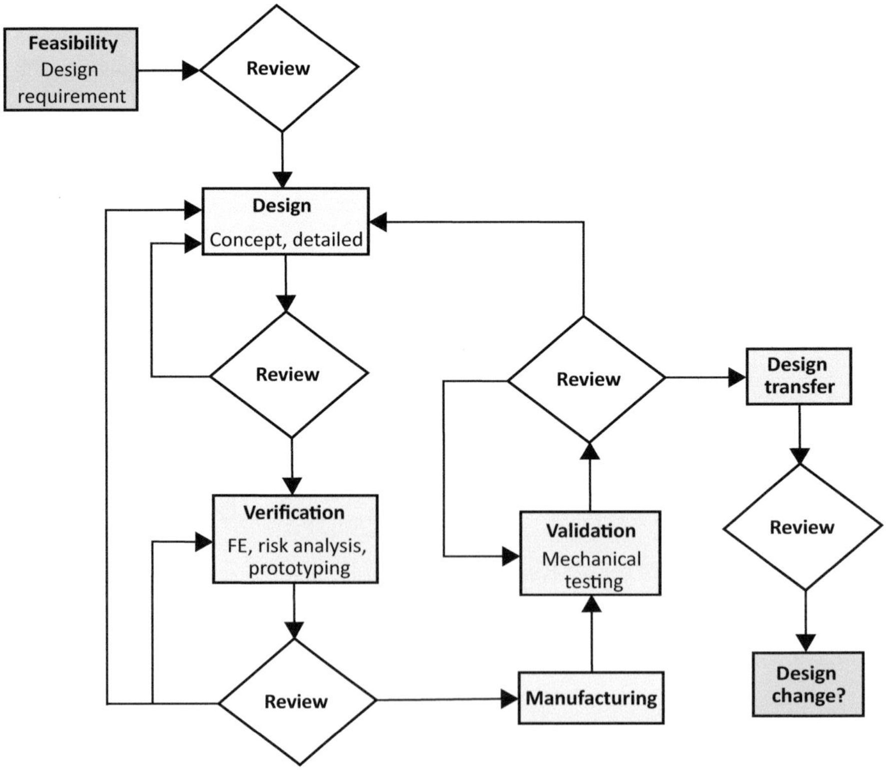

Fig. 8.19 A flowchart highlighting the major steps in the design and development process of orthopaedic implants

commercially viable. A well-performing similar product may exist that captures a large market share, hindering the success of a new product. Therefore, it is essential to find out if any potentially conflicting designs are already in the market (Aitchinson et al. 2009). Similarly, the prior arts (intellectual property—IP) in the field must also be considered to avoid infringing any existing IPs and to protect the new design. The entire project is framed by proper identification and break-up of the required design activities, with special attention to the time, budget, and an alternative plan for risk management for each activity. A newly designed implant requires regulatory clearance before it can be allowed for clinical use. Several regulatory guidelines have highlighted the requirements that orthopaedic implants must satisfy in order to avoid risk to patients and medical professionals. For example, the European Union Medical Device Regulation (EU MDR) and the United States Food and Drug Administration (FDA) have produced regulatory guidelines for medical devices to promote and protect public health in their respective territory. The regulatory guidelines suggest that the risk level for any implant design should be established early. The design specifications, such as its intended functions, range of movements, selection of materials, fixation method, manufacturing method, verification and validation techniques, requirements of tools and instrumentations, potential surgical approach, sterilization and packaging, and other design- and patient-related variables need to be identified (Pal and Gupta 2020). Standards are available to facilitate the general and specific design requirements (ISO 21535: 2023; ISO 21536: 2023). A comprehensive design review activity is necessary for a methodical evaluation of the design requirements, any foreseeable problem, and its potential solution.

8.4.2 Concept and Detailed Design

The ideas that potentially satisfy the design requirements are known as concepts. Through brainstorming, the designer needs to come up with multiple ideas (Pal and Gupta 2020). Sketch and CAD models of such ideas need to be prepared. Other important aspects to consider are the choice of best materials and manufacturing methods. The proposed ideas are then assessed for their suitability with regard to manufacturability, cost-effectiveness, intended function, instrumentation, sterilization and packaging, and potential intellectual property value. The most appropriate design is then selected from the list of concept designs that were suitably rated and compared with each other. Thereafter, the best concept design is further used to develop a detailed design. This step encompasses the preparation of a CAD model, selection of materials, drafting of engineering drawings including dimensions and tolerances, evaluation of the range of movements, and checking the appropriateness of the design for manufacture and assembly.

8.4.3 Design Verification

In the early design process, it is important to examine whether the design satisfies the prescribed design requirements. As mentioned in the design feasibility section, a verification method must be considered when identifying the design requirements. Finite element analysis, risk analysis, and prototyping are usually used to verify the design. Load transfer, strength, and potential failure mechanisms can be investigated using FE analysis, saving valuable time and cost. Another important technique in this process is the analysis of risks. The associated risks should be minimised. Careful comparison should be made between any remaining risks and the potential benefits. Guidelines for risk analysis methods, such as Failure Mode Effect Analysis and Fault Tree Analysis, are reported in the standards. Associated surgical instruments, packaging, and sterilization methods should also be considered in the risk analysis. Prototyping is a useful technique for verifying new designs. Also, it facilitates visualization of the product features, helping the exchange of ideas between the various stakeholders. The designer can use several methods for making prototypes, including various 3D printing techniques. The implant prototypes can be used for visual inspection, sizing, demonstration, and trial surgery using artificial bone models.

8.4.4 Manufacturing

The technique employed to manufacture the implants must be reliable and repeatable. The choice of manufacturing method depends on several factors, such as quantities, desired surface finish, risk of contamination during manufacturing, and cleaning, packaging, and sterilization. Casting and machining techniques have been traditionally used for implant manufacturing. Computerized numerical control (CNC), computer-aided design, and computer-aided manufacturing (CAD/CAM) can be used for manufacturing implants. Drag finishing, tumbling, and polishing operations are performed depending on the required surface finish. Coating, plating, heat treatment, and anodizing may also be required. Cleaning and sterilization of the final product are important to avoid the risk of contamination. The advancement of 3D-printing technology has enabled the manufacturing of implants with complex geometry. Medical regulatory bodies impose strict requirements for the quality and compliance of the implants. A stringent quality management system should be in place for any implant manufacturing unit, including material verification, process validation, and inspection.

8.4.5 *Design Validation*

Design validation is very important in the process of implant design. The mechanical behaviour of a design must be assessed under conditions closely replicating the physiological condition. Verification is about checking whether the product being developed is right. In contrast, validation checks whether the right product is being developed. The most common technique used for validation is mechanical testing of the manufactured device. The mechanical performance of the design and its allied risk of failure must be investigated by physical testing. Several professional bodies have set standards for such testing, such as the International Organisation for Standardization (ISO), the British Standard Institution (BSI) and the European Standard (EN). Both composite bones and cadaver bones have been used in the mechanical testing of implants. Clinical investigations with due ethical and regulatory approval and existing clinical evidence of similar design are also useful tools. Having clinical collaborators and an industrial partner in a team is very useful in conducting the required clinical investigations.

8.4.6 *Design Transfer*

Upon achieving promising results from the verification and validation steps, and on regulatory approval, the design is recommended to be transferred to an interested industry to manufacture it commercially for clinical use. A filed patent application and/or a granted patent of the design may attract a suitable industrial partner. Hence, it is advisable not to disclose the ideas publicly at the design stage and to seek protection of the design through patenting at a suitable stage of the design process. All supporting evidence and models generated throughout the design process should be well-recorded and preserved, including instructions for use and the proposed surgical approach. Once the commercial manufacturing of the design and its clinical use is initiated, a process needs to be planned for collecting patients' and surgeons' feedback on the implant performance. Any major concerns should be addressed by suitable modifications of the implant design.

Multiple Choice Questions (MCQ)

1. Which of the following procedures is not related to image processing of bone?
   (A) Thresholding
   (B) Region growing
   (C) Meshing
   (D) Contour detection

2. In medical image processing, the process of separating the bone from the surrounding soft-tissue structures is known as
   (A) Segmentation
   (B) Region growing
   (C) Smoothening
   (D) Contour detection

3. In which of the following steps does an initial selection of a small region get propagated if the adjoining pixels have similar characteristics?
   (A) Smoothening
   (B) Region growing
   (C) Segmentation
   (D) Contour detection

4. Virtual surgery of the bone model is performed right after
   (A) Generation of CAD model of bone
   (B) Meshing of the bone model
   (C) Implant positioning in the bone model
   (D) Segmentation of bone

5. A geometry-based meshing technique is commonly used to generate a finer mesh in the desired areas. This statement is
   (A) True
   (B) False

6. How do population-based models address the issue of variability in bone geometry and material properties?
   (A) By using a single subject-specific FE model for all predictions.
   (B) By statistically representing the geometry and material properties of a population.
   (C) By focusing only on the material properties and ignoring bone geometry.
   (D) By using a limited number of subject-specific models without incorporating variability.

7. In CT-scan grey value based heterogeneous material property assignment of bone, the relationships used between the apparent density and CT grey value (in HU) and the Elastic modulus and apparent density are ________ and ___________, respectively.
   (A) Linear, power law
   (B) Quadratic, power law
   (C) Linear, exponential
   (D) Quadratic, exponential

8. The partial volume effect can affect the distribution of material properties in the FE model of bones. This statement is _____.

(A) True
(B) False

9. What is the typical technique used for design validation?

(A) Prototyping
(B) Mechanical testing
(C) Finite element analysis

10. Verification of a numerical model involves ________________.

(A) Solving the equations correctly
(B) Solving the right equations
(C) Comparing the computational predictions to experimental data

11. In an implant-bone structure, linear variable differential transformer (LVDT) can be used for the measurement of

(A) Longitudinal strain
(B) Principal stress
(C) Implant-bone relative displacement

12. Which of the following is a full-field surface strain measurement technique?

(A) Digital image correlation
(B) Digital volume correlation
(C) Strain gauge/rosette

13. Which of the following techniques is used for measuring strain inside a loaded specimen?

(A) Digital image correlation
(B) Digital volume correlation
(C) Strain gauge/rosette
(D) LVDT

14. The speckle pattern used on the surface of the specimen in a Digital Image Correlation (DIC) experiment should be ______ and ______.

(A) Non-repetitive, high contrast
(B) Repetitive, high contrast
(C) Repetitive, low contrast
(D) Non-repetitive, low contrast

15. In which of the following stages, the design requirements of a new or improved implant design are identified?

(A) Design feasibility
(B) Detailed design
(C) Verification
(D) Validation

Answer Key to MCQs

1. C; 2. A; 3. B; 4. C; 5. A; 6. B; 7. A; 8. A; 9. B; 10. A; 11. C; 12. A; 13. B; 14. A; 15. A

References

Aitchison GA, Hukins DW, Parry JJ, Shepherd DE, Trotman SG (2009) A review of the design process for implantable orthopedic medical devices. Open Biomed Eng J 3:21–27

Anderson AE, Ellis BJ, Weiss JA (2007) Verification, validation and sensitivity studies in computational biomechanics. Comput Methods Biomech Biomed Eng 10(3):171–184

Anderson AE, Peters CL, Tuttle BD et al (2005) Subject-specific finite element model of the pelvis: development, validation and sensitivity studies. Trans ASME J Biomech Eng 127(3):364–373

Bergmann G, Deuretzbacher G, Heller M, Graichen F, Rohlmann A, Strauss J, Duda GN (2001) Hip contact forces and gait patterns from routine activities. J Biomech 34:859–871

Brown TD, Shaw DT (1983) In vitro contact stress distribution in the natural human hip. J Biomech 16:373–384

Bryan R, SuryaMohan P, Galloway F, Taylor M, Nair PB (2010) Statistical modeling of the whole human femur incorporating geometric and material properties. Med Eng Phys 32(1):57–65

Byrne N, Velasco-Forte M, Tandon A, Valverde I, Hussain T (2016) A systematic review of image segmentation methodology used in the additive manufacture of patient-specific 3D printed models of the cardiovascular system. JRSM Cardiovasc Dis. https://doi.org/10.1177/2048004016645467

Chanda S, Dickinson AS, Gupta S, Browne M (2015) Full-field in vitro measurements and in silico predictions of strain shielding in the implanted femur after total hip arthroplasty. Proc Inst Mech Eng Part H J Eng Med 229(8):549–559

Dall'Ara E, Peña-Fernández M, Palanca M, Giorgi M, Cristofolini L, Tozzi G (2017) Precision of digital volume correlation approaches for strain analysis in bone imaged with micro-computed tomography at different dimensional levels. Front Mater 4:31

Dally JW, Riley WF (1991) Experimental stress analysis, 3rd edn. McGraw Hill

Dickinson AS, Taylor AC, Ozturk H, Browne M (2011) Experimental validation of a finite element model of the proximal femur using digital image correlation and a composite bone model. Trans ASME J Biomech Eng 133(1):1–6

Duda GN, Brand D, Freitag S, Liersel W, Schneider E (1996) Variability of femoral muscle attachments. J Biomech 29(9):1185–1190

Galloway F, Kahnt M, Ramm H, Worsley P, Zachow S, Nair P, Taylor M (2013) A large scale finite element study of a cementless osseointegrated tibial tray. J Biomech 46(11):1900–1906

Ghosh R, Gupta S, Dickinson A, Browne M (2012) Experimental validation of finite element models of intact and implanted composite hemi-pelvises using digital image correlation. Trans ASME J Biomech Eng 134(8):081003

Ghosh R, Mukherjee K, Gupta S (2013a) Bone remodelling around uncemented metallic and ceramic acetabular components. Proc Inst Mech Eng Part H J Eng Med 227(5):490–502

Ghosh R, Gupta S, Dickinson A, Browne M (2013b) Experimental validation of numerically predicted strain and micromotion in intact and implanted composite hemi-pelvises. Proc Inst Mech Eng Part H J Eng Med 227(2):162–174

Ghosh R, Pal B, Ghosh D (2015) Finite element analysis of a hemi-pelvis: the effect of inclusion of cartilage layer on acetabular stresses and strain. Comput Methods Biomech Biomed Eng 18(7):697–710

Grassi L, Väänänen SP, Yavari SA et al (2014) Full-field strain measurement during mechanical testing of the human femur at physiologically relevant strain rates. Trans ASME J Biomech Eng 136(11):111010

Gupta S, van der Helm FCT, van Keulen F (2004a) Stress analysis of cemented glenoid prostheses in total shoulder arthroplasty. J Biomech 37(11):1777–1786

Gupta S, van der Helm FCT, Sterk JC, van Keulen F, Kaptein BI (2004b) Development and experimental validation of a three-dimensional finite element model of the human scapula. Proc Inst Mech Eng Part H J Eng Med 218(2):127–142

Heller M, Bergmann G, Deuretzbacher G, Durselen L, Pohl M, Claes L, Duda GN (2001) Musculoskeletal loading conditions during walking and stair climbing. J Biomech 34:883–893

Hvid I, Bentzen SM, Linde F, Mosekilde L, Pongsoipetch B (1989) X-ray quantitative computed tomography: the relations to physical properties of proximal tibial trabecular bone specimens. J Biomech 22(8–9):837–844

ISO 21535:2023 (2023) Non-active surgical implants—Joint replacement implants—Specific requirements for hip-joint replacement implants. International Organization for Standardization, Geneva

ISO 21536:2023 (2023) Non-active surgical implants—Joint replacement implants—Specific requirements for knee-joint replacement implants. International Organization for Standardization, Geneva

Keyak JH, Meagher JM, Skinner HB, Mote CD (1990) Automated three-dimensional finite element modelling of bone: a new method. J Biomed Eng 12:389–397

Laz PJ, Stowe JQ, Baldwin MA et al (2007) Incorporating uncertainty in mechanical properties for finite element based evaluation of bone mechanics. J Biomech 40(13):2831–2836

Laz PJ, Browne M (2010) A review of probabilistic analysis in orthopaedic biomechanics. Proc Inst Mech Eng Part H J Eng Med 224(8):927–943

Lengsfeld M, Schmitt J, Alter P, Kaminsky J, Leppek R (1998) Comparison of geometry-based and CT voxel-based finite element modelling and experimental validation. Med Eng Phys 20:515–522

Madi K, Tozzi G, Zhang QH, Tong J, Cossey A, Au A, Hollis D, Hild F (2013) Computation of full-field displacements in a scaffold implant using digital volume correlation and finite element analysis. Med Eng Phys 35(9):1298–1312

Mathai B, Gupta S (2019) Numerical predictions of hip joint and muscle forces during daily activities: a comparison of musculoskeletal models. Proc Inst Mech Eng Part H J Eng Med 233(6):636–647

Morgan EF, Bayraktar HH, Keaveny TM (2003) Trabecular bone modulus-density relationships depend on anatomic site. J Biomech 36:897–904

Pal B, Gupta S (2020) The relevance of biomechanical analysis in joint replacements: a review. J Inst Eng India Ser C 101:913–927

Pal B, Gupta S, New AM (2009) A numerical study of failure mechanisms in the cemented resurfaced femur: effects of interface characteristics and bone remodelling. Proc Inst Mech Eng Part H J Eng Med 223:471–484

Pal B, Gupta S, New AMR, Browne M (2010) Strain and micromotion in intact and resurfaced composite femurs: experimental and numerical investigations. J Biomech 43:1923–1930

Peña Fernández M, Barber AH, Blunn GW, Tozzi G (2018) Optimization of digital volume correlation computation in SR-microCT images of trabecular bone and bone-biomaterial systems. J Microsc 272(3):213–228

Poelert S, Valstar E, Weinans H, Zadpoor AA (2012) Patient-specific finite element modeling of bones. Proc Inst Mech Eng Part H J Eng Med 227(4):464–478

Radcliffe IAJ, Taylor M (2007a) Investigation into the effect of varus-valgus orientation on load transfer in the resurfaced femoral head: a multi-femur finite element analysis. Clin Biomech 22(7):780–786

Radcliffe IAJ, Taylor M (2007b) Investigation into the effect of cementing techniques on load transfer in the resurfaced femoral head: a multi-femur finite element analysis. Clin Biomech 22(4):422–430

Rapagna S, Berahmani S, Wyers CE, Van den Bergh JP, Reynolds KJ, Tozzi G, Janssen D, Perilli E (2019) Quantification of human bone microarchitecture damage in press-fit femoral knee implantation using HR-pQCT and digital volume correlation. J Mech Behav Biomed Mater 97:278–287

Ridzwan MIZ, Sukjamsri C, Pal B, van Arkel RJ, Bell A, Khanna M, Baskaradas A, Abel R, Boughton O, Cobb J, Hansen UN (2018) Femoral fracture type can be predicted from femoral structure: a finite element study validated by digital volume correlation experiments. J Orthop Res 36(3):993–1001

Sas A, Tanck E, Sermon A et al (2020) Finite element models for fracture prevention in patients with metastatic bone disease: a literature review. Bone Rep 12:100286

Schileo E, Taddei F, Malandrino A, Cristofolini L, Viceconti M (2007) Subject-specific finite element models can accurately predict strain levels in long bones. J Biomech 40:2982–2989

Sukjamsri C, Geraldes DM, Gregory T, Ahmed F, Hollis D, Schenk S, Amis A, Emery R, Hansen U (2015) Digital volume correlation and micro-CT: an in-vitro technique for measuring full-field interface micromotion around polyethylene implants. J Biomech 48(12):3447–3454

Taddei F, Pancanti A, Viceconti M (2004) An improved method for the automatic mapping of computed tomography numbers onto finite element models. Med Eng Phys 26:61–69

Talukdar RG, Mukhopadhyay KK, Dhara S, Gupta S (2021) Numerical analysis of the mechanical behaviour of intact and implanted lumbar functional spinal units: effects of loading and boundary conditions. Proc Inst Mech Eng Part H J Eng Med 235(7):792–804

Taylor M, Prendergast PJ (2015) Four decades of finite element analysis of orthopaedic devices: where are we now and what are the opportunities? J Biomech 48:767–778

Taylor M, Bryan R, Galloway F (2013) Accounting for patient variability in finite element analysis of the intact and implanted hip and knee: a review. Int J Numer Methods Biomed Eng 29(2):273–292

Viceconti M, Bellingeri L, Cristofolini L, Toni A (1998) A comparative study on different methods of automatic mesh generation of human femurs. Med Eng Phys 20:1–10

Viceconti M, Olsen S, Nolte L, Burton K (2005) Extracting clinically relevant data from finite element simulations. Clin Biomech 20(5):451–454

Yosibash Z, Padan R, Joskowicz L et al (2007) A CT-based high-order finite element analysis of the human proximal femur compared to in-vitro experiments. Trans ASME J Biomech Eng 129(3):297–309

Zannoni C, Mantovani R, Viceconti M (1999) Material properties assignment to finite element models of bone structures: a new method. Med Eng Phys 20(10):735–740

Chapter 9
Adaptive Bone Remodelling

Bone is a highly adaptive material and is sensitive to disuse, immobilisation, surgical intervention, or activities involving high mechanical loads. Bone tissue may change its properties in response to changes in the biomechanical and biochemical environment. In this chapter, the cellular activities involved in bone remodelling, the theory of adaptive bone remodelling and the numerical frameworks developed in the last few decades to quantify this adaptive phenomenon have been discussed.

9.1 Biology of Bone Remodelling

The adult skeleton undergoes a continuous renewal through a process known as remodelling, which persists throughout life. Bone resorption (bone loss) and bone apposition (bone formation) are coupled processes, owing to the cellular activities of osteoclast and osteoblast, respectively. The osteoblast and osteoclast cells belong to a unique temporary structure known as a Basic Multicellular Unit (BMU). In other words, BMU is a specialised compartment where bone remodelling occurs. The process of bone remodelling involves the initiation of 3–4 million BMUs annually, of which 1 million are actively engaged in bone remodelling at any given time (Manolagas 2000). The BMUs engage in a series of cellular processes, resulting in the continual renewal of bone tissue to maintain its structural strength and integrity. Let us elaborate on the activities of various bone cells involved in bone remodelling.

Osteoclasts: Osteoclasts are large multinucleated cells and are differentiated from monocytes. The osteoclasts attach to bone surfaces, creating a tight seal around their contact area. Subsequently, these cells secrete acid to dissolve the bone beneath them, followed by the ingestion and expulsion of the resulting debris into the bloodstream. Osteoclasts help in regulating calcium homeostasis, since calcium is present in the debris. Osteoclasts resorb the bone, resulting in depressions on the bone surface. These depressions are known as Howship's lacunae.

S. Gupta et al., *Biomechanics of Joints and Implants*,
https://doi.org/10.1007/978-981-96-0586-6_9

Osteoblasts: Unlike osteoclasts, osteoblasts are mono-nucleated cells and they function as groups of connected cells. The main function of osteoblast is to synthesise various components that constitute the extracellular matrix of bone. Moreover, these cells promote the mineralisation of the organic bone matrix. Osteoblasts secrete osteoid that facilitates the formation and maintenance of the bone structure. Osteoid is a protein mixture that forms the un-mineralised and organic bone matrix, which is later mineralised to form the mature bone tissue. The inactive osteoblasts are known as bone-lining cells. These cells are regarded as post-proliferative osteoblasts, which protect the bone surface from resorption activity. The bone lining cells may later differentiate into reversal cells.

Reversal cells: These are mono-nucleated cells that exist on the eroded surface after the departure of the osteoclasts. The reversal phase links the osteoclastic bone resorption to osteoblastic bone apposition (Baron and Saffar 1977). These reversal cells are considered to differentiate to become osteoblasts (Dempster 2017).

Osteocytes: After the bone-forming phase, the osteoblasts get buried in the extracellular bone matrix. These cells represent the most mature differentiation stage of osteoblasts and are known as osteocytes. Osteocytes comprise 90–95% of the total bone cells (Franz-Odendaal et al. 2006). They are regarded as the mechanoreceptors of the bone tissue since they can sense mechanical stimuli. In response to these mechanical stimuli, osteocytes control the onset of both bone apposition and resorption (Schaffler and Kennedy 2012). Osteocytes communicate with nearby osteocytes and other bone cells through a meshwork of cell processes, located inside the canaliculi within the bone matrix.

Bone remodelling is a dynamic process that involves coordinated activities of osteoclasts and osteoblasts within the BMUs (Ryser et al. 2009). This coordination is necessary to maintain the balance in the remodelling occurring at each focal point (local remodelling site). Focal coupling in bone remodelling refers to the transfer of signals across different cells responsible for bone remodelling. This transfer of signals enables the coupling of cellular events necessary for bone remodelling. Coupling means that the receptors on the osteoclast membrane, as well as the regulatory factors released by osteoclasts, are coupled with sequential recruitment and differentiation of osteoblast lineage cells toward the mature phenotype, laying down the bone matrix and inducing mineralisation (Bolamperti et al. 2022). Osteoclasts not only resorb bone but also secrete coupling factors that recruit osteoblast precursors to regulate the overall bone remodelling process. The balance between bone resorption and formation is crucial, since an imbalance can lead to improper bone turnover (i.e. bone loss) and diseases like osteoporosis.

9.1.1 Different Phases of Bone Remodelling

Bone remodelling is a complex process consisting of four phases: activation, resorption, reversal, and formation, where the osteoclasts and osteoblasts work sequentially within the same BMU (Raggatt and Partridge 2010).

Activation phase: This first stage of bone remodelling involves the detection of remodelling signal. This signal can be of any type, for example (a) mechanical strain resulting in structural damage, or (b) action of hormones, such as estrogen or parathyroid hormone, on bone cells, owing to changes in homeostasis.

Resorption phase: In this phase, the osteoclast precursor cells are recruited to the remodelling site. These cells are differentiated into the osteoclasts, which create an isolated microenvironment called a sealed zone. Mineralised matrix is dissolved within this sealed acidic space, resulting in Howship's lacunae.

Reversal phase: Following bone resorption, the remaining undigested demineralised collagen matrix covers the Howship's lacunae. The reversal cells remove these collagen remnants and prepare the bone surface for bone formation. Moreover, the reversal cells produce coupling signals that allow the transition from bone resorption to bone formation within the BMU (Raggatt and Partridge 2010).

Formation phase: The formation phase is considerably longer than the first three phases and it involves the production of a matrix by progressive waves of osteoblasts. The osteoblasts secrete osteoid, which is later mineralised to form bone. Following mineralisation, the osteoblasts may either (a) undergo apoptosis (cell death), (b) become inactive cells (bone lining cells), or (c) get embedded in the mineralised matrix and differentiate into osteocytes.

After reaching peak bone mass, a balanced bone remodelling maintains the stable bone mass for a decade or two, before age-related bone loss begins. This loss is primarily due to increased resorption and decreased bone formation (Liang et al. 2022). Abnormalities in bone remodelling can lead to conditions of low or high bone mass, forming the basis of various bone-mass syndromes such as Paget's disease, osteopenia, and osteopetrosis. Various types of remodelling cycles have been presented in Fig. 9.1. The remodelling cycle, which prevails in the physiological conditions, exhibits a completely filled cavity indicating no loss (Fig. 9.1a). In contrast, Fig. 9.1b, c correspond to pathways of bone loss (lack of bone reconstruction) during bone remodelling. Although coupling occurs in the first pathway of bone loss (imbalance), the osteoblasts do not synthesise as much bone matrix as compared to that resorbed by osteoclasts (Fig. 9.1b). In the case of the second pathway, uncoupling/delayed coupling, osteoblasts do not appear in the remodelling site (Fig. 9.1c) (Delaisse et al. 2020). Imbalanced and uncoupled bone remodelling correspond to the cases where the cavity is partially filled and completely unfilled, respectively (Fig. 9.1b, c). The reversal bone phase in the uncoupled bone remodelling cycle is called the arrested reversal phase (Andersen et al. 2013).

9.1.2 *Bone Turnover*

Bone turnover refers to the proportional bone volume replacement per unit time and is a key aspect of bone remodelling (Parfitt 2002). Moreover, bone turnover is regulated by the number of BMUs and the focal balance within each BMU (Bolamperti et al. 2022). It has been further reported that geometric and biological factors influence

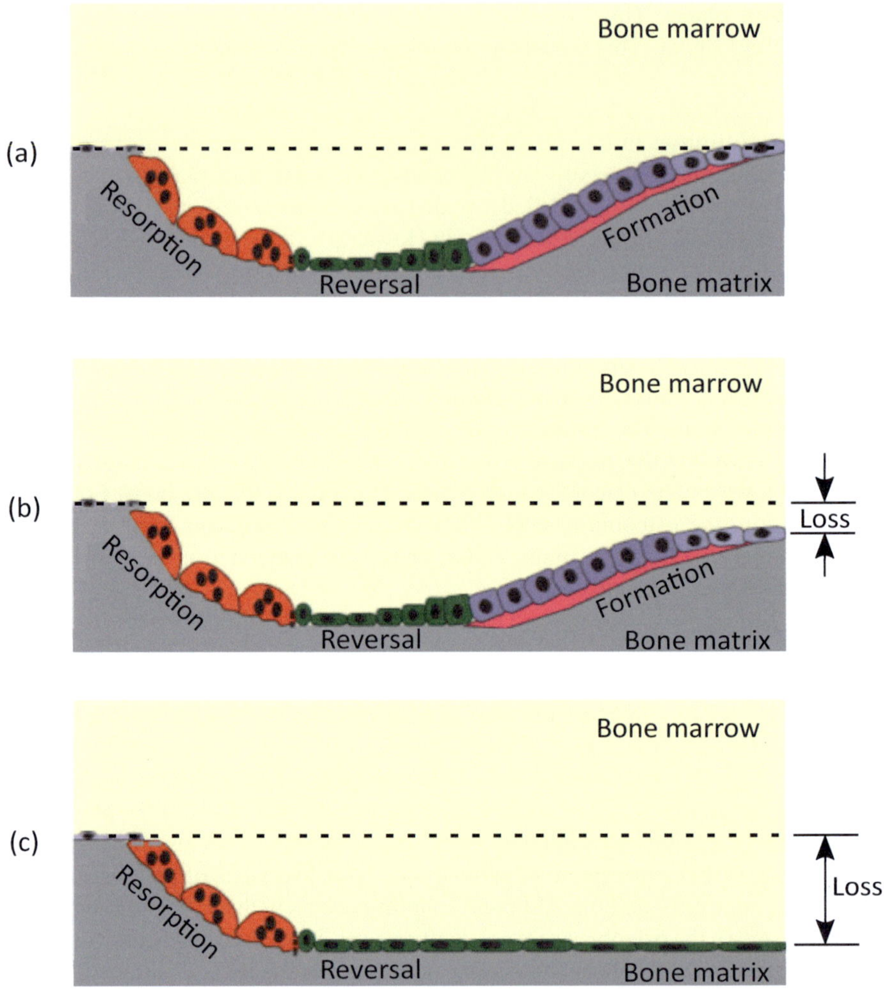

Fig. 9.1 Three types of bone remodelling cycles; **a** the cavity is completely refilled, **b** the cavity is partially filled, **c** the cavity is completely unfilled. The bone formation is not even initiated in the case of completely unfilled cavity. Bone remodelling on the surface of the trabecular bone is illustrated here. Adapted from Delaisse et al. (2020)

bone turnover (Parfitt 2002) (Table 9.1). The bone formation rate to bone surface ratio (BFR/BS) and bone surface to bone volume ratio (BS/BV) are the biological and geometric factors, respectively. Bone turnover is usually expressed in terms of bone volume replacement (BFR/BV). The popular belief is that the cancellous bone turnover is higher than that of cortical bone. However, this statement is valid only for site-specific comparison. As evident in Table 9.1, if one considers either central or peripheral anatomy, the determinants (factors) of bone turnover are in such a way that supports the popular belief mentioned earlier. However, if we consider

Table 9.1 Determinants of bone turnover in peripheral and central anatomy.

Anatomy	Bone type	BS/BV	BFR/BS
Peripheral	Cortical	Low	Low
	Cancellous	High	Low
Central	Cortical	Low	High
	Cancellous	High	High

Source Parfitt (2002)

the peripheral cancellous and central cortical, the popular belief is not valid (Parfitt 2002). Moreover, in this case, the biological factors outweigh the geometric ones, resulting in higher turnover of cortical bone than cancellous bone.

With aging, the relative importance of cortical remodelling increases due to the loss of cancellous bone and increased remodelling activity (Rauch et al. 2007). In cortical bone, remodelling occurs at periosteal and endo-cortical surfaces, as well as within the intra-cortical region (Haversian). Surface-based remodelling processes that occur at the endo-cortical and periosteal surfaces are similar to cancellous bone remodelling. Whereas, Haversian remodelling involves the development of the Haversian system or osteon, characterised by concentric layers of bone surrounding a central canal.

9.1.3 Haversian Bone Remodelling

The secondary osteons are formed as a result of the coordinated actions of cells within the BMU. Secondary osteons differ from primary osteons as the former is formed by the replacement of the existing bone, whereas the latter is likely formed during bone growth through the mineralisation of cartilage, where bone was not previously present. Figure 9.2 presents an overview of the Haversian bone remodelling, which results in the formation of secondary osteons. Section 1–1 corresponds to the completed Haversian system, section 2–2 corresponds to the forming Haversian system, and section 3–3 corresponds to the forming resorption cavity (Fig. 9.2). The osteoblastic apposition of concentric lamellae continues until the diameter of a typical Haversian canal is attained. This culminates in the formation of a new secondary Haversian system or osteon (Zoetis et al. 2003).

9.1.4 Trabecular Bone Remodelling

The remodelling of trabecular bone is not as organised as it is within the Haversian system of cortical bone. As evident in Fig. 9.3, the bone marrow envelope (BME) is a thin layer of cells surrounding bone marrow. After the bone remodelling cycle is initiated, the osteoclasts lift the BME. Thereby, BME forms a canopy that covers the

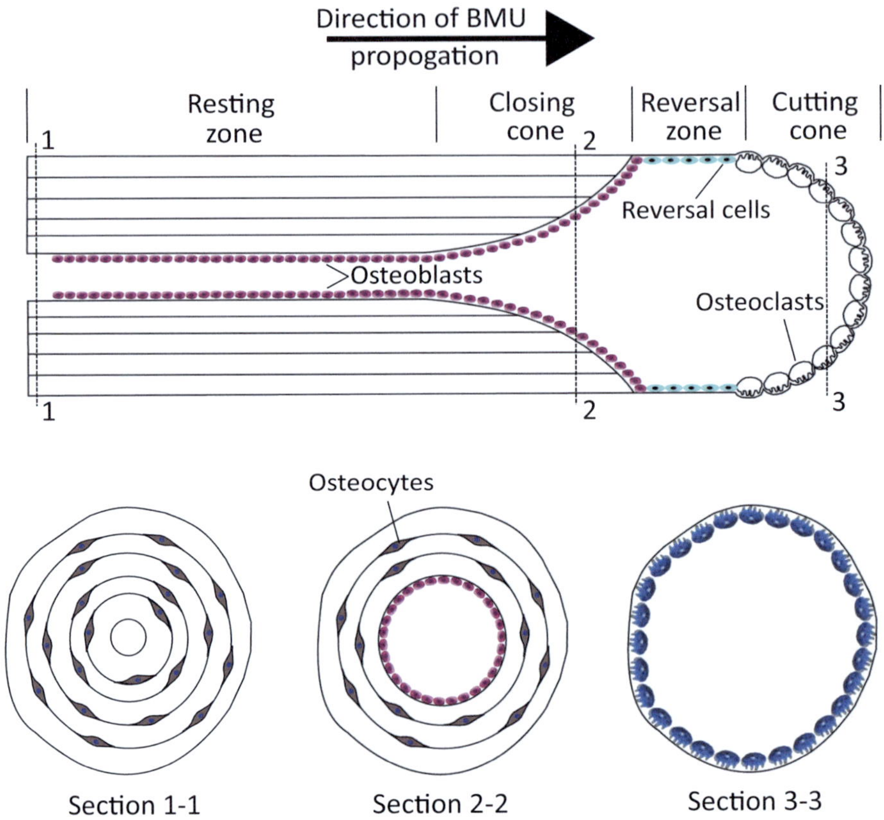

Fig. 9.2 Schematic overview of the Haversian bone remodelling, indicating coordinated action of osteoblasts and osteoclasts. The cross-sectional views of various stages are also shown. Adapted from Bolamperti et al. (2022). The cutting cone consists of osteoclasts, which is followed by reversal zone consisting of reversal cells. The closing zone consists of active osteoblasts, whereas quiescent (inactive) osteoblasts are present in the resting zone

cells involved in remodelling. Figure 9.3 illustrates trabecular remodelling, wherein an osteoclast is observed to erode the bone surface, while it moves over it. Initially, the osteoclast approaches the bone lining cell (BLC), which is inactive between the BME and inactive bone surface. Thereafter, the osteoclast moves forward, and the BLC differentiates to a reversal cell on the eroded surface. Here, the BME cell turns into a canopy cell in contact with the capillary (Delaisse et al. 2020). Subsequently, the reversal cell density increases as a result of the differentiation of BLCs and the recruitment of cells from the canopy. Finally, these reversal cells differentiate into mature bone osteoblasts which deposit osteoid. Figure 9.3 simplifies the remodelling cycle considering only one osteoclast cell. However, it should be noted that a wave of osteoclasts may be involved in bone remodelling, resulting in erosion of an already eroded bone surface.

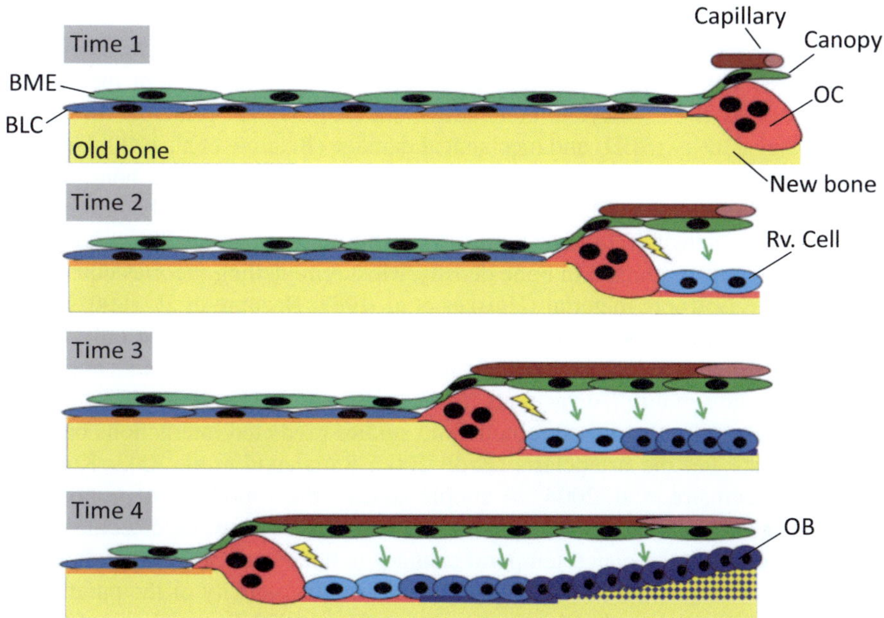

Fig. 9.3 Evolutionary trabecular bone remodelling. BME: Bone Marrow Envelope, BLC: Bone Lining Cell, Rv. Cell: Reversal Cell, OC: Osteoclast, OB: Osteoblast. Adapted from Delaisse et al. (2020)

9.2 Theory of Adaptive Bone Remodelling

The capability of bone to adapt its structure in response to alterations in the mechanical environment by bone apposition and resorption is known as adaptive bone remodelling. In addition to the changes in the mechanical environment, alterations in the biochemical environment also influence bone remodelling. As discussed earlier, an equilibrium between bone apposition and bone resorption exists in the natural state. However, if the mechanical environment is altered, e.g. owing to the insertion of a prosthesis, the equilibrium is disturbed, triggering bone remodelling. This process occurs in two ways.

- External bone remodelling—changes in the geometry of cortical bone.
- Internal bone remodelling—changes in density and morphology of cancellous bone.

According to Roux's (1881) functional adaptation theory, bone apposition and resorption are two biological processes that are governed by the local state of mechanical stress. Based on this theory, Wolff (1986) further proposed that every change in a bone's form or function causes definite changes in the internal architecture of the bone as well as changes in its external geometry. This law of bone transformation later came to be known as Wolff's law of bone remodelling.

In order to quantify the mechanical influence on bone remodelling, identifying the most suitable mechanical stimulus has been a subject of ongoing debate. Therefore, the proposed bone remodelling algorithms followed several different mechanical stimuli, such as effective cyclic bone stress, strains along the principal directions, strain energy density (SED) and mechanical damage (Beaupre et al. 1990; Geraldes et al. 2016; Levenston and Carter 1998; Huiskes et al. 1987). The bone remodelling algorithm based on SED has been quite popular in predicting changes in periprosthetic bone density (Huiskes et al. 1987). This algorithm related the changes in SED with the rate of change in bone density. Most remodelling theories considered bone to be an isotropic material (Huiskes et al. 1987; Beaupre et al. 1990; Levenston and Carter 1998). Since these remodelling algorithms accounted only for the mechanical stimulus, a lack of mechanical loading may result in complete bone loss, which contradicts reality (Avval et al. 2014).

Biochemical stimulus-based remodelling studies predicted interactions between the osteoblasts and the osteoclasts (Kroll 2000; Rattanakul et al. 2003; Komarov et al. 2003; Lemaire et al. 2004). A simple mathematical model was proposed by Komarov et al. (2003) that could predict the interactions between osteoblasts and osteoclasts and the associated temporal changes in bone mass. Earlier studies have also attempted to mathematically model the differential activity of the parathyroid hormone (PTH), associated with bone adaptation (Kroll 2000; Rattanakul et al. 2003). A more complex dynamic interaction model was proposed considering the interaction between pre(responding)-osteoblast, osteoblast, and osteoclast populations (Lemaire et al. 2004). It may be noted, however, that very few attempts have been made to couple mechanical stimulus with biochemical models (Moroz et al. 2006; Bonfoh et al. 2011; Hambli 2014).

A remodelling model that introduced coupling between biochemical reactions and mechanical loading was proposed by Bourgherara et al. (2010). Moreover, an earlier study employed this model on bone remodelling around a femoral implant and predicted bone resorption of 52%, when mechanical loads were absent (Avval et al. 2014). However, the effect of bone anisotropy was ignored in this mechanobiochemical model (Avval et al. 2014). Although attempts have been made to incorporate bone anisotropy in bone remodelling, biochemical stimulus was never accounted for in those models.

Anisotropic material properties can be used to define the orientation of trabecular bone at the continuum level. A study by Jacobs et al. (1997) proposed an anisotropic remodelling model in which bone apparent density and stiffness tensor were considered mutually independent variables. Moreover, damage-based theoretical models have been proven to accurately predict certain aspects of bone remodelling (Prendergast and Taylor 1994; Prendergast and Huiskes 1995). Based on the damage repair theory, an alternate anisotropic bone remodelling hypothesis was proposed by Doblare and Garcia (2002), which considered microdamage in the bone surface as the remodelling stimulus. Later, McNamara and Prendergast (2007) investigated bone adaptation using a combination of strain and microdamage as the remodelling stimuli. The bone remodelling algorithm proposed by Geraldes and Philips (2014) assumed bone to be orthotropic. In this algorithm, the strains along the principal direction were considered as the mechanical stimuli. Other algorithms based on structural topology optimisation have also been attempted to predict implant-induced bone remodelling (Jang et al. 2009; Mathai et al. 2022).

9.3 Bone Remodelling Models: Mathematical Formulations and Computational Schemes

The mathematical formulations of bone remodelling algorithms were based on various remodelling stimuli, e.g. SED, orthotropic strain, and orthotropic mechanobiochemical (MBC) stimulus, discussed elaborately in the following section. The computational framework associated with these numerical models is also presented.

9.3.1 SED-Based Bone Remodelling Model

The SED-based bone remodelling model employed a site-specific formulation considering strain energy as the mechanical stimulus responsible for the remodelling (Cowin and Hegedus 1976; Huiskes et al. 1987). The element-specific reference stimulus (S_{ref}) and the actual stimulus (S) for intact and implanted bone, respectively, were equal to the elastic strain energy density (U) per unit apparent bone density (ρ) averaged over a loading history (n). The stimulus was averaged over the loading history to account for the differences in the loading conditions (Eq. 9.1).

$$S = \frac{1}{n} \sum_{i=1}^{n} \frac{U_i}{\rho} \tag{9.1}$$

Bone is not responsive to small deviations in mechanical stimulus (Frost 1964). Later, the bone remodelling theory incorporated this minimal threshold of the inhibitory stimulus (s) as the “lazy zone” or “dead zone” (Huiskes et al. 1987; Van Rietbergen et al. 1993) (Fig. 9.4). The dead zone (s) is often taken as ± 0.75 of the reference stimulus S_{ref} for human (Huiskes et al. 1992). It is suggested that the available free surface area ($A(\rho)$) affects the rate of adaptive changes in bone (Martin 1984). Martin’s assumption was used to compute the internal free surface area per unit volume of bone, $a(\rho) = A(\rho)/V$ (Martin 1984). The $a(\rho)$ was estimated as a function of the apparent density, using the following empirical relationship:

$$\begin{aligned} a(\rho) = &-0.0293 + 9.5124\rho - 4.887\rho^2 - 1.568\rho^3 \\ &+ 3.7182\rho^4 - 1.6352\rho^5 \end{aligned} \tag{9.2}$$

The above relation is presented graphically in Fig. 9.5.

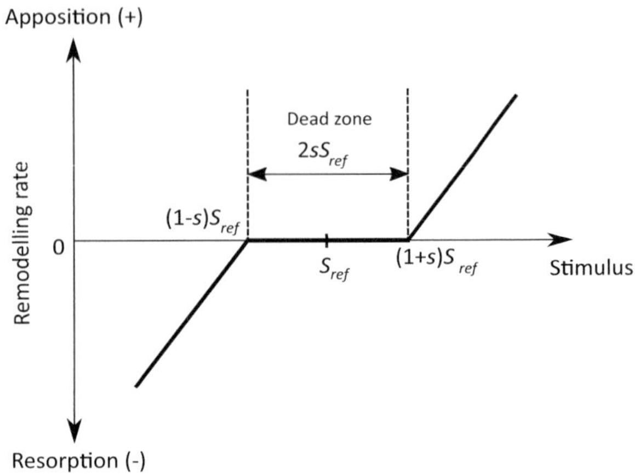

Fig. 9.4 The SED-based bone remodelling rule indicating the dead zone. The reference stimulus is denoted by S_{ref}

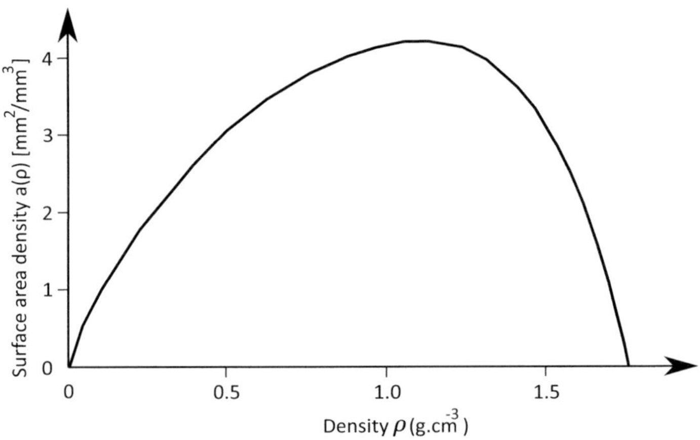

Fig. 9.5 Relationship between internal free surface area density and apparent density (Martin 1984)

The value of $a(\rho)$ is 0.0 for $\rho = \rho_{\text{max}} = 1.73$ g cm^{-3}, which means no internal remodelling occurs in the cortical bone with an apparent density of 1.73. However, when the ρ is equal to 0, owing to the non-existence of bone tissue, there is no free surface area. The changes in bone morphology are caused by cellular activities on the free surface. The greater the available free surface area, the more predominant the osteoblastic activity. If the apparent density is very small and the stimulus is favouring the remodelling, the internal adaptation results in the development of rod-like trabeculae, which later transition to plate-like trabeculae (dense trabeculae). The plate-like trabeculae result in a decrease in the available surface area for further remodelling, which is indicated by the decrease in slope (Fig. 9.5). Therefore, the change in bone density will be higher in the porous bone than in the dense bone. Hence, bone adaptation is proportional to the free surface area density at a specific location. The rate of change in bone mass, representing the bone remodelling process, was expressed as follows:

$$\frac{dM}{dt} = \begin{cases} A(\rho)\{S - (1 \pm s)S_{\text{ref}}\}\tau, & \text{if } S \leq (1 - s)S_{\text{ref}} \text{ or } S \geq (1 + s)S_{\text{ref}} \\ 0, & \text{if } S_{\text{ref}}(1 - s) < S < S_{\text{ref}}(1 + s) \end{cases} \tag{9.3}$$

The parameter τ is a time constant of 129.6 g/(m^2 (J/g) month). The rate of change of bone mass was expressed in terms of the stimulus in Eq. 9.3. This equation was rewritten in terms of rate of change of the apparent bone density, as follows (Eq. 9.4).

$$\frac{d\rho}{dt} = \begin{cases} a(\rho)\{S - (1 \pm s)S_{\text{ref}}\}\tau, & \text{if } S \leq (1 - s)S_{\text{ref}} \text{ or } S \geq (1 + s)S_{\text{ref}} \\ 0, & \text{if } S_{\text{ref}}(1 - s) < S < S_{\text{ref}}(1 + s) \end{cases} \tag{9.4}$$

Based on the time increment value, Δt, Eq. 9.4 can be iteratively solved using Euler's forward integration method to determine site-specific, i.e. element-specific in finite element (FE) analysis, adaptive changes in bone density (Eq. 9.5).

$$\Delta\rho^t = \tau a(\rho)\left\{S - (1 \pm s)S_{ref}\right\}\Delta t \tag{9.5}$$

The updated apparent density (ρ^{t+1}) of each element was found using Eq. 9.6.

$$\rho^{t+1} = \rho^t + \Delta\rho^t \tag{9.6}$$

The above integration was carried out in steps of $\tau\Delta t$, which represents bone adaptation at a simulation time scale (Weinans et al. 1993). The time increment (Δt) was determined in each iteration of the integration procedure such that the maximum change in of bone density does not exceed 0.865 g cm^{-3}.

$$\Delta t = \frac{0.865}{\tau[a(\rho_i)\{S - (1 \pm s)S_{\text{ref}}\}]_{\max}} \tag{9.7}$$

Upon estimating the time step (Δt), the updated apparent density was calculated using Eqs. 9.5 and 9.6.

Steps involved in the computational framework: Let us now discuss the computational framework with reference to bone remodelling around an uncemented femoral implant (Mathai and Gupta 2020). This framework is presented as a flow diagram in Fig. 9.6.

(1) The first step is to develop CAD models of the intact and implanted femurs. The CAD models are generally developed using CT-scan dataset, as discussed earlier in Chap. 8. The implanted femur model is obtained by carrying out virtual implantation on the intact femur.
(2) Based on these CAD models, the FE models of the intact and implanted femurs are generated. The FE analyses are undertaken for both intact and implanted

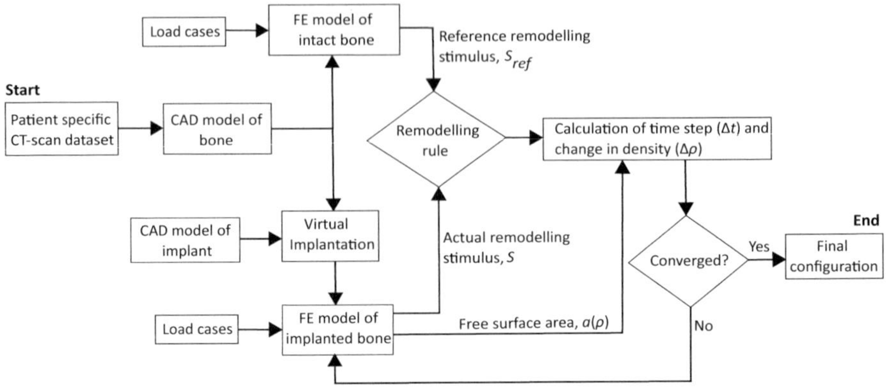

Fig. 9.6 Computational framework of SED-based bone remodelling framework

femurs to obtain the reference and actual remodelling stimuli (Mathai and Gupta 2020).

(3) The remodelling stimulus corresponding to the bone elements of almost identical spatial locations in the implanted and intact FE models, are further compared (Fig. 9.4). If the remodelling stimulus value is beyond the threshold range of the dead zone, bone remodelling process is triggered.

(4) The time increment (Δt) is calculated such that the maximum density change in any cancellous bone element does not exceed 0.865 g cm^{-3}. The time increment is a variable in the simulation, since it varies with bone remodelling iterations.

(5) The change in element-specific apparent bone density is calculated using Eq. 9.4. The elastic properties of the bone elements are updated according to the anatomical site-specific empirical equations. In the case of femur, the elastic modulus of each bone element is updated according to Eq. 9.8. In this computational framework, the bone material is assumed to be linearly elastic and isotropic (Mathai and Gupta 2020). The Poisson's ratio is assumed to be 0.2.

$$E = 7281\rho^{1.52} \tag{9.8}$$

(6) The process of iterative bone remodelling is continued until an equilibrium is attained. The equilibrium or convergence of the remodelling simulation is assumed when the maximum change in bone density for any element is not more than 0.005 g cm^{-3} per time increment ($\tau \Delta t$).

9.3.2 *Orthotropic Strain-Based Bone Remodelling Model*

In this section, the orthotropic strain-based bone remodelling model proposed by Mathai et al. (2021) is discussed. This bone remodelling formulation assumed that directional material properties change according to local principal stress (Frost 1964;

Miller et al. 2002). Two major sequential steps involved in the algorithm were (a) estimation of material orientation, and (b) estimation of material properties along the principal directions (orthogonal directions).

(a) Estimation of Material Orientation: This step primarily involves the determination of the principal direction (element specific) based on Eigen analysis of the stress tensor. The following is an elaborative description of this step. The FE model of the implanted bone was solved during each iteration and the stress tensor ($\boldsymbol{\sigma}$), and strain tensor ($\boldsymbol{\varepsilon}$) were extracted for each bone element. Since multiple load cases were considered in this study (Mathai et al. 2021), it was necessary to choose a particular load case for the bone remodelling rule. This selection was carried out element-wise, using the following procedure.

A guiding frame corresponding to the load case that produced the absolute maximum normal strain value $\left(\varepsilon_n^{\max}\right)$, was identified by Eq. 9.9.

$$\varepsilon_n^{\max} = \max(|\varepsilon_{ii}|_k) \tag{9.9}$$

where k represents the load case under consideration. The guiding stress tensor ($\boldsymbol{\sigma}^g$) was defined based on this guiding frame (Geraldes et al. 2016). Eigen analysis of the guiding stress tensor ($\boldsymbol{\sigma}^g$) was carried out to determine the principal stress values, $\boldsymbol{\sigma}_p^g$ and the corresponding element principal directions, $\boldsymbol{\sigma}_{pv}^g$ (Eq. 9.10). The principal directions were considered as the target material orientations.

$$eig\left(\sigma_{ij}^g\right) = \left[\boldsymbol{\sigma}_{pv}^g, \boldsymbol{\sigma}_p^g\right] = \left[\begin{bmatrix} \sigma_{pv1}^{\min} & \sigma_{pv1}^{\text{med}} & \sigma_{pv}^{\max} \\ \sigma_{pv2}^{\min} & \sigma_{pv2}^{\text{med}} & \sigma_{pv2}^{\max} \\ \sigma_{pv3}^{\min} & \sigma_{pv3}^{\text{med}} & \sigma_{pv3}^{\max} \end{bmatrix}, \begin{bmatrix} \sigma_p^{\min} & 0 & 0 \\ 0 & \sigma_p^{\text{med}} & 0 \\ 0 & 0 & \sigma_p^{\max} \end{bmatrix}\right] \tag{9.10}$$

(b) Estimation of material properties along the principal directions: Based on the assumption that the bone structure adapts itself to the maximum strain (Judex and Zernicke 2000; Geraldes et al. 2016), a strain envelope denoted by $\varepsilon_{ij}^{\max}$ was estimated using Eq. 9.11.

$$\varepsilon_{ij}^{\max} = \max\left(\left|\varepsilon_{ij}\right|_k\right) \tag{9.11}$$

The strain envelope was obtained from maximum absolute strain components resulting from every load case under consideration. The strain stimuli along the target directions (ε^*) were later calculated from σ_{pv}^g and $\varepsilon^{\max}$ using Eq. 9.12 (Skedros and Baucom 2007).

$$\boldsymbol{\varepsilon}^* = \left(\boldsymbol{\sigma}_{pv}^g\right)^T \boldsymbol{\varepsilon}^{\mathbf{max}}\left(\boldsymbol{\sigma}_{pv}^g\right) \tag{9.12}$$

In this study, strain, ε^*, was considered as the remodelling stimuli. The site-specific (element-wise) reference stimulus was obtained from the FE model of intact bone. The von Mises strain was determined for each bone element of the intact model and mapped to the corresponding bone element of the implanted model. The Young's modulus (E) was updated iteratively using Eqs. 9.13–9.14.

$$\Delta E_{ii} = \begin{cases} a(\rho)\left\{\left|\varepsilon_{ii}^{*}\right| - (1 \pm s)\varepsilon_{\text{ref}}\right\}\tau_{\text{ortho}}\Delta t, & \text{if } \left|\varepsilon_{ii}^{*}\right| \leq (1-s)\varepsilon_{\text{ref}} \\ & \text{or } \left|\varepsilon_{ii}^{*}\right| \geq (1+s)\varepsilon_{\text{ref}} \\ 0, & \text{if } (1-s)\varepsilon_{\text{ref}} < \left|\varepsilon_{ii}^{*}\right| < \varepsilon_{\text{ref}}(1+s) \end{cases} \quad (9.13)$$

$$E_{ii}^{t+1} = E_{ii}^{t} + \Delta E_{ii} \quad (9.14)$$

where E_{ii}^{t+1} stands for updated Young's modulus along the i-direction. The threshold value of the dead zone, denoted by s, was assumed to be 0.5 (Mathai et al. 2021). The remodelling was carried out computationally in the simulation scale denoted by $\tau_{\text{ortho}}\Delta t$ (Van Rietbergen et al. 1993). The increment in the time step, Δt, was variable and was calculated during every iteration using Eq. 9.15 (Van Rietbergen et al. 1993; Weinans et al. 1993).

$$\Delta t = \frac{\Delta E_{\max}}{\tau_{\text{ortho}}\left[a(\rho)\left\{\left|\varepsilon_{ii}^{*}\right| - (1 \pm s)\varepsilon_{\text{ref}}\right\}\right]_{\max}} \quad (9.15)$$

The constant $\Delta E_{\max}$ was calculated corresponding to a maximum change in apparent density of 0.865 g cm^{-3}, using Eq. 9.8. An arbitrary value of 2 MPa mm^{-2} mm^{3} per month was assumed as time constant, τ_{ortho} (Mathai et al. 2021). As discussed in Sect. 9.3.1, trabecular bone remodelling is considered to be a surface phenomenon, and its rate was assumed to be proportional to internal free surface area density ($a(\rho)$) (Martin 1984).

The modulus of rigidity (G_{ij}^{t+1}) was estimated using Eq. 9.16, wherein a target strain, ε_{st}, was considered and its value was assumed to be 1443 με (Geraldes et al. 2016).

$$G_{ij}^{t+1} = G_{ij}^{t}\left(\frac{\left|\varepsilon_{ij}^{*}\right|}{\varepsilon_{st}}\right) \quad (9.16)$$

The Poisson's ratios, ν_{ij}^{t+1}, of bone elements were estimated such that it satisfies the thermodynamic restrictions on the elastic constants (Eq. 9.17) (Cowin and van Buskirk 1986; Miller et al. 2002).

$$\nu_{ij}^{t+1} < \left(\frac{E_{ii}^{t+1}}{E_{jj}^{t+1}}\right)^{0.5} \quad (9.17)$$

Once the material properties were determined, the final step was aimed at the determination of updated apparent density (ρ^{t+1}), corresponding to the updated Young's modulus. In the case of femur, the apparent density was calculated using Eq. 9.18.

$$\rho^{t+1} = \left(\frac{E_{\text{rep}}^{t+1}}{7281}\right)^{-1.52} \tag{9.18}$$

where the representative Young's modulus for a bone element, E_{rep}^{t+1}, was calculated according to Eq. 9.19 (Miller et al. 2002; Geraldes et al. 2016).

$$E_{\text{rep}}^{t+1} = \frac{E_{11}^{t+1} + E_{22}^{t+1} + E_{33}^{t+1}}{3} \tag{9.19}$$

Steps involved in the computational framework: Let us now discuss the steps involved in the computational framework associated with implant-induced orthotropic remodelling around a femoral implant. This framework is presented as a flow diagram in Fig. 9.7.

(1) As the first step, the orthotropic material property of the intact femur is obtained by employing the orthotropic remodelling rule in the intact FE model. The initial orthotropic material orientations are aligned with the directions of the global coordinate system, Whereas the simplified orthotropic material properties (dummy) are based on the isotropic material properties obtained from the CT-scan dataset ($E_{11} = E_{22} = E_{33} = E$). After convergence in bone remodelling, we obtain the orthotropic material properties and orientations of the intact femur. These properties and orientations are further mapped to the implanted femur model.
(2) Similar to the computational framework of the SED-based model, the reference and remodelling stimuli are obtained from the FE analysis of intact and implanted femur models. However, the reference stimulus and remodelling stimuli are the equivalent (von Mises) strain and strain along the principal directions, respectively.
(3) The time increment and the change in elastic modulus are calculated using the stimuli values (Eqs. 9.13–9.15).
(4) The bone density is updated based on the updated elastic modulus (Eqs. 9.18–9.19).
(5) The bone remodelling simulation is considered to have converged when the bone density changes were limited to 0.005 g cm^{-3} in successive iterations.

9.3.3 *Mechanobiochemical Bone Remodelling Model*

So far, we have been discussing remodelling algorithms based solely on mechanical stimulus. In this section, a remodelling framework is proposed that coupled biochemical and mechanical stimulus. In the mechanobiochemical (MBC) bone remodelling algorithm, proposed by Bougherara et al. (2010), bone was considered to be an open thermodynamic system. Various biochemical reactions involving mononuclear cells (*MCELL*), multinucleated osteoclasts (*MNOC*), old bone (*OldB*),

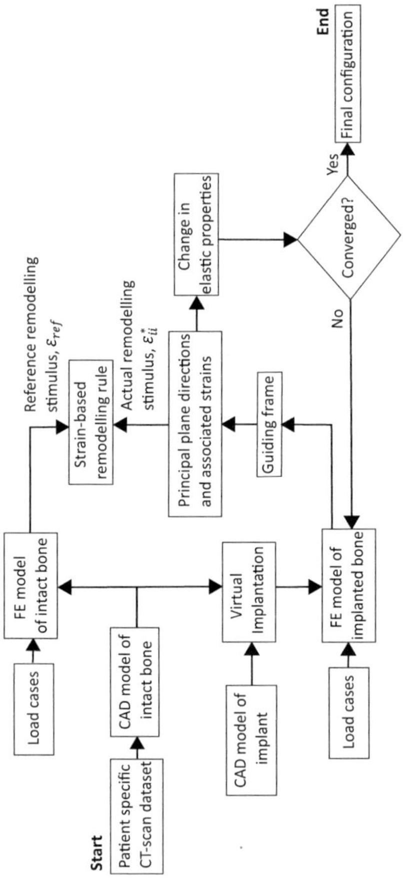

Fig. 9.7 Computational framework of orthotropic strain-based bone remodelling

activator osteoblasts (*ActivOB*), osteoblasts (*OB*), osteoids (*Osteoid*) were considered in the MBC model. These biochemical reactions followed the general form of the Michaelis–Menten enzyme reaction (Eq. 9.20) (Michaelis and Menten 1913):

$$E + S \rightleftharpoons SE \rightarrow E + P \tag{9.20}$$

The complex (SE) is formed as the enzyme (E) binds with the substrate (S). Later, the complex is decomposed into enzyme (E) and product (P).

Biochemical reactions: The reactions considered in the MBC model are summarised as follows. In this model, the first reaction ($\alpha = 1$) involves the formation of the *MNOC*.

$$N_1 + \boldsymbol{MCELL} \rightleftharpoons \boldsymbol{MNOC} + N_4$$

The mixture of chemical substances (N_1) initiates reaction with *MCELL* for the formation of *MNOC*. In the subsequent reaction ($\alpha = 2$), *MNOC* reacts with the *OldB* and resorbs the bone matrix resulting in the chemical substances (N_6 and N_7).

$$\boldsymbol{MNOC} + \boldsymbol{OldB} \rightleftharpoons N_6 + \boldsymbol{N_7}$$

The N_7 reacts with *OldB* to produce *ActivOB* ($\alpha = 3$), which is responsible for activating the osteoblasts.

$$\boldsymbol{N_7} + \boldsymbol{OldB} \rightleftharpoons \boldsymbol{ActivOB} + N_9$$

The *ActivOB* reacts with the *OB*, resulting in the production of the *Osteoid* (unmineralised bone, Collagen type 1) ($\alpha = 4$):

$$\boldsymbol{ActivOB} + \boldsymbol{OB} \rightleftharpoons \boldsymbol{Osteoid} + N_{12}$$

The last reaction ($\alpha = 5$) involves the mineralisation of *Osteoid* by N_{13} resulting in *NewB*.

$$\boldsymbol{N_{13}} + \boldsymbol{Osteoid} \rightleftharpoons \boldsymbol{NewB} + N_{15}$$

In order to achieve the coupling between the mechanical stimulus and the concentrations of the constituents in the biochemical reactions, the standard law of mass action was modified (Eq. 9.21).

$$\begin{aligned} r_\alpha &= l_{\alpha\alpha} A_\alpha + l_{\alpha v} d_{(1)} \\ &= k_{+\alpha} \prod_{i=1}^{14} [n_i]^{\nu_{\alpha i}} - k_{-\alpha} \prod_{i=1}^{14} [n_i]^{\nu'_{\alpha i}} + l_{\alpha v} d_{(1)} \end{aligned} \tag{9.21}$$

where A_α and r_α represented the affinity and the rate of the αth reaction (α varied from 1 to 5), respectively. The rate of volume change was denoted using $d_{(1)}$. The k and l represented the reaction rate and phenomenological coefficients, respectively. The molar concentration of different substances was represented by $[n_i]$. The $\nu_{\alpha i}$ and $\nu'_{\alpha i}$ were the stoichiometric coefficients of the mixture of n_i entering and leaving the αth reaction, respectively. The rate of change of molar concentrations of the chemical substances, $[\dot{n}_i]$ were expressed in terms of the reaction rate (r_α), flux of chemical substances (j_i), and the stoichiometric coefficients as follows (Klika 2006; Bougherara et al. 2010).

$$[\dot{n}_i] = \sum_{\alpha=1}^{5} \left(\nu'_{\alpha i} - \nu_{\alpha i}\right) \cdot r_\alpha + j_i \tag{9.22}$$

Governing differential equations: Using Eqs. 9.21 and 9.22, the kinetics equations were obtained (Klika 2006; Bougherara et al. 2010). Assuming forward reactions to dominate the backward reactions, the kinetics equations were simplified and expressed in dimensionless form as follows:

$$\dot{N}_{MCELL} = -\delta_1(\beta_1 + N_{MCELL})N_{MCELL} + J_3 + J_{14} - D_1 \tag{9.23}$$

$$\begin{aligned}\dot{N}_{OldB} = &-(\beta_3 - N_{MCELL} + N_{OldB} + N_{ActivOB} \\ &+ N_{Osteoid} + N_{NewB})N_{OldB} - D_2 \\ &- \delta_3(\beta_7 - N_{OldB} - 2(N_{ActivOB} + N_{Osteoid} + N_{NewB})) \\ &N_{OldB} + 2J_{14} - D_3\end{aligned} \tag{9.24}$$

$$\begin{aligned}\dot{N}_{ActivOB} = &\,\delta_3(\beta_7 - N_{OldB} - 2(N_{ActivOB} + N_{Osteoid} + N_{NewB}))N_{OldB} - \delta_4 \\ &(\beta_{10} - N_{Osteoid} - N_{NewB})N_{ActivOB} + D_3 - D_4\end{aligned} \tag{9.25}$$

$$\begin{aligned}\dot{N}_{Osteoid} = &\,\delta_4(\beta_{10} - N_{Osteoid} - N_{NewB})N_{ActivOB} \\ &- \delta_5(\beta_{13} - N_{NewB})N_{Osteoid} + D_4 - D_5\end{aligned} \tag{9.26}$$

$$\dot{N}_{NewB} = \delta_5(\beta_{13} - N_{NewB})N_{Osteoid} - J_{14} + D_5 \tag{9.27}$$

where $\dot{N}_i$ represented the normalised rate of change of the molar concentration, J_i denoted dimensionless flux of the ith substance. The ratio of the rate of αth reaction to the second reaction was expressed as δ_α. D_α represented the effect of mechanical loading on the αth biochemical reaction. The model parameters, denoted by β_i, were related to the normalised initial molar concentrations of chemical substances as follows:

$$\beta_1 = \overline{N}_{1_0} - \overline{N}_{MCELL_0} \tag{9.28}$$

$$\beta_3 = \overline{N}_{MNOC_0} + \overline{N}_{MCELL_0} - \overline{N}_{OldB_0} - \overline{N}_{ActivOB_0} - \overline{N}_{Osteiod_0} - \overline{N}_{NewB_0} \tag{9.29}$$

$$\beta_7 = \overline{N}_{7_0} + \overline{N}_{OldB_0} + 2\left(\overline{N}_{ActivOB_0} + \overline{N}_{Osteiod_0} + \overline{N}_{NewB_0}\right) \tag{9.30}$$

$$\beta_{10} = \overline{N}_{OB_0} + \overline{N}_{Osteiod_0} + \overline{N}_{NewB_0} \tag{9.31}$$

$$\beta_{13} = \overline{N}_{13_0} + \overline{N}_{NewB_0} \tag{9.32}$$

Here, $\overline{N}_{i_0}$ denoted the normalised initial molar concentration of various substances involved in the biochemical reactions. The values of β_i were assumed to be constant (Avval et al. 2014, 2015; Bougherara et al. 2010).

Estimation of bone density change: Upon setting the rate of change of molar concentration values (Eqs. 9.23–9.27) to zero and solving the resulting algebraic equations, the stationary solutions corresponding to the kinetic equations were obtained. The dimensionless positive stationary solutions were expressed as follows:

$$\overline{N}_{MCELL} = \frac{1}{2}\left(\beta_1 + \sqrt{\beta_1^2 + 4\frac{J_3 + J_{14} - D_1}{\delta_1}}\right) \tag{9.33}$$

$$\overline{N}_{OldB} = \frac{1}{2}\Bigg(-(\beta_7 + 2\beta_3 - 2\overline{N}_{MCELL}) + \sqrt{(\beta_7 + 2\beta_3 - 2\overline{N}_{MCELL})^2 + 4(2J_{14} - 2D_1 + \frac{J_3 - D_3}{\delta_3})}\Bigg) \tag{9.34}$$

$$\overline{N}_{ActivOB} = \frac{1}{2}\left(-\left(\beta_{10} + \frac{1}{2}\left(\overline{N}_{OldB} - \beta_7 + \frac{J_{14} - D_3}{\delta_3\overline{N}_{OldB}}\right)\right) + \sqrt{\left(\beta_{10} + \frac{1}{2}\left(\overline{N}_{OldB} - \beta_7 + \frac{J_{14} - D_3}{\delta_3\overline{N}_{OldB}}\right)\right)^2 + 4\frac{J_{14} - D_4}{\delta_4}}\right) \tag{9.35}$$

$$\overline{N}_{Osteoid} = \frac{1}{2}\left(-\left(\beta_{13} - \beta_{10} + \frac{J_{14} - D_4}{\delta_4\overline{N}_{ActivOB}}\right) + \sqrt{\left(\beta_{13} - \beta_{10} + \frac{J_{14} - D_4}{\delta_3\overline{N}_{ActivOB}}\right)^2 + 4\frac{J_{14} - D_5}{\delta_5}}\right) \tag{9.36}$$

$$\overline{N}_{NewB} = -\overline{N}_{Osteoid} + \beta_{10} - \frac{J_{14} - D_4}{\delta_4\overline{N}_{ActivOB}} \tag{9.37}$$

where $\overline{N}_i$ was the normalised molar concentration of the ith substance. The D_α was the model parameter which was related to the mechanical stimulus and was expressed as

$$D_\alpha = \frac{|\varepsilon_1| D_{\alpha_{\text{ref}}}}{\Delta t \dot{S}_{\text{ref}}} \tag{9.38}$$

The influence of the reference strain rate ($\dot{S}_{\text{ref}}$) on the αth reaction was denoted using a constant, $D_{\alpha_{\text{ref}}}$. The ε_1 represented the first invariant of the strain tensor, which was considered as the remodelling stimulus obtained from the implanted model (Bougherara et al. 2010). The $\dot{S}_{\text{ref}}$ was approximated as S_{ref} divided by the change in time, assuming the initial strain to be zero, which simplified Eq. 9.38 as

$$D_\alpha = \frac{|\varepsilon_1| D_{\alpha_{\text{ref}}}}{S_{\text{ref}}} \tag{9.39}$$

The element-specific equivalent (von Mises) strain of the intact model was considered as the reference strain. The values of constants $\left(J_i,\ D_{\alpha_{\text{ref}}}\right)$ were obtained from earlier studies (Avval et al. 2014, 2015; Bougherara et al. 2010). The concentration levels of $\overline{N}_{OldB}$ and $\overline{N}_{NewB}$ were obtained after solving the set of Eqs. 9.33–9.37. According to the rule of mixtures, the updated apparent bone density (ρ^t) was evaluated as the product of the bone density corresponding to the previous iteration (t) and the sum of the $\overline{N}_{OldB}$ and $\overline{N}_{NewB}$ (Eq. 9.40).

$$\rho^{t+1} = \rho^t \left(\overline{N}_{OldB} + \overline{N}_{NewB}\right) \tag{9.40}$$

An anatomical site-specific (pelvic bone) empirical relation was used to evaluate the effective elastic modulus (E_{eff}^{t+1}) (Dalstra et al. 1993):

$$E_{\text{eff}}^{t+1} = 2017.3(\rho^{t+1})^{2.46} \tag{9.41}$$

Orthotropic MBC model: In the earlier part of this subsection, we dealt with various biochemical reactions considered in the model and the procedure with which we obtained the apparent density and the corresponding isotropic material property. However, recently, Saviour et al. (2024) proposed a framework that could couple the MBC model and the bone orthotropy. The major sequential steps involved in the framework are (1) estimation of bone density changes and (2) estimation of the orthotropic material directions and properties. The estimation of the density was carried out similarly to the MBC model (Eq. 9.40). However, the orthotropic MBC model accounted for the anisotropic trabecular adaptation by determining strains along the principal directions.

Similar to the orthotropic strain-based model, Eqs. 9.9–9.12 resulted in the determination of the strains along the principal directions $\left(\varepsilon_{ii}^*\right)$. However, the orthotropic material properties were updated using Eq. 9.42. Assuming that the trabeculae are oriented along the dominant direction, E_{rep}^{t+1} was distributed such that the local elastic modulus along the dominant principal direction was greater than that along other principal directions

$$E_{ii}^{t+1} = E_{\text{rep}}^{t+1}\left(\frac{|\varepsilon_{ii}^{*}|}{\epsilon_{\text{ref}}}\right) \tag{9.42}$$

Here, E_{ii}^{t+1} denoted the updated elastic modulus along the principal directions, corresponding to the $t + 1$th iteration. The ϵ_{ref} denotes the von Mises strain obtained from the intact model. A dead zone value of 0.4 was taken while updating the elastic modulus. In order to avoid any sudden jumps in the values of material property, Young's modulus was averaged over four iterations (Eq. 9.43).

$$E_{ii}^{t+1} = \frac{1}{4}\sum_{j=0}^{3} E_{ii}^{t+1-j} \tag{9.43}$$

The modulus of rigidity and the Poison's ratio were updated using Eqs. 9.16 and 9.17.

Steps involved in the computational framework: The steps involved in the computational framework, associated with implant-induced bone remodelling around an acetabular component, are presented sequentially. This computational framework is presented as a flow diagram in Fig. 9.8.

(1) Similar to the computational framework of the SED-based and orthotropic strain-based model, the reference and remodelling stimuli are obtained from the FE analysis of intact and implanted hemipelvis models.
(2) The reference stimulus and remodelling stimuli are the equivalent (von Mises) strain and first invariant of the strain tensor, respectively.
(3) The concentration values of different chemical substances are calculated using the stimuli values (Eqs. 9.32–9.37).
(4) Based on the concentration values of the *NewB* and *OldB*, the change in density is calculated (Eq. 9.40).

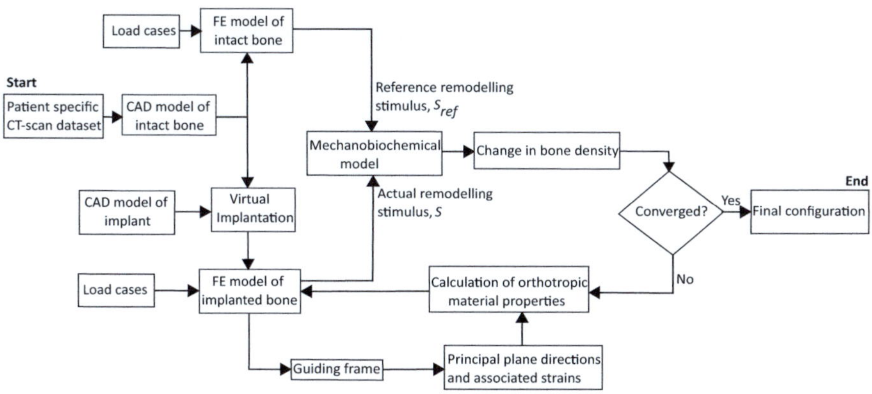

Fig. 9.8 Computational framework of orthotropic mechanobiochemical bone remodelling

(5) Based on the updated density values and the strain along the principal directions, the orthotropic elastic properties are calculated (Eq. 9.42).
(6) The bone remodelling is assumed to have reached equilibrium, when the bone density changes were limited to 0.005 g cm^{-3} in successive iterations.

It should be noted that the bone remodelling algorithms discussed in this chapter have their unique set of assumptions. Moreover, one should be aware of these assumptions before utilising these algorithms for predicting bone remodelling.

Multiple Choice Questions (MCQ)

1. External bone remodelling results in a change in _____.

   (A) Bone apparent density
   (B) Cortical bone thickness
   (C) Cancellous architecture

2. What is the primary role of osteoclasts in bone remodelling?

   (A) Forming new bone tissue
   (B) Creating bone depressions through resorption
   (C) Acting as mechanoreceptors for bone tissue
   (D) Protecting the bone surface from resorption

3. Which cells are responsible for synthesising and mineralising the extracellular matrix of bone?

   (A) Osteoclasts
   (B) Osteocytes
   (C) Osteoblasts
   (D) Reversal cells

4. Which cells are responsible for synthesising and mineralising the extracellular matrix of bone?

   (A) Osteoclasts
   (B) Osteocytes
   (C) Osteoblasts
   (D) Reversal cells

5. Which type of bone remodelling is characterised by a partial or complete lack of bone matrix formation despite resorption?

   (A) Balanced remodelling
   (B) Imbalanced remodelling
   (C) Uncoupled/delayed coupling

6. The remodelling and reference stimulus for bone adaptation are obtained from intact and implanted FE models, respectively.

   (A) True
   (B) False

7. What typically causes age-related bone loss?

   (A) Increased osteoblast activity
   (B) Increased bone formation
   (C) Increased bone resorption and decreased bone formation
   (D) Increased osteocyte activity

8. According to SED-based remodelling rule, ____ is proportional to the difference in remodelling and reference stimulus.

   (A) Strain energy density
   (B) Change in bone mass
   (C) Change in bone density
   (D) Rate of bone loss or gain

9. Free surface area per unit volume of bone is higher for ______.

   (A) cortical bone
   (B) cancellous bone of density 0.4 g cm^{-3}
   (C) cancellous bone of density 1.2 g cm^{-3}
   (D) cancellous bone of density 0.6 g cm^{-3}

10. According to the orthotropic strain-based remodelling rule, trabecular orientation in bone is assumed to coincide with the direction of_____.

   (A) principal stresses
   (B) principal strain
   (C) normal strain

11. Bone does not react to small deviations in mechanical stimulus known as "dead zone".

   (A) True
   (B) False

12. In which scenario is the popular belief that cancellous bone turnover is higher than cortical bone not valid?

   (A) When comparing central cancellous and cortical bone
   (B) When comparing peripheral cancellous and cortical bone
   (C) When considering bone remodelling at the periosteal surface
   (D) When examining the Haversian system

13. Which factor is a biological determinant of bone turnover?

   (A) Bone surface to bone volume ratio (BS/BV)

(B) Bone formation rate to bone surface ratio (BFR/BS)
(C) Bone mineral density
(D) Cortical bone thickness

14. What role do osteocytes play in bone remodelling?
    (A) They secrete acid to dissolve bone matrix
    (B) They form new bone tissue by synthesising osteoid
    (C) They act as stress/strain sensors and regulate bone formation and resorption
    (D) They remove collagen remnants and prepare the bone surface for formation
15. What distinguishes secondary osteons from primary osteons in bone?
    (A) Secondary osteons form by the mineralisation of cartilage, while primary osteons replace existing bone
    (B) Secondary osteons form by replacing existing bone, while primary osteons form by the mineralisation of cartilage
    (C) Secondary osteons are associated with trabecular bone, while primary osteons are found in cortical bone.

Answer Key to MCQs
1. B; 2. B; 3. C; 4. C; 5. B; 6. B; 7. C; 8. D; 9. C; 10. A; 11. A; 12. B; 13. B; 14. C; 15. B.

References

Andersen TL, Abdelgawad ME, Kristensen HB, Hauge EM, Rolighed L, Bollerslev J, Kjærsgaard-Andersen P, Delaisse JM (2013) Understanding coupling between bone resorption and formation: are reversal cells the missing link? Am J Pathol 183(1):235–246

Avval PT, Klika V, Bougherara H (2014) Predicting bone remodelling in response to total hip arthroplasty: computational study using mechanobiochemical model. ASME J Biomech Eng 136(5):051002

Avval PT, Samiezadeh S, Klika V, Bougherara H (2015) Investigating stress shielding spanned by biomimetic polymer-composite vs. metallic hip stem: a computational study using mechano-biochemical model. J Mech Behav Biomed Mater 41:56–67

Baron R, Saffar JL (1977) A quantitative study of the effects of prolonged calcitonin treatment on alveolar bone remodelling in the golden hamster. Calcif Tissue Res 22(3):265–274

Beaupre GS, Orr TE, Carter DR (1990) An approach for time-dependent bone modeling and remodelling-application: a preliminary remodelling simulation. J Orthop Res 8(5):662–670

Bolamperti S, Villa I, Rubinacci A (2022) Bone remodelling: an operational process ensuring survival and bone mechanical competence. Bone Res 10(1):48

Bonfoh N, Novinyo E, Lipinski P (2011) Modeling of bone adaptative behavior based on cells activities. Biomech Model Mechanobiol 10(5):789–798

Bougherara H, Klika V, Marsík F, Marík IA, Yahia L (2010) New predictive model for monitoring bone remodelling. J Biomed Mater Res A 95(1):9–24

Cowin S, Hegedus D (1976) Bone remodelling i: theory of adaptive elasticity. J Elast 6:313–326

Cowin SC, Van Buskirk WC (1986) Thermodynamic restrictions on the elastic constants of bone. J Biomech 19:85–87

Dalstra M, Huiskes R, Odgaard A, van Erning L (1993) Mechanical and textural properties of pelvic trabecular bone. J Biomech 26:523–535

Delaisse JM, Andersen TL, Kristensen HB, Jensen PR, Andreasen CM, Søe K (2020) Re-thinking the bone remodelling cycle mechanism and the origin of bone loss. Bone 141:115628

Dempster DW (2017) Tethering formation to resorption: reversal revisited. J Bone Miner Res 32(7):1389–1390

Doblare M, García J (2002) Anisotropic bone remodelling model based on a continuum damage-repair theory. J Biomech 35(1):1–17

Franz-Odendaal TA, Hall BK, Witten PE (2006) Buried alive: how osteoblasts become osteocytes. Dev Dyn 235(1):176–190. https://doi.org/10.1002/dvdy.20603

Frost HM (1964) The laws of bone structure. Charles C. Thomas, Springfield, IL

Geraldes DM, Phillips A (2014) A comparative study of orthotropic and isotropic bone adaptation in the femur. Int J Numer Methods Biomed Eng 30(9):873–889

Geraldes DM, Modenese L, Phillips AT (2016) Consideration of multiple load cases is critical in modelling orthotropic bone adaptation in the femur. Biomech Model Mechanobiol 15:1029–1042

Hambli R (2014) Connecting mechanics and bone cell activities in the bone remodelling process: an integrated finite element modeling. Front Bioeng Biotechnol 2:6

Huiskes R, Weinans H, Grootenboer HJ, Dalstra M, Fudala B, Slooff J (1987) Adaptive bone-remodelling theory applied to prosthetic-design analysis. J Biomech 20(11–12):1135–1150

Huiskes R, Weinans H, van Rietbergen B (1992) The relationship between stress shielding and bone resorption around total hip stems and the effects of flexible materials. Clin Orthop Relat Res 274:124–134

Jacobs CR, Simo JC, Beaupré GS, Carter DR (1997) Adaptive bone remodelling incorporating simultaneous density and anisotropy considerations. J Biomech 30(6):603–613

Jang IG, Kim IY, Kwak BM (2009) Analogy of strain energy density based bone remodelling algorithm and structural topology optimization with compliance minimization. Trans ASME J Biomech Eng 131(1):011012

Judex S, Zernicke RF (2000) High-impact exercise and growing bone: relation between high strain rates and enhanced bone formation. J Appl Physiol 88:2183–2191

Klika V (2006) Mathematical and numerical analysis of differential equations of bone remodeling. Diploma thesis, Czech Technical University, Czech Republic

Komarova SV, Smith RJ, Dixon SJ, Sims SM, Wahl LM (2003) Mathematical model predicts a critical role for osteoclast autocrine regulation in the control of bone remodelling. Bone 33:206–215

Kroll MH (2000) Parathyroid hormone temporal effects on bone formation and resorption. Bull Math Biol 61(1):163–188

Liang B, Burley G, Lin S, Shi YC (2022) Osteoporosis pathogenesis and treatment: existing and emerging avenues. Cell Mol Biol Lett 27(1):72. https://doi.org/10.1186/s11658-022-00371-3

Lemaire V, Tobin FL, Greller LD, Cho CR, Suva LJ (2004) Modeling the interactions between osteoblast and osteoclast activities in bone remodelling. J Theor Biol 229:293–309

Levenston ME, Carter DR (1998) An energy dissipation-based model for damage stimulated bone adaptation. J Biomech 31(7):579–586

Manolagas SC (2000) Birth and death of bone cells: basic regulatory mechanisms and implications for the pathogenesis and treatment of osteoporosis. Endocr Rev 21(2):115–137

Martin RB (1984) Porosity and specific surface of bone. Crit Rev Biomed Eng 10(3):179–222

Mathai B, Gupta S (2020) The influence of loading configurations on numerical evaluation of failure mechanisms in an uncemented femoral prosthesis. Int J Numer Methods Biomed Eng 36(8):e3353

Mathai B, Dhara S, Gupta S (2021) Orthotropic bone remodelling around uncemented femoral implant: a comparison with isotropic formulation. Biomech Model Mechanobiol 20(3):1115–1134

Mathai B, Dhara S, Gupta S (2022) Bone remodelling in implanted proximal femur using topology optimization and parameterized cellular model. J Mech Behav Biomed Mater 125:104903

McNamara LM, Prendergast PJ (2007) Bone remodelling algorithms incorporating both strain and microdamage stimuli. J Biomech 40:1381–1391

Michaelis L, Menten ML (1913) Die Kinetik der Invertinwirkung. Biochem Z 49:333–369

Miller Z, Fuchs MB, Arcan M (2002) Trabecular bone adaptation with an orthotropic material model. J Biomech 35:247–256

Moroz A, Crane MC, Smith G, Wimpenny DI (2006) Phenomenological model of bone remodelling cycle containing osteocyte regulation loop. Biosystems 84(3):183–190

Parfitt AM (2002) Misconceptions (2): turnover is always higher in cancellous than in cortical bone. Bone 30(6):807–809

Prendergast PJ, Huiskes R (1995) Mathematical modelling of microdamage in bone remodelling and adaptation. In: Odgaard A, Weinans H (eds) Bone structure and remodeling. World Scientific Publishers, Singapore, pp 213–224

Prendergast PJ, Taylor D (1994) Prediction of bone adaptation using damage accumulation. J Biomech 27(8):1067–1076

Raggatt LJ, Partridge NC (2010) Cellular and molecular mechanisms of bone remodelling. J Biol Chem 285(33):25103–25108

Rattanakul C, Lenbury Y, Krishnamara N, Wollkind DJ (2003) Modeling of bone formation and resorption mediated by parathyroid hormone: response to estrogen/PTH therapy. Biosystems 70:55–72

Rauch F, Travers R, Glorieux FH (2007) Intracortical remodeling during human bone development—A histomorphometric study. Bone 40(2):274–280. https://doi.org/10.1016/j.bone.2006.09.012

Roux W (1881) Der ziichtende kampf der teile, oder die 'tei/auslese' imorganismus (theorie der 'finktionellen unpassung'). Wilhelm Engelmann, Leipzig

Ryser MD, Nigam N, Komarova SV (2009) Mathematical modeling of spatio-temporal dynamics of a single bone multicellular unit. J Bone Miner Res 24(5):860–870. https://doi.org/10.1359/jbmr.081229

Saviour CM, Mathai B, Gupta S (2024) Mechanobiochemical bone remodelling around an uncemented acetabular component: influence of bone orthotropy. Med Biol Eng Comput 62(6):1717–1732

Schaffler MB, Kennedy OD (2012) Osteocyte signaling in bone. Curr Osteoporos Rep 10(2):118–125

Skedros JG, Baucom SL (2007) Mathematical analysis of trabecular trajectories in apparent trajectorial structures: the unfortunate historical emphasis on the human proximal femur. J Theor Biol 244:15–45

Van Rietbergen B, Huiskes R, Weinans H, Sumner D, Turner T, Galante J (1993) The mechanism of bone remodelling and resorption around press-fitted THA stems. J Biomech 26:369–382

Weinans H, Huiskes R, van Reitbergen B, Sumner DR, Turner TM, Galante JO (1993) Adaptive bone remodelling around bonded noncemented total hip arthroplasty: a comparison between animal experiments and computer simulation. J Orthop Res 11:500–513

Wolff J (1986) The law of bone remodelling (trans by Maquet P and Furlong R). Springer, New York

Zoetis T, Tassinari MS, Bagi C, Walthall K, Hurtt ME (2003) Species comparison of postnatal bone growth and development. Birth Defects Res Part B: Dev Reprod Toxicol 68(2):86–110

Chapter 10
Bone Fracture Healing

In this chapter, we will explore the details of bone fracture healing. The causes of bone fractures can be broadly categorised into two types: pathological and traumatic. Some of the causes of pathological fractures are osteoporosis, osteonecrosis, and bone tumours. Conversely, traumatic fractures arise from events like vehicle accidents or falls from heights. In the later part of the chapter, we focus on the mechanobiology of bone fracture healing. Mechanobiology is an emerging field of science at the interface of biology, engineering, and physics. It focuses on how cells detect and respond to mechanical stimuli (Lee et al. 2011). The ability of bone tissue to sense and convert external mechanical stimuli into a biochemical response, which eventually alters the cell phenotype and function of the cell, is known as mechanotransduction.

10.1 Types of Bone Fractures

Bone fractures can be categorised into different types based on several factors, such as the cause, location, and pattern of the fracture (Fig. 10.1). The following are the common types of bone fractures:

(a) **Closed fracture** (Simple fracture): In a closed fracture, the fractured bone does not penetrate the skin and therefore, the skin remains intact (Fig. 10.1a).
(b) **Open fracture** (Compound fracture): In an open fracture, the fractured fragments of bone pierce through the skin and get exposed to the external environment. This type of fracture carries a higher risk of infection due to open wound (Fig. 10.1b).

In addition to the open and closed fractures, the fractured bones can be further classified into the following types:

S. Gupta et al., *Biomechanics of Joints and Implants*,
https://doi.org/10.1007/978-981-96-0586-6_10

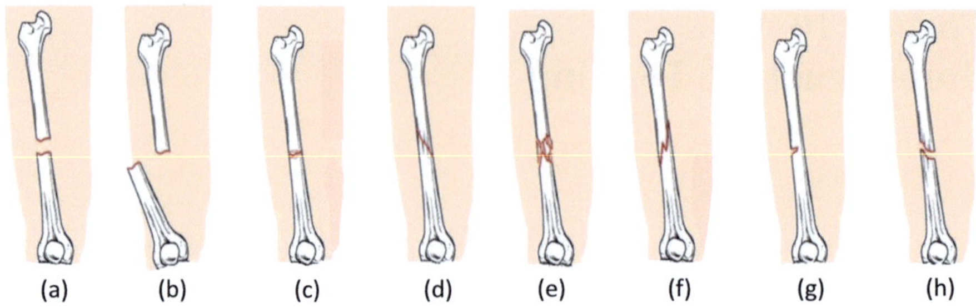

Fig. 10.1 Types of bone fractures; **a** closed, **b** open, **c** transverse, **d** spiral, **e** comminuted, **f** impacted, **g** greenstick, **h** oblique

(c) **Transverse fracture**: A transverse fracture occurs when the fracture line is perpendicular to the long axis of bone and this fracture results in two bone fragments (Fig. 10.1c).
(d) **Spiral fracture**: Spiral fractures result from a twisting or torsional load applied to the bone. The fracture line makes a spiral about the axis of the bone (Fig. 10.1d).
(e) **Comminuted fracture**: In a comminuted fracture, the bone is fractured into multiple pieces or fragments. This type of fracture causes a void or gap which needs to be filled by bone grafts. Therefore, treating this type of fracture can be more complicated (Fig. 10.1e).
(f) **Impacted fracture**: An impacted fracture occurs when the bone fragments are driven into each other, causing compression and shortening of the bone (Fig. 10.1f).
(g) **Greenstick fracture**: Greenstick fractures are more common in children. This type of fracture occurs when the bone is partially cracked, instead of breaking into fragments (Fig. 10.1g).
(h) **Oblique fracture**: An oblique fracture is characterised by a diagonal fracture across the bone. It typically occurs at an angle to the long axis of the bone (Fig. 10.1h).

10.2 Stages of Bone Fracture Healing

A bone fracture entails a breach in the structural continuity of the bone cortex, with a degree of injury to the adjacent soft tissues. Bone fracture healing is an intricate, sequential regenerative process that aims to restore the damaged bone to its original (pre-injury) state and cellular composition. The overall process of bone fracture healing can be subdivided into the following four stages: inflammation and hematoma formation, subsequent development of a soft callus, transformation into a hard callus, and bone remodelling (Fig. 10.2).

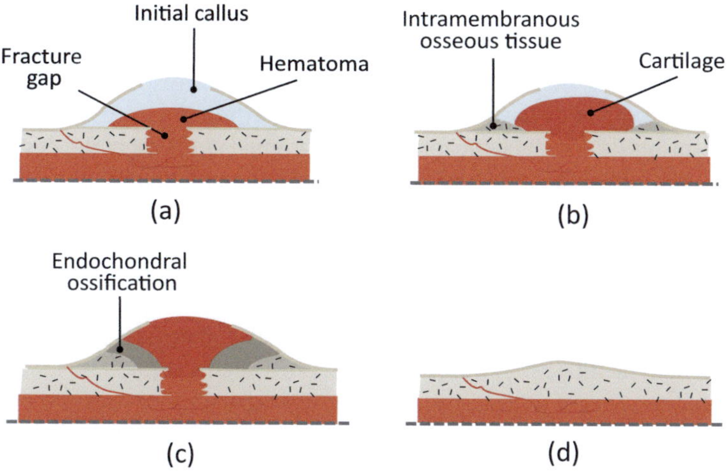

Fig. 10.2 Stages of fracture healing; **a** hematoma formation **b** granular tissue formation **c** formation of soft and hard callus **d** remodelling (formation of lamellar/compact bone)

Inflammation and hematoma formation: The initial stage of fracture healing is the inflammatory phase, which lasts for several days. When a bone is fractured, the body initiates an immediate response. This response aims to minimise further damage and prepare the fractured site for healing. Blood vessels around the fracture constrict, reducing blood flow and limiting the bleeding. A blood clot or hematoma forms at the fracture site. This clot serves as a temporary seal to stop bleeding and contains various immune cells and growth factors necessary for healing. Inflammatory cells and white blood cells migrate to the injured area. These cells play a crucial role in clearing debris, bacteria, and damaged tissue. Following the inflammatory phase, fibroblasts, a type of cell found in connective tissue, start to produce collagen. Collagen is a structural protein that forms the framework for new tissue formation. Granulation tissue is a new connective tissue with microscopic blood vessels that are developed at the wound site during the healing process. Granulation tissue acts as a temporary scaffold for the subsequent stages of healing. Blood vessels also begin to grow into the granulation tissue to provide nutrients. This stage spans approximately 2 weeks.

Soft callus formation: Fibrous tissue and new cartilage begin to form along with revascularisation. The soft callus is mainly composed of this cartilaginous tissue. It is softer and more flexible than the mature bone tissue but serves to hold the fractured bone segments together. This stage occurs during the second to fifth weeks, post fracture.

Hard callus formation: The soft callus formation is followed by the hard callus formation phase, where woven bone begins to form. Woven bone is a less organised type of bone tissue. Osteoblasts deposit bone matrix (osteoid) directly within the soft callus. The hard callus is stronger than the soft callus and provides more stability to the fractured bone. At this stage, the bone has made significant progress towards healing. This stage occurs during the sixth to twelfth weeks, post fracture.

Remodelling phase: The final phase of fracture healing is the remodelling phase, which can extend for years. This phase also involves the replacement of woven bone with mature lamellar bone. Osteoclasts, specialised cells responsible for breaking down and resorbing bone tissue, become active. They remove excess callus material and shape the newly formed bone to match the original structure of the bone. During this phase, the bone gradually regains its original strength and shape. Unlike other stages, bone remodelling is a slow process that may span for a few years.

The success of the fracture healing process and the associated time duration depends on various factors like the type and location of the fracture, age and health of the patient, and the procedure used for fracture fixation. In many cases, surgical interventions using intramedullary nails, plates, and screws, may be required to stabilise the fracture and facilitate the healing process.

10.3 Cellular Mechanisms Involved in Fracture Healing

Several biological events involving the migration, proliferation, and differentiation of various cell types play pivotal roles in the overall bone fracture healing process. These events are tightly orchestrated and occur in a sequential manner. A brief summary of these key events is provided below.

Migration: Cell migration is characterised as the movement of cells in response to various signals, such as chemical and mechanical (Fig. 10.3). For example, immediately after bone fracture, immune cells, primarily neutrophils and macrophages, migrate to the fracture site. Initially, the neutrophils arrive to address the immediate infection risk, while the macrophages play a longer-term role in tissue clean-up and signalling. The Mesenchymal Stem Cells (MSCs) migrate to the fracture site in response to chemotactic signals produced by the immune cells and local microenvironment. MSCs are multipotent cells found in bone marrow and other tissues. These cells are attracted to the fracture site by the signalling molecules released during the inflammatory phase.

Chemotaxis and haptotaxis are cellular behaviours related to cell migration, and they often involve directional movement. Chemotaxis involves the directed movement of cells in response to chemical gradients or concentration gradients of specific molecules, known as chemoattractants (for positive chemotaxis) or chemorepellents (for negative chemotaxis). Cells exhibiting chemotaxis actively move towards or away from the source of the chemical gradient, guided by the concentration of the signalling molecules. Angiogenesis, a form of chemotaxis involving endothelial cells, is vital for delivering nutrients and oxygen to the healing tissue. Vascular endothelial growth factor (VEGF) acts as a chemoattractant, guiding endothelial cells to form new blood vessels within a fracture site.

Haptotaxis, on the other hand, involves the directed movement of cells in response to gradients in adhesive properties, rather than chemical gradients. Within the extracellular matrix (ECM) of the wounded tissue, there may be variations in the composition and density of adhesive molecules, such as fibronectin and collagen, as compared

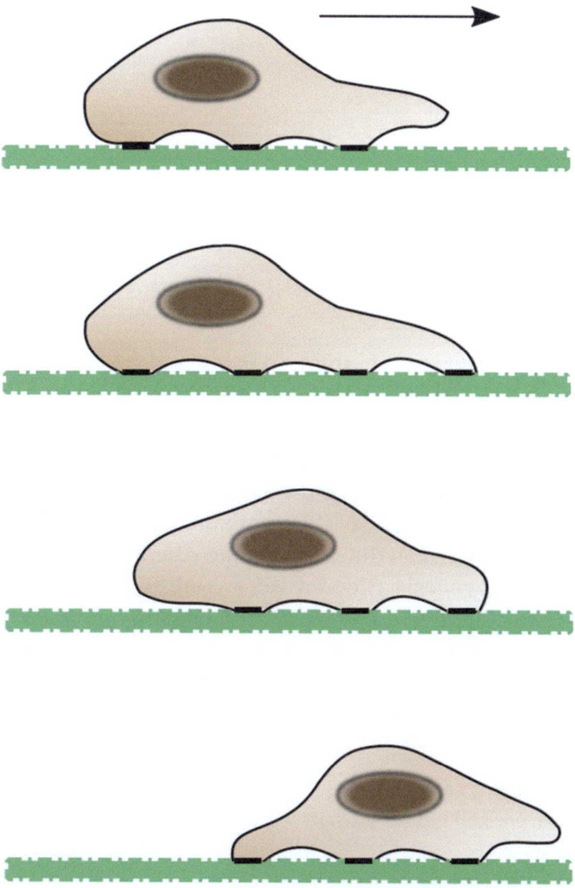

Fig. 10.3 The process of cell migration. Arrow indicates the direction of migration

to that of a healthy tissue. These differences create gradients in adhesive properties. When fibroblasts encounter regions of higher adhesive molecule density (increased adhesion), they may slow down their motility, adhere more firmly to the ECM, and become more active in matrix production.

Proliferation: Cell proliferation is a process by which the parent cell undergoes division, resulting in the production of two or more daughter cells (Fig. 10.4). For example, fibroblasts proliferate during the inflammatory phase and the early stages of soft callus formation. They are involved in producing collagen, a critical component of the extracellular matrix. Chondrocytes, responsible for cartilage formation, proliferate during soft callus formation. They multiply and produce cartilage matrix. Osteoblasts, bone-forming cells, proliferate during the hard callus formation phase. They multiply and actively deposit bone matrix (osteoid).

Differentiation: Cell differentiation is the process by which a cell changes its cell type to another, often involving changes in gene expression, cellular morphology and function (Fig. 10.5). Fibroblastic differentiation refers to the process by which MSCs

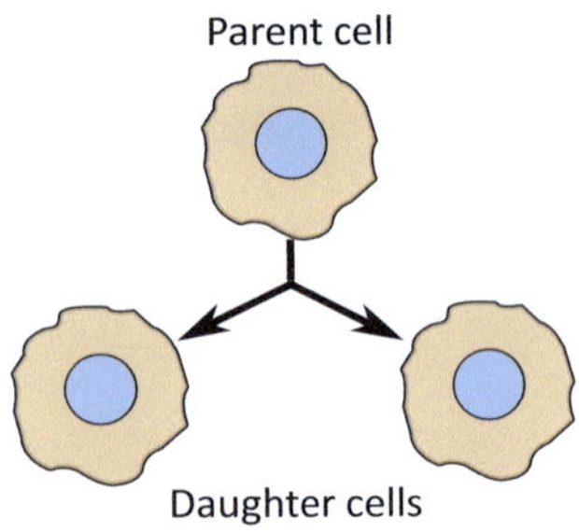

Fig. 10.4 Proliferation (cell division) of parent cell into daughter cells

differentiate into fibroblasts. Fibroblasts are associated with the formation of fibrous tissue (Fig. 10.6). In chondrogenic differentiation, MSCs differentiate into chondrocytes, which facilitates cartilage matrix formation. Whereas in osteogenic differentiation, osteoblasts are derived from the MSCs. These osteoblasts are responsible for the formation of woven bone during the hard callus formation phase. Additionally, fibroblast is capable of differentiating into chondrocyte and osteoblast. Whereas, the chondrocytes can differentiate into osteoblast (Fig. 10.6).

Angiogenesis: Blood vessel growth, or angiogenesis, is a critical event throughout the fracture healing process. Angiogenesis involves the migration and proliferation of endothelial cells to form new blood vessels. Adequate blood supply (vascularity) is essential for delivering nutrients, oxygen, and immune cells to the healing tissue. It supports the metabolic needs of the proliferating and differentiating cells.

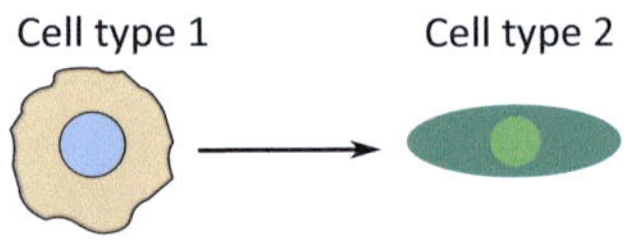

Fig. 10.5 Differentiation from one cell type to another

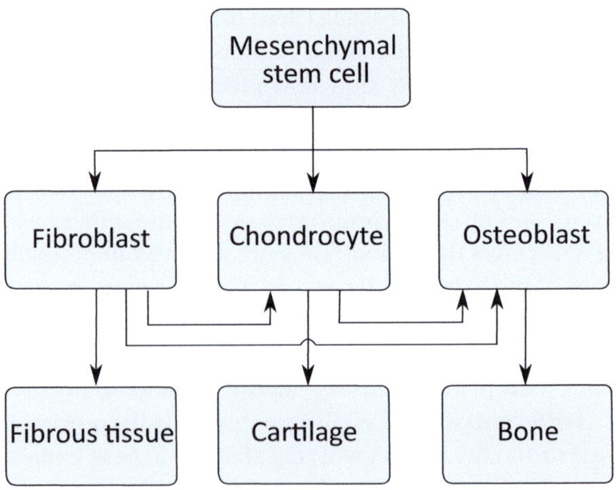

Fig. 10.6 A flowchart of the tissue differentiation process from the mesenchymal stem cells

10.4 MSC Differentiation Pathway

MSCs are multipotent cells found in bone marrow and other tissues. These cells are attracted to the fracture site by the signalling molecules released during the inflammatory phase. These recruited MSCs are undifferentiated cells capable of differentiating into various cell types, including osteoblasts, chondrocytes, fibroblasts, myoblasts, and stromal cells, as needed for the healing process. Depending on the signals they receive from their microenvironment, MSCs differentiate into specific cell phenotypes to initiate tissue repair and regeneration. These differentiations follow a specific lineage described as mesengenesis (Fig. 10.7). Osteoblasts produce the bone matrix and initiate the formation of new bone tissue. Chondrocytes produce a cartilage matrix, which helps in stabilising the fracture site. Myoblasts are precursors to muscle cells. In cases of muscle repair, MSCs may differentiate into myoblasts to regenerate muscle tissue. Stromal cells are supportive cells found in the bone marrow. They contribute to the formation of bone marrow environment and support haematopoiesis (formation of blood cells). Fibroblasts are known for their ability to synthesise and secrete components of the Extracellular matrix (ECM), including collagen, fibronectin, and proteoglycans. These ECM components provide the structural framework for tissues and facilitate tissue remodelling and repair. The precursor cells of tendon and ligaments are known as tendon and ligament fibroblasts, respectively.

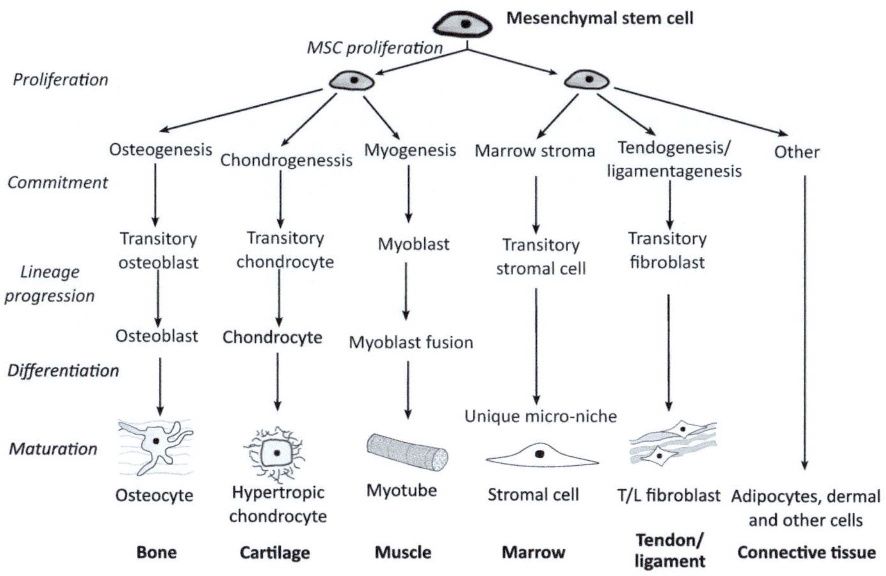

Fig. 10.7 The mesengenic process indicating the MSC differentiation pathways (adapted from Caplan 2005)

The MSCs create a supportive microenvironment within the fracture site by releasing various growth factors, including transforming growth factor-beta (TGF-β), bone morphogenetic proteins (BMPs), vascular endothelial growth factor (VEGF), and fibroblast growth factors (FGFs), among others. These growth factors have specific roles in the healing process. TGF-β and BMPs induce the differentiation of MSCs into osteoblasts, promoting bone formation. VEGF stimulates angiogenesis (formation of new blood vessels), ensuring a sufficient blood supply to the healing tissue. FGFs play a vital role in tissue repair, including angiogenesis and recruitment of other cells involved in the healing process. In bone healing, chemotaxis and haptotaxis play essential roles in guiding various cell types to the fracture site and promoting the formation and remodelling of bone tissue.

10.5 Primary and Secondary Bone Healing

Primary bone healing, also known as direct bone healing or anatomical healing, occurs when fractured bone ends are brought into close contact without any notable gap, and there is no formation of a visible callus during the healing process. This type of healing is typically seen in cases where a fracture is adequately fixed through precise reduction and rigid immobilisation.

Primary bone healing is commonly observed in surgically managed fractures that are anatomically reduced and rigidly stabilised, using common fracture fixation techniques discussed later in Sect. 10.9. The mechanical environment created by rigid fixation is conducive to direct bone union, without the formation of a callus. In primary bone healing, the fractured bone is aligned nearly perfectly, with only a minimal gap between the broken ends. This close alignment allows the ends of the fragments to fuse together, restoring mechanical continuity. Primary healing can either occur through gap healing or contact healing (Marsell and Einhorn 2011). However, both these methods require alignment of the fractured bone, which ensures that the original anatomical structure is restored. In gap healing, the gap must be less than 1 mm (Marsell and Einhorn 2011). Whereas in contact healing, the gap must be less than 0.01 mm (Marsell and Einhorn 2011). The rigid fixation minimises the micromotion, thereby allowing for direct bone-to-bone contact and leading to direct osteonal healing, without the need for callus formation (Schenk and Willenegger 1977; Claes et al. 1997).

Secondary bone healing, also known as indirect bone healing or callus formation, is a process by which a fractured bone heals through the formation of callus and subsequent bone remodelling. This type of healing occurs when there is a significant gap between the fractured bone ends, and the goal is to promote callus formation to bridge the gap. The characteristics of the fracture or the difficulty in achieving precise anatomical alignment often lead to a gap between the fractured bone ends. The fracture callus initially consists of granulation tissue, cartilage, and woven bone, and serves as a temporary bridge stabilising the fracture. Over time, the callus undergoes a remodelling process where the cartilage is replaced by mature bone tissue through endochondral ossification. This results in eventual restoration of the normal structure of bone. The mechanical environment at the fracture site in secondary bone healing is characterised by some degree of micromotion. This dynamic environment promotes the formation and remodelling of the callus. Secondary bone healing is commonly observed in fractures that are treated conservatively or those that cannot be anatomically reduced and rigidly stabilised.

10.6 Need for Immobilisation of Fracture Site

Immobilisation of the fracture site is a key aspect of fracture management since it plays a crucial role in bone fracture healing. When a bone is fractured, the continuity of the bone is disrupted, and the bone fragments may be displaced. Immobilisation, through the use of casts, braces, or splints, helps to hold the bone fragments together and restores proper anatomical alignment. This is particularly important for fractures that are displaced or unstable. Osteoblasts require a relatively stable mechanical environment to produce and deposit new bone matrix, a process known as osteogenesis. Excessive micromotion can disrupt the early stages of healing, delay the formation of a bridging callus, and potentially result in non-union (failure of the fracture to heal).

Immobilisation is particularly important for endochondral ossification. In this process, cartilage initially forms at the fracture site. Immobilisation is necessary to allow for the gradual conversion of cartilage to bone. Proper immobilisation reduces the risk of complications associated with bone fractures, such as malunion (improper alignment), non-union, and infection. It is important to note that the specific method of immobilisation and treatment, surgical or conservative, depends on various factors. These include the type and location of the fracture, the patient's age and overall health, and the intended outcome of the treatment. These factors need to be carefully assessed by an orthopaedic surgeon in order to determine the most appropriate approach for fracture treatment.

10.7 Factors Affecting Bone Fracture Healing

Bone fracture healing is a complex process influenced by various factors, both intrinsic (related to the patient's health) and extrinsic (related to treatment and environmental factors). Several key factors affect the bone fracture healing process. Younger individuals generally have a higher potential for rapid bone healing owing to their more robust regenerative capabilities. Older individuals experience slower healing and may be at a higher risk of complications. Adequate blood supply to the fracture site is crucial for providing oxygen and nutrients necessary for healing. Fractures with compromised blood supply (e.g., due to vascular damage) may experience delayed healing. Infections at or near the fracture site can disrupt the healing process and may require additional treatment. Proper fixation and immobilisation of the fractured bone are critical for successful healing. Stable fixation reduces movement at the fracture site, permitting improved bone-to-bone contact. The type of fracture (e.g., simple, comminuted, open, or closed) can significantly impact healing. The location of the fracture within the body, as well as its proximity to major blood vessels and nerves, can affect healing rates.

10.8 Common Fracture Fixation Techniques

Fracture fixation is a surgical procedure used to stabilise fractured bones, facilitating proper alignment and healing. There are several types of fracture fixation methods, and the choice of a method depends on factors such as the type and location of the fracture, the age and health of the patient, and the goal of the treatment. The fracture fixation is commonly classified as follows:

Open reduction: Open reduction involves making a larger incision or creating a surgical exposure to directly access and visualize the fractured bone. This approach allows for precise reduction, fixation, and stabilisation of the fracture.

- Plates and screws: Plates made of metal (usually titanium or stainless steel) are attached to the surface of the fractured bone using screws (Fig. 10.8a). This method provides a stable fixation and precise alignment.

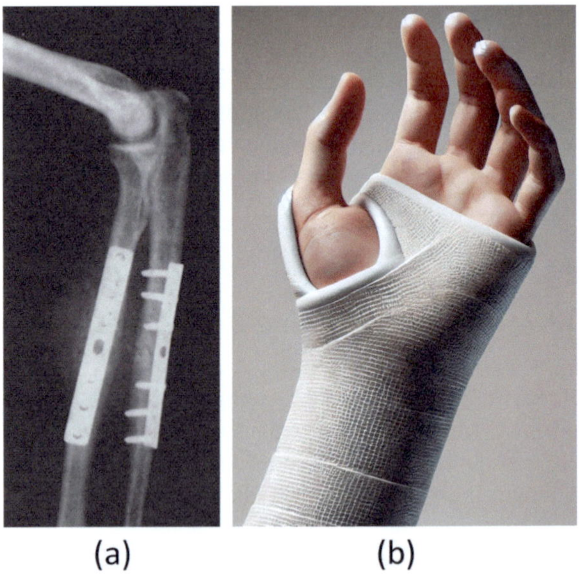

Fig. 10.8 Types of fracture fixation techniques: **a** open reduction of tibia and fibula using plates and screws, **b** closed reduction of wrist using plaster cast

- Intramedullary nails: Intramedullary nails, also known as rods or nails, are inserted into the hollow centre (medullary canal) of long bones. They are used for fractures of long bones, like the femur and the tibia.
- Intramedullary pins: Intramedullary pins are thin metal rods that are inserted into the medullary canal to provide stabilisation for small bone fractures, such as those in the fingers or toes.
- Kirschner wires (K-wires): K-wires are thin, smooth metal wires that can be used to stabilise fractures temporarily or in conjunction with other fixation methods. K-wires are frequently used to stabilise fractures, especially in cases where the fracture is small, simple, or involves the extremities (such as finger, hand, or toe fractures).

Closed Reduction: Closed reduction method is minimally invasive and does not require extensive surgical exposure or large incisions. These methods rely on percutaneous or small incisions to insert implants for fracture stabilisation.

- Percutaneous Fixation: This minimally invasive technique involves the insertion of pins or wires percutaneously (through the skin) to stabilise fractures, often in the extremities. This approach offers several advantages, including reduced tissue damage, shorter recovery times, and lower infection risk as compared to traditional open surgical procedures.
- Functional Bracing: Certain fractures, especially those that are stable and do not require surgical intervention, can be treated with orthopaedic braces or splints. These devices provide external support and immobilisation to promote healing.
- Casting: Plaster casts are made by wrapping the injured limb in layers of wet plaster (Fig. 10.8b). They harden over time and provide support and immobilisation for the healing bone. Fiberglass casts, which are lighter and more durable than plaster casts, are also used for immobilisation.
- External Compression: Some fractures, such as certain wrist fractures, can be treated with external compression devices like compression sleeves or wraps. These devices help to stabilise the fracture by applying pressure to the injured area.

In certain situations, a combination of open and closed fixation methods might be employed to ensure the best possible stabilisation and alignment, particularly for complex fractures or those located near the joints.

10.9 Secondary Fracture Healing Models

The secondary fracture healing models is classified into mechanoregulatory, bioregulatory, mechanobioregulatory models, based on the type of stimulus. Let us explore these models individually in the following subsections.

10.9.1 Mechanoregulatory Models

Mechanoregulatory models focused on predicting bone formation based on local mechanical stimuli. Several mechanoregulatory models were proposed for various cases, ranging from normal fracture healing to more complex processes like distraction osteogenesis and bone regeneration. In this section, we will explore the theoretical foundation of key models.

Pauwels' theoretical framework was a crucial advancement in comprehending how mechanical stimuli influence tissue differentiation (Pauwels 1960). The study found that MSCs experience stress and deformation owing to physical factors, and these mechanical stimuli are instrumental in regulating the differentiation of these cells into different types of tissue. It was proposed that strain stimulates the formation of fibrous tissue (Pauwels 1960). This suggests that mechanical forces that induce strain can promote the development of collagenous fibres and other fibrous components. Pauwels (1960) also suggested that hydrostatic compression stimulates the formation of cartilage (Fig. 10.9). This implied that specific mechanical conditions, possibly related to pressure, are conducive to cartilage development. However, this theory did not provide specific details on the limiting values of strain and stress (Pauwels 1960).

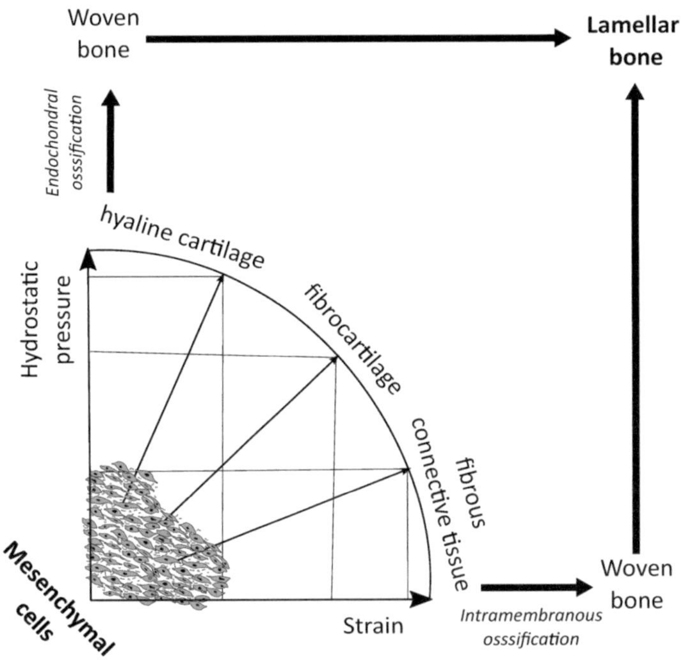

Fig. 10.9 Schematic representation of mechanoregulatory algorithm proposed by Pauwels (1960). Adapted from Pauwels (1960)

Building on Pauwels' work, the study by Perren and Cordey (1980) introduced the concept of interfragmentary strain (IFS). This measure of strain was proposed to better understand how the mechanical environment influences bone fracture healing. They stated that a tissue is formed or remains between the fractured bone ends, only if it can tolerate the local mechanical strain owing to the interfragmentary movement. This concept was useful for evaluating treatment strategies for fractures. It is important to acknowledge, however, that IFS has limitations, especially regarding its application to bone fracture healing. It simplified the complex and heterogeneous nature of tissue structure and mechanics in fracture callus, which includes different tissue types. Subsequently, Carter et al. (1988) proposed two-dimensional models based on the theory of Pauwel to explain how mechanical stimuli influence tissue differentiation during the healing process (Fig. 10.10). Thereafter, they developed a more generalized mechanotransduction model, which provides a broader framework for understanding how different tissue types were formed in response to mechanical stimuli. Fibrous tissue tends to form in regions where the tissue experienced high tensile strain, specifically above the tension line. Areas characterised by the hydrostatic compressive stress, which are to the left of the pressure line, promote chondrogenesis. Bone formation occurred in areas with low hydrostatic compressive stress, particularly to the right of the pressure line. It was, therefore, concluded that mechanical conditions characterised by low hydrostatic compression are favourable for bone formation (Carter et al. 1988).

It is important to note that the model proposed by Carter et al. (1988) does not specify exact limits for the tension or pressure lines. Instead, it provided a conceptual framework for understanding how different mechanical conditions influence tissue differentiation. This generalized mechanotransduction model has been used to study various aspects of tissue differentiation, such as pseudoarthrosis formation, asymmetric fractures, and distraction osteogenesis (Carter et al. 1988). Using a two-dimensional finite element model, Blenman et al. (1989) analysed the fracture callus at different ossification phases, including the initial ossification phase, the newly ossified callus tissue phase, and a near-completion phase of healing. The analysis was based on the mechanoregulatory principle developed by Carter et al. (1988)

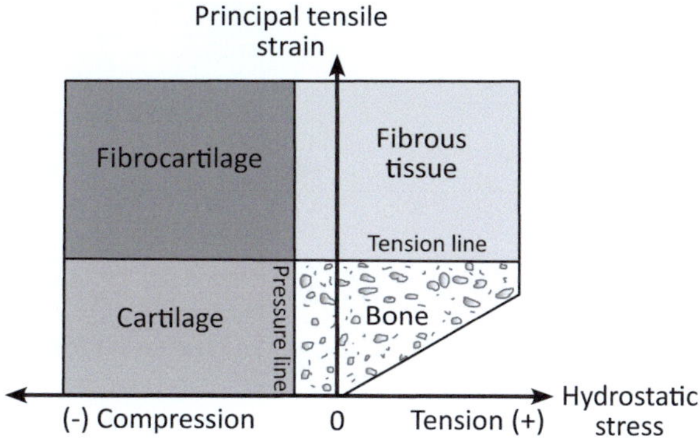

Fig. 10.10 Bone fracture healing algorithm proposed by Carter et al. (1988) (adapted from Carter et al. 1988)

that investigated the influence of mechanical factors on tissue differentiation during fracture healing.

The study of Claes and Heigele (1999) introduced limiting values of mechanical stimuli to regulate new tissue formation (Fig. 10.11). Using finite element analysis, they determined the specific thresholds. Following are the key findings of their study (Claes and Heigele 1999).

- When local strain is less than 5% and hydrostatic pressure is between ±0.15 MPa, intramembranous bone formation occurs. This means that under these mechanical conditions, bone forms directly without a cartilaginous intermediate stage. This study, however, mentioned local strain without being specific about it. Additionally, we assume that they might have considered the maximum principal strain to represent the local strain.
- When strain is smaller than 15% and hydrostatic pressure is greater than −0.15 MPa, endochondral ossification occurs. Endochondral ossification involves the formation of cartilage, followed by its replacement with bone tissue.
- All other conditions, not meeting the above criteria, result in the formation of fibrous tissue or fibrocartilage.

The theory of bone formation was based on the idea that new bone only forms on the surface of existing bone or calcified tissue, emphasising the importance of mechanical factors in guiding tissue differentiation during the healing process. This marked a significant step towards understanding the role of mechanical stimuli on specific outcomes of tissue formation during fracture healing.

The aforementioned models of fracture healing considered the tissues to be solid elastic materials. However, tissues consist of a solid phase and a fluid phase. Biphasic finite element models consider the interaction between these two phases. The first biphasic finite element model of fracture healing was reported by Prendergast et al. (1997). This model utilised the shear strain in the solid phase and the fluid velocity

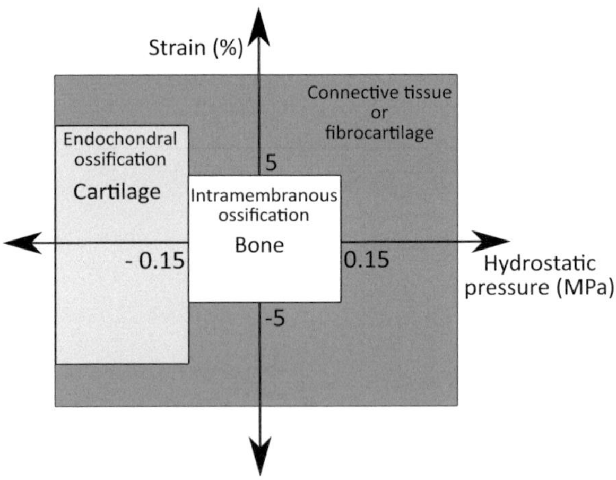

Fig. 10.11 Mechanoregulatory algorithm proposed by Claes and Heigele (1999). Adapted from Claes and Heigele (1999)

in the interstitial fluid phase as the biophysical stimuli for tissue differentiation (Fig. 10.12). The model predicted that fibrous tissue would develop under high values of both solid shear strain and fluid velocity, whereas cartilage tissue would form under intermediate values. Only low values promoted bone formation. However, none of the above mentioned studies accounted for the evolutionary bone regeneration. Lacroix et al. (2002) investigated the time-course of bone regeneration, by incorporating the net effect of MSC migration and proliferation into a biphasic poroelastic finite element model (Fig. 10.13).

This net effect was modelled using the following diffusion equation:

$$\frac{\mathrm{d}n}{\mathrm{d}t} = D\nabla^2 n \tag{10.1}$$

where D is a diffusion coefficient in m^2/day and n represents the concentration of the MSC. The flowchart for predicting tissue differentiation is presented in Fig. 10.13. The biophysical stimuli were evaluated using a poroelastic FE simulation, which is known to regulate cell differentiation. Once the new tissue phenotype is determined, the material properties are updated using the rule of mixtures (Eq. 10.2).

$$E_{n+1} = \left(\frac{c_{\max} - c_{\text{tissue}}}{c_{\max}}\right)E_{\text{granulation}} + \left(\frac{c_{\text{tissue}}}{c_{\max}}\right)E_{\text{tissue}} \tag{10.2}$$

where $c_{\max}$ and c_{tissue} are the maximum cell concentration, equal to one, and the actual concentration of cells, respectively. E_{tissue} and $E_{\text{granulation}}$ are the material properties of newly formed tissue and granulation tissue, respectively. A temporal smoothing procedure was employed to eliminate numerical instability that mightoccur due to sudden changes in material properties. Lacroix et al. (2002) found that the origin site of MSCs considerably influenced the healing pattern. They also observed that

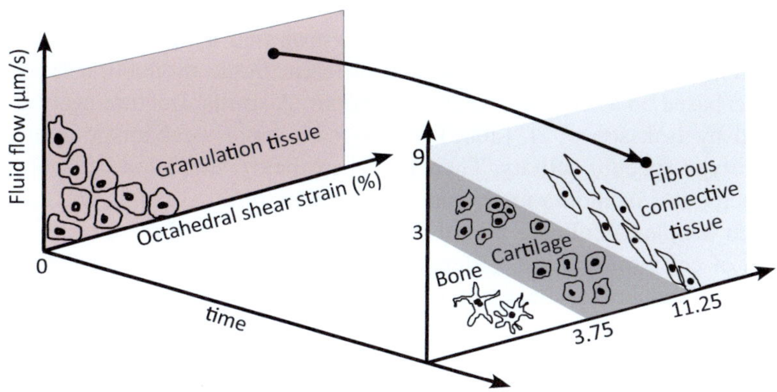

Fig. 10.12 Mechanoregulatory algorithm based on biphasic stimulus proposed by Prendergast et al. (1997). Adapted from Prendergast et al. (1997)

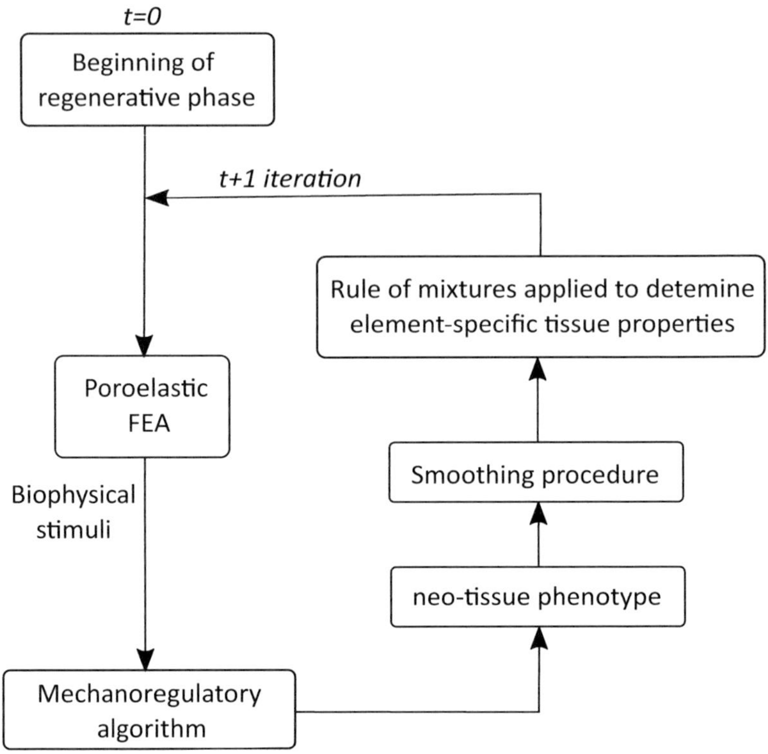

Fig. 10.13 Flow chart for predicting evolutionary tissue differentiation (adapted from Lacroix et al. 2002). Here the biophysical stimuli are obtained from a poroelastic finite element analysis (FEA)

increased gap size resulted in slower healing and increased interfragmentary strain promoted connective tissue formation (Lacroix et al. 2002).

Another study by Andreykiv et al. (2008) proposed a fracture healing model that explicitly modelled proliferation and different tissue production rates. Their model was based on a set of differential equations. A similar fracture healing model developed by Isaksson et al. (2008), employed a more mechanistic approach to cell-phenotype specific activity. Their model explicitly simulated cell-phenotype-specific activities, including migration, proliferation, differentiation, and apoptosis, with seven coupled nonlinear partial differential equations (Eqs. 10.3–10.8). The evolution of cell proliferation, differentiation, and apoptosis was regulated by the magnitudes of octahedral shear strain and fluid velocity (Prendergast et al. 1997).

$$\frac{\partial c_j}{\partial t} = \nabla D_i(c)\nabla c_i + f_i^{PR}(\psi)c_i\left(1 - \frac{c_i}{c_{\text{space}}}\right) - F_i^D(\psi, c) - f_i^{AP}(\psi)c_i \qquad (10.3)$$

where t represents time and c_i is the concentration of cell type i, where $i = 1$ (MSC), 2 (Fibroblast, FB), 3 (Chondrocyte, CC), or 4 (Osteoblast, OB). The four parts of the equation describe migration, proliferation, differentiation, and apoptosis of each of the cell types i. A diffusive process was used to model cell migration. D_i is the concentration-dependent diffusivity for cell type i. The proliferation and apoptosis terms are represented by f_{PR} and f_{AP}, respectively. These are either turned 'off' or 'on' depending on the mechanical stimulation (Ψ) (Prendergast et al. 1997). The differentiation term, F_i^D, is calculated for each cell type, i, and is dependent on the mechanical stimulation (Eqs. 10.4 and 10.5).

$$\text{if } \psi = i, \quad F_i^D = \sum_{k=1,k\neq i}^{4} \left(-f_k^D c_k\right) \tag{10.4}$$

$$\text{if } \psi \neq i, \quad F_i^D = f_i^D c_i \tag{10.5}$$

where f_k^D are the rates for each cell type k. Thus, if cell phenotype i is stimulated mechanically, cells of other phenotypes except i, will differentiate into cell phenotype i. The value of F^D is dependent on the individual differentiation potential of each cell type.

The maximum rates of f_{PR}, f_D, and f_{AP} for each cell type are shown in Table 10.1. When $\psi = i$, then proliferation of cell type i is 'on' (Table 10.1), and if $\psi \neq i$, then proliferation of cell type i is 'off'. The c_{space} is the 'available space' in the element, and is calculated using Eq. 10.6.

$$c_{\text{space}} = c - \sum_{\text{i}=1}^{4} c_i \tag{10.6}$$

These cells can produce skeletal tissue types, i.e. fibrous tissue (FT), cartilage (C), and bone (B), as described in Eq. 10.7.

$$\frac{\partial m_j}{\partial t} = f_j^{PM}(\Psi)C_{j-3}\left(1 - \frac{m_j}{m_{\text{space}}}\right) - f_j^{DM}(\Psi)C_{j-3}m_j \tag{10.7}$$

where m_j represents the concentration of matrix type j, here $j = 5$ (FT), 6 (C), 7 (B), and $j - 3$ is the corresponding cell type to the matrix type j, e.g., chondrocyte to cartilage ($i = j - 3$). The equation can be split into production and degradation

Table 10.1 Various model parameters corresponding to different cellular activities

Cell	D (mm^2/day)	f^{PR} (day^{-1})	f^D (day^{-1})	f^{AP} (day^{-1})
MSC	0.65	0.60	0.30	0.05
FB	0.50	0.55	0.20	0.05
CC	0.00	0.20	0.10	0.10
OB	0.20	0.30	0.15	0.15

Source Isaksson et al. (2008)

of matrix, where f^{PM} and f^{DM} regulate the rates of production and degradation of matrix, respectively. Matrix production and degradation are turned 'on' or 'off' depending on ψ. For example, if Ψ is 3, stimulating chondrogenic activity, that also results in maximum matrix production of cartilage tissue, and minimal degradation of cartilage tissue.

The maximum values of f^{PM} and f^{DM} are shown in Table 10.2. The m_{space} denotes the 'available space' in the element, in Eq. 10.8.

$$m_{\text{space}} = m_{\max} - \sum_{j=5}^{7} m_j \tag{10.8}$$

The models discussed above did not consider angiogenesis during bone fracture healing. Angiogenesis is the process of new blood vessel formation and plays a crucial role in fracture healing, since it provides the necessary oxygen and nutrients for cells to repair and regenerate bone tissue. A number of recent fracture healing models have incorporated angiogenesis in their simulations. These models have demonstrated that angiogenesis is crucial for osteogenesis and tissue differentiation. One example of an angiogenesis model is the lattice model developed by Checa and Prendergast (2009), which included three fundamental events in angiogenesis: branching, sprout growth, and anastomosis.

The studies discussed, so far in this chapter, assumed a callus with a constant volume. However, callus growth is a crucial aspect of the fracture healing process and is affected by several factors, including mechanical loading, blood supply, and cellular activity. The callus growth model developed by Gomez-Benito et al. (2005) included MSC proliferation and chondrocyte hypertrophy, during endochondral ossification. The cells could proliferate only with an increase in callus size, when the MSC concentration reached saturation. Chondrocyte hypertrophy, characterised by an increase in cell volume, also leads to callus growth. Hypertrophic chondrocyte is the terminal state of chondrocyte, prior to apoptosis. Later, this model was employed in a three-dimensional geometry with thermoelastic analysis (Gomez-Benito et al. 2006). This study reported the effect of the stiffness of the external fixator on the bone fracture healing pattern (Gomez-Benito et al. 2006). Moreover, they observed that lower stiffness causes a larger callus volume and delays fracture healing (Gomez-Benito et al. 2006). The study by Garcia-Aznar et al. (2007) investigated the effect of interfragmentary movement on callus growth and shape. In their model, the second invariant of the deviatoric strain tensor (i.e. octahedral shear strain) was considered to

Table 10.2 Various model parameters corresponding to production and degradation of different matrix (*Source* Isaksson et al., 2008)

Matrix	f^{PM} (day^{-1})	f^{DM} (day^{-1})
FT	0.20	0.05
C	0.05	0.05
B	0.10	0.05

be the mechanical stimulus, which regulated the cell proliferation, tissue differentiation process, and chondrocyte hypertrophy, in order to determine the callus shape and growth. They found that increase in interfragmentary movement led to an increase in the callus size (diameter).

In order to compare the dependence of different mechanical stimuli on model predictions, Geris et al. (2003) considered the models of Prendergast et al. (1997) and Claes and Heigele (1999) to simulate bone healing inside a rabbit bone chamber. The model of Prendergast et al. (1997) considered biphasic stimuli, octahedral shear strain, and relative fluid velocity. In contrast, the model proposed by Claes and Heigele (1999) considered hydrostatic stress and the maximum principal strain. Based on the comparative study, Geris et al. (2003) concluded that the biphasic stimuli played a vital role in predicting more realistic patterns of tissue differentiation.

10.9.2 Bioregulatory Models

Biochemical factors, such as growth factors and oxygen tension, also play an important role in bone fracture healing. Bioregulatory healing models simulate the fracture healing process considering only the role of biochemical factors, excluding mechanical factors. Bailon-Plaza and van der Meulen were the first to incorporate the influence of growth factors into the healing models (Bailon-Plaza and van der Meulen 2001). Seven partial differential equations were considered to model cellular activities like migration, proliferation, and differentiation in the callus region. Their model demonstrated the importance of growth factors during the healing process (Bailon-Plaza and van der Meulen 2001). Geris et al. (2008) extended the model presented by Bailon-Plaza and van der Meulen by incorporating angiogenesis. In their model, the process of angiogenesis was modelled using a partial differential equation regulated by endothelial cells. While the model suggested by Geris et al. (2008) was a great improvement over the previous one, it did not consider the discrete nature of the vasculature. A hybrid bioregulatory model was proposed by Peiffer et al. (2011) that for the discrete nature of the vasculature. Additionally, individual vessel growth was directed by endothelial tip cell migration. The processes of branching and anastomoses were also included in the model.

The study by Carlier et al. (2012) extended the bioregulatory framework proposed by Peiffer et al. (2011). This study presented a multiscale model of osteogenesis and sprouting angiogenesis. Their multiscale model was found to be useful in comprehending the influence of molecular mechanisms on angiogenesis and associated bone tissue formation. Later, Carlier et al. (2015) incorporated the effect of oxygen into their multiscale model. In their model, cellular activities such as cell proliferation, cell differentiation, cell death, oxygen consumption, and the production of growth factors were modelled with the incorporation of oxygen-related parameters. This model was able to predict non-unions caused by insufficient oxygen supply (Carlier et al. 2015).

10.9.3 Mechanobioregulatory Models

The mechanobioregulatory model considers the effect of both mechanical and biological stimuli on bone formation. Bailon-Plaza and van der Meulen improved their earlier bioregulatory model by incorporating mechanical stimuli (Bailon-Plaza and van der Meulen 2001, 2003). In their mechanobioregulatory model (Bailon-Plaza and van der Meulen 2003), two types of mechanical strains were considered:

(1) Deviatoric strain, which had a promoting effect on ossification (bone formation).
(2) Dilatational strain, which had an inhibitory effect on ossification.

Through numerical simulations, they demonstrated that moderate early loading had beneficial effects on bone healing as compared to excessive delayed loading. This suggested that applying the right amount of mechanical stimulation at the right time can improve the healing process.

The study by Geris et al. (2010) employed a similar approach by integrating the mechanical stimuli as parametric values into their existing model (Geris et al. 2008). In their model, these mechanical parameters were linked to different biological activities, including angiogenesis (formation of new blood vessels), matrix and growth factor production, as well as cell proliferation and differentiation. However, a much simpler mathematical model was proposed by Burke and Kelly (2012), which hypothesised that substrate stiffness and oxygen tension regulate stem cell differentiation during fracture healing. It was assumed that the mechanical stimulus (deviatoric strain) regulates angiogenesis, which in turn affects oxygen tension, since blood vessels transport oxygen. The model assumed hypoxia (reduced oxygen tension) to facilitate chondrogenic differentiation, whereas in well-vascularised regions, a soft local substrate was assumed to facilitate adipogenesis and stiff substrate to facilitate osteogenesis. The cell migration and proliferation were modelled using a simple diffusion equation. Figure 10.14 presents the flowchart of the model proposed by Burke and Kelly (2012).

These studies emphasised the need to consider both mechanical factors and biological responses to comprehend tissue healing and regeneration. Moreover, they showcased how computational models can be used to investigate the intricate interactions between mechanics and biology in tissue repair and the challenge of non-unions (when bone healing fails due to overloading).

10.10 Bone Ingrowth Around Porous Coated Implants

The success of uncemented implants is dependent on the degree of osseointgration, which is in turn dependent on the amount of peri-prosthetic bone ingrowth. The steps involved in the simulation of bone ingrowth are similar to bone fracture healing, as discussed earlier. Moreover, the local micromechanical environment is known to influence this biological phenomenon. Numerical evaluations of evolutionary bone

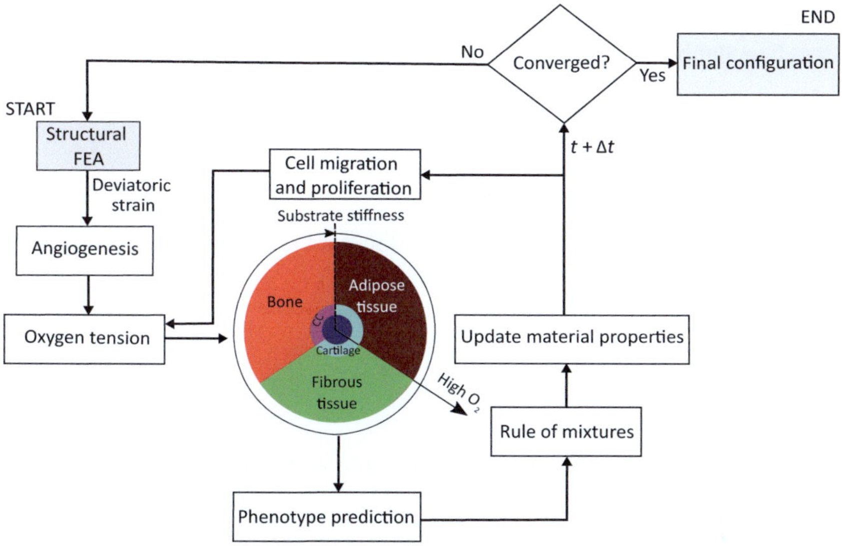

Fig. 10.14 Flowchart of mechanobioregulatory model proposed by Burke and Kelly (2012). Here CC denotes calcified cartilage

ingrowth require microscale finite element analysis in combination with a tissue differentiation algorithm. Over the past decade, several numerical investigations on the spatial distribution of evolutionary tissue ingrowth around porous coated implants have been reported (Simmons et al. 2001; Checa and Prendergast 2009; Mukherjee and Gupta 2016).

Multiple Choice Questions (MCQ)

(1) What is the primary characteristic of a comminuted fracture?

(A) The bone is fractured into multiple pieces
(B) The fracture line is diagonal
(C) The fracture causes a single clean break
(D) The bone fragments are driven into each other

(2) Which stage of bone fracture healing involves the formation of granulation tissue?

(A) Inflammation and hematoma formation
(B) Soft callus formation
(C) Hard callus formation
(D) Remodelling phase

(3) During which phase does the bone undergo a transition from woven bone to lamellar bone?

(A) Inflammation and hematoma formation
(B) Soft callus formation
(C) Hard callus formation
(D) Remodelling phase

(4) Which cell type is primarily responsible for the production of collagen during the initial stages of fracture healing?

(A) Osteoblasts
(B) Chondrocytes
(C) Fibroblasts
(D) Macrophages

(5) What process involves the replacement of cartilage with bone during fracture healing?

(A) Endochondral ossification
(B) Intramembranous ossification

(6) What term describes the migration of cells in response to chemical gradients?

(A) Haptotaxis
(B) Chemotaxis
(C) Thermotaxis
(D) Phototaxis

(7) Which type of cell differentiation involves MSCs becoming chondrocytes?

(A) Osteogenic differentiation
(B) Chondrogenic differentiation
(C) Myogenic differentiation
(D) Fibrogenic differentiation

(8) What is the key characteristic of primary bone healing?

(A) Formation of a visible callus
(B) Presence of a significant gap between bone ends
(C) Close contact between bone ends without callus formation
(D) Slow and gradual bone remodelling

(9) In primary bone healing, what is required for gap healing?

(A) Gap must be less than 1 mm
(B) Gap must be less than 0.1 mm
(C) Immediate surgical intervention
(D) Use of external fixators

(10) What type of bone healing is characterised by callus formation and subsequent bone remodelling?

(A) Primary bone healing
(B) Secondary bone healing

(11) Which cell type is responsible for resorbing excess callus material during the remodelling phase?

(A) Osteoclasts
(B) Chondrocytes
(C) Osteoblasts
(D) Fibroblasts

(12) What is a key characteristic of percutaneous fixation as compared to open reduction techniques?

(A) Longer recovery times
(B) Higher risk of infection
(C) Reduced tissue damage
(D) More invasive procedure

(13) Which model incorporates the effects of both mechanical and biological stimuli on bone formation?

(A) Mechanoregulatory model
(B) Mechanobioregulatory model
(C) Bioregulatory model
(D) Angiogenesis model

(14) In mechanotransduction model proposed by Carter et al. (1988), which mechanical condition is associated with bone formation?

(A) High tensile strain
(B) High hydrostatic compressive stress
(C) Low hydrostatic compressive stress
(D) High interfragmentary movement

(15) Pauwels' theoretical framework primarily focuses on the influence of which type of stimuli on bone formation?

(A) Biochemical
(B) Mechanical
(C) Chemical

Answer Key to MCQs

1. A; 2. A; 3. D; 4. C; 5. A; 6. B; 7. B; 8. C; 9. A; 10. B; 11. A; 12. C; 13. B; 14. C; 15. B.

References

Andreykiv A, van Keulen F, Prendergast PJ (2008) Simulation of fracture healing incorporating mechanoregulation of tissue differentiation and dispersal/proliferation of cells. Biomech Model Mechanobiol 7(5):443–461

Bailon-Plaza A, van der Meulen MCH (2001) A mathematical framework to study the effects of growth factor influences on fracture healing. J Theor Biol 212(2):191–2010

Bailon-Plaza A, van der Meulen MCH (2003) Beneficial effects of moderate, early loading and adverse effects of delayed or excessive loading on bone healing. J Biomech 36(8):1069–1077

Blenman PR, Carter DR, Beaupré GS (1989) Role of mechanical loading in the progressive ossification of a fracture callus. J Orthop Res 7(3):398–407

Burke DP, Kelly DJ (2012) Substrate stiffness and oxygen as regulators of stem cell differentiation during skeletal tissue regeneration: a mechanobiological model. PLoS ONE 7(7):e40737

Caplan AI (2005) Review: mesenchymal stem cells: cell-based reconstructive therapy in orthopedics. Tissue Eng 11(7–8):1198–1211

Carlier A, Geris L, Bentley K, Carmeliet G, Carmeliet P, Van Oosterwyck H (2012) MOSAIC: a multiscale model of osteogenesis and sprouting angiogenesis with lateral inhibition of endothelial cells. PLoS Comput Biol 8(12):e1002724

Carlier A, Geris L, van Gastel N, Carmeliet G, Van Oosterwyck H (2015) Oxygen as a critical determinant of bone fracture healing: a multiscale model. J Theor Biol 365:247–264

Carter DR, Blenman PR, Beaupré GS (1988) Correlations between mechanical stress history and tissue differentiation in initial fracture healing. Clin Orthop Relat Res 6:736–748

Checa S, Prendergast PJ (2009) A mechanobiological model for tissue differentiation that includes angiogenesis: a lattice-based modeling approach. Ann Biomed Eng 37(1):129–145

Claes LE, Heigele CA (1999) Magnitudes of local stress and strain along bony surfaces predict the course and type of fracture healing. J Biomech 32(3):255–266

Claes L, Augat P, Suger G, Wilke HJ (1997) Influence of size and stability of the osteotomy gap on the success of fracture healing. J Orthop Res 15(4):577–584

Garcia-Aznar JM, Kuiper JH, Gomez-Benito MJ, Doblare M, Richardson JB (2007) Computational simulation of fracture healing: influence of interfragmentary movement on the callus growth. J Biomech 40(7):1467–1476

Geris L, Van Oosterwyck H, Vander Sloten J, Duyck J, Naert I (2003) Assessment of mechanobiological models for the numerical simulation of tissue differentiation around immediately loaded implants. Comput Methods Biomech Biomed Eng 6(4):277–288

Geris L, Gerisch A, Sloten JV, Weiner R, Van Oosterwyck H (2008) Angiogenesis in bone fracture healing: a bioregulatory model. J Theor Biol 251(1):137–158

Geris L, Sloten JV, Van Oosterwyck H (2010) Connecting biology and mechanics in fracture healing: an integrated mathematical modeling framework for the study of nonunions. Biomech Model Mechanobiol 9(6):713–724

Gomez-Benito MJ, Garcia-Aznar JM, Kuiper JH, Doblare M (2005) Influence of fracture gap size on the pattern of long bone healing: a computational study. J Theor Biol 235(1):105–1110

Gomez-Benito MJ, Garcia-Aznar JM, Kuiper JH, Doblare M (2006) A 3D computational simulation of fracture callus formation: influence of the stiffness of the external fixator. J Biomech Eng 128(2):290–2910

Isaksson H, van Donkelaar CC, Huiskes R, Ito K (2008) A mechano-regulatory bone-healing model incorporating cell-phenotype specific activity. J Theor Biol 252(2):230–246

Lacroix D, Prendergast PJ, Li G, Marsh D (2002) Biomechanical model to simulate tissue differentiation and bone regeneration: application to fracture healing. Med Biol Eng Comput 40(1):14–21

Lee DA, Knight MM, Campbell JJ, Bader DL (2011) Stem cell mechanobiology. J Cell Biochem 112(1):1–10

Marsell R, Einhorn TA (2011) The biology of fracture healing. Injury 42(6):551–555

Mukherjee K, Gupta S (2016) Bone ingrowth around porous-coated acetabular implant: a three-dimensional finite element study using mechanoregulatory algorithm. Biomech Model Mechanobiol 15(2):389–403

Pauwels F (1960) A new theory on the influence of mechanical stimuli on the differentiation of supporting tissue. The tenth contribution to the functional anatomy and causal morphology of the supporting structure. Z Anat Entwickl Gesch 121(5):478–515

Peiffer V, Gerisch A, Vandepitte D, Van Oosterwyck H, Geris L (2011) A hybrid bioregulatory model of angiogenesis during bone fracture healing. Biomech Model Mechanobiol 10(3):383–395

Perren SM, Cordey J (1980) The concept of interfragmentary strain. In: Uhthoff HK (ed) Current concepts of internal fixation of fractures. Springer, pp 63–77

Prendergast PJ, Huiskes R, Søballe K (1997) Biophysical stimuli on cells during tissue differentiation at implant interfaces. J Biomech 30(6):539–548

Schenk RK, Willenegger HR (1977) Zur Histologie der primaren Knochenheilung (Modifikationen und Grenzen der Spaltheilung in Abhagigkeit von der DefektgroRe). Unfallheilkunde 80(3):155–160

Simmons CA, Meguid SA, Pilliar RM (2001) Differences in osseointegration rate due to implant surface geometry can be explained by local tissue strains. J Orthop Res 19(2):187–194

Appendix
Artificial Intelligence in Orthopaedic Biomechanics

Artificial Intelligence (AI) involves developing computer systems or programs that can perform tasks that typically require human intelligence, such as making decisions, comprehending natural language, and learning from information. AI has applications across diverse fields, including the arts, commerce, physical sciences, and life sciences. Clinical methods are often complex, time-consuming, and pose ethical challenges. However, rapid advancements in computer hardware have allowed AI and its allied technologies to contribute to research across various domains, including orthopaedic biomechanics. For example, comprehensive pre-clinical evaluations of implant designs are computationally expensive, since they involve simulation of a sequence of complex biological events. In such cases, AI is useful in offering rapid design solutions. This appendix provides a brief overview of AI-based methods relevant to joint biomechanics and implants, followed by a discussion of AI applications in specific areas of biomechanics.

A.1 Introduction to the AI

The goal of AI is to develop computer systems capable of emulating human intelligence, enabling them to automate tasks, operate more efficiently, and augment human capabilities. Some key characteristics of AI include (a) Ability to learn from experience, (b) Ability to adapt to new situations, and (c) Ability to make decisions based on available data and information. AI is categorised into two broad types (Fig. A.1).

Type I: Based on capabilities

- *Narrow AI*: Designed to perform very specific tasks or solve specific problems. Examples include Apple's Siri, Cortana, and Google Assistant.
- *General AI*: Capable of understanding, learning, and performing intellectual tasks similar to a human. A self-driving car is an example of General AI.

S. Gupta et al., *Biomechanics of Joints and Implants*,
https://doi.org/10.1007/978-981-96-0586-6

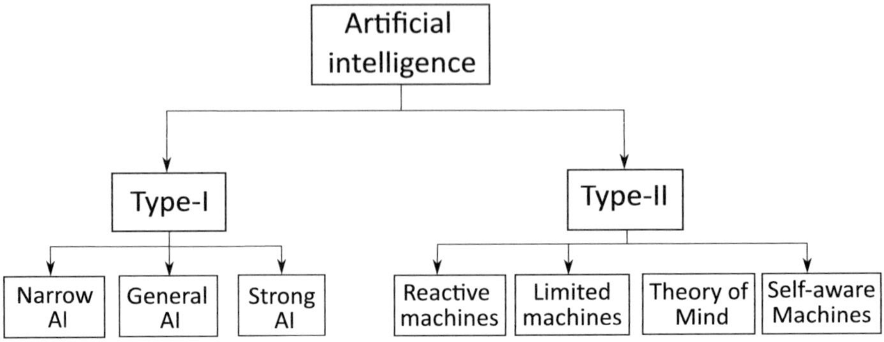

Fig. A.1 Different categories of AI

- *Strong AI*: Hypothetical AI that would exhibit human-level, general intelligence, which has not yet been realized.

Type II: Based on functionality

- *Reactive machines*: The simplest form of AI, which responds to specific inputs without memory of past events (e.g., IBM's Deep Blue).
- *Limited memory*: Machines that use limited memory of past events to inform their decisions, like driverless cars.
- *Theory of mind*: An early form of general intelligence that would allow machines to understand other entities and their own environment, but these systems are still in development.
- *Self-aware machines*: The most advanced theoretical form of AI, which would possess consciousness and self-awareness; this type has not yet been developed.

A.1.1 Machine Learning (ML)

Machine Learning (ML) is a subset of AI focused on developing algorithms and models that enable computers to learn and make predictions or decisions without explicit programming. We may consider a popular video streaming platform as an example, it learns about an individual's movie preferences and makes recommendations accordingly. Each time a user watches a video or adds it to their library, the service updates its algorithm to improve recommendations. Three main approaches are used in ML as described as follows (Fig. A.2).

Rule-based approach: This approach is usually based on the information and rules fed into the machine. The machine delivers the output reacting to the rules which are generally based on the features. Let us consider a file containing 50 images of apples and 50 images of oranges. In order to train the machine, one needs to feed these 100 images into the machine and label each image as either apple or orange.

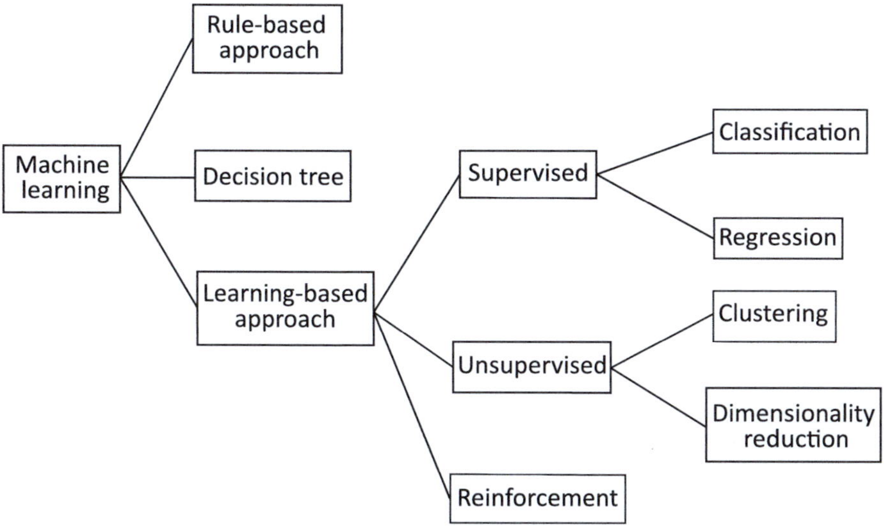

Fig. A.2 Classifications of Machine Learning (ML)

Here, the rule can be based on the colour (feature) of the object. Now if one tests the machine with the image of an orange, the machine will employ the predefined rule on the image. This is known as a rule-based approach in ML.

Decision tree: This approach follows a tree-like structure of the decisions with all possible results. The decision tree, which follows a top-down approach, is made up of two nodes: decision node and leaf node. The top most node in a decision tree is known as root. Decision nodes are used to make decisions and they typically have multiple branches, whereas leaf nodes are the output of these decisions and they do not contain any further branches. Decision trees in ML can either be classification trees or regression trees, which together are known as Classification And Regression Tree (CART). Decision Tree allows developers to evaluate the possible consequences of a decision. By accessing more data, the algorithm of decision trees can envision outcomes for future data.

Learning-based approach: Here, the machine is provided with input data and, in some cases, the desired output. The machine creates its own algorithms to map the data to the output or infer patterns. A relevant example would be the classification of stray dogs using image-based learning. Suppose there is a collection of a sizable number of images of random stray dogs from a locality. However, one barely has any clue about the trend in this dataset with regard to the colour, breed, or any other feature that may be missing. Now, if this data is fed into a learning-based AI machine, it may extract many patterns found in the features of these images. It might as well cluster the data on the basis of size, colour, and other criteria.

Prior to classifying the learning-based approach, let us discuss the difference between labelled data and unlabelled data. Suppose you have a dataset containing

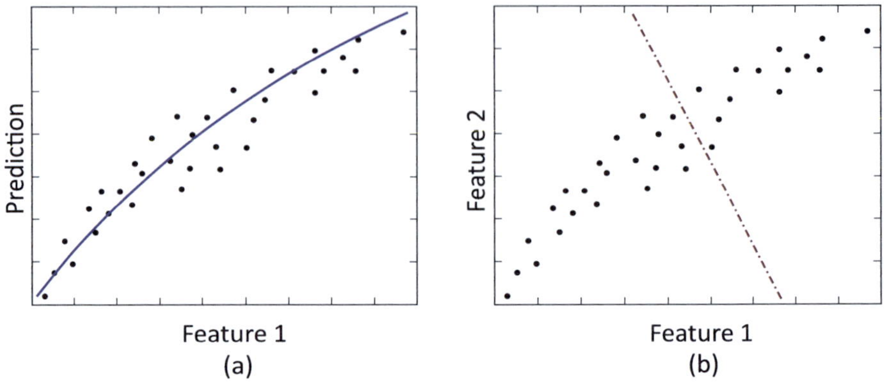

Fig. A.3 Major prediction problems within supervised learning: **a** regression finds the best fit for the data, **b** classifier finds the boundary to classifying the dataset into different classes based on two features

information about different types of fruits. The dataset includes features like colour, size, and weight, and each data point is labelled with the type of fruit it represents (e.g., apple, orange, banana). In this case, the labels (Apple, Orange, Banana) convey the type of fruit each data point represents. This type of data containing labels are called as labelled data since each data point is associated with a specific category or outcome. Now let us consider a dataset with the same features (colour, size, weight) but without any labels indicating the type of fruit. In this scenario, you have the features of the fruits but no information about what type of fruit each data point denotes. This type of data is known as unlabelled data since there are no predefined categories or outcomes associated with the data points.

The learning-based approach can further be divided into three types:

(a) ***Supervised learning***—Involves training a model on labelled data. The model learns to map inputs to outputs for tasks like regression (predicting continuous data) and classification (categorising data into discrete classes).

Regression: Such models typically work on continuous data. For example, if such a model is required to predict the next salary, then the data of the previous salary alongside any increments and perks need to be provided as input to the system. Here, the input data which has been fed to the machine is continuous having a specific trend (Fig. A.3a). Regression involves mapping of the input variable and the output variable.

Classification: Such a model can work on data that is not continuous (Fig. A.3b). For example, in a grading system, students are classified according to their grades based on their examination scores.

Support Vector Machine (SVM): SVM is a widely used supervised learning algorithm that addresses both classification and regression tasks. It leverages

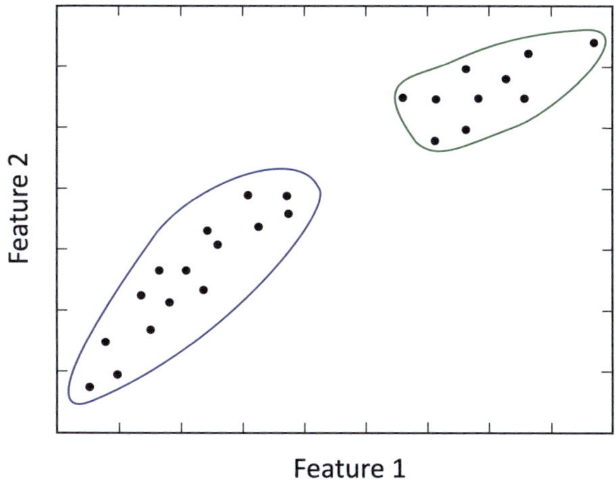

Fig. A.4 Patterns and separation of two identified clusters

supervised learning models to tackle complex problems in classification, regression, and outlier detection. SVM works by performing optimal data transformations to establish boundaries between data points based on predefined classes, labels, or outputs.

(b) ***Unsupervised learning***—This refers to a type of machine learning in which a machine can learn from data on its own without external supervision. The unannotated dataset can be employed to train the unsupervised models. In order to address the association and the clustering problems, such models are used. There are two broad types of unsupervised learning models in AI:

Clustering: This approach can cluster the unlabelled data according to patterns or similarities found within it (Fig. A.4). It assesses similarity across the data using metrics such as Euclidean distance, Cosine similarity, or Manhattan distance, and then groups points together based on their highest similarity scores.

Dimensionality Reduction: This approach is the transformation of data from a high-dimensional space into a low-dimensional space (for the sake of computational efficiency) so that the low-dimensional representation retains some meaningful properties of the original data. As we reduce the dimension of an entity, the information tends to become distorted. Upon further reduction of the dimensions, more and more information will be lost. Hence, meaningful removal of redundant features is necessary for reducing the number of features in data while preserving important information. Principal Component Analysis (PCA) is a widely used dimensionality reduction algorithms.

A.1.2 Neural Networks and Deep Learning

Let us explore the biological neural network, prior to the artificial neural networks (ANNs). Neural network is the process of how a human brain works, behaves, and interprets the situation. Biological neural networks consist of neurons that transmit information, using electrical impulses and chemical signals. These neurons comprise a cell body, dendrites (receiving signals), and an axon (transmitting signals) and they communicate through synapses using neurotransmitters.

ANN is a machine learning algorithm that mimics biological neural networks. ANNs are composed of layers, with each layer containing nodes (or neurons) that process data (Fig. A.5). Each node completes its own tasks and passes on the derivatives to the next layer. The first layer of an ANN, input layer, acquires data and feeds it to the neural network. Each node in the hidden layers has its own machine learning algorithm which processes the data received from the input layer. There can be multiple hidden layers in the architecture of ANN. The last hidden layer passes the final processed data to the output layer, which is available to the user as the final output (Fig. A.5). In the following, we will try to answer the question: "what really happens in the hidden layer?". There are three major learning paradigms used to teach the machine so that it can mimic biological systems. The three paradigms are supervised learning, unsupervised learning, and reinforcement learning. Supervised and unsupervised learning have already been discussed in Section A.1.1. In reinforcement learning, neural networks learn to make decisions by interacting with an environment. Unlike supervised and unsupervised learning, the data is not provided here as a fixed set of examples. The machine learns upon the data acquired from the external system, known as the environment, through dynamic interactions. The learning process is driven by feedback in the form of rewards or penalties, based on its performance, which guides the network to improve its decision-making strategy over time. Reinforcement learning is often employed in robots to learn specific tasks. In this case, the robot learns based on the observations from the environment which is the feedback to its actions.

Let us now explore the architecture of ANN in detail. In a neural network, the fundamental unit is a node. Each node processes inputs from other nodes or from the external input data, computes a weighted sum of these inputs, adds a bias term,

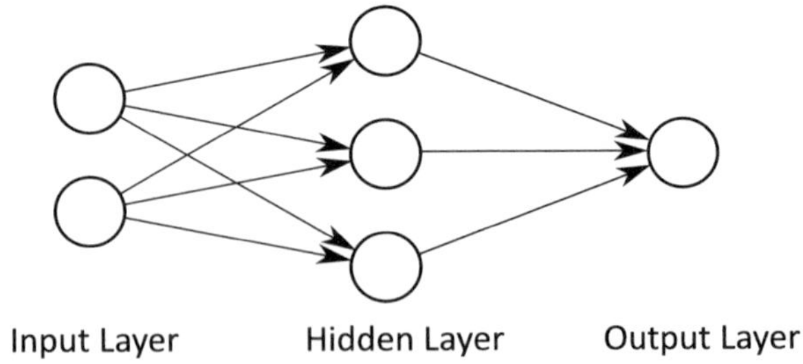

Fig. A.5 The architecture of ANN involving input layer, hidden layer and output layer. Nodes are denoted using circles

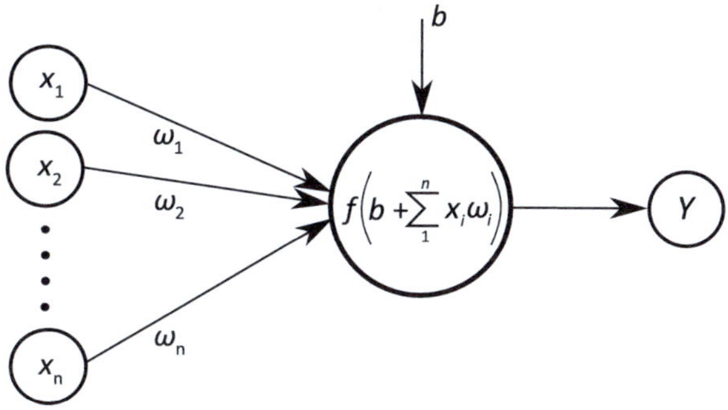

Fig. A.6 Activation function within ANN; bias (b) is added to the weight sum of inputs ($x_i\omega_i$)

and then applies an activation function to produce an output. This process occurs in multiple layers, each playing a distinct role, as described in the following.

Input Layer: The input layer consists of neurons that receive the raw data. Each node in this layer corresponds to a feature (x_i) of the input data. The inputs are directly passed to the next layer without applying any modifications or activation functions.

Hidden Layers: Hidden layers are intermediate layers between the input and output layers. Each node in a hidden layer performs the following steps: (i) The weighted sum of its inputs is computed, where each input is multiplied by a corresponding synaptic weight (ω_i). (ii) A bias term, b, is added to the weighted sum to adjust the output. (iii) An activation function, f, is employed on the resulting net input to introduce non-linearity (Fig. A.6). This non-linearity enables the network to learn and model complex patterns. The output from each neuron in a hidden layer is passed as input to the neurons in the subsequent layer.

Output Layer: The output layer, consisting of neurons, generates the final output (Y) of the network. Similar to hidden layers, each neuron in the output layer computes a weighted sum of its inputs, adds a bias, and applies an activation function. The choice of activation function in the output layer depends on the specific task. The output obtained from the output layer represents the prediction or decision based on the input data.

In the following, we discuss some of the types of ANN, starting with the simplest among them.

A.1.2.1 Feedforward Neural Network

The feedforward neural network (FNN) is a type of ANN which contains multiple nodes arranged in layers as shown in Fig. A.6. Nodes from adjacent layers have connections or edges between them and these connections have weights associated with them.

The two types of FNN are: (a) Single-Layer Perceptron (SLP) and (b) Multi-Layer Perceptron (MLP). SLP is the simplest form of feedforward neural network and does not contain any hidden layer. MLP, on the other hand, has one or more hidden layers, alongside one input and one output layer. While SLP can only learn linear functions, MLP can also learn non-linear functions. FNNs are typically used for tasks like classification and regression, where the relationship between inputs and outputs is pretty much straightforward.

A.1.2.2 Backward Propagation Neural Network

Backpropagation Neural Networks (BPNNs) build upon the basic structure of Feed-forward Neural Networks by introducing a crucial learning algorithm known as backpropagation. During backpropagation, the network calculates the total error at the output nodes and propagates these errors backward through the layers to compute the gradients. These gradients are then used to adjust the weights in the network in order to minimise the error. It's important to note that, initially, all weights in the network are typically assigned randomly before training begins. BPNNs are more powerful as compared to FNN because they allow the network to learn complex, non-linear mappings between input and output. They are widely used in tasks such as speech recognition, image classification, and time series forecasting. Despite their advantages, BPNNs can be computationally intensive, especially when the network is deep or the dataset is large. Nevertheless, BPNNs are a fundamental component of many modern neural networks and have significantly advanced the field of machine learning.

A.1.2.3 Deep Learning Neural Networks

Before we move to the next section, it is important to understand Deep Learning (DL) in the context of ANN. What sets Deep Learning Neural Networks (DLNNs) apart is their deep architecture, characterised by having many hidden layers between the input and output layers. These additional layers enable the network to learn hierarchical representations of data, where higher-level layer capture more abstract features.

For example, in image recognition, the initial layers might detect simple features like edges and corners, while subsequent layers combine these features to identify more complex structures such as shapes and objects. Although a machine learning algorithm can learn from small datasets, deep learning algorithms require large datasets containing heterogeneous and unstructured data. The relationship between AI, ML, and DL can be best described by a Venn diagram (Fig. A.7). Convolutional Neural Networks (CNNs) and Recurrent Neural Networks (RNNs) are specialised types of Deep Learning Neural Networks (DLNNs).

I. Convolutional Neural Networks (CNN): CNNs are particularly well-suited for processing grid-like data, such as images (Fig. A.8). They utilise a special operation called convolution, which applies filters across the input data to capture

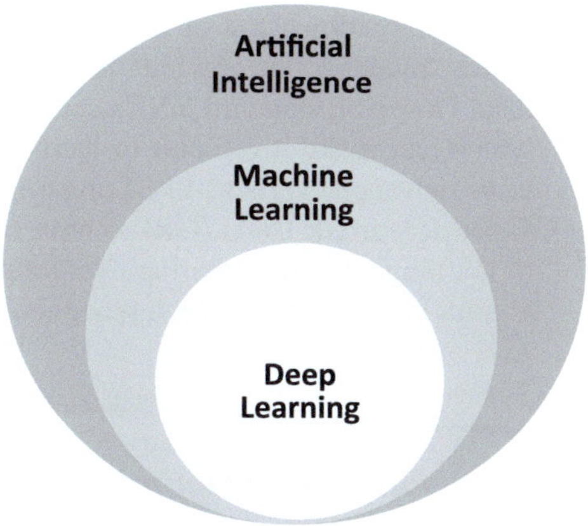

Fig. A.7 A Venn diagram of AI presenting the link across AI, ML, and DL

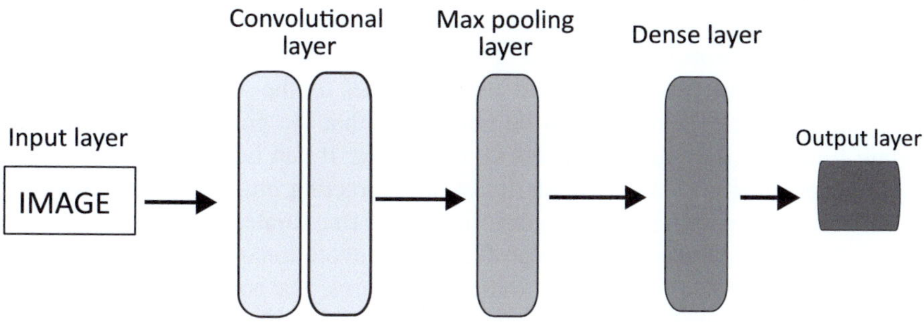

Fig. A.8 CNN architecture depicting various layers involved

spatial hierarchies of features. This allows CNN to automatically and efficiently recognize patterns, such as edges, textures, and shapes, making them ideal for tasks like image classification, object detection, and facial recognition.

A typical CNN architecture includes several layers, including convolutional layers, pooling layers, and dense layers. The convolutional layer extracts features from the input data, pooling layers reduce the dimensionality while retaining important information, and dense layers make the final prediction based on the extracted features. CNNs are primarily used in state-of-the-art computer vision applications.

II. Recurrent Neural Networks (RNNs): RNNs are designed to handle sequential data, where the order of the data points is important. What makes RNNs special is that they have loops that allow information to be passed from one step of the sequence to the next. This means RNN can remember what happened earlier in the sequence and use that information to make decisions later on. This makes them particularly effective for tasks where context or temporal dynamics are important, such as natural language processing, speech recognition, and time series prediction. One of the key features of RNNs is their ability to maintain a "memory" of previous inputs, which helps in processing sequences of data. However, standard RNNs can suffer from issues like vanishing gradients, which can make it difficult to learn long-term dependencies. To address this issue, advanced variants like Long Short-Term Memory (LSTM) networks and Gated Recurrent Units (GRUs) have been developed, which are more capable of learning and remembering over long sequences. RNNs are primarily used in natural language and speech recognition applications.

A.1.3 Evolutionary Optimisation

Evolutionary optimisation is a specially curated discipline in optimisation that does not tread the path of classical gradient search. This black box scheme deals with a derivative-free approach to search for optimality. Thus, it does not make any assumption of the objective function $f(x)$ and rather searches for the solution point x heuristically in the search space. It may be noted here that the gradient information on certain functions may be unavailable or unreliable. It can happen when the function is discreet or possesses singularity. Also, extracting gradient information can be tedious in some cases. This makes derivative-free strategies robust and more practical to address complex design problems. The evolutionary algorithms (EA) are also commonly known as population-based algorithms. The popular EAs are Particle Swarm Optimisation (PSO), Differential Evolution Algorithm (DEA), and Genetic Algorithms (GA).

- PSO simulates the social behaviour of birds or fish to explore solutions. This approach uses a group of candidate solutions, called particles, which move around in the solution space. Each particle adjusts its position based on its own experience and the experience of neighbouring particles, sharing information about the best solutions found. The particles collectively converge towards the optimal solution by learning from both their individual experiences and those of the swarm.
- DEA focuses on generating new candidate solutions by combining and perturbing existing solutions using differential mutations, which are based on the differences between randomly selected pairs of solutions. This approach emphasizes, making changes proportional to the differences among solutions, allowing for the exploration of the solution space in a more directed manner. By incorporating these differences, DEA can efficiently explore diverse regions of the solution space and adaptively refine solutions.
- GA mimics biological evolution, using processes like selection, crossover, and mutation to evolve solutions over generations. This approach uses selection to choose the best solutions based on their fitness, crossover to combine parts of

these solutions to create new candidate solutions, and mutation to introduce variability and maintain genetic diversity. GA relies on these evolutionary operations to explore and refine the solution space over multiple generations, gradually improving the quality of solutions.

In a nutshell, PSO leverages collective intelligence, DEA uses differential information for evolution, and GA relies on evolutionary principles to refine solutions.

A.1.4 Fuzzy Logic

Fuzzy logic represents logical reasoning where the truth of a statement is not simply 'true' or 'false'. Instead, it encompasses a range of truth values from 0 (indicating complete falsehood) to 1 (indicating complete truth). Widely considered a legacy AI system, fuzzy logic helps in encoding experience-based knowledge into a format that computers can comprehend through logical rules. Thus, when binary logic proves insufficient, fuzzy logic becomes essential. Fuzzy logic enables us to create artificial intelligence by designing a fuzzy inference system, which is a function that translates a set of inputs into outputs using rules that are understandable to humans, rather than relying on more abstract mathematical concepts. The advantage is that developing such a fuzzy inference system does not require a model. Therefore, fuzzy logic is effective for complex systems whose underlying mechanisms are not fully understood. With some experience and intuition about these complex systems, rules can be framed and implemented to manage them effectively. Fuzzy logic transforms precise input values, known as crisp inputs, into fuzzy variables through a process called fuzzification. Then, the fuzzy inference system applies the rules to these fuzzified variables to generate new fuzzy outputs, and subsequently, these fuzzy outputs are converted back into crisp, specific outputs through a process called defuzzification. Fuzzy logic is widely used in control systems for automotive and home appliances, industrial automation, healthcare diagnostics, and treatment planning, financial risk assessment and investment analysis, image and signal processing, decision-making systems in insurance and smart technologies, and environmental management for pollution control and resource optimisation.

A.2 Application of AI in Orthopaedic Research

The section presents a brief overview of few specific applications of AI in the domain of orthopaedic research, such as estimation of joint kinematics and kinetics, orthopaedic image analysis, design of load-bearing implants, and prediction of bone fracture healing.

A.2.1 AI in Estimation of Joint Kinematics and Kinetics

Estimating joint kinematics and muscle forces remains a critical issue in both rehabilitation and implant biomechanics. More recently, several researchers have been increasingly employing state-of-the-art AI tools to simulate the kinematic parameters related to the gait cycle. In order to predict the effect of muscle activation on joint angles and moments, BPNN was effectively employed by Sepulveda et al. (1993). Moreover, this ANN was trained using the data adapted from Winter (1987). In another study (Prentice et al. 2001), data associated with limb movement was used as input to the FNN model, which predicted the EMG activity of muscles of the lower limb and trunk. In a study by Matijevich et al. (2020), an ML regression algorithm was employed to extract 13 discrete features from wearable sensor signals in order to predict tibial force, thereby offering insights into bone damage and injury risk. A study by Rapp et al. (2021) compared two DLNNs, CNN and LSTM, in predicting joint kinematics in walking and running (Rapp et al. 2021). The input data used in their study was angular velocity and linear acceleration from two adjacent segments. Additionally, Rapp et al. (2021) reported that CNN exhibited better performance.

In an SVM-based classification study by Conforti et al. (2020), the kinematic data from a biomechanical model was used as the input. Based on features, e.g. range of the motion of the lower limb and lumbosacral joint, the differences between correct and incorrect postures in lifting loads were estimated, thereby resulting in a classification of postures. Favre et al. (2012) demonstrated the use of the MLP model to predict knee adduction moment during walking, based on anthropometric measurements and ground reaction force. Some of the technical challenges of AI in joint kinematics include high variability of medical data, anthropometric differences among subjects, requirement of manual interference in combining sensor data with AI technology and lack of transparency in AI models. In order to address patient variability, the prognosis of bone and joint disorders can be treated as a classification problem in the field of AI. These classifiers can function as interpreters, simulating potential clinical outcomes based on the data. A more detailed review of the ML applications in human movement science is available in Halilaj et al. (2018).

A.2.2 AI in Orthopaedic Image Analysis

Orthopaedic imaging plays a crucial role in the diagnosis and treatment of joint disorders and bone fractures. In joint reconstruction, AI has been used to analyse images for automated diagnostic purposes, implant validation, successful prediction of clinical outcomes and to optimise the process of pre-surgical stages for patient-specific implants. In a study by Kruse et al. (2017), fracture risk levels were predicted by undertaking clustering analysis on dual-energy X-ray absorptiometry scan dataset. The clustering analysis was carried out using hierarchical agglomerative clustering (HAC), an unsupervised machine learning technique. HAC is based on a bottom-up approach. This procedure starts with each data point as its own cluster and iteratively merges the closest clusters until all points are in one cluster or until a desired number of clusters is attained.

AI has also been effectively employed to analyse imaging datasets resulting in diagnostic accuracies that often match or exceed those of trained radiologists and orthopaedic surgeons (Merkely et al. 2021; Borjali et al. 2020; Xue et al. 2017). In a study by Borjali et al. (2020), mechanical loosening of total hip replacement implants was detected from X-ray images using CNN. The application of CNN in improving the diagnostic performance of MRI in the detection of tibiofemoral cartilage defect was reported by Merkely et al. (2021). In another study, hip osteoarthritis has been diagnosed effectively using CNN (Xue et al. 2017). This study reported that the effectiveness of the diagnosis was comparable to that of an attending physician having 10 years of experience (Xue et al. 2017). A recent study has highlighted the use of RNN as a diagnostic technique to identify anterior cruciate ligament injuries, using automated video analysis of in-game motion patterns (Schulc et al. 2024). In this study, videos of different sports activities, involving high-risk movements, were used (Schulc et al. 2024).

The application of AI not only expedites the diagnostic process but also reduces human error, thereby improving patient outcomes (Ramkumar et al. 2022; Lalehzarian et al. 2021; Hetherington et al. 2017). In a study by Hetherington et al. (2017), a CNN model was employed to identify lumbar vertebral levels from a sequence of ultrasound images. Percutaneous spinal needle insertion procedures typically require accurate identification of the vertebral level to ensure effective and safe delivery of analgesic agents (Hetherington et al. 2017). The application of AI extends beyond mere image interpretation; it encompasses predictive analytics that can forecast post-operative outcomes and resource utilisation, thereby facilitating personalized patient care (Huffman et al. 2024; Hui et al. 2022; Navarro et al. 2018). It is evident from the literature that there exists a positive reception among orthopaedic surgeons towards AI integration in the clinical workflow (Rupp et al. 2024; Ormond et al. 2023). As AI continues to evolve, its role in orthopaedic image analysis is expected to expand, offering innovative solutions that can enhance both the efficiency and quality of healthcare in orthopaedic practice (Federer and Jones 2021).

A.2.3 AI in Design of Load-Bearing Orthopaedic Implants

Load-bearing orthopaedic implants pose a significant design challenge since they require both biological compliance as well as structural robustness. These two aspects are often conflicting, necessitating a trade-off between them. Over the past decade, there has been a paradigm shift in the design approach for the load-bearing implants. While the biomechanical failure theories are being addressed in a new light, traditional failure-based design strategies are now being replaced by smart optimisation schemes integrated with data-driven models. One such example is the integration of optimisation algorithms using AI trained on finite element analysis (FEA) (Saviour and Gupta 2024; Dammak et al. 2020; Cilla et al. 2017; Chanda et al. 2016; Chanda et al. 2015).

A recent study by Dammak et al. (2020) proposed a multi-objective reliability-based design optimisation method for optimising the design of cementless hip prostheses by integrating FEA with a kriging model and GA. Kriging is a statistical method used for spatial interpolation and modelling. It relies on understanding spatial correlations and patterns within data. In another study, FE-based mechanoregulatory scheme coupled with BPNN and GA evaluated optimum surface macro-textures from implant models in an attempt to enhance bone ingrowth (Ghosh et al. 2021). The study by Cheng et al. (2018) employed a kriging model coupled with genetic algorithm (GA) to optimise the shape of the implant, which minimised implant-bone micromotion. In another study by Chanda et al. (2016) on design optimisation of hip stem, a hybrid framework (NN-GA-FEA) was implemented in order to minimise the chances of interfacial debonding, bone resorption owing to stress-shielding, and implant-bone micromotion. The contribution of AI in the design of orthopaedic implant is expected to expand, leading to enhanced patient outcomes and more effective surgical interventions.

A.2.4 AI in Prediction of Bone Fracture Healing

Most fracture healing models are based on the theory of mechanoregulation, which has been discussed in Chap. 10. However, these mechanobiological rules for cellular activities and tissue differentiation are rigid and hence, have limitations in modelling such complex biological phenomenon. AI systems help overcome such limitations through heuristic assessment capabilities that are intrinsic to such algorithms. Moreover, classification models assist clinicians in identifying both major and minor healing parameters by analysing medical data.

Earlier studies have simulated the callus formation by controlling the addition of new elements iteratively using a framework which couples the mesh growth algorithm with the FE model (Naveiro et al. 2021, 2023). This algorithm was able to reproduce the generation of primary callus in long bones, irrespective of the fracture type and complexity. In another study by Lau et al. (2022), an ML model was implemented to predict the healing time of upper limb fractures in children. In this study, SVR and Random Forest (RF) were employed to predict and identify variables that were related to the fracture healing time. RF is an ensemble learning method that combines multiple decision trees to improve accuracy and reduce overfitting. By aggregating predictions from various trees, it enhances the robustness and allows better generalization on new data. Malek et al. (2016) evaluated paediatric fracture healing by classifying the healing time using a BPNN. In another study by Malek et al. (2018), lower limb fractures in children were examined using RF and Self Organising feature Map (SOM). The SOM is an unsupervised neural network that clusters data, by reducing the high-dimensional inputs to visualize them on a lower-dimensional grid (Vesanto 1999). In this study, RF was used to rank the most important variables influencing fracture healing time, whereas SOM was later implemented to relate the selected variables with healing time (Malek et al. 2018). It was reported that a combination of these two algorithms effectively assisted in the analysis of paediatric fracture healing time (Malek et al. 2018).

In a recent study by Liu et al. (2023), classification ML algorithms were employed to design optimal distal radius fractures (DRF) rehabilitation at different stages of healing. In this study, the FE model predictions were used as the input data to train the ML algorithm. The computational model considered the cellular processes, such as cell differentiation and angiogenesis (Liu et al. 2023). Various Fuzzy logic-based frameworks have also been utilised to comprehend bone healing (Nag et al. 2022; Wehner et al. 2010; Shefelbine et al. 2005; Simon et al. 2003). In a study by Nag et al. (2022), a 3D dynamic fracture healing scheme was modelled on a subtrochanteric femur fracture, regulated by FEA and Fuzzy logic control to investigate the spatio-temporal healing phenomena of extramedullary plates. A combined micro-FE and ANN framework was proposed by Minku and Ghosh (2024) to predict site-specific bone ingrowth around tibial implants after total ankle replacement. This model enabled rapid prediction of the bone-stem relative micromotion as compared to that of FEA (Minku and Ghosh 2024).

In a summary, the integration of advanced machine learning (ML) techniques with traditional fracture healing models marks a significant advancement in the field of orthopaedics. By leveraging heuristic algorithms and classification models, researchers and clinicians are now able to gain insights into the complex dynamics of bone healing within a reasonable timeframe.

References

Borjali A, Chen AF, Muratoglu OK, Morid MA, Varadarajan KM (2020) Detecting total hip replacement prosthesis design on plain radiographs using deep convolutional neural network. J Orthop Res 38(7):1465–1471

Chanda S, Gupta S, Pratihar DK (2015) A genetic algorithm based multi-objective shape optimization scheme for cementless femoral implant. J Biomech Eng 137:034502(1–12)

Chanda S, Gupta S, Pratihar DK (2016) Effects of interfacial conditions on shape optimization of cementless hip stem: an investigation based on a hybrid framework. Struct Multidiscip Optim 53(5):1143–1155

Cheng YC, Jiang CP, Lin DH (2018) Finite element based optimization design for a one-piece zirconia ceramic dental implant under dynamic loading and fatigue life validation. Struct Multidiscip Optim 59(3):835–849

Cilla M, Borgiani E, Martínez J, Duda GN, Checa S (2017) Machine learning techniques for the optimization of joint replacements: application to a short-stem hip implant. PLoS ONE 12(9)

Conforti I, Mileti I, del Prete Z, Palermo E (2020) Measuring biomechanical risk in lifting load tasks through wearable system and machine-learning approach. Sensors (Switz) 20(6)

Dammak K, Hami A (2020) Multi-objective reliability based design optimization using Kriging surrogate model for cementless hip prosthesis. Comput Methods Biomech Biomed Eng 23(12):854–867

Favre J, Hayoz M, Erhart-Hledik JC, Andriacchi TP (2012) A neural network model to predict knee adduction moment during walking based on ground reaction force and anthropometric measurements. J Biomech 45(4):692–698
Federer SJ, Jones GG (2021) Artificial intelligence in orthopaedics: A scoping review. PLoS ONE 16(11):e0260471
Ghosh R, Chanda S, Chakraborty D (2021) Qualitative predictions of bone growth over optimally designed macro-textured implant surfaces obtained using NN-GA based machine learning framework. Med Eng Phys 95:64–75
Halilaj E, Rajagopal A, Fiterau M, Hicks JL, Hastie TJ, Delp SL (2018) Machine learning in human movement biomechanics: Best practices, common pitfalls, and new opportunities. J Biomech 81:1–11
Hetherington J, Lessoway V, Gunka V, Abolmaesumi P, Rohling R (2017) SLIDE: Automatic spine level identification system using a deep convolutional neural network. Int J Comput Assist Radiol Surg 12(7):1189–1198
Huffman N, Pasqualini I, Khan ST, Klika AK, Deren ME, Jin Y, Kunze KN, Piuzzi NS (2024) Enabling personalized medicine in orthopaedic surgery through artificial intelligence: a critical analysis review. JBJS Rev 12(3):e23.00232
Hui AT, Alvandi LM, Eleswarapu AS, Fornari ED (2022) Artificial intelligence in modern orthopaedics: current and future applications. JBJS Rev 10(10):e22.00086
Kruse C, Eiken P, Vestergaard P (2017) Clinical fracture risk evaluated by hierarchical agglomerative clustering. Osteoporos Int 28(3):819–832
Lalehzarian SP, Gowd AK, Liu JN (2021) Machine learning in orthopaedic surgery. World J. Orthop. 12(9):685–699
Lau CF, Malek S, Gunalan R, Chee WH, Saw A, Aziz F (2022) Paediatric upper limb fracture healing time prediction using a machine learning approach. All Life 15(1):490–499
Liu X, Miramini S, Patel M, Ebeling P, Liao J, Zhang L (2023) Development of numerical model-based machine learning algorithms for different healing stages of distal radius fracture healing. Comput Methods Programs Biomed 233
Malek S, Gunalan R, Kedija SY, Lau CF, Mosleh MAA, Milow P, Amber H, Saw A (2016) A primary study on application of artificial neural network in classification of pediatric fracture healing time of the lower limb. Adv Intell Syst Comput 477:23–30
Malek S, Gunalan R, Kedija SY, Lau CF, Mosleh MAA, Milow P, Lee SA, Saw A (2018) Random forest and self-organizing maps application for analysis of pediatric fracture healing time of the lower limb. Neurocomputing 272:55–62
Matijevich ES, Scott LR, Volgyesi P, Derry KH, Zelik KE (2020) Combining wearable sensor signals, machine learning and biomechanics to estimate tibial bone force and damage during running. Hum Mov Sci 74
Minku, Ghosh R (2024) A macro–micro FE and ANN framework to assess site-specific bone ingrowth around the porous beaded-coated implant: an example with BOX® tibial implant for total ankle replacement. Med Biol Eng Comput 62(6):1639–1654
Nag P, Chanda S (2022) A preclinical model of post-surgery secondary bone healing for subtrochanteric femoral fracture based on fuzzy interpretations. PLoS ONE 17
Navarro SM, Wang EY, Haeberle HS, Mont MA, Krebs VE, Patterson BM, Ramkumar PN (2018) Machine learning and primary total knee arthroplasty: Patient forecasting for a patient-specific payment model. J Arthroplasty 33(12):3617–3623
Naveiro JM, Gracia L, Roces J, Albareda J, Puértolas S (2023) Three-dimensional computational model simulating the initial callus growth during fracture healing in long bones: application to different fracture types. Bioengineering 10(2)
Naveiro JM, Puértolas S, Rosell J, Hidalgo A, Ibarz E, Albareda J, Gracia L (2021) A new approach for initial callus growth during fracture healing in long bones. Comput Methods Programs Biomed 208
Ormond MJ, Clement ND, Harder BG, Farrow L, Glester A (2023) Acceptance and understanding of artificial intelligence in medical research among orthopaedic surgeons. Bone Joint Open 4(9):696–703

Prentice SD, Patla AE, Stacey DA (2001) Artificial neural network model for the generation of muscle activation patterns for human locomotion. J Electromyogr Kinesiol 11(1):19–30
Ramkumar PN, Luu BC, Haeberle HS, Karnuta JM, Nwachukwu BU, Williams RJ (2022) Sports medicine and artificial intelligence: A primer. Am J Sports Med 50(4):1166–1174
Rapp E, Shin S, Thomsen W, Ferber R, Halilaj E (2021) Estimation of kinematics from inertial measurement units using a combined deep learning and optimization framework. J Biomech 116
Rupp MC, Moser LB, Hess S, Angele P, Aurich M, Dyrna F, Nehrer S, Neubauer M, Pawelczyk J, Izadpanah K, Zellner J, Niemeyer P, AGA-Komitee Innovation und Translation, (2024) Orthopaedic surgeons display a positive outlook towards artificial intelligence: A survey among members of the AGA Society for Arthroscopy and Joint Surgery. J. Exp. Orthop. 11(3):e12080
Saviour CM, Gupta S (2024) Towards an optimal design of a functionally graded porous uncemented acetabular component using genetic algorithm. Med Eng Phys 126:104159
Schulc A, Leite CBG, Csákvári M, Lattermann L, Zgoda MF, Farina EM, Lattermann C, Tősér Z, Merkely G (2024) Identifying anterior cruciate ligament injuries through automated video analysis of in-game motion patterns. Orthop J Sports Med 12(3):23259671231221580
Sepulveda F, Wells DM, Vaughan CL (1993) A neural network representation of electromyography and joint dynamics in human gait. J Biomech 26(2):101–109. https://doi.org/10.1016/0021-9290(93)90041-c
Shefelbine SJ, Augat P, Claes L, Simon U (2005) Trabecular bone fracture healing simulation with finite element analysis and fuzzy logic. J Biomech 38(12):2440–2450
Simon U, Augat P, Utz M, Claes L (2003) Simulation of tissue development and vascularization in the callus healing process. Trans Orthop Res Soc 28
Vesanto J (1999) SOM-based data visualization methods. Intell. Data Anal. 3(2):111–126
Wehner T, Claes L, Niemeyer F, Nolte D, Simon U (2010) Influence of the fixation stability on the healing time: A numerical study of a patient-specific fracture healing process. Clin Biomech 25(6):606–612
Winter DA (1987) The biomechanics and motor control of human gait, 2nd edn. University of Waterloo Press
Xue Y, Zhang R, Deng Y, Chen K, Jiang T (2017) A preliminary examination of the diagnostic value of deep learning in hip osteoarthritis. PLoS ONE 12(6)

Solution to Practice Problems

Solutions to Practice Problems of Chap. 2

Problem 2.1

Given data:

Reaction force, $R = 80$ kg (subject standing on one foot)
The weight of limb (W_L) = 13 kg

Unknown data:

Hip joint reaction force ($\boldsymbol{F_J}$) acting at an angle θ with the horizontal.
Considering the force equilibrium along x- and y-directions (Fig. S2.1),

$$\begin{aligned} &\sum F_X = 0 \\ &F_A \cos 70 = F_J \cos \theta \end{aligned} \tag{S2.1}$$

$$\begin{aligned} &\sum F_Y = 0 \\ &F_A \sin 70 + R = \mathrm{W_L} + F_J \sin \theta \\ &F_A \sin 70 + 80 = 13 + F_J \sin \theta \\ &F_A \sin 70 + 67 = F_J \sin \theta \end{aligned} \tag{S2.2}$$

Taking a moment about centre of pressure,

$$\begin{aligned} &\sum M = 0 \\ &0.082 \times F_A \sin 70 + 0.890 \times F_A \cos 70 - 0.008 \times W_L \\ &-0.014 \times F_J \sin \theta - 0.9 \times F_J \cos \theta = 0 \end{aligned} \tag{S2.3}$$

Substituting Eqs. (S2.1) and (S2.2) in Eq. (S2.3)

S. Gupta et al., *Biomechanics of Joints and Implants*,
https://doi.org/10.1007/978-981-96-0586-6

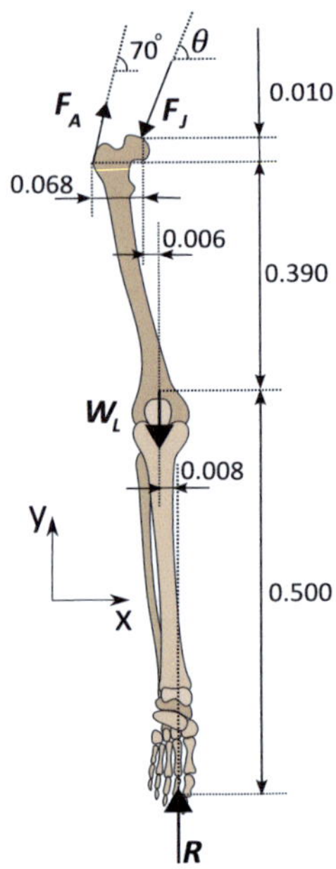

Fig. S2.1 Forces acting on a lower limb while standing

$$0.082 \times F_A \sin 70 + 0.89 \times F_A \cos 70 - 0.008 \times 13$$
$$-0.014 \times (F_A \sin 70 + 67) - 0.9 \times F_A \cos 70 = 0$$
$$F_A(0.082 \sin 70 + 0.89 \cos 70 - 0.014 \sin 70 - 0.9 \cos 70)$$
$$-0.014 \times 67 = 1.04$$
$$F_A(0.082 \sin 70 + 0.89 \cos 70 - 0.014 \sin 70 - 0.9 \cos 70) = 1.042$$
$$F_A = 17.229kg$$

Using Eqs. (S2.1) and (S2.2),

$$\frac{F_J \sin \theta}{F_J \cos \theta} = \frac{(F_A \sin 70 + 67)}{F_A \cos 70}$$
$$\tan \theta = \frac{(\mathrm{F_A} \sin 70 + 67)}{\mathrm{F_A} \cos 70} \tag{S2.4}$$

Upon substituting the value of F_A in Eq. (S2.4),

$$\tan \theta = \frac{(17.229 \sin 70 + 67)}{17.229 \cos 70}$$
$$\theta = \tan^{-1} 14.116$$
$$\theta = 85.94^\circ$$

Upon substituting the value of F_A and θ in Eq. (S2.1),

$$17.229 \cos 70 = F_J \cos 85.94$$
$$F_J = 83.23 \text{ kg}$$
$$F_J = 83.23 \times 9.81$$

Therefore, the hip joint reaction force $F_J = 816.48$ N.

Problem 2.2

Given Data:

Dumbbell weight is 10 kg
Head weighs 5 kg
Trunk weighs 35 kg

Unknown data:

Erector spinae muscle force (F_E) at a distance of 0.04 m from the centre of vertebra
Upon considering moment equilibrium about point C (Fig. S2.2),

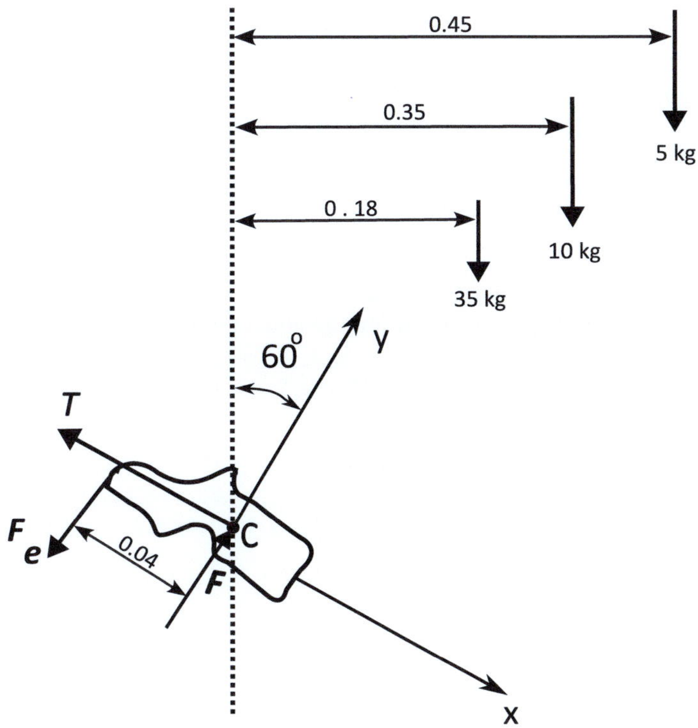

Fig. S2.2 Forces acting on a vertebra

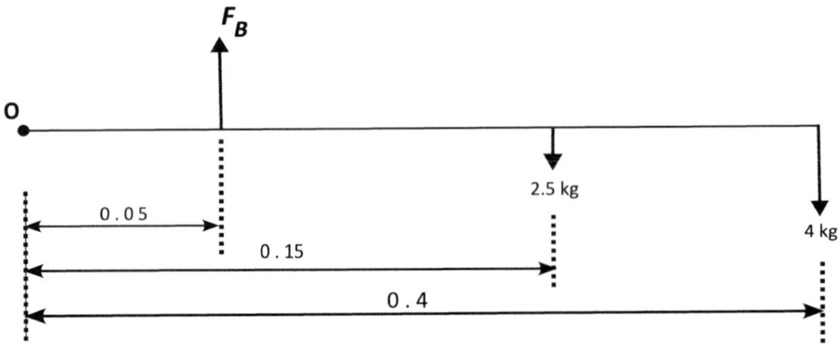

Fig. S2.3 Forces acting on a flexed elbow holding a ball

$$F_E \times 0.04 = 35 \times 0.18 + 10 \times 0.35 + 5 \times 0.45$$
$$F_E = 301.25 \text{ kg}$$
$$F_E = 301.25 \times 9.81 \text{ N}$$

The erector spinae muscle force (F_E) is obtained as 2955.26 N.

Problem 2.7.3

Given Data:

A flexed elbow is holding a ball weighing 4 kg
The weight of the forearm is 2.5 kg

Unknown data:

The force exerted by the biceps muscle (F_B)
Upon considering the moment equilibrium about point O (Fig. S2.3),

$$F_B \times 0.05 = 0.15 \times 2.5 + 4 \times 0.4$$
$$F_B = 39.5 \text{ kg}$$
$$F_B = 39.5 \times 9.81 \text{ N}$$

The biceps muscle force is calculated as, $F_B = 387.495$ N.

Problem 2.7.4

Given Data:

Whole body of weighing 70 kg
Lower leg weighing 5 kg

Unknown Data:

Tensile force exerted by the patellar tendon (***T***)

Considering that the body weight is supported by both the lower limbs, the reaction force on each foot can be assumed to be equal to half the body weight. Therefore, reaction force (R) = 35 kg.

Upon considering moment equilibrium about point O (Fig. S2.4),

$$R \times 0.22 = 5 \times 0.10 + T \times 0.05$$
$$35 \times 0.22 = 5 \times 0.10 + T \times 0.05$$
$$T = 144 \text{ kg}$$
$$T = 144 \times 9.81 \text{ N}$$

The patellar tendon force is obtained as
$T = 1412.64$ N

Problem 2.7.5

Given Data:
Deltoid muscle force (F_D) = 400 N
Supraspinatus muscle force (F_s) = 200 N
Weight of the arm (W_A) = 50 N

Unknown Data:
Glenohumeral joint reaction force (F_G)

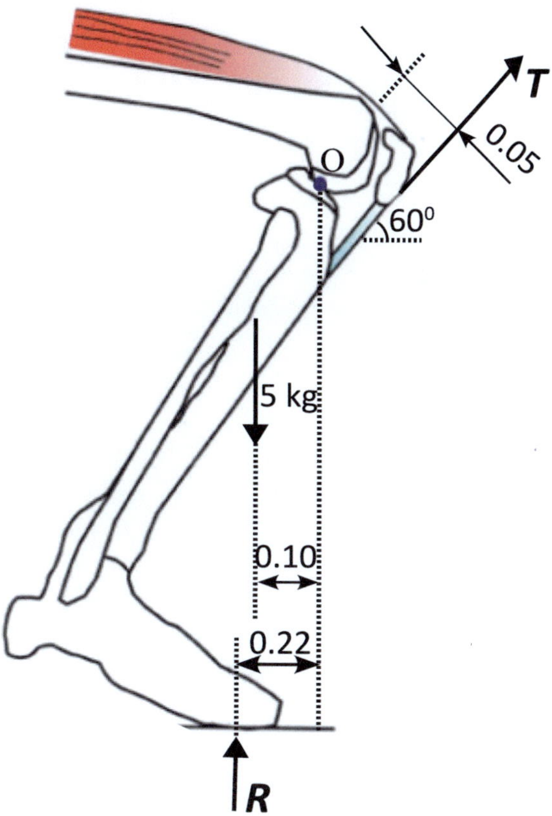

Fig. S2.4 Forces acting on the shank when a subject is about to rise from squatting position

Upon considering force equilibrium along the X- and Y-direction (Fig. S2.5),

$$\begin{aligned} &\sum F_X = 0 \\ &-F_G \cos\theta + F_S \cos 35 + F_D \sin 5 = 0 \\ &-F_G \cos\theta + 200 \cos 35 + 400 \sin 5 = 0 \\ &F_G \cos\theta = 200 \cos 35 + 400 \sin 5 \end{aligned} \tag{S2.5}$$

$$\begin{aligned} &\sum F_Y = 0 \\ &-W_A - F_G \sin\theta + F_S \sin 35 + F_D \cos 5 = 0 \\ &-50 - F_G \sin\theta + 200 \sin 35 + 400 \cos 5 = 0 \\ &F_G \sin\theta = 200 \sin 35 + 400 \cos 5 - 50 \end{aligned} \tag{S2.6}$$

Using Eqs. (S2.6) and (S2.5)

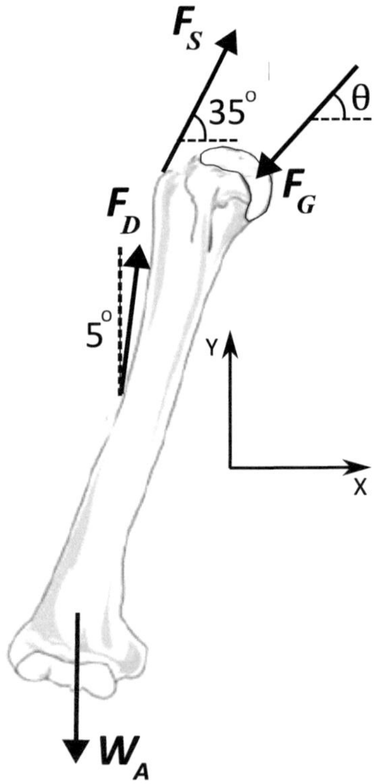

Fig. S2.5 Forces acting on the arm during an instance of humeral abduction

$$\frac{F_G \sin\theta}{F_G \cos\theta} = \frac{200\sin 35 + 400\cos 5 - 50}{200\cos 35 + 400\sin 5}$$
$$\tan\theta = \frac{463.19}{198.69}$$
$$\theta = \tan^{-1} 2.33$$
$$\theta = 66.78^\circ$$

Upon substituting θ in Eq. (S2.5)

$$F_G \cos 66.78 = 200\cos 35 + 400\sin 5$$
$$F_G = \frac{198.69}{\cos 66.78}$$

The glenohumeral joint force is determined as, $F_G = 504$ N

Solutions to Practice Problems of Chap. 5

Problem 5.1

Given data:

Hip joint reaction force (F_{HJR}) is 2000 N
Femoral neck width (y) is 30 mm
Area of cross-section (A) is 800 mm^2
Second moment of area (I) is 40000 mm^4

Unknown data:

Maximum normal stress acting at the section X–X (σ)

The combined stresses act at the femoral neck (X–X) i.e., bending stress and axial stress (Fig. S5.1).

The tangential force component (along the line x–x) of the F_{HJR} can be expressed as

$$F_t = F_{HJR}\cos(180^\circ - 50^\circ - 75^\circ)$$
$$F_t = 2000\cos 55^\circ$$
$$F_t = 1147.15 \text{ N}$$

The normal force component (perpendicular to the line x–x) of the F_{HJR} can be expressed as

$$F_n = F_{HJR}\sin 55^\circ$$
$$F_n = 2000\sin 55^\circ$$
$$F_n = 1638.304 \text{ N}$$

Fig. S5.1 A femur fixed distally and subjected to joint reaction force (F_{HJR})

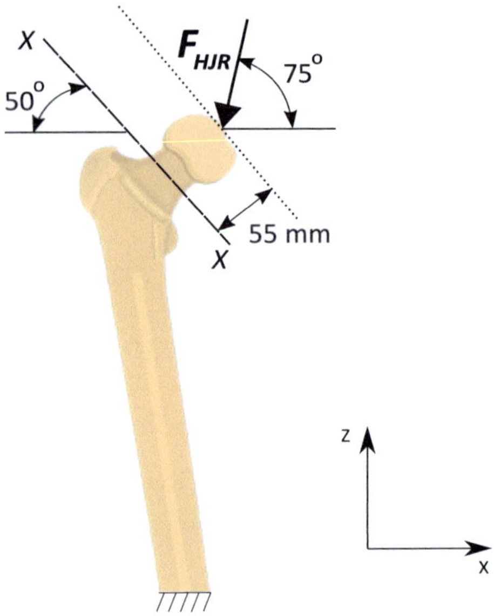

Bending moment (M) produced by F_t about the axis X–X can be written as

$$M = F_t \times 55$$
$$M = 1147.15 \times 55$$
$$M = 63093.25 \text{ N mm}$$

Using the flexural formula, we can obtain the maximum bending stress ($(\sigma_b)_{\max}$).

$$(\sigma_b)_{\max} = \frac{M}{I} y_{\max}$$

Here, $y_{\max}$ is equal to half the width of the femoral cross-section.

$$(\sigma_b)_{\max} = \frac{63098.25}{40000} \times 15$$

$$(\sigma_b)_{\max} = 23.66 \text{ N/mm}^2$$

The axial stress owing to the normal component (F_n) can be calculated as

$$\sigma_a = \frac{F_n}{\text{A}}$$
$$\sigma_a = \frac{1638.304}{800}$$
$$\sigma_a = 2.05 \text{ N/mm}^2$$

The total stress, which occurs at the outermost layer, is the sum of the maximum bending stress ($(\sigma_b)_{\max}$) and the axial stress (σ_a), as discussed in Sect. 5.1.4.

$$\sigma = \sigma_b + \sigma_a$$
$$\sigma = 23.66 + 2.05$$

The stress acting at the cross section X–X, $\sigma = 25.71\ \text{N/mm}^2 \sigma = 25.71\ \text{N/mm}^2$

Problem 5.2

Given data:

$$\sigma = \begin{bmatrix} 5\ 1\ 0 \\ 1\ 3\ 2 \\ 0\ 2\ 6 \end{bmatrix} kPa\,;\ \in = \begin{bmatrix} 0.001 & 0.002 & 0 \\ 0.002 & 0.004 & 0.001 \\ 0 & 0.001 & 0.006 \end{bmatrix}$$

Unknown data:

As we know that the orthotropic remodelling rule assumes that the material orients itself along the principal direction, we need to find the Eigenvectors of the stress tensor.

$$\sigma = \begin{bmatrix} 5\ 1\ 0 \\ 1\ 3\ 2 \\ 0\ 2\ 6 \end{bmatrix} \text{kPa}$$

$$\begin{bmatrix} \sigma_{xx} & \tau_{xy} & \tau_{xz} \\ \tau_{xy} & \sigma_{yy} & \tau_{yz} \\ \tau_{xz} & \tau_{yz} & \sigma_{zz} \end{bmatrix} = \begin{bmatrix} 5\ 1\ 0 \\ 1\ 3\ 2 \\ 0\ 2\ 6 \end{bmatrix}$$

The stress invariants are as follows:

$$I_1 = \sigma_{xx} + \sigma_{yy} + \sigma_{zz} = 5 + 3 + 6 = 14$$

$$I_2 = \begin{vmatrix} \sigma_{xx} & \tau_{xy} \\ \tau_{xy} & \sigma_{yy} \end{vmatrix} + \begin{vmatrix} \sigma_{yy} & \tau_{yz} \\ \tau_{yz} & \sigma_{zz} \end{vmatrix} + \begin{vmatrix} \sigma_{xx} & \tau_{xz} \\ \tau_{xz} & \sigma_{zz} \end{vmatrix} = \begin{vmatrix} 5\ 1 \\ 1\ 3 \end{vmatrix} + \begin{vmatrix} 3\ 2 \\ 2\ 6 \end{vmatrix} + \begin{vmatrix} 5\ 0 \\ 0\ 6 \end{vmatrix} = 58$$

$$I_3 = \begin{vmatrix} 5\ 1\ 0 \\ 1\ 3\ 2 \\ 0\ 2\ 6 \end{vmatrix} = 64$$

The characteristics equations can be written as

$$\sigma^3 - I_1\sigma + I_2\sigma - I_3 = 0$$

Upon substituting the invariants in the above equation:

$$\sigma^3 - 14\sigma + 58\sigma - 64 = 0$$

Upon solving the above equation, we obtain the principal stresses as

$$\sigma_1 = 7.103 \text{ MPa}$$
$$\sigma_2 = 5.146 \text{ MPa}$$
$$\sigma_3 = 1.751 \text{ MPa}$$

The principal directions corresponding to σ_1 can be found out by substituting the stress value into the Eigenvalue problem. Upon solving it using the Gauss elimination method and using the constraint of unit vector, as mentioned earlier, we obtain the principal stress direction corresponding to σ_1:

$$\begin{bmatrix} n_1 \\ n_2 \\ n_3 \end{bmatrix} = \begin{bmatrix} 0.224 \\ 0.471 \\ 0.854 \end{bmatrix}$$

Similarly, one can find the principal directions corresponding to other principal stresses. The principal direction corresponding to σ_2:

$$\begin{bmatrix} n_1 \\ n_2 \\ n_3 \end{bmatrix} = \begin{bmatrix} 0.937 \\ 0.137 \\ -0.321 \end{bmatrix}$$

The principal direction corresponding to σ_3:

$$\begin{bmatrix} n_1 \\ n_2 \\ n_3 \end{bmatrix} = \begin{bmatrix} 0.268 \\ -0.872 \\ 0.410 \end{bmatrix}$$

Solutions to Practice Problems of Chap. 6

Problem 6.1

Given Data:

The top of the vertebral body is loaded with a central compressive force (***P***)
Diameter of the cancellous region is d
Thickness of the outer cortical region is t
cortical and cancellous bone has a height of h
Young's modulus of the cancellous region is E_{canc}
Area of cross-section of cancellous is A_{canc}

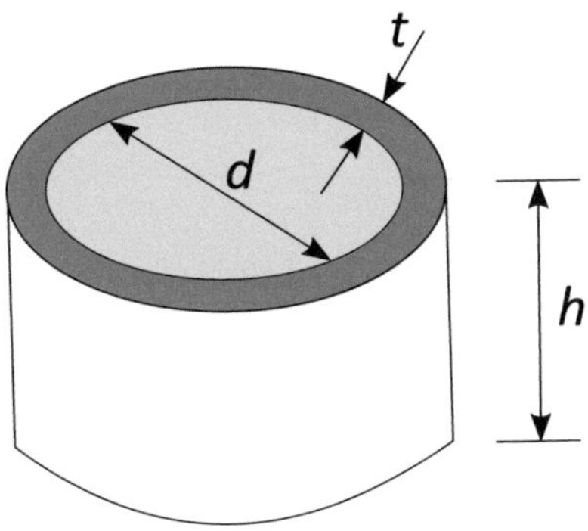

Fig. S6.1 Simplified representation of cortical bone and cancellous bone

Young's modulus of the cortical region is E_{cort}
Area of cross-section of cortical is A_{cort}

Unknown Data:
Find the load shared by the cortical (F_{cort}) and the cancellous bone (F_{canc}).
The area of cross-section of cancellous bone is expressed as (Fig. S6.1):

$$(A_{canc}) = \frac{\pi}{4}d^2$$

The area of cross-section of cortical bone is expressed as

$$(A_{cort}) = \frac{\pi}{4}\left((d+2t)^2 - d^2\right) = \frac{\pi}{4}(4t(t+d))$$

Using the concept of composite bar subjected to axial load, we can write

$$\epsilon = \frac{F}{(E_{canc}A_{canc} + E_{cort}A_{cort})}$$

The stresses generated in Material 1 and 2 is expressed as:

$$\sigma_1 = E_{canc}\epsilon_{canc} = E_{canc}\epsilon = \frac{PE_{canc}}{(E_{canc}A_{canc} + E_{cort}A_{cort})}$$
$$\sigma_{cort} = E_{cort}\epsilon_{cort} = E_{cort}\epsilon = \frac{PE_{cort}}{(E_{canc}A_{canc} + E_{cort}A_{cort})}$$

Following the concept of load transfer in a composite bar subjected to axial load, the load carried by the cancellous (P_{canc}) and the cortical (P_{cort}) is expressed as follows:

$$P_{canc} = \sigma_{canc}A_{canc} = \frac{PE_{canc}A_{canc}}{(E_{canc}A_{canc} + E_{cort}A_{cort})}$$

$$P_{cort} = \sigma_{cort}A_{cort} = \frac{PE_{cort}A_{cort}}{(E_{canc}A_{canc} + E_{cort}A_{cort})}$$

Area of cross-section of cancellous and cortical is expressed in terms of thickness (t) and diameter (d). The load carried by cancellous and cortical bone is rewritten as

$$P_{canc} = \frac{PE_{canc}A_{canc}}{(E_{canc}A_{canc} + E_{cort}A_{cort})} = \frac{PE_{canc}\frac{\pi}{4}d^2}{\left(E_{canc}\frac{\pi}{4}d^2 + E_{cort}\frac{\pi}{4}(4t(t+d))\right)}$$

$$P_{canc} = \frac{PE_{canc}d^2}{\left(E_{canc}d^2 + E_{cort}(4t(t+d))\right)}$$

$$P_{cort} = \frac{PE_{cort}A_{cort}}{(E_{canc}A_{canc} + E_{cort}A_{cort})} = \frac{PE_{cort}\frac{\pi}{4}(4t(t+d))}{\left(E_{canc}\frac{\pi}{4}d^2 + E_{cort}\frac{\pi}{4}(4t(t+d))\right)}$$

$$P_{cort} = \frac{PE_{cort}(4t(t+d))}{\left(E_{canc}d^2 + E_{cort}(4t(t+d))\right)}$$

Problem 6.2

Given Data:

The state of stress at a specific location on the implant-bone interface is expressed as

$$\sigma = \begin{bmatrix} 45 & -15 \\ -15 & 25 \end{bmatrix} \text{MPa}$$

The implant-bone interface is inclined at 45°

The bone density of the nearest bone element (ρ) = 1 g cm^{-3}.

Unknown:

Hoffman failure index (FL)

The Hoffman number (FL) at that location is calculated using the following expression:

$$\text{FL} = \frac{\sigma_n^2}{S_t S_c} + \left(\frac{1}{S_t} - \frac{1}{S_c}\right)\sigma_n + \frac{\tau^2}{S_s^2}$$

The strength of the bone depends on the apparent density as expressed below:

$$S_t = 14.5\rho^{1.71};\ S_c = 32.4\rho^{1.85};\ S_s = 21.6\rho^{1.65}$$

Since the bone density of the nearest bone element is equal to 1 g cm^{-3}, we get

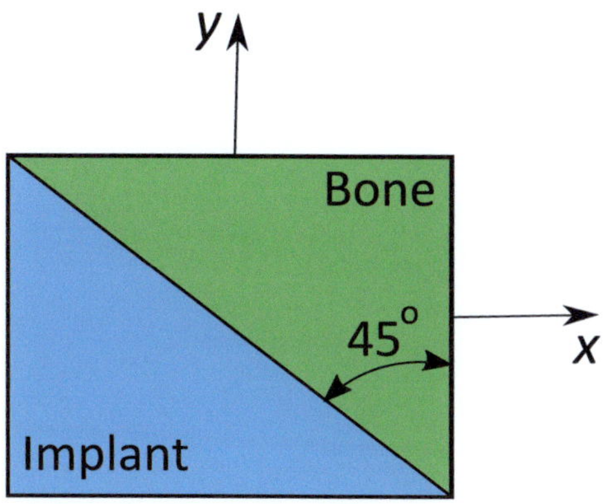

Fig. S6.2 Simplified representation of the implant-bone interface

$$S_t = 14.5\rho^{1.71} = 14.5(1)^{1.71} = 14.5\ \text{MPa}$$
$$S_c = 32.4\rho^{1.85} = 32.4(1)^{1.85} = 32.4\ \text{MPa}$$
$$S_s = 21.6\rho^{1.65} = 21.6(1)^{1.65} = 21.6\ \text{MPa}$$

Now, we need to find the normal and tangential stress acting at the interface using the stress transformation matrix (Fig. S6.2).

$$[\sigma'] = \begin{bmatrix} \cos(45^\circ) & \sin(45^\circ) \\ -\sin(45^\circ) & \cos(45^\circ) \end{bmatrix} \begin{bmatrix} 45 & -15 \\ -15 & 25 \end{bmatrix} \begin{bmatrix} \cos(45^\circ) & -\sin(45^\circ) \\ \sin(45^\circ) & \cos(45^\circ) \end{bmatrix}$$
$$[\sigma'] = \frac{1}{2}\begin{bmatrix} 1 & 1 \\ -1 & 1 \end{bmatrix} \begin{bmatrix} 45 & -15 \\ -15 & 25 \end{bmatrix} \begin{bmatrix} 1 & -1 \\ 1 & 1 \end{bmatrix} = \begin{bmatrix} 20 & -10 \\ -10 & 50 \end{bmatrix} \text{MPa}$$

From $[\sigma']$, it is clear that $\sigma_n = 20$ MPa and $\tau = -10$ MPa

Using the strength and stress values, the FL number is found as follows:

$$\text{FL} = \frac{\sigma_n^2}{S_t S_c} + \left(\frac{1}{S_t} - \frac{1}{S_c}\right)\sigma_n + \frac{\tau^2}{S_s^2} = \frac{20^2}{14.5 \times 32.4} + \left(\frac{1}{14.5} - \frac{1}{32.4}\right) \times 20 + \frac{(-10)^2}{21.6^2}$$
$$\text{FL} = 1.83$$

Since the FL number is higher than 1, the interface is likely to fail at that specific location.